Advancing Embedded Systems and Real-Time Communications with Emerging Technologies

Seppo Virtanen
University of Turku, Finland

A volume in the Advances in Systems
Analysis, Software Engineering, and High
Performance Computing (ASASEHPC)
Book Series

An Imprint of IGI Global

Managing Director:	Lindsay Johnston
Production Editor:	Jennifer Yoder
Development Editor:	Erin O'Dea
Acquisitions Editor:	Kayla Wolfe
Typesetter:	Thomas Creedon
Cover Design:	Jason Mull

Published in the United States of America by
    Information Science Reference (an imprint of IGI Global)
    701 E. Chocolate Avenue
    Hershey PA 17033
    Tel: 717-533-8845
    Fax: 717-533-8661
    E-mail: cust@igi-global.com
    Web site: http://www.igi-global.com

Library of Congress Cataloging-in-Publication Data

Advancing embedded systems and real-time communications with emerging technologies / Seppo Virtanen, editor.
    pages cm
 Includes bibliographical references and index.
   ISBN 978-1-4666-6034-2 (hardcover) -- ISBN 978-1-4666-6037-3 (print & perpetual access) -- ISBN 978-1-4666-6035-9 (ebook) 1. Embedded computer systems--Technological innovations. 2. Wireless communication systems--Equipment and supplies. I. Virtanen, Seppo, 1972- editor of compilation.
 TK7895.E42A383 2014
 006.2'2--dc23
                        2014004940

This book is published in the IGI Global book series Advances in Systems Analysis, Software Engineering, and High Performance Computing (ASASEHPC) (ISSN: 2327-3453; eISSN: 2327-3461)

British Cataloguing in Publication Data
A Cataloguing in Publication record for this book is available from the British Library.

For electronic access to this publication, please contact: eresources@igi-global.com.

Advances in Systems Analysis, Software Engineering, and High Performance Computing (ASASEHPC) Book Series

Vijayan Sugumaran
Oakland University, USA

ISSN: 2327-3453
EISSN: 2327-3461

MISSION

The theory and practice of computing applications and distributed systems has emerged as one of the key areas of research driving innovations in business, engineering, and science. The fields of software engineering, systems analysis, and high performance computing offer a wide range of applications and solutions in solving computational problems for any modern organization.

The **Advances in Systems Analysis, Software Engineering, and High Performance Computing (ASASEHPC) Book Series** brings together research in the areas of distributed computing, systems and software engineering, high performance computing, and service science. This collection of publications is useful for academics, researchers, and practitioners seeking the latest practices and knowledge in this field.

COVERAGE

- Computer Graphics
- Computer Networking
- Computer System Analysis
- Distributed Cloud Computing
- Enterprise Information Systems
- Metadata and Semantic Web
- Parallel Architectures
- Performance Modeling
- Software Engineering
- Virtual Data Systems

IGI Global is currently accepting manuscripts for publication within this series. To submit a proposal for a volume in this series, please contact our Acquisition Editors at Acquisitions@igi-global.com or visit: http://www.igi-global.com/publish/.

Titles in this Series

For a list of additional titles in this series, please visit: www.igi-global.com

Handbook of Research on Innovations in Systems and Software Engineering
Vicente García Díaz (University of Oviedo, Spain) Juan Manuel Cueva Lovelle (University of Oviedo, Spain) and
B. Cristina Pelayo García-Bustelo (University of Oviedo, Spain)
Information Science Reference • copyright 2015 • 723pp • H/C (ISBN: 9781466663596) • US $515.00 (our price)

Handbook of Research on Emerging Advancements and Technologies in Software Engineering
Imran Ghani (Universiti Teknologi Malaysia, Malaysia) Wan Mohd Nasir Wan Kadir (Universiti Teknologi Malaysia, Malaysia) and Mohammad Nazir Ahmad (Universiti Teknologi Malaysia, Malaysia)
Engineering Science Reference • copyright 2014 • 478pp • H/C (ISBN: 9781466660267) • US $395.00 (our price)

Advancing Embedded Systems and Real-Time Communications with Emerging Technologies
Seppo Virtanen (University of Turku, Finland)
Information Science Reference • copyright 2014 • 308pp • H/C (ISBN: 9781466660342) • US $235.00 (our price)

Handbook of Research on High Performance and Cloud Computing in Scientific Research and Education
Marijana Despotović-Zrakić (University of Belgrade, Serbia) Veljko Milutinović (University of Belgrade, Serbia) and Aleksandar Belić (University of Belgrade, Serbia)
Information Science Reference • copyright 2014 • 476pp • H/C (ISBN: 9781466657847) • US $325.00 (our price)

Agile Estimation Techniques and Innovative Approaches to Software Process Improvement
Ricardo Colomo-Palacios (Østfold University College, Norway) Jose Antonio Calvo-Manzano Villalón (Universidad Politécnica De Madrid, Spain) Antonio de Amescua Seco (Universidad Carlos III de Madrid, Spain) and Tomás San Feliu Gilabert (Universidad Politécnica De Madrid, Spain)
Information Science Reference • copyright 2014 • 399pp • H/C (ISBN: 9781466651821) • US $215.00 (our price)

Enabling the New Era of Cloud Computing Data Security, Transfer, and Management
Yushi Shen (Microsoft, USA) Yale Li (Microsoft, USA) Ling Wu (EMC, USA) Shaofeng Liu (Microsoft, USA) and Qian Wen (Endronic Corp, USA)
Information Science Reference • copyright 2014 • 336pp • H/C (ISBN: 9781466648012) • US $195.00 (our price)

Theory and Application of Multi-Formalism Modeling
Marco Gribaudo (Politecnico di Milano, Italy) and Mauro Iacono (Seconda Università degli Studi di Napoli, Italy)
Information Science Reference • copyright 2014 • 314pp • H/C (ISBN: 9781466646599) • US $195.00 (our price)

www.igi-global.com

701 E. Chocolate Ave., Hershey, PA 17033
Order online at www.igi-global.com or call 717-533-8845 x100
To place a standing order for titles released in this series, contact: cust@igi-global.com
Mon-Fri 8:00 am - 5:00 pm (est) or fax 24 hours a day 717-533-8661

Table of Contents

Section 1
Technology Review

Liang Guang, University of Turku, Finland
Juha Plosila, University of Turku, Finland
Hannu Tenhunen, University of Turku, Finland & Royal Institute of Technology, Sweden

Fabrício A. B. da Silva, Oswaldo Cruz Foundation (FIOCRUZ), Brazil
David F. C. Moura, Brazilian Army Technological Center, Brazil
Juraci F. Galdino, Brazilian Army Technological Center, Brazil & Military Institute of
Engineering, Brazil

Fabio Dovis, Politecnico di Torino, Italy
Luciano Musumeci, Politecnico di Torino, Italy
Nicola Linty, Politecnico di Torino, Italy
Marco Pini, Istituto Superiore Mario Boella, Italy

Section 2
Innovations for Wireless Networks and Positioning

Teemu Laukkarinen, Tampere University of Technology, Finland
Lasse Määttä, Tampere University of Technology, Finland
Jukka Suhonen, Tampere University of Technology, Finland
Marko Hännikäinen, Tampere University of Technology, Finland

Section 3
System Specification and Modeling

Section 4
Technologies for Network-On-Chip

Detailed Table of Contents

Section 1
Technology Review

    Liang Guang, University of Turku, Finland
    Juha Plosila, University of Turku, Finland
    Hannu Tenhunen, University of Turku, Finland & Royal Institute of Technology, Sweden

Dependability is a primary concern for emerging billion-transistor SoCs (Systems-on-Chip), especially when the constant technology scaling introduces an increasing rate of faults and errors. Considering the time-dependent device degradation (e.g. caused by aging and run-time voltage and temperature variations), self-adaptive circuits and architectures to improve dependability is promising and very likely inevitable. This chapter extensively surveys existing works on monitoring, decision-making, and reconfiguration addressing different dependability threats to Very Large Scale Integration (VLSI) chips. Centralized, distributed, and hierarchical fault management, utilizing various redundancy schemes and exploiting logical or physical reconfiguration methods, are all examined. As future research directions, the challenge of integrating different error management schemes to account for multifold threats and the great promise of error resilient computing are identified. This chapter provides, for chip designers, much needed insights on applying a self-adaptive computing paradigm to approach dependability on error-prone, cost-sensitive SoCs.

    Fabrício A. B. da Silva, Oswaldo Cruz Foundation (FIOCRUZ), Brazil
    David F. C. Moura, Brazilian Army Technological Center, Brazil
    Juraci F. Galdino, Brazilian Army Technological Center, Brazil & Military Institute of
        Engineering, Brazil

In recent years, the development of radio communication technology solutions has experienced a huge paradigm change – the Software-Defined Radio (SDR) technology upspring, in which previously hardware-based features became software-defined and users may also introduce new application waveforms on-the-fly. Given its growing importance for SDR application vendors and developers in different project domains, one of first steps in engineering a secure SDR system is the identification of classes of attacks on a SDR, along with the associated threats and vulnerabilities. Therefore, the identification of classes of attacks is necessary for the definition of realistic and relevant security requirements. One contribution of this chapter is to identify classes of attacks that Software Communications Architecture (SCA) compliant Software-Defined Radios (SDR) can suffer. It is noteworthy that, with the advancement of technology, new vulnerabilities emerge every day, and with them, new forms of threats and attacks on systems. The authors intend, however, to highlight in this chapter the classes of attacks that are more relevant for tactical software-defined radios, taking into account expected losses for legitimate radio network users. They also discuss, in this chapter, mitigation strategies for several identified attacks and how attack mitigation strategies can affect a SCA-compliant operating environment. Finally, the authors present several case studies, along with simulation results, illustrating the identified attack classes.

Chapter 3

Fabio Dovis, Politecnico di Torino, Italy
Luciano Musumeci, Politecnico di Torino, Italy
Nicola Linty, Politecnico di Torino, Italy
Marco Pini, Istituto Superiore Mario Boella, Italy

This chapter deals with one of the major concerns for reliable use of Global Navigation Satellite Systems (GNSS), providing a description of intentional and unintentional threats, such as interference, jamming, and spoofing. Despite the fact that these phenomena have been studied since the early stages of Global Positioning System (GPS), they were mainly addressed for military applications of GNSS. However, a wide range of recent civil applications related to user safety or featuring financial implications would be deeply affected by interfering or spoofing signals intentionally created. For such a reason, added value processing algorithms are being studied and designed in order to embed in the receiver an interference reporting capability so that they can monitor and possibly mitigate interference events.

Section 2
Innovations for Wireless Networks and Positioning

Chapter 4

Teemu Laukkarinen, Tampere University of Technology, Finland
Lasse Määttä, Tampere University of Technology, Finland
Jukka Suhonen, Tampere University of Technology, Finland
Marko Hännikäinen, Tampere University of Technology, Finland

Wireless Sensor Networks (WSNs) require automated over the air software updates for fixing errors or adding new features. Reprogramming nodes manually is often impractical or even impossible. Current update methods require an external memory, additional computation, and/or external WSN transport protocol. In this chapter, the authors propose Program Image Dissemination Protocol (PIDP) for WSNs. Combining PIDP with an application description language provides a complete method for WSN firmware management. PIDP is reliable, lightweight, and supports multi-hopping. PIDP does not require external memory, is independent of the WSN implementation, transfers firmware reliably, and reprograms the whole program memory. In addition, PIDP allows several levels of WSN node heterogeneity. PIDP was implemented on an 8-bit node platform with a 2.4 GHz radio. Implementation requires 22 bytes of data memory and less than 7 kilobytes of program memory. PIDP updates 178 nodes within 5 hours. One update consumes under 1% of the energy of two AA batteries.

Chapter 5

Pramita Mitra, Ford Research and Innovation Center, USA
Christian Poellabauer, University of Notre Dame, USA

Recent experimental research has revealed that the link conditions in realistic wireless networks vary significantly from the ideal disk model, and a substantial percentage of links are asymmetric. Many existing geographic routing protocols fail to consider asymmetric links during neighbor discovery and thus discount a significant number of potentially stable routes with good one-way reliability. This chapter provides a detailed overview of a number of location-aware routing protocols that explicitly use asymmetric links in routing to obtain efficient and shorter (low latency) routes. An asymmetric link routing protocol, called Asymmetric Geographic Forwarding (A-GF) is discussed in detail. A-GF discovers asymmetric links in the network, evaluates them for stability (e.g., based on mobility), and uses them to improve the routing efficiency.

Chapter 6

Maitane Barrenechea, University of Mondragón, Spain
Mikel Mendicute, University of Mondragón, Spain
Andreas Burg, École Polytechnique Fédérale de Lausanne (EPFL), Switzerland
John Thompson, University of Edinburgh, UK

The multiuser MIMO environment enables the communication between a base-station and multiple users with several antennas. In such a scenario, the use of precoding techniques is required in order to detect the signal at the users' terminals without any cooperation between them. This contribution presents various designs and hardware implementations of a high-capacity precoder based on vector perturbation. To this aim, three tree-search techniques and their associated user-ordering schemes are investigated in this chapter: the well-known K-Best precoder, the fixed-complexity Fixed Sphere Encoder (FSE), and the variable complexity Single Best-Node Expansion (SBE). All of the aforementioned techniques aim at finding the most suitable perturbation vector within an infinite lattice without the high computational complexity of an exhaustive search.

Chapter 7

Liang Chen, Finnish Geodetic Institute, Finland
Heidi Kuusniemi, Finnish Geodetic Institute, Finland
Yuwei Chen, Finnish Geodetic Institute, Finland
Ling Pei, Shanghai Jiao Tong University, China
Jingbin Liu, Finnish Geodetic Institute, Finland
Jian Tang, Wuhan University, China
Laura Ruotsalainen, Finnish Geodetic Institute, Finland
Ruizhi Chen, Finnish Geodetic Institute, Finland & Texas A&M University – Corpus Christi, USA

This chapter studies wireless positioning using a network of Bluetooth signals. Fingerprints of Received Signal Strength Indicators (RSSI) are used for localization. Due to the relatively long interval between the available consecutive Bluetooth signal strength measurements, the authors applied an information filter method with speed detection, which combines the estimation information from the RSSI measurements with the prior information from the motion model. Speed detection is assisted to correct the outliers of position estimation. The field tests show the effectiveness of the information filter-assisted positioning method, which improves the horizontal positioning accuracy of indoor navigation by about 17% compared to the static fingerprinting positioning method, achieving a 4.2 m positioning accuracy on the average, and about 16% improvement compared to the point Kalman filter. In RSSI fingerprinting localization, building a fingerprint database is usually time-consuming and labour-intensive. In the final section, a self-designed autonomous SLAM robot platform is introduced to be able to carry out the Bluetooth RSS data collecting.

Chapter 8

Francesco Sottile, Istituto Superiore Mario Boella, Italy
Zhoubing Xiong, Istituto Superiore Mario Boella, Italy
Claudio Pastrone, Istituto Superiore Mario Boella, Italy

This chapter analyzes some hybrid and cooperative GNSS-terrestrial positioning algorithms that combine both pseudorange measurements from satellites and terrestrial range measurements based on radio frequency communication to improve both position accuracy and availability. A Simulation Tool (ST) is also presented as a viable tool able to test and evaluate the performance of these hybrid positioning algorithms in different scenarios. In particular, the ST simulates devices belonging to a Peer-to-Peer (P2P) wireless network where peers, equipped with a wireless interface and a GNSS receiver, cooperate among them by exchanging positioning aiding data in order to enhance the overall performance. Different hybrid and cooperative algorithms, based on Bayesian and least squares approaches proposed in the literature, have been implemented in the ST and simulated in different simulation scenarios including the vehicular urban one. Moreover, all these algorithms are compared in terms of computational complexity to better understand their feasibility to achieve a real-time implementation. Finally, the sensitivity of the hybrid and cooperative algorithms when pseudorange measurements are affected by large noise and in presence of malicious peers in the P2P network is also assessed by means of the ST.

Section 3
System Specification and Modeling

Chapter 9

Valentin Olenev, St. Petersburg State University of Aerospace Instrumentation, Russia
Irina Lavrovskaya, St. Petersburg State University of Aerospace Instrumentation, Russia
Pavel Morozkin, St. Petersburg State University of Aerospace Instrumentation, Russia
Alexey Rabin, St. Petersburg State University of Aerospace Instrumentation, Russia
Sergey Balandin, Open Innovations Association FRUCT, Finland
Michel Gillet, Nokia Research Center, Finland

This chapter gives an overview of a modeling application in the general embedded systems design flow and presents two general approaches for the embedded networks simulation: network modeling and protocol stack modeling. The authors select two widely used modeling languages, which are SDL and SystemC.

The analysis shows that both languages have a number of advantages that could be combined by the joint use of SystemC and SDL. Thus, the authors propose an approach for the SystemC and SDL co-modeling. This approach can be used in practice to perform protocol stack simulation as well as simulation of network operation. Therefore, the authors give examples of co-modeling practical applications.

Detlef Streitferdt, Technische Universität Ilmenau, Germany
Florian Kantz, ABB Corporate Research, Germany
Philipp Nenninger, ABB Automation Products, Germany
Thomas Ruschival, Datacom Telematica, Brazil
Holger Kaul, ABB Corporate Research, Germany
Thomas Bauer, Fraunhofer IESE, Germany
Tanvir Hussain, The Mathworks GmbH, Germany
Robert Eschbach, ITK Engineering AG, Germany

This chapter reports the results of a cycle computer case study and a previously conducted industrial case study from the automation domain. The key result is a model-based testing process for highly configurable embedded systems. The initial version of the testing process was built upon parameterizeable systems. The cycle computer case study adds the configuration using the product line concept and a feature model to store the parameterizable data. Thus, parameters and their constraints can be managed in a very structured way. Escalating demand for flexibility has made modern embedded software systems highly adjustable. This configurability is often realized through parameters and a highly configurable system possesses a handful of those. Small changes in parameter values can often account for significant changes in the system's behavior, whereas in some other cases, changed parameters may not result in any perceivable reaction. The case studies address the challenge of applying model-based testing to configurable embedded software systems in order to reduce development effort. As a result of the case studies, a model-based testing process was developed. This process integrates existing model-based testing methods and tools such as combinatorial design and constraint processing as well as the product line engineering approach. The testing process was applied as part of the case studies and analyzed in terms of its actual saving potentials, which turned out to reduce the testing effort by more than a third.

Tapio Pahikkala, University of Turku, Finland
Antti Airola, University of Turku, Finland
Thomas Canhao Xu, University of Turku, Finland
Pasi Liljeberg, University of Turku, Finland
Hannu Tenhunen, University of Turku, Finland
Tapio Salakoski, University of Turku, Finland

This chapter considers parallel implementation of the online multi-label regularized least-squares machine-learning algorithm for embedded hardware platforms. The authors focus on the following properties required in real-time adaptive systems: learning in online fashion, that is, the model improves with new data but does not require storing it; the method can fully utilize the computational abilities of modern embedded multi-core computer architectures; and the system efficiently learns to predict several labels simultaneously. They demonstrate on a hand-written digit recognition task that the online algorithm converges faster, with respect to the amount of training data processed, to an accurate solution than a stochastic gradient descent based baseline. Further, the authors show that our parallelization of the method scales well on a quad-core platform. Moreover, since Network-on-Chip (NoC) has been proposed as a promising candidate for future multi-core architectures, they implement a NoC system consisting of 16 cores. The proposed machine learning algorithm is evaluated in the NoC platform. Experimental results show that, by optimizing the cache behaviour of the program, cache/memory efficiency can improve significantly. Results from the chapter provide a guideline for designing future embedded multi-core machine learning devices.

Chapter 12

Hervé Yviquel, IRISA, France
Emmanuel Casseau, IRISA, France
Matthieu Wipliez, Synflow SAS, France
Jérôme Gorin, Telecom ParisTech, France
Mickaël Raulet, IETR/INSA, France

This chapter reviews dataflow programming as a whole and presents a classification-based methodology to bridge the gap between predictable and dynamic dataflow modeling in order to achieve expressiveness of the programming language as well as efficiency of the implementation. The authors conduct experiments across three MPEG video decoders including one based on the new High Efficiency Video Coding standard. Those dataflow-based video decoders are executed onto two different platforms: a desktop processor and an embedded platform composed of interconnected and tiny Very Long Instruction Word-style processors. The authors show that the fully automated transformations presented can result in a 80% gain in speed compared to runtime scheduling in the more favorable case.

Chapter 13

Sergey Ostroumov, Åbo Akademi University, Finland & TUCS - Turku Centre for Computer Science, Finland
Leonidas Tsiopoulos, Åbo Akademi University, Finland
Marina Waldén, Åbo Akademi University, Finland
Juha Plosila, University of Turku, Finland

A Network-On-Chip is a paradigm that tackles limitations of traditional bus-based interconnects. It allows complex applications that demand many resources to be deployed on many-core platforms effectively. To satisfy requirements on dependability, however, a NoC platform requires dynamic monitoring and reconfiguration mechanisms. In this chapter, the authors propose an agent-based management system that monitors the state of the platform and applies various reconfiguration techniques. These techniques aim at enabling uninterruptable execution of applications satisfying dependability requirements. The authors develop the proposed system within Event-B that provides a means for stepwise and correct-by-construction specification supported by mathematical proofs. Furthermore, the authors show the mechanism of decomposition of Event-B specifications such that a well-structured and hierarchical agent-based management system is derived.

Section 4
Technologies for Network-On-Chip

Diandian Zhang, RWTH Aachen University, Germany
Jeronimo Castrillon, Dresden University of Technology, Germany
Stefan Schürmans, RWTH Aachen University, Germany
Gerd Ascheid, RWTH Aachen University, Germany
Rainer Leupers, RWTH Aachen University, Germany
Bart Vanthournout, Synopsys Inc., Belgium

Efficient runtime resource management in heterogeneous Multi-Processor Systems-on-Chip (MPSoCs) for achieving high performance and energy efficiency is one key challenge for system designers. In the past years, several IP blocks have been proposed that implement system-wide runtime task and resource management. As the processor count continues to increase, it is important to analyze the scalability of runtime managers at the system-level for different communication architectures. In this chapter, the authors analyze the scalability of an Application-Specific Instruction-Set Processor (ASIP) for runtime management called OSIP on two platform paradigms: shared and distributed memory. For the former, a generic bus is used as interconnect. For distributed memory, a Network-on-Chip (NoC) is used. The effects of OSIP and the communication architecture are jointly investigated from the system point of view, based on a broad case study with real applications (an H.264 video decoder and a digital receiver for wireless communications) and a synthetic benchmark application.

Tim Wegner, University of Rostock, Germany
Martin Gag, University of Rostock, Germany
Dirk Timmermann, University of Rostock, Germany

With the progress of deep submicron technology, power consumption and temperature-related issues have become dominant factors for chip design. Therefore, very large-scale integrated systems like Systems-on-Chip (SoCs) are exposed to an increasing thermal stress. On the one hand, this necessitates effective mechanisms for thermal management and task mapping. On the other hand, application of according thermal-aware approaches is accompanied by disturbance of system integrity and degradation of system performance. In this chapter, a method to predict and proactively manage the on-chip temperature distribution of systems based on Networks-on-Chip (NoCs) is proposed. Thereby, traditional reactive approaches for thermal management and task mapping can be replaced. This results in shorter response times for the application of management measures and therefore in a reduction of temperature and thermal imbalances and causes less impairment of system performance. The systematic analysis of simulations conducted for NoC sizes up to 4x4 proves that under certain conditions the proactive approach is able to mitigate the negative impact of thermal management on system performance while still improving the on-chip temperature profile. Similar effects can be observed for proactive thermal-aware task mapping at system runtime allowing for the consideration of prospective thermal conditions during the mapping process.

Chapter 16

Alessandro Strano, Intel Mobile Communications, Germany
Carles Hernández, Barcelona Supercomputing Center, Spain
Federico Silla, Universitat Politècnica de València, Spain
Davide Bertozzi, Università degli studi di Ferrara, Italy

In the context of multi-IP chips making use of internal communication paths other than the traditional buses, source synchronous links for use in multi-synchronous Networks-on-Chip (NoCs) are becoming the most vulnerable points for correct network operation and therefore need to be safeguarded against intra-link delay variations and signal misalignments. The intricacy of matching link net attributes during placement and routing and the growing role of process parameter variations in nanoscale silicon technologies, as well as the deterioration due to the ageing of the chip, are the root causes for this. This chapter addresses the challenge of designing a timing variation and layout mismatch tolerant link for synchronizer-based GALS NoCs by implementing a self-calibration mechanism. A timing variation detector senses the misalignment, due to process variation and wearout, between data lines with themselves and with the transmitter clock routed with data in source synchronous links. Then, a suitable delayed replica of the transmitter clock is selected for safe sampling of misaligned data. This chapter proves the robustness of the link in isolation with respect to a detector-less link, also addressing integration issues with the downstream synchronizer and switch architecture, proving the benefits in a realistic experimental setting for cost-effective NoCs.

Chapter 17

Milos Krstic, IHP, Germany
Xin Fan, IHP, Germany
Eckhard Grass, IHP, Germany
Luca Benini, University of Bologna, Italy
M. R. Kakoee, University of Bologna, Italy
Christoph Heer, Intel Mobile Communications, Germany
Birgit Sanders, Intel Mobile Communications, Germany
Alessandro Strano, Intel Mobile Communications, Germany
Gabriele Miorandi, University of Ferrara, Italy
Alberto Ghiribaldi, University of Ferrara, Italy
Davide Bertozzi, University of Ferrara, Italy

The GALS methodology has been discussed for many years, but only a few relevant implementations in silicon have been done. This chapter describes the implementation and test of the Moonrake Chip – complex GALS demonstrator implemented in 40 nm CMOS technology. Two novel types of GALS interface circuits are validated: point-to-point pausible clocking GALS interfaces and GALS NoC interconnects. Point-to-point GALS interfaces are integrated within a complex OFDM baseband transmitter block, and for NoC switches special test structures are defined. This chapter discloses the full structure of the respective interfaces, the complete GALS system, as well as the design flow utilized to implement them on the chip. Moreover, the full set of measurement results are presented, including area, power, and EMI results. Significant benefits and robustness of our applied GALS methodology are shown. Finally, some outlook and vision of the future role of GALS are outlined.

Preface

The rapid development of Information and Communication Technology (ICT) is enabling the penetration of embedded computing systems to places where they have not traditionally been seen. For some time already it has been possible to vote in elections using computerized ballot machines; we are traveling using passports that contain our personal information stored electronically and made accessible to officials by wireless communication; we now drive cars that have automatic cruise speed adjustment to meet the speed of the vehicle in front of us; we have medical implants in our bodies that use advanced electronics to control a bodily function like heartbeat with the possibility of having a physician adjust the settings wirelessly and with the ability to store lengthy history logs of the controlled bodily function, for example the patient's heartbeat history for the past year. These are examples of everyday applications of embedded systems and embedded communication systems introduced to our lives recently in application areas where we traditionally may not have been taking advantage of information and communication technologies. We can only expect that the development will rapidly go further and we will continue to see advanced ICTs appear in environments where we are not used to seeing them.

As embedded communication systems enter new application domains where they have traditionally not been seen, the system-level and hardware/software design processes are becoming more and more challenging. The processes need to cater for all the different kinds of functionality required from the devices, consisting of ever more complex operations and tasks while the devices still are most often required to operate on battery power. Depending on the application domain, the battery may be required to provide power to the device for a matter of hours or days (for example smartphones and tablets), weeks or months (sensor networks), or even years (medical implants). Each different type of power requirement requires different kinds of optimizations to the system and the way it uses power, and the design processes are expected to extensively support the designer in making these optimizations. Additionally, for lengthy battery lifetimes, the device is often required to recharge its battery using the energy of some available physical phenomenon in its operating environment, for example sunlight or motion, causing further pressure on the device design processes. Of course meeting the strict performance and power requirements serves little purpose if the manufacturing costs of the device are too high for the target consumer group and potential clientele: the manufacturers need to constantly keep a balance between adequate performance of the device, low enough manufacturing and design costs, and a short enough time-to-market.

As the name suggests, each embedded communication system utilizes some form of communication substrate to provide interoperability. Depending on the type of device in question, the substrate may be intended for communicating with the outside world, it may be intended for communication within the device, or both. For interoperability with the outside world, the devices may use technologies like

Wi-Fi and cellular data (3G and LTE) for fast Internet access (for example, smartphones and tablets), low-power and low-speed wireless transmission where the application needs to limit the amount of communication and the amount of power used for communication to a minimum (sensor networks), and Radio Frequency Identification (RFID) or Near Field Communication (NFC) for short-range wireless information exchange (for example smartcards, modern passports, and medical ICT implants). The more complex the embedded communication system is, the greater a number of these communication technologies it is usually expected to support.

Communication within the device is the target of System-on-Chip (SoC) and Network-on-Chip (NoC) research. In today's more complex devices like for example smartphones and tablets, the de facto approach is to utilize system circuitry consisting of two or more processor cores, that is, utilizing multiprocessor implementations. Within the system circuitry, data transportation most often needs to be extremely fast between the different system components, for example, between the cores inside the Graphics Processing Unit (GPU) or the communication between the GPU, the multicore system processor and system memory. Device interoperability at the on-chip level is typically increased by utilizing standardized buses and interfaces. In device interoperability, the key focus is moving from inter-component issues onto component level suitability to different types of platforms. Furthermore, due to physical effects producing errors, fault-free pre-tested circuits will be increasingly difficult to manufacture and the process will be cost intensive. Instead of this very costly and time-consuming approach, the key problem to be solved is how to build robust, error-free, and highly scalable systems when their basic building blocks can be defective due to static and dynamic errors or failures.

When considering the multiparty interaction and network connections to the external world, an issue of utmost importance is the protection of information security and privacy. Security functionality should be an integral built-in part of future embedded systems. The dramatically increased importance of system security (for example, security of the data stored in the device), implementation security (for example, the hardware and software components are formally verified to function correctly), and communication security (for example, connections through each available network need to be secure) must drive the processes of modeling and designing communication systems towards security enabled embedded systems very rapidly to ensure future-proof devices also from the point-of-view of security and privacy. With the rapid development of embedded communication systems, their security and ability to secure user privacy has not received such scholarly research and industrial R&D attention as it should. People, processes, and technology are all needed to adequately secure systems and facilities. Design processes have not adequately addressed security and privacy concerns arising from the characteristics and abilities of the systems and by far not enough attention, effort and resourcing has been allocated to explorative scientific research in this important topic area. Currently, there is no comprehensive systematic methodology for security design of embedded systems. Security is typically implemented either as an add-on property or a separate security software package. In addition, information security is not in the area of core competence of system designers today. As embedded communication systems are often the link between an unsecure dynamic network and the user, it is not enough that the system gives users the ability to execute applications. The users have to be able to use their applications in a secure and safe manner no matter when and where they are using them.

In the designing of embedded communication systems, clearly the challenges and development trends discussed above must converge in the design processes to facilitate the requirements and demands placed on today's state-of-the-art systems as well as those of tomorrow and the foreseeable future. The converging development is made possible only through the active research performed in the field of embedded and real-time communication systems. The disciplines of computer science, computer engineering, telecommunication, and communication engineering are well established and scientists, researchers, and industry professionals in these disciplines are numerous in all continents. The convergence of these disciplines into embedded and real-time communication systems is a natural development as can be seen for example in the smartphone and tablet industries today. The field is interdisciplinary in scope, binding together research from the mentioned disciplines with focus on how they converge to embedded and real-time systems for the communication application domain.

Forthcoming research in embedded and real-time communication systems needs to target the challenges in future complex converged wireless systems by adventurous development and technological exploration, and experimentation with novel technologies, systems, and system design methodologies. Key research areas in this respect are embedded system design, communication system design, system-wide security, and hardware/software co-design, producing results that converge into novel technologies usable in future secure embedded communication systems. Advances in research in some of these key areas are covered in this book. This book is a summation volume of research originally published in the *International Journal of Embedded and Real-Time Communication Systems* (IJERTCS) in its second and third volume years, 2011 and 2012. The chapters of this book have been extended, enhanced, and updated from the original journal articles during late 2013 and early 2014. The journal has an interdisciplinary scope, binding together research from different disciplines with focus on how the disciplines converge to embedded and real-time systems for the communication application domain. The subject coverage of the journal is broad, which enables a clear presentation of how the research results presented in the journal benefit the convergence of embedded systems, real-time systems and communication systems. The journal is aimed to benefit scientists, researchers, industry professionals, educators, and junior researchers like PhD students in the embedded systems and communication systems sector. An important aim is to provide the target audience with a forum to disseminate and obtain information on the most recent advances in embedded and real-time communication systems research: to give the readers the opportunity to take advantage of the research presented in the journal in their scientific, industrial, or educational purposes.

As a journal in the focal point of disciplines such as computer science, computer engineering, telecommunication, and communication engineering, the *International Journal of Embedded and Real-Time Communication Systems* is positioned well to provide its readership with interesting and well-focused articles based on recent high-quality research. The journal's coverage in topics from embedded systems, real-time systems, and communications system engineering, and especially how these disciplines interact in the field of embedded and real-time systems for communication, offers its readership both theoretical and practical research results facilitating the convergence of embedded systems, real-time computing, and communication system technologies and paradigms.

This book is organized into four thematic sections. The first section is titled "Technology Review" and it consists of three chapters that each provides an insightful and extensive literary review based coverage of a key topic area in the discipline of embedded and real-time communication systems: System-on-Chip dependability through self-adaptive technologies, Software-Defined Radio (SDR) security issues with focus on tactical applications, and mitigation of interference and spoofing attacks on satellite navigation.

The second section, titled "Innovations for Wireless Networks and Positioning," consists of five chapters. This section focuses on novel technological advances for embedded and resource-constrained networks as well as both indoor and outdoor positioning technologies. Many different aspects of technological advances are covered, for example protocol design for remotely updating sensor network firmware, improving the reliability and performance of ad-hoc routing in wireless sensor networks, high-performance precoders for wireless communication, filter assisted indoor positioning, and real-time terrestrial positioning algorithms. The selection of chapters in this section clearly highlights the multi-disciplinary nature of this research area: embedded and real-time communication systems are positioned in the focal points of research of several different information technology disciplines.

The third section is titled "System Specification and Modeling," and it consists of five chapters. Developing new design and verification methodologies and flows is essential to efficiently engineer more and more complex embedded and real-time communication systems. The classical way of first developing a system and then testing it at the end of a development cycle does not meet the requirements put on system quality and reliability in a short development time. This section addresses the design challenges with far-reaching research in using several description languages in parallel for co-modeling, applying model-based testing to verification of highly configurable systems, using parallel online learning for system-level adaptability, optimization of dataflow programs, and formal agent-based monitoring for dynamic Network-on-Chip reconfiguration.

The fourth thematic section is titled "Technologies for Network-on-Chip," and it consists of four chapters. In Networks-on-Chip, research focuses on problems regarding reliable communication inside a chip, for example between the different on-chip cores that together form a Multi-Processor System-on-Chip (MPSoC). An on-chip communication implementation itself can be seen as an embedded and real-time communication system, but very often today, for example, MPSoCs are an essential building part of some larger embedded communication system. The efficient and reliable implementation of on-chip communication is essential for ensuring excellence in device performance and variety in its operating capabilities, and it is thus in the heart of the research areas needing scholarly focus for future communication systems and environments.

The first thematic section of this book, "Technology Review," consists of three chapters (1-3). Chapter 1 is titled "Self-Adaptive SoCs for Dependability: Review and Prospects" and is written by Liang Guang, Juha Plosila, and Hannu Tenhunen from University of Turku, Finland. The chapter introduces existing and proposed monitoring strategies in parallel embedded systems, particularly, in adaptive on-chip networks. The chapter extensively discusses and reviews research on monitoring, decision-making and reconfiguration addressing a variety of dependability threats faced by modern silicon chips. Chip designers receive extremely valuable insights on applying the self-adaptive computing paradigm for attaining improved dependability for Systems-on-Chips (SoCs).

Chapter 2 is titled "Security Issues in Tactical Software-Defined Radios: Analysis of Attacks and Case Studies" and it is written by Fabrício A. B. da Silva, David F. C. Moura, and Juraci F. Galdino from the Brazilian Army Technological Center and the Military Institute of Engineering in Rio de Janeiro, Brazil. The chapter presents an extensive review and classification of attacks against tactical Software Defined Radios (SDR). The authors identify several classes of possible attacks against Software Communications Architecture (SCA) compliant SDRs based on an extensive literature survey. The authors also identify several sources of vulnerabilities and discuss related research directions. The authors note that security engineering should be incorporated in system design as early as possible in the process, ideally during or right after the initial specification. The conclusion is that the attack classes identified in the chapter define a point of departure for the threat analysis and the requirements definition related to the security engineering process of SCA compliant radios.

Chapter 3 ends the first section of this book. The chapter is titled "Interference and Spoofing: New Challenges for Satellite Navigation Receivers," and it is written by Fabio Dovis, Luciano Musumeci, and Nicola Linty from the Department of Electronics and Telecommunications, Politecnico di Torino,

Italy and Marco Pini from the Navigation Lab, Istituto Superiore Mario Boella, Italy. A major concern in satellite-based positioning is the security of the weak signals from satellites in space that are used in positioning. This chapter classifies intentional and unintentional threats, such as interference, jamming, and spoofing, and discusses some of the recent trends concerning techniques for their detection and mitigation. The topic is very important as satellite positioning is vital to a variety of widely used applications related to for example safety or finances: disruptions in the signal coverage or falsification of the signals would render the applications unusable and in the worst case cause severe hazards or financial losses. The authors present several techniques to detect and countermeasure even very sophisticated spoofing attempts. The complexity of a spoofing/interference monitoring unit may be challenging and the requirements for the computational capabilities and memory size make, at present, these techniques more suitable for high-end receivers.

The second thematic section of this book, "Innovations for Wireless Networks and Positioning," consists of five chapters (4-8). Chapter 4 is titled "A Multi-Hop Software Update Method for Resource Constrained Wireless Sensor Networks" and is written by Teemu Laukkarinen, Lasse Määttä, Jukka Suhonen, and Marko Hännikäinen from Tampere University of Technology (Finland). As wireless sensor networks are a very resource-constrained low-cost device environment, they pose high demands for energy efficiency throughout the distributed system. Due to the nature of the systems, it is also a big challenge to update the software in the network nodes. Current update protocols rely on a large external memory or on external transport protocols. The authors present the design, implementation, and experiments of a lightweight and reliable Program Image Dissemination Protocol (PIDP) for autonomous multihop wireless sensor networks. The proposed PIDP does not require external memory; it is independent of the wireless sensor network implementation; it transfers the program image; and it reprograms the whole program image. As energy is an important issue in wireless sensor networks, one update only consumes under 1% of energy of two AA batteries in a typical network node, and updates a deployment of 178 nodes in 5 hours.

Chapter 5 is titled "Asymmetric Link Routing in Location-Aware Mobile Ad-Hoc Networks," and it is written by Pramita Mitra from Ford Research and Innovation Center, USA and Christian Poellabauer from Department of Computer Science and Engineering, University of Notre Dame, USA. This chapter introduces a location aware routing approach called A-GF (Asymmetric Geographic Forwarding). The approach takes advantage of asymmetric links in increasing the reliability and performance of ad-hoc routing. Link asymmetry in wireless transmission occurs when the transmission ranges of communicating parties are asymmetric: the station with a wider transmission range is able to successfully reach its peer station, but the peer station's transmission range is not wide enough to reach the other station. In this situation, only one-way transmissions are possible. Existing geographic forwarding algorithms assume that their environment consists of symmetric links. By discovering and exploiting also the asymmetric links the routing hop count is reduced, latencies decrease and routing reliability is improved. In making routing decisions, A-GF uses stability and minimum latency as metrics.

Chapter 6 is titled "Implementation Strategies for High-Performance Multiuser MIMO Precoders," and it is written by Maitane Barrenechea and Mikel Mendicute from the Department of Electronics and Computer Science, University of Mondragón, Spain, Andreas Burg from the Institute for Electrical Engineering, École Polytechnique Fédérale de Lausanne (EPFL), Switzerland, and John Thompson from the Institute for Digital Communications, University of Edinburgh, UK. This chapter evaluates a hardware implementation of a high-throughput vector precoder that can be used, for example, to separate data streams in multiuser Multiple-Input Multiple-Output (MIMO) systems. In addition, vector-preceding efficiency is currently experiencing high demand to bolster the convergence of personal computing devices as well as the Internet of Things (IoT) concept. Both IoT and personal device networks perform a lot of sensor processing, but communicating the unprocessed data is much more common for IoT applications due to energy efficiency concerns of IoT nodes on the one hand, and security issues related to personal devices on the other. In both of these multiuser environments, the separation of data streams in an energy efficient manner is of interest, although the energy envelopes of IoT applications are much more restrictive.

Chapter 7 is titled "Information Filter-Assisted Indoor Bluetooth Positioning," and it is written by Liang Chen, Heidi Kuusniemi, and Laura Ruotsalainen from the Department of Navigation and Positioning, Finnish Geodetic Institute, Finland, Yuwei Chen and Jingbin Liu from the Department of Remote Sensing and Photogrammetry, Finnish Geodetic Institute, Finland, Ling Pei from the School of Electronic Information and Electrical Engineering, Shanghai Jiao Tong University, China, Jian Tang from the GNSS Research Center, Wuhan University, Hubei, China, and Ruizhi Chen from the Conrad Blucher Institute for Surveying & Science, School of Engineering and Computer Science, Texas A&M, University Corpus Christi, USA. In this chapter, fingerprints of Received Signal Strength Indicators (RSSI) are used for localization. Due to the relatively long interval between the available consecutive Bluetooth signal strength measurements, a method of information filtering with speed detection is proposed. The method combines estimation information from RSSI measurements with prior information from the motion model. Speed detection is further assisted to correct the outliers of position estimation. Field tests show that the proposed new algorithm improves horizontal positioning accuracy of indoor navigation, achieving a 4.2 m positioning accuracy on average. The accuracy of the method compares favorably with the Bayesian static estimation method and the point Kalman filter.

Chapter 8 is the last chapter of section 2 of this book. The title of the chapter is "Analysis of Real-Time Hybrid-Cooperative GNSS-Terrestrial Positioning Algorithms," and it is written by Francesco Sottile, Zhoubing Xiong, and Claudio Pastrone from Istituto Superiore Mario Boella, Italy. The authors address the need for verifying co-operative positioning methods combined with traditional satellite-based positioning or terrestrial signals. In particular, the presented simulation tool simulates devices belonging to a Peer-to-Peer (P2P) wireless network where peers, also equipped with a GNSS receiver, co-operate among each other by exchanging aiding data in order to improve both positioning accuracy and availability. The authors also propose a method to increase the robustness of co-operative algorithms based on the estimated position covariance matrix. In particular, the proposed approach assures a faster estimation convergence and improved accuracy while lowering computational complexity and network traffic. The authors also present results from tests on the sensitivity of the implemented positioning algorithms in two different scenarios, first in presence of high level of pseudo-range noise and then in presence of a malicious peer in the P2P network.

The third thematic section of this book is titled "System Specification and Modeling," and it consists of five chapters (9-13). Chapter 9 is titled "Co-Modeling of Embedded Networks Using SystemC and SDL: From Theory to Practice," and it is written by Valentin Olenev, Irina Lavrovskaya, Pavel Morozkin, and Alexey Rabin from St. Petersburg State University of Aerospace Instrumentation, Russia, Sergey Balandin from the Open Innovations Association FRUCT, Finland, and Michel Gillet from Nokia Research Center, Nokia Devices, Finland. This chapter focuses on the problem of modeling embedded systems using SDL and SystemC. Both modeling environments have their own benefits and specialties that may provide important advantages in one project whereas may play a miniscule role in some other project. Hence, the research in this chapter aims to combine SDL and SystemC into a co-modeling environment, where the benefits of both would be available to designers. Specific attention is given to integration of SDL and SystemC modeling environments, exchanging the data and control information between the SDL and SystemC sub-modules, and the real-time co-modeling aspects of the integrated SDL/SystemC system. The co-modeling environment is put to test by finding a solution for embedded network protocols simulation. The authors reach the conclusion that of the different ways of co-modeling using SDL and SystemC, a solution where SDL modules are included in SystemC is the most viable modeling option. The authors state that SDL and SystemC co-modeling is a very perspective and interesting area of research and can result in improved quality in facilitation of the specification, modeling, and verification work.

Chapter 10 is titled "Model-Based Testing of Highly Configurable Embedded Systems," and it is written by Detlef Streitferdt from the Technische Universität Ilmenau, Germany, Florian Kantz and Holger Kaul from ABB Corporate Research, Ladenburg, Germany, Philipp Nenninger from ABB Automation Products, Goettingen, Germany, Thomas Ruschival from Datacom Telematica, Porto Alegre, Brazil, Thomas Bauer from Fraunhofer IESE, Kaiserlautern, Germany, Tanvir Hussain from the Mathworks GmbH, Ismaning, Germany, and Robert Eschbach from ITK Engineering AG, Ruelzheim, Germany. The authors present a practical instantiation of the model-based testing paradigm based on an industrial case study in the automotive domain. The chapter addresses the challenge of applying model-based testing to configurable embedded software systems in order to reduce the development efforts, while measuring the benefits (development quality, test case quality, and test effort) against traditional testing. The model-based testing support described in the chapter integrates existing model-based testing methods and tools such as combinatorial design and constraint processing. The main conclusion of the study was that by using model-based testing techniques the testing effort could be reduced by more than a third compared to the traditional testing process.

Chapter 11 is titled "On Parallel Online Learning for Adaptive Embedded Systems," and it is written by Tapio Pahikkala, Antti Airola, Thomas Canhao Xu, Pasi Liljeberg, Hannu Tenhunen, and Tapio Salakoski from the Turku Centre for Computer Science (TUCS) and Department of Information Technology, University of Turku, Finland. The authors propose a machine-learning algorithm that enables embedded systems to autonomously learn about emerging system behaviours, which may not have been foreseen at design time. The proposed machine-learning algorithm is evaluated in a Network-on-Chip platform consisting of 16 cores. The authors state, based on their experimental results that by optimizing the cache behaviour of the program, cache/memory efficiency can be significantly improved. Results presented in this chapter can be utilized as a guideline for designing future embedded multi-core machine learning devices.

Chapter 12 is titled "Classification-Based Optimization of Dynamic Dataflow Programs," and it is written by Hervé Yviquel and Emmanuel Casseau from IRISA, France, Matthieu Wipliez from Synflow SAS, France, Jérôme Gorin from Telecom ParisTech, France, and Mickaël Raulet from IETR/INSA, France. The authors study dynamic dataflow actor classification in different models of computation through abstract interpretation of an intermediate representation. Signal processing applications designed for parallel execution using conventional models of computation often exhibit excessive sequentialization in order to guarantee correctness under all potential run-time scenarios. This limits the performance increase potentially achievable by application level parallelization. Modern day parallel and dynamic signal processing applications are well matched with the dataflow model of computation. Hence, these applications can gain significant speedups from dataflow analysis when compiled to massively parallel execution platforms.

Chapter 13 is the last chapter in section 3 of this book. It is titled "Hierarchical Agent-Based Monitoring Systems for Dynamic Reconfiguration in NoC Platforms: A Formal Approach," and it is written by Sergey Ostroumov, Leonidas Tsiopoulos, and Marina Waldén from Åbo Akademi University, Finland and Juha Plosila from the Department of Information Technology, University of Turku, Finland. The chapter proposes formal models, a stepwise development methodology, and tool support for an agent-based system whose function is to dynamically monitor the state of a multi-core platform and to perform a reconfiguration when faults occur. The authors focus on certain specific parameters for monitoring and manipulating, namely temperature, faults, frequency, and voltage supply. They consider an obvious direction for future work to be to explore and integrate into the proposed approach other characteristics of many-core platforms such as traffic and activity of routers.

The fourth thematic section of this book is "Technologies for Network-on-Chip," and it consists of four chapters (14-17). Chapter 14 is titled "System-Level Analysis of MPSoCs with a Hardware Scheduler," and it is written by Diandian Zhang, Jeronimo Castrillon, Stefan Schürmans, Gerd Ascheid, and Rainer Leupers from RWTH Aachen University, Germany and Bart Vanthournout from Synopsys Inc., Belgium. Efficient runtime resource management in MPSoCs for achieving high performance and low energy consumption is one of the key challenges for system designers. The authors of this chapter base their proposed solution to this problem on OSIP, an operating system application-specific instruction-set processor, together with its well-defined programming model. They highlight the vital importance of the communication architecture for OSIP-based systems and optimize the communication architecture. Furthermore, they investigate the effects of OSIP and the communication architecture jointly from the system point of view, based on a broad case study for a real life application (H.264) and a synthetic benchmark application.

Chapter 15 is titled "Efficiency Analysis of Approaches for Temperature Management and Task Mapping in Networks-on-Chip," and it is written by Tim Wegner, Martin Gag, and Dirk Timmermann from the University of Rostock, Germany. Temperature management has become a major issue in System-on-Chip (SoC) as very large-scale circuits are increasingly exposed to thermal stress. This is addressed in the context of Networks-on-Chips (NoCs) in this chapter. On the one hand, the heat generated by the operating chip necessitates effective mechanisms for thermal management. On the other hand, application of thermal management is accompanied by disturbance of system integrity and degradation of system performance. The authors propose to precompute and proactively manage on-chip temperature of systems based on NoCs. With this proposed solution, traditional reactive approaches utilizing NoC infrastructure to perform thermal management can be replaced. This results not only in shorter response times for applying the management measures and therefore in reduced temperature and thermal imbalances, but also in less impairment of system integrity and performance. The systematic analysis of simulations conducted for NoC sizes ranging from 2x2 to 4x4 proves that under certain conditions the proactive approach is able to mitigate the negative impact of thermal management on system performance while still improving the on-chip temperature profile.

Chapter 16 is titled "Wearout and Variation Tolerant Source Synchronous Communication for GALS Network-on-Chip Design," and it is written by Alessandro Strano from Intel Mobile Communications, Germany, Carles Hernández from Barcelona Supercomputing Center, Spain, Federico Silla from Universitat Politècnica de València, Spain, and Davide Bertozzi from Università degli studi di Ferrara, Italy. Multiprocessor SoCs are largely relying on networked interconnects. The authors are addressing the vulnerability of source synchronous links in Networks-on-Chip (NoC). In particular, they are discussing the challenge of designing a process variation and layout mismatch tolerant link for synchronizer based Globally Asynchronous, Locally Synchronous (GALS) NoCs by implementing a self-calibration mechanism. They present a variation detector, which guarantees the reliability of NoC source synchronous interfaces under high variability. The variability detector is placed in front of the regular synchronizers in the communication architecture. They consider different cases of mismatch: misalignment between data and clock, misalignment between wires of the data link, and even random process variability of the detector's own logic cells. Analysis of the comprehensive metric of skew tolerance for a mesochronous synchronizer has revealed that the insertion of the variation detector has advantages in most cases and for the main synchronization architectures of practical interest (loosely vs. tightly coupled with the NoC). Even for non-variability-dominated links, the architecture with a variability detector can better cope with the layout constraints of the link.

Chapter 17 concludes the fourth thematic section and is also the last chapter of this book. The chapter is titled "Silicon Validation of GALS Methods and Architectures in State-of-the-Art CMOS Process," and it is written by Milos Krstic, Xin Fan, and Eckhard Grass from IHP, Germany, Luca Benini and M. R. Kakoee from University of Bologna, Italy, Christoph Heer, Birgit Sanders, and Alessandro Strano from Intel Mobile Communications, Germany, and Gabriele Miorandi, Alberto Ghiribaldi, and Davide Bertozzi from the University of Ferrara, Italy. One paradigm for constructing the communication infrastructure for complex systems on a single large die is to rely on the Globally Asynchronous, Locally Synchronous (GALS) scheme. There, the local computing subsystems are clocked synchronously, but this requirement is relaxed on chip level by letting the clock islands to run independently. This requires synchronization at the domain edges. The authors address the GALS methods in modern scaled-down microelectronics circuits based on a complex GALS implementation in 40 nm technology. The design analysis, measurement, and test results confirm the potential of the GALS approach for the scaled technologies, showing the significant benefits with regard to area, power, and EMI when it comes to the complex system implementation. Furthermore, 91% of the tests performed on the GALS network-on-chip test structures completed successfully, validating the timing robustness of new area and latency-efficient synchronization schemes and proving that the design flow for GALS synchronization technology can be implemented by means of mainstream industrial tools. This is crucial for the adoption of the methodology for commercial applications.

Seppo Virtanen
University of Turku, Finland
January 31, 2014

Acknowledgment

As the outgoing Editor-in-Chief of the *International Journal of Embedded and Real-Time Communication Systems* (IJERTCS), I wish to extend my sincerest thanks to all individuals who have contributed to the scientific content of the journal during my five years (January 2009 – December 2013) as Editor-in-Chief: all contributing authors, all current and past members of IJERTCS editorial review board, and all guest reviewers. It has been a pleasure and privilege to work with you in launching and establishing this journal.

This book is based on and extended from research published in volumes 2 (2011) and 3 (2012) of IJERTCS, and for these volume years, I wish to specifically extend my warmest thanks and appreciation to Professor Jari Nurmi (Tampere University of Technology, Finland and IJERTCS Associate Editor), Dr. Ina Schieferdecker (Technical University Berlin, Germany), Dr. Colin Willcock (Nokia Siemens Networks, Germany and former IJERTCS Editorial Review Board member), Dr. Dragos Truscan (Åbo Akademi University, Finland and former IJERTCS Associate Editor), Dr. Tapani Ahonen (Tampere University of Technology, Finland and IJERTCS Editorial Review Board member), Dr. Juha Plosila (University of Turku, Finland and IJERTCS Associate Editor), Dr. Waltenegus Dargie (Technical University of Dresden, Germany), Dr. Elena-Simona Lohan (Tampere University of Technology, Finland), Dr. Stephan Sand (German Aerospace Center, Germany), and Dr. John Raquet (Air Force Institute of Technology, USA) for their hard work and effort as Guest Editors or Guest Co-Editors of journal issues during the second and third volume years of IJERTCS.

Seppo Virtanen
University of Turku, Finland

Section 1
Technology Review

Chapter 1
Self–Adaptive SoCs for Dependability:
Review and Prospects

Liang Guang
University of Turku, Finland

Juha Plosila
University of Turku, Finland

Hannu Tenhunen
University of Turku, Finland & Royal Institute of Technology, Sweden

ABSTRACT

Dependability is a primary concern for emerging billion-transistor SoCs (Systems-on-Chip), especially when the constant technology scaling introduces an increasing rate of faults and errors. Considering the time-dependent device degradation (e.g. caused by aging and run-time voltage and temperature variations), self-adaptive circuits and architectures to improve dependability is promising and very likely inevitable. This chapter extensively surveys existing works on monitoring, decision-making, and reconfiguration addressing different dependability threats to Very Large Scale Integration (VLSI) chips. Centralized, distributed, and hierarchical fault management, utilizing various redundancy schemes and exploiting logical or physical reconfiguration methods, are all examined. As future research directions, the challenge of integrating different error management schemes to account for multifold threats and the great promise of error resilient computing are identified. This chapter provides, for chip designers, much needed insights on applying a self-adaptive computing paradigm to approach dependability on error-prone, cost-sensitive SoCs.

INTRODUCTION

Dependability is a primary requirement for computer-based systems. Communication systems, embedded control systems and even entertainment systems all demand certain level of dependability as to be satisfying to the application requirements (Kobayashi & Onodera, 2008). Dependability broadly encompasses the attributes of availability, reliability, safety, integrity, confidentiality and maintainability (Avizienis, Laprie, Randell, Landwehr, 2004; Knight, 2002).

DOI: 10.4018/978-1-4666-6034-2.ch001

SoC (System-on-Chip) is a modern form of VLSI (Very Large Scale Integration) chips. Diverse system components, e.g. processors, memories, accelerator, sensors and actuators, and different interconnections between these components, can be designed and implemented on a single chip, due to the constant scaling of semiconductor devices. The increasing usage of SoC in modern and emerging computing systems, e.g. health monitoring and industrial automation (Eshraghian, 2006), demands more attentive research on the dependability of SoCs. Especially the shrinking transistor feature size introduces significant process variations (Unsal et al., 2006), crosstalk, leakage and other submicron effects. Process variations refer to the deviation of actual physical parameters from nominal values due to the manufacturing process or post-manufacture aging. In particular time-dependent device degradation (Collet, Zajac, Psarakis, Gizopoulos, 2011), e.g. due to aging and material wearout, run-time voltage and temperature variations (Henkel, Ebi, Amrouch, & Khdr, 2013), and even radiations (Duzellier, 2005), pose more profound challenges to the design of SoCs.

Conventional techniques to provide dependability are either too costly or very limited in applicability. For example, TMR (triple modular redundancy) makes three-fold replicas of the circuits to detect and correct errors, whose area penalty is large (Kobayashi & Onodera, 2008). Worst-case-based design, which applies a conservatively large safety margin (e.g. a high supply voltage to account for voltage variations), incurs significantly performance or power overhead (Ernst et al., 2003). Besides, effects such aging cannot be detected by conventional duplication-based techniques, as replicas may suffer from aging simultaneously.

Self-adaptive computing is an emerging paradigm to achieve functional diversity, energy and power efficiency and dependability in systems of different scales (Salehie & Tahvildari, 2009). By monitoring the internal (e.g. power consumption) and external (e.g. ambient temperature) status, a self-adaptive system can tune its own configurations to approach various objectives (specified by user, system administrator, or derived from these external sources) (Guang, 2012). Without overly relying on static pre-assumption of circuit and environmental status, self-adaptive techniques can better address the run-time dependability concerns, and apply diverse temporal or spatial recovery techniques to maintain the system function and performance. Such techniques have been widely proposed on various processing, communication and memory components and the overall systems (e.g. Das et al., 2009; Collet et al., 2011; Boesen, Keymeulen, Madsen, Lu, & Tien-Hsin Chao, 2011b). Great potential can be observed in this direction.

There exist survey papers either on self-adaptive softwares, e.g. (Salehie & Tahvildari, 2009), or dependability of specific subsystems, e.g. (Radetzki, Feng, Zhao, & Jantsch, 2013), and overview papers on the dependability issues of embedded systems in general, e.g. (Henkel et al., 2011; Knight, 2002). This chapter gives a dedicated study on self-adaptive architectures and techniques for dependability provision on SoCs. Particular focus lies on time-dependent hardware errors and degradation, e.g. due to aging, thermal stress or radiation, and run-time techniques to predict, detect, recover and tolerate these dependability threats.

The rest of the chapter is organized as follows. Section "Dependability and Self-Adaptivity: Overview" provides a brief conceptual overview, and describes how the combined study of the two features will be performed. Section "Threats to SoC's Dependability" presents the technological perspective on the increasing criticality of dependability issues of on-chip systems. Section "Self-Adaptive Architectures and Techniques" extensively reviews recent works on self-adaptive architectures or techniques addressing the foreseen dependability concerns. Section "Summary and Future Perspectives," based on summarizing existing achievement, identifies remaining challenges and future directions. Section "Conclusion" concludes this chapter.

DEPENDABILITY AND SELF-ADAPTIVITY: OVERVIEW

In general, dependability is "the ability to deliver service that can justifiably be trusted" (Avizienis et al., 2004). The concept of dependability can be elaborated by six attributes (Knight, 2002): availability- operational when needed, reliability-continuation of correct operation, safety- without dangerous output, confidentiality- without unauthorized disclosure, integrity- without unauthorized modification, and maintainability- the ability to be modified and repaired. Dependability can be jeopardized by software faults, e.g. design errors or programming errors, or hardware defects, e.g. open or short wires, aging or thermal stress. The scope of dependability is much wider than the conventional fault-tolerance, which is, strictly, one of the means to achieve dependability (others being fault prevention, fault removal and forecasting) (Avizienis et al., 2004). In SoCs, many dependability considerations, e.g. thermal stress, the monitoring of aging before failures, e.g. (Agarwal, Paul, Ming Zhang & Mitra, 2007), or error resilience architectures (Leem, Cho, Bau, Jacobson, & Mitra, 2010), are beyond the realm of classic fault tolerance discussions. This survey does not intend to delve into the detailed differentiation of various dependability attributes. Instead it focuses on the major threats to dependability on SoCs, esp. those causing run-time hardware defects.

Self-adaptivity is a newcomer of system features, but very important in future systems. The concept of self-adaptivity was initially applied to software design- "Self-adaptive software evaluates its own behavior and changes behavior when the evaluation indicates that it is not accomplishing what the software is intended to do, or when better functionality or performance is possible" (Laddaga, 1997). Self-adaptivity has a different flavor of focus from autonomic computing (Horn, 2001). While autonomic computing has proposed the key aspects of adaptive functions- self-configuring, self-healing, self-optimizing and self-protecting (Horn, 2001), self-adaptive computing focuses on the features and constructions of the self-adaptation process, e.g. the monitoring, decision-making and reconfiguration loop (Guang, 2012), or the "Where-When-What-Why-Who-How" questions for adaptation requirement elicitation (Salehie & Tahvildari, 2009).

To review how self-adaptive computing can be utilized to provide dependability, the survey examines the extensive amount of existing works related to monitoring, processing and recovering run-time errors based on the self-adaptive process framework (Figure 1). To account for the different types of errors (functional or timing-related), monitoring techniques need to be specifically designed, whose outputs are then analyzed via the decision-making processes before reconfiguration of the system to tolerate the fault or recover the function. The decision-making process can be performed in a distributed, centralized or hybrid manner. And both the monitoring and reconfiguration processes have diverse design and implementation techniques. We will explore the broad design space in the following sections.

THREATS TO SOC's DEPENDABILITY

With technology scaling and diverse application needs, dependability has more concerns beyond manufacturing defects and yield. For instance, aging and thermal effects influence the dependability during the lifetime reliability of chips (Henkel et al., 2011). These effects lead to errors, faults and failures of various features. This section presents the sources to hardware errors after the manufacturing process, and the common forms of errors, faults and failures on SoC components and systems.

Figure 1. Studying self-adaptation processes for dependability on SoCs

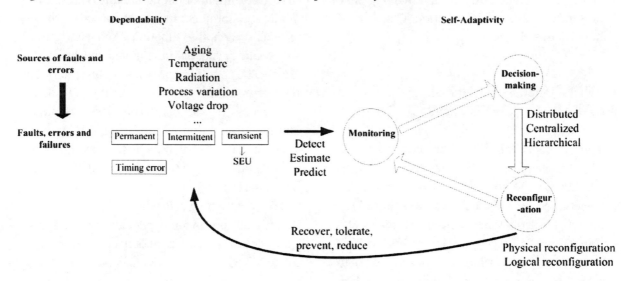

Sources

After chips have been manufactured, tested, delivered and deployed, their function and performance can still be negatively influenced by aging, thermal effect and radiations.

Aging is a primary source of errors. It is the degradation of transistor performance over time (Agarwal et al., 2007) due to atomic or subatomic-level physical activities within the semiconductor material. Common physical mechanisms include NBTI (negative bias temperature instability), HCI (hot carrier injection), TDDB (time dependent dielectric breakdown) and electromigration (Kim, Wang, & Choi, 2010; Keane, & Kim, 2011). NBTI is the dominating mechanism to cause the increase of threshold voltage (thus lowering the switching speed) of PMOS transistors (Agarwal et al., 2007). HCI causes lower saturation current and higher threshold voltage in NMOS transistors (Kim et al., 2010). TDDB and electromigration both can degrade the interconnection between transistors (Noguchi et al., 2001; Keane, & Kim, 2011). Aging can impact the performance of processors, memories (Joshi, Kanj, Adams, & Warnock, 2013) and interconnects.

In addition to aging, thermal effects is also a major concern for chips' dependability. The peak temperature, the thermal gradients (as the temperature is uneven across the chip), and the thermal cycling (period phases of heating and cooling) all affect the chips (Henkel et al., 2013). The temperature influences the dependability mainly through two manners - causing timing errors and worsening the aging effects (Henkel et al., 2013). It is claimed by Narayanan and Xie (2006) that the interconnect delay increases by 5% with 10°C increase in temperature. Major mechanisms of aging, e.g. electromigration and NBTI, are worsened by the thermal effects. In fact, it is considered the dominant stress factor in electronic devices, e.g. automotive systems (Actel, 2003).

Radiation is also posing further challenges on chips' dependability. Radiation-induced high-energy particles strike the semiconductor material and induce many observable effects (Duzellier, 2005). One most common effect is SEU (single event upset; see Section "Faults, Errors and Failures"), a transient soft error strongly affecting latches and SRAMs (Actel, 2003). Specific materials, e.g. flash and antifuse, can be used to alleviate such effects (Actel, 2003). This concern is more profound in, e.g. space applications where the radiation is stronger in the external environment.

In addition to thermal, radiation and aging effects, other deep submicron effects e.g. crosstalk and voltage variations, also affect the dependability. Crosstalk, as the noise between adjacent wires, is considered as one main source of glitches and delay errors on interconnection (Radetzki et al., 2013). Voltage variation refers to the difference of voltage values, in particular the supply voltage, on different places in the power delivery network (Gupta, Oatley, Joseph, Wei, & Brooks, 2007). The voltage variation will affect the delay of processing elements and interconnect. With technology scaling, circuits are more susceptible to crosstalk and voltage variations (Gupta et al., 2007; Chandra & Aitken, 2008).

Faults, Errors, and Failures

In strict senses, faults, errors and failures are referred to dependability threats of different features. A failure, or a service failure is "an event that occurs when the delivered service deviates from correct service," which indicates "one (or more) external state of the system deviates from the correct service state"- an error (Avizienis et al., 2004). A fault is the "adjudged or hypothesized cause of an error" (Avizienis et al., 2004). A fault does not necessarily lead to an error- a dormant fault is one that has not produced an error (Mushtaq, Al-Ars, & Bertels, 2011), otherwise active. For instance, a fault in a redundant component is dormant until the component is activated. This survey does not emphasize on the taxonomic differences of the three terms, which will be used interchangeably unless specifically stated.

Faults can be categorized, in terms of persistence, as permanent, intermittent and transient (Radetzki et al., 2013). Permanent faults can be caused by manufacturing defects, aging (e.g. electromigration may lead to open and short circuits (Constantinescu, 2003)), or thermal stress. Intermittent faults are repetitively appearing faults,

often in bursts, which are caused by unstable or marginal hardware, e.g. partial metal separation in the via (Constantinescu, 2003). Intermittent faults can develop into permanent faults. Transient faults are caused by temporary interference, e.g. crosstalk, voltage noises or radiation.

Two types of faults are of particular interest under the shrinking technology. SEU (single event upset) is a common type of soft error, which can be corrected as the hardware device is not permanently defected (Actel, 2003). It is a transient error caused by particle hits which are strong enough to flip the content of a state-saving element (Chandra & Aitken, 2008), e.g. latches, SRAM. The occurrence of SER is increasing due to the decrease in node capacitance and supply voltage (Chandra & Aitken, 2008; Constantinescu, 2003). And it is more often at higher altitudes (Actel, 2003), which needs to be especially considered for, e.g. space applications (Duzellier, 2005). Due to its mechanism, SEU affects various types of memory circuits, e.g. latches, SRAM and DRAM, thus affects the dependability of register files (in processors), on-chip caches and memory (Wang, Hu, & Ziavras, 2008). Common techniques to deal with SEUs include using technologies less vulnerable to soft errors, e.g. fully depleted SOI devices, parity coding and ECC (error correction coding) (Chandra & Aitken, 2008), and replicas of memory content to recover from errors (Wang et al., 2008). In particular, adaptive techniques against SEU will be elaborated in Section "Self-Adaptive Architectures and Techniques."

Timing error is another common type of error. Timing errors refer to the late arrival of data signals compared to the clock signal, so that the data cannot be sampled correctly (Guang, Nigussie, Plosila, Isoaho, & Tenhunen, 2012a). Timing errors can be caused by process or voltage variations (Das et al., 2009), aging (Agarwal et al., 2007) and thermal variations (Narayanan & Xie, 2006). Process or voltage variations can lead to timing errors due to the decrease of supply voltage and the increase of threshold voltage, which slow down the transistors' switching speed. Aging causes the timing error in a similar manner, e.g. the NBTI effect increases the threshold voltage of PMOS transistors. High temperature also delays the interconnect and transistor switching (Narayanan & Xie, 2006). Timing errors can influence processing, communication and memory subsystems (Das et al., 2009; Simone, Lajolo, & Bertozzi, 2008; Karl, Sylvester, Blaauw, 2005). Timing errors can be detected or predicted by inserting a delayed clock, and comparing the outputs sampled by the original and the delayed clock (Das et al., 2009; Fuketa, Hashimoto, Mitsuyama, Onoye, 2009). They can be recovered via online tuning of threshold voltage, supply voltage and clock frequencies. Extensive examples for timing error detection and correction will be studied.

The categorization of fault types is helpful to the identification of causes and proper countermeasures. For instance, permanent faults are often due to manufacturing defects, process variations, aging, thermal breakdown etc., while transient faults can be caused by, e.g., crosstalk, noise, radiation etc. In terms of countermeasures, generally speaking, spatial redundancy (e.g. spare component) best addresses permanent faults, temporal redundancy (e.g. retransmission) can recover transient faults, while information redundancy (e.g. ECC) is suitable for both permanent and transient faults (Radetzki et al., 2013). Most works do not treat intermittent faults separately. If a fault persists after repeating the operation, it is considered as permanent, otherwise treated as a transient fault (Lehtonen, Wolpert, Liljeberg, Plosila, & Ampadu, 2010). Surely these general guidelines are not exclusive. The next section will elaborate on these techniques.

SELF-ADAPTIVE ARCHITECTURES AND TECHNIQUES

This section extensively surveys self-adaptive techniques and architectures, which have been proposed to address the above mentioned depend-ability concerns. The techniques are discussed, based on the generic self-adaptation processes (Figure 1), for monitoring, decision-making and reconfiguration processes respectively. Overall system architectures, which integrate the three self-adaptation processes into a unique paradigm, are researched at the end.

Monitoring

Common techniques for run-time monitoring of faults and errors include self-test, duplication, error coding, delayed sampling (for timing errors), stability checkers, thermal sensors and other application-specific sensors (e.g. aging sensors).

Self-test (Kamran & Navabi, 2013; Lehtonen et al., 2010; Sylvester, Blaauw, & Karl, 2006) is a well-thought-of technique for permanent errors (e.g. short circuit or open circuit) due to manufacturing defects or aging. It is a natural application of the classic functional test for on-line detection of similar errors. To prevent interruption of normal circuit operations, measures such as switching to a peer component (Lehtonen et al., 2010) or periodic triggering of testing operation (Kamran & Navabi, 2013) can be taken. Self-test can be applied for any on-chip component, in a centralized or distributed manner.

Duplication (Blome, Gupta, Feng, Mahlke, & Bradley, 2006) is also a common technique following the classic DMR (double modular redundancy) to detect permanent, intermittent or transient errors. TMR is an enhanced version of DMR, which can both detect and correct errors. The basic principle of duplication is to replicate a circuit and compare the outputs of the two circuits (Figure 2). Different outputs will trigger an error signal. Blome et al. (2006) presents a RVC (register value cache) architecture, where the mostly used register file entries are replicated in a separate storage (Figure 2). By comparing the contents of the original entry and its replica, errors can be detected. As both the original copy and the replica can be wrong, cyclic redundancy check (an error coding technique) is performed on both

Figure 2. From basic DMR technique to register file duplication for error detection and correction (Blome et al., 2006)

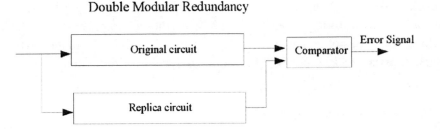

Double Modular Redundancy

Register Value Cache architecture

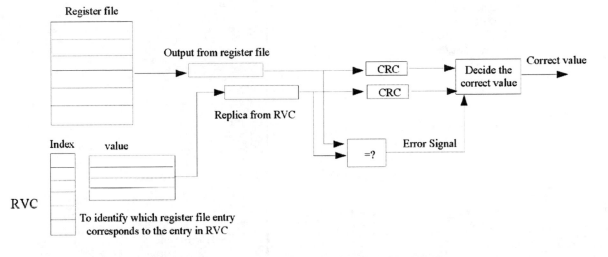

copies. The value to be adopted is the one passing the check. The work assumes the probability of both copies being corrupted in the same time period is negligible. This example also shows that duplication can be used for error correction (see Section Reconfiguration). Compared to self-test, duplication can detect transient errors as the error detection is performed in parallel with the normal operation. In addition, it can be extended for error correction (e.g. TMR). However duplication will introduce area overheads, thus measures are taken to predict circuits which are mostly likely to suffer from errors, e.g. the register file.

Error detection coding is a widely used technique for permanent, intermittent and transient error detection, which is often combined with duplication (Wang et al., 2008; Blome et al., 2006). Error detection coding adds tags to the data,

which can be used to verify the correctness of the data after being operated or transmitted. There are different coding schemes. The simplest one is parity checking, including even parity and odd parity. Other codings, e.g. CRC (cyclic redundancy coding), are also often used (Blome et al., 2006). Error detection coding can be applied to different subsystems. Wang et al. (2008) applied parity coding to detect cache error, where a cache line is replicated with both copies added parity checking. If either copy passes the parity checking, this copy will be considered as the correct value. Lehtonen et al. (2010) applies syndrome decoding (also an error correction coding) for interconnection errors. Error detection coding can be enhanced to be error correction coding, which will be explained further in Section Reconfiguration.

Delayed sampling is a popular detection technique for timing errors (Ernst et al., 2003; Karl et al., 2005; Simone et al., 2008). If a data signal propagates slower than expected, i.e. timing error, the correct value of the signal will not arrive at the input of the register before the clock signal. In this case a wrong value of the data signal will be sampled. To detect this error, a delayed sampling of the signal can be performed, e.g. by using a delayed clock, when the data signal has surely arrived. The difference of the original sample and the delayed sample indicates a timing error (Figure 3). This

Figure 3. Delayed Sampling and its Application in Timing Error Detection (Ernst et al., 2003; Simone et al., 2008)

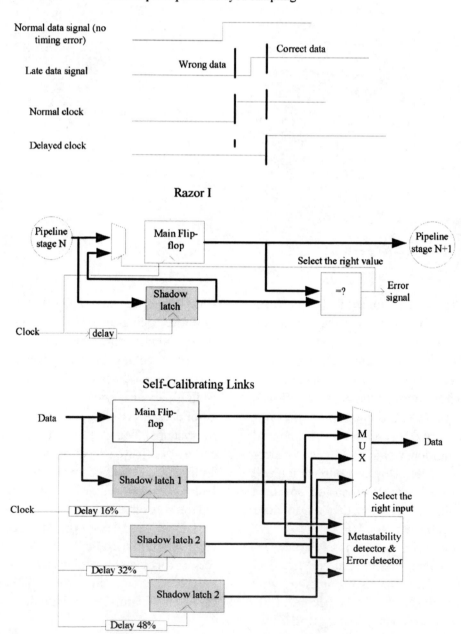

technique is widely used in different subsystems, including on-chip interconnect (Simone et al., 2008), processor (Ernst et al., 2003) and memory (Karl et al., 2005), to detect timing errors due to various causes, such as aging, process, voltage and the temperature variations. Razor I (Ernst et al., 2003) is a pioneering work applying delayed sampling for timing error detection and correction in processor pipelines. A shadow latch samples the input data with a delayed clock (Figure 3). The worst-case delay of the input data is guaranteed to meet the setup time of the shadow latch (Ernst et al., 2003), so that the shadow latch can always obtain the correct value. Karl et al. (2005) apply a similar design for on-chip SRAM-based memory. It implements a shadow sense amplifier to detect and correct timing errors on memory access. Simone et al. (2008) design several shadow latches to quantify the amount of timing delay on self-calibrating links. The three shadow latches as illustrated in Figure 3 are sampled by clocks with, e.g., 16%, 32% and 48% cycle delay. By comparing the outputs of the original sample and the three delayed samples, the timing error and its approximate delay range can be identified. Several design challenges were raised for shadow latch-based timing error detection. Metastability, as the consequence of simultaneous clock and data transition, will disable the timing error detection. Thus metastability detectors are often integrated in such designs (Ernst et al., 2003; Simone et al., 2008). In addition, the area overhead of the shadow latch can increase the design cost, thus the new Razor II (Das et al., 2009) uses a lightweight transition detector (described later in the same section) instead of a shadow latch.

Temperature sensor is specifically used for temperature monitoring (Memik, Mukherjee, Min Ni, & Jieyi Long, 2008; Clabes et al., 2004). In current technology, it is highly feasible to integrate many temperature sensors on a single chip (e.g. IBM Power5 microprocessor integrates 24 digital temperature sensors (Clabes et al., 2004)), to monitor overheating or build the temperature profile of the chip. The location of temperature sensors should be optimized based on the likely hotspots and thermal gradients (Memik et al., 2008).

In addition to the above mentioned common monitoring techniques, other application-specific error detection or prediction circuits have also been proposed for individual purposes. For instance, to sense aging, instead of waiting for its consequence in timing errors, Agarwal et al. (2007) have designed a stability checker to predict aging-induced timing error. A guard band, which is an interval before the circuit actually has a timing error, is set based on aging analysis of the chip during a short period after the chip is manufactured. At the run-time, if the data transits during the guard band, it is considered that a timing error may occur in the near future (with continued aging). A stability detector, which is an analog circuit-based component, is designed to detect such transition (Agarwal et al., 2007). To target more specific causes of aging, Kim et al. (2010) have designed sensor circuits to detect NBTI and HCI-induced aging effects respectively. The design compares the threshold voltages of stressed (with aging) and unstressed (with high voltage input) devices to detect increase of the threshold voltage, to capture the consequence of aging.

Decision-Making

The decision-making process analyzes the monitored information and decides on the proper reconfiguration, if needed. The decision-making process can be categorized based on its temporal and spatial features.

In terms of the temporal feature, the process can be performed locally, centrally, cooperatively or hierarchically. Distributed decision-making can be used for errors or effects with local influence, e.g. error checking and correction in each interconnect channel (Simone et al. 2008). Such decision does not require the enabling or support of peer components, e.g. the error detector in nearby channels. Cooperative decision-making is the process where multiple peer components work together to make decisions. For instance, in the eDNA (Boesen, Madsen, & Keymeulen, 2011a) system model for self-organization and self-healing hardware architecture (further elaborated in Section "Overall Architectures"), three eCells work together to identify the location of a single fault, similar to the TMR approach but on a higher structural level. Collet et al. (2011) also utilizes a cooperative process to identify possible routes on a multi-core array when interconnects, routers or cores fail. The source node sends routing requests to other nodes via flooding protocol (i.e. messages sent to all possible nodes except the sender), and all proper nodes receiving the request will add the local index before acknowledging and forwarding the messages. In this case, all peer nodes help to establish a possible route for sender nodes to reach expected destinations. Centralized decision-making is often applied to parameters which have been global influence or cannot be dealt properly only with local conditions. Temperature is one of such threats. When a hotspot appears, in order to prevent overheating, not only the power dissipation of the hotspot but its neighbors or the whole chip needs to be considered. For instance, on Power5 processors, once an overheating is sensed by one of the 24 thermal sensors, the centralized core control logic will perform frequency reduction or thermal throttling (reducing the processor activities) to deal with the overheating (Clabes et al., 2004). With SoCs becoming massively parallel (e.g. with many-core architectures), the temperature management is often handled in a distributed or cooperative manner. For instance, Shang, Peh, Kumar, and Jha (2004) apply distributed thermal throttling on each router to reduce local hotspot and traffic to enter other hotspots, on a network-on-chip platform. Hierarchical decision-making is a hybrid of distributed, cooperative and centralized processes. While the local component makes decisions for its individual configuration, the system (or a subsystem) decides on the global configuration which is applied to all of its components. For instance, to deal with soft errors in data cache, Wang et al. (2008) integrates different monitoring and recovery schemes locally (e.g. replicas of cache lines), while globally deciding on which scheme to apply based on the collective error rate. Guang (2012) proposes a hierarchical agent-based control architecture, where the local agents detect and report the errors, while the central agent allocates and assigns spare components upon permanent failures. Both works will be elaborated in Section "Overall Architectures."

In terms of the spatial feature, the decision-making process can be present-based reactive action or prediction-based proactive configuration. Present-based decision making is more straightforward, which decides on the configuration to respond to current system or circuit status. It is most applicable to errors or defects which are random or difficult to identify any temporal connection. For instance, Razor II processor (Das et al., 2009) performs detection and correction for timing error and soft error, whose correlation to future incidents is hard to identify. The thermal management on Power 5 processor (Clabes et al., 2004) is also reactive when a sensor detects overheating. When the future of a parameter can be predicted e.g. via a quantitative model, prediction-based proactive decision-making can be performed. For instance, aging is usually considered cumulative with estimation model of its process, e.g., (Paul, Kunhyuk Kang, Kufluoglu, Alam, & Roy, 2005), thus a guardband to predict aging-induced failures can be estimated (Agarwal et al., 2007). Similarly, Nakura, Nose and Mizuno (2007) observe path delay increase to predict failures caused by latent defects. Shang et al. (2004) estimate and predict network traffic for thermal throttling.

To choose proper temporal and spatial features for decision-making, the application requirement, the dependability of the decision-making process itself, the power and area overhead, as well as the availability of quantitative models need to be considered. For dependability threats with disastrous consequences, e.g. thermal breakdown, or leading to unsafe conditions, prediction-based proactive measures are preferrable. Centralized decision-making, though low in overhead or conflicts, suffers from single point of failure (Boesen et al., 2011). Distributed decision-making, though more dependable, requires careful consideration of power and area overheads. In addition, for specific decision-making processes, esp. prediction-based approaches, the availability and accuracy of quantitative models for analysis and estimation are essential, which sometimes require extensive research (Paul et al., 2005).

Reconfiguration

To prevent, tolerate or recover from the faults and errors, diverse techniques with different overheads have been designed. In addition to the conventional duplication and error correction coding schemes, this section focuses more on the techniques or methods specifically proposed for self-adaptive systems.

Duplication, spares, error correction coding and architectural replay are the classic methods for error correction and recovery. TMR is the most traditional form of duplication for automatic error recovery, where three replicas vote for the correct output. As TMR incurs large (two-fold) overhead, lower-overhead spatial redundancy schemes are usually designed. For instance Blome et al. (2006) make duplicate of most recently accessed register file entries, instead of the whole register file to reduce overhead. Spares provide spatial redundancy, e.g. in the form spare processing unit, memory, communication channels or routers. For instance, Lehtonen et al. (2010) use a small number of spare wires to recover from permanent interconnect errors. Error correction coding, as the enhanced version of error detection coding, can automatically correct limited bits of errors as it has higher information redundancy. Different error correction codings, e.g., Hamming code, are used in processors, e.g. Intel Itanium processors (Quach, 2000), memory and on-chip

networking. Architectural replay, when used in pipelined processors, refers to the process that processor rollbacks to the point before the error occurs, and reperforms the pipeline stages. It is a common and well-studied technique in processors with speculative operations or fault-tolerance. For instance, Razor II architecture (Das et al., 2009) adopts the technique for error recovery. Architectural replay's counterpart in networking is ARQ (automatic repeat request), when the receiver requests the sender to retransmit the data upon errors. While error correction coding is suitable for either permanent or transient faults, spares are more suitable for permanent errors. Duplication and architectural replay are used mainly for transient errors.

Physical reconfiguration techniques, in particular adaptive voltage and frequency scaling, are suitable for timing errors caused by various sources (aging, voltage, temperature or process variations). The principle of voltage and frequency scaling for timing tuning comes from the relation between the CMOS circuit delay and these physical or electrical parameters, e.g. Equation 1 and Equation 2.

$$t = \frac{1}{f} \tag{1}$$

$$t = K \frac{C_L * V_{DD}}{\beta \left(V_{DD} - V_{TH}\right)^{\alpha}} \tag{2}$$
$$\left(Nose \ \& \ Sakurai, \ 2000\right)$$

t is the circuit delay. f is the operating frequency. V_{DD} is the supply voltage. V_{TH} is the threshold voltage. K is a fitting parameter. α (typically $\alpha >$ 1.1) is the velocity saturation index dependent on the technology. β is a parameter dependent on α but independent from V_{DD}. From Equation 1, we can see that tuning frequency lower can accommodate for the increase of delay. From Equation 2, we can see that by increasing the supply voltage, or reducing the threshold voltage, timing errors can be recovered. Physical reconfiguration can be applied to memory, interconnect and processors. For instance, Razor I (Ernst et al., 2003) is an early work utilizing adaptive voltage scaling to tradeoff power consumption and error rate. When the timing error rate is too high, the supply voltage can be increased to lower the error occurrence. Karl et al. (2005) apply similar tradeoff for SRAM-based memory system. Simone et al. (2008) utilize adaptive voltage swing (the range of voltage on the wire) for on-chip interconnects to compensate for the timing delay in switches. Current technology has provided low-overhead hardware circuits for coarse and fine-grained voltage and frequency scaling (Guang, Nigussie, Plosila, Isoaho, & Tenhunen, 2012b).

Mapping, task migration and self-organization are system-level techniques, relying on spatial redundancies, to recover from faulty processing elements or communication channels (Collet et al., 2011; Boesen et al., 2011a). Self-organization is the combined application of task and communication mapping schemes, to rebuild a functional system upon permanent failures. As illustrated in Figure 4, when a processing element fails, a replacement can be chosen from a list of available spares based on specific criteria, e.g. minimum Manhattan distance (Boesen et al., 2011a). For instance, the eDNA architecture (Boesen et al., 2011a; Boesen et al., 2011b) utilizes cooperative decision-making in choosing spares and updating system organization (to be elaborated in Section "Overall Architectures"). If certain communication channels fail, possible routes need to be discovered (if fully adaptive routing is not enabled) to maintain the reachability between processing elements, e.g. the cooperative route discovery with flooding protocol (Collet et al., 2011). To enable run-time self-organization, the information of current system structure, e.g. the location of spares or the status of each component, needs to be

Figure 4. Illustrating task migration and route remapping for self-organization

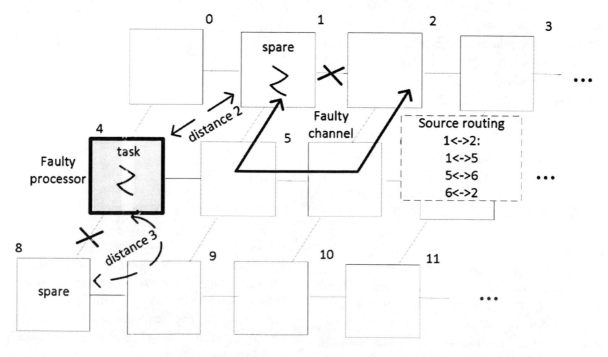

stored and made available to decision-makers. In the next Section, we will see how eDNA (Boesen et al., 2011a) architecture and H2A architecture (Guang, 2012) have designed supporting structures for this purpose.

Thermal throttling, temperature-aware mapping, and power and clock gating are techniques, in terms of dependability provision, to address thermal stress on SoCs (Shang et al., 2004; Pedram & Nazarian, 2006; Henkel et al., 2013). Thermal throttling is the logic process of lowering the workload on the processing element or the interconnect in order to avoid overheating, e.g. by distracting traffics from the hotspot (Shang et al.) or migrating a low-power application to a processor which originally runs a high performance application (Henkel et al., 2013). Temperature aware mapping considers the location of processing elements, the processing workload and the communication to avoid thermal hotspot (Addo-Quaye, 2005). Power gating and clock gating are conventional techniques to reduce the temperature by reducing the power consumption (Pedram & Nazarian, 2006). As the temperature directly influences other effects, e.g. aging and timing errors, temperature management can also be used to deal with these related issues. For instance, Henkel et al. (2013) apply thermal throttling to alleviate aging.

Overall Architectures

Previous sections have examined techniques for individual self-adaptation processes. This section highlights four works, which propose integrated monitoring, decision-making, and reconfiguration system architectures for dependable computing that can be applied to SoCs. Each of these works is representative in a particular design aspect, either in design paradigm, system model, adaptation methods, or microarchitectural implementation. They will be presented in the chronological order, to reflect the architectural evolution.

ElastIC (Sylvester et al., 2006) is an early work positioning an integrated self-adaptive system

architecture. It strongly motivates the necessity and feasibility of maximized inherent adaptivity of the system, to deal with the constantly increasing variability. ElastIC is composed of a centralized DAP (diagnostic and adaptivity processing) unit, and a scalable group of modularized processing elements. The processing elements are embedded with tunable parameters, which support, on the hardware level, run-time performance monitoring and reconfiguration schemes such as self-test and dynamic voltage and frequency scaling. With processing elements' hardware support, DAP unit performs run-time monitoring, self-testing, self-healing and adaptive power tuning with a system-level scheduler. The memory units utilize error coding and redundancy to tolerate functional errors and parametric variations. Even though ElastIC did not physically implement the visioned micro-architectural designs, its proposals- 1) systematic structure of self-adaptivity, with centralized supervisor and distributed components; 2) processing element design with modularized monitoring and tuning interfaces; 3) hybrid application of dependability provision schemes, e.g. self-test, voltage and frequency scaling, error coding, for specific purposes on different components - have inspired follow-up researches.

SA-RDC (self-adaptive reliable data cache) (Wang et al., 2008) specifically addresses soft errors on data caches. It employs distributed architectural-level protection schemes and system-level policy switching, which effectively constitute two cycles of self-adaptation (Figure 5). In the distributed cache structures, error detection, reduction and recovery schemes, such as parity coding, in-cache replication (replicas of most used cache entries in other cache lines) and dead-time early write back (to the dirty cache lines), are implemented. These schemes have different error tolerance, overhead and performance trade-offs. For instance, the basic parity coding with software-based error recovery has the lowest overhead when the error rate is low. But with higher error rate, the penalty for softwre-based

Figure 5. Reliable data cache with system-level and architectural-level error-driven adaptation

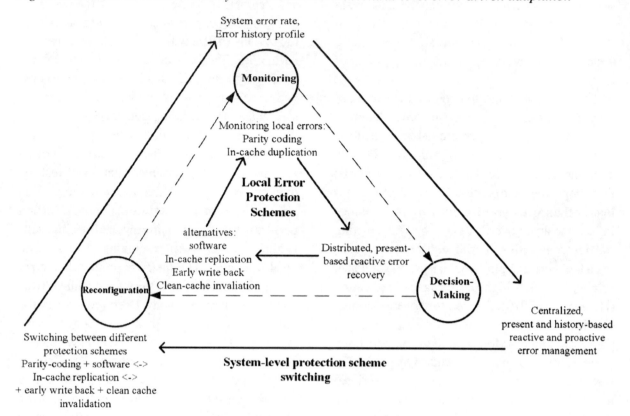

recovery increases significantly. Thus the other adaptation cycle, at the system-level, switches the error management schemes based on the collective error rate. For instance, if the current protection scheme is parity coding and the error rate is e-4, the system switches to in-line cache replication to tolerate the high error rate. Similar strategies, but with history-based error profiles, are employed to switch from higher-protection schemes to lower-protection schemes.

SA-RDC is highly noteworthy especially due to its novel combination of dynamic system-level policy adaptation and architectural-level software and hardware support of various error protection schemes. Such architecture can handle environments with different levels of error rates, with adaptive tuning of error tolerance, performance and energy overhead. Besides, these choices of protection schemes are tailored to the targeted

subsystem- data cache, which is particularly vulnerable to the addressed dependability threat-soft error.

eDNA (Boesen et al., 2011a; Boesen et al., 2011b) is a brand new system architecture utilizing fully distributed fault management and adaptation to realize a self-organizing and self-healing hardware platform. Its system model, interpreted with biological analogy, consists of eDNA (the program to be computed) and eCells (the distributed hardware components to execute the eDNA). Each eCell implements a part of the eDNA- gene (as the gene in biological terms). The authors have proposed a specific language to program the eDNA, which can be automatically translated to genes. Each eCell then is assigned with a particular gene based on its unique identifier. In terms of hardware, each eCell contains an ALU to process the gene and a microprocessor to perform diagnosis and self-healing functions (termed the ribosomal DNA considering its function in biological cells). The system contains spare eCells to replace faulty ones.

The highlight of the eDNA hardware architecture is the self-organization and self-healing processes, which are fully performed in a distributed and cooperative manner. The initial organization, i.e. the mapping of genes to eCells, is done directly by relating the index of the gene to the unique identifier of the eCells. Routing tables in each eCell will record the mapping of all eCells for communication and future modification. A fault is detected and located with two active eCells (with genes mapped) and a spare eCell (the fault detector) running the same test program and comparing the outputs. When the eCell is identified as erroneous, a spare eCell will be chosen (base on simple criteria as minimum Manhattan distance) to replace the faulty one. The routing table of each eCell is updated to reflect the changes of organization.

The beauty of eDNA architecture lies its modularization, simplicity and scalability. With a built-from-scratch system architecture, it maximizes the capabilities of purely distributed and cooperative adaptation processes with straightforward and simple rules, e.g. relying on peer spare components for error recovery, only handling one fault at one time to resolve potential contention, and clear criteria to choose fault detectors and eCell for replacement. Its potential in safety-critical, high-redundancy applications, e.g. space applications, is promising.

H2A (Hierarchical Agent-based Adaptation) (Guang, 2012) proposes a generic agent-based architecture for monitoring, decision-making and reconfiguring parallel systems. The main thrust of this systematic architecture is the hierarchical partitioning of adaptive services and software/hardware-cosynthesis of agents, to provide scalability in terms of service functions and implementation overhead. One platform agent performs global configuration for the whole system, e.g.

mapping. The system is dynamically organized into clusters, each with a cluster agent supervising the application running in the cluster. Each cluster is composed of modularized cells, embedded with a hardware-based cell agent to monitor cell status and actuate reconfiguration. The cluster and platform agents are implemented as software threads running on processing elements, to allow for maximized flexibility and scalable physical overhead (compared to hardware-based implementation).

In terms of dependability provision, the current version of H2A focuses on the dynamic clusterization process (Figure 6). When cell agents detect errors (permanent) and report to the platform agent, the latter decides on proper spares to replace the broken cells. The decision will trigger the recording of new cells into a cluster (with the faulty cell), which in effect leads to a new system organization. Similar in principle to the routing tables proposed in eDNA architecture, three levels of LUT (look-up table)-based supporting structures are designed for each level of agent respectively, to support the dynamic clusterization. As illustrated in Figure 6, on the platform agent, a RLT (resource look-up table) with as many entries as the number of cells in the system is implemented, where the status of each cell (faulty, used or spare) is stored. On each cluster agent, a CLT (cluster look-up table) with as many entries as the maximum allowed cluster size is implemented, where the indexes of currently affiliated cells are stored. In CLT, all faulty cells will be removed, with only currently used cells occupying the entries. On each cell agent, a RT (re-routing table) is implemented with as many entries as the maximum allowed spares in each cluster, where the locations of spares to replace faulty cells are stored. The hierarchical design of supporting structures on each level of agent is to minimize the physical overhead, esp. on the distributed cells (linear area overhead with system expansion), while enabling self-organization in case of permanent errors.

Figure 6. Hierarchical Supporting Structures for Dynamic Clusterization on H2A architecture

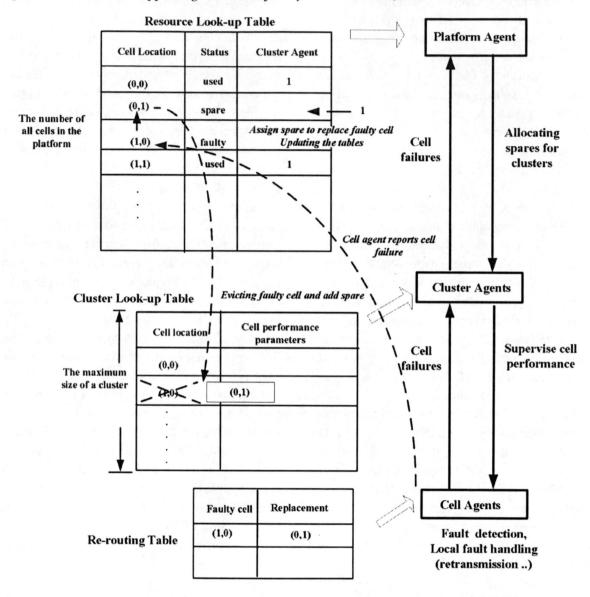

SUMMARY AND FUTURE PERSPECTIVES

The extensive study of existing literature has demonstrated major progress in all of the monitoring, decision-making and reconfiguration processes for dependability improvement. Monitoring and sensing circuits and designs for different types of faults and errors due to various sources have been proposed. Reconfiguration methods are so diverse that all architectural levels have been exploited to tolerate, reduce and recover from errors. In terms of the decision-making process, the development of parallel computing systems has incubated adaptation algorithms or methods running by centralized, distributed or hierarchical decision-makers. Novel systematic approaches have projected promising paradigms addressing the dependability issues, for instance the eDNA architecture as a fully distributed, cooperative approach for self-organization and self-healing and the H2A architecture focusing on the functional and overhead scalability of adaptation processes. Surely the self-adaptation for dependability is closely related to the optimization and tradeoff with other concerns, e.g. the power consumption (Fuketa et al., 2009; Ernst et al, 2003). Such discussion is beyond the scope of this chapter.

What is worth noticing is that a SoC is in the potential risk of many types of dependability threats, e.g. temperature, SEU, aging, process variations, etc. While there are techniques addressing individual concerns, it remains a challenge on how to deploy all these techniques on a chip with affordable overhead. Firstly, a specific type of error, e.g. the timing error, may require dedicated error detection circuits, e.g. shadow latches (Simone et al., 2008) or transition detector (Das et al., 2009). Secondly, one type of errors may imply different sources, e.g. the increasing timing error rate may be due to supply voltage variations, temperature increase or aging effects (increase of threshold voltage). Without identifying the most contributing source, the reconfiguration measures may be ineffective or too costly, e.g. increasing the supply voltage when the temperature is already so high to cause timing errors. In addition, as all types of reconfiguration techniques introduce some level of area, performance or power overhead, the deployment of reconfiguration techniques needs to be made considering the most critical threats and the targeted system, for instance the tailored error recovery techniques addressing the strong impact of soft errors on data cache (Wang et al., 2008).

A new research direction- error resilient computing (Lammers, 2010; Henkel et al., 2011), is particularly promising. Insteading of trying to detect and correct every incidence of error, the error resilient architecture (also called probabilistic computing or stochastic processing) tolerates a reasonable level of error occurrence to reduce the error management overhead or optimize other metrics (e.g. power), as long as the quality of result is acceptable (Lammers, 2010). The ini-tial work of Razor I (Ernst et al., 2013) already shares a similar perspective, where the processor works with a controlled level error rate to minimize the power consumption. This paradigm is more relevant in current technology where the error rate seems to be constantly increasing due to transistor scaling, voltage reduction, thermal issues etc. Thus, the researchers have started to exploit the application and algorithm resilience to improve dependability. Many types of applications, e.g. audio and video processings, are tolerant of low error rate that barely affects the output quality (Henkel et al., 2011). Algorithms, such are statistical or aggregative, also have inherent resilience, e.g. errors can be averaged out or self-corrected in later iterations (Chippa, Mohapatra, Raghunathan, Roy, & Chakradhar, 2010). The higher-level resilience can be strongly utilized to alleviate the error management overhead in the hardware architecture. The starting efforts along this direction demonstrate major potentials (e.g. Chippa et al., 2010; Leem et al., 2010).

CONCLUSION

Self-adaptive measures are effective in the system monitoring and reconfiguration for dependability. Existing literature has presented a significant amount and diversity of fault and error management techniques against every concerned threat to dependability on SoCs. These works significantly expands the design choices available to circuit and system developers, much beyond the conventional duplication-based or error coding-based schemes. This chapter reviews the common faults, errors and failures on SoCs, in particular those caused by time-dependent device degradation, radiation, and run-time voltage and thermal variations, and examines representative published works based on the monitoring, decision-making and reconfiguration self-adaptation processes. The survey is expected to inform circuit and system designers the wide range of choices for integrating self-adaptive measures in all architectural levels of chip design. The decision should be made based on the most influencing dependability threats and the requirements of the targeted systems or subsystem.

Albeit our intention to be comprehensive, the broad issues related to the self-adaptive approach to dependability are by no means fully discussed. For example, the performance, area and power overheads of adaptive techniques and architectures, though concisely presented, require more in-depth study and comparison. Also the quantitative comparison between conventional and self-adaptive archiectures, in terms of error tolerance and design overhead, is also needed to convince more designers switching to the new paradigm.

REFERENCES

Actel. (2003). *Reliability Considerations for Automotive FPGAs*. Retrieved Dec. 19, 2013 from http://www.actel.com/documents/AutoWP.pdf

Addo-Quaye, C. (2005). Thermal-aware mapping and placement for 3-D NoC designs. In *Proceedings of IEEE International SOC Conference*, (pp. 25-28). IEEE.

Agarwal, M., & Paul, B. C. Zhang, & Mitra, S. (2007). Circuit Failure Prediction and Its Application to Transistor Aging. In *Proceedings of 25th IEEE VLSI Test Symposium* (pp. 277-286). IEEE.

Avizienis, A., Laprie, J.-C., Randell, B., & Landwehr, C. (2004). Basic Concepts and Taxonomy of Dependable and Secure Computing. *IEEE Transactions on Dependable and Secure Computing*, *1*(1), 11–33. doi:10.1109/TDSC.2004.2

Blome, J., Gupta, S., Feng, S., Mahlke, S., & Bradley, D. (2006). Cost-efficient Soft Error Protection for Embedded Microprocessors. In *Proceedings of the 2006 International Conference on Compilers, Architecture and Synthesis for Embedded System*, (pp. 421-431). IEEE.

Boesen, M. R., Keymeulen, D., Madsen, J., Lu, T., & Chao, T.-H. (2011b). Integration of the reconfigurable self-healing eDNA architecture in an embedded system. In *Proceedings of 2011 IEEE Aerospace Conference*, (pp. 1-11). IEEE.

Boesen, M. R., Madsen, J., & Keymeulen, D. (2011a). Autonomous distributed self-organizing and self-healing hardware architecture - The eDNA concept. In *Proceedings of Aerospace Conference*. IEEE.

Chandra, V., & Aitken, R. (2008). Impact of technology and voltage scaling on the soft error susceptibility in nanoscale CMOS. In *Proceedings of IEEE International Symposium on Defect and Fault Tolerance of VLSI Systems* (pp. 114-122). IEEE.

Chippa, V. K., Mohapatra, D., Raghunathan, A., Roy, K., & Chakradhar, S. T. (2010). Scalable effort hardware design: Exploiting algorithmic resilience for energy efficiency. In *Proceedings of 47th ACM/IEEE Design Automation Conference (DAC)*, (pp. 555-560). ACM/IEEE.

Clabes, J., Friedrich, J., Sweet, M., DiLullo, J., Chu, S., & Plass, D. … Dodson, S. (2004). Design and implementation of the POWER5 microprocessor. In *Proceedings of the 41st annual Design Automation Conference*, (pp. 670--672). IEEE.

Collet, J. H., Zajac, P., Psarakis, M., & Gizopoulos, D. (2011). Chip Self-Organization and Fault Tolerance in Massively Defective Multicore Arrays. *IEEE Transactions on Dependable and Secure Computing*, *8*(2), 207–217. doi:10.1109/TDSC.2009.53

Constantinescu, C. (2003). Trends and Challenges in VLSI Circuit Reliability. *IEEE Micro*, *23*(4), 14–19. doi:10.1109/MM.2003.1225959

Das, S., Tokunaga, C., Pant, S., Ma, W.-H., Kalaiselvan, S., & Lai, K. … Blaauw, D.T. (2009). Razor II: In Situ Error Detection and Correction for PVT and SER Tolerance. JSSC, 44(1), 32-48.

Duzellier, S. (2005). Radiation effects on electronic devices in space. *Aerospace Science and Technology, 9*(1), 93–99. doi:10.1016/j.ast.2004.08.006

Ernst, D. Kim, Das, S., Pant, S., Rao, R., Pham, … Mudge, T. (2003). Razor: A low-power pipeline based on circuit-level timing speculation. In *Proceedings of MICRO-36*, (pp. 7--18). MICRO.

Eshraghian, K. (2006). SoC Emerging Technologies. *Proceedings of the IEEE, 94*(6), 1197–1213. doi:10.1109/JPROC.2006.873615

Fuketa, H., Hashimoto, M., Mitsuyama, Y., & Onoye, T. (2009). Trade-off analysis between timing error rate and power dissipation for adaptive speed control with timing error prediction. In *Proceedings of ASP-DAC*, (pp. 266-271). ASP-DAC.

Guang, L. (2012). *Hierarchical Agent-Based Adaptation for Self-Aware Embedded Computing Systems*. (Doctoral Dissertation). University of Turku, Turku, Finland. Retrieved from http://www.doria.fi/handle/10024/86210

Guang, L., Nigussie, E., Plosila, J., Isoaho, J., & Tenhunen, H. (2012a). Survey of Self-Adaptive NoCs with Energy-Efficiency and Dependability. *International Journal of Embedded and Real-Time Communication Systems, 3*(2), 1–22. doi:10.4018/jertcs.2012040101

Guang, L., Nigussie, E., Plosila, J., Isoaho, J., & Tenhunen, H. (2012b). Coarse and fine-grained monitoring and reconfiguration for energy-efficient NoCs. In *Proceedings of 2012 International Symposium on System on Chip (SoC)*, (pp. 1-7). SoC.

Gupta, M. S., Oatley, J. L., & Joseph, R. Wei, & Brooks, D.M. (2007). Understanding Voltage Variations in Chip Multiprocessors using a Distributed Power-Delivery Network. In Proceedings of Design, Automation & Test in Europe Conference & Exhibition (pp. 1--6). Academic Press.

Henkel, J., Bauer, L., Becker, J., Bringmann, O., Brinkschulte, U., & Chakraborty, S. … Wunderlich, H. (2011). Design and architectures for dependable embedded systems. In *Proceedings of the 9th International Conference on Hardware/Software Codesign and System Synthesis (CODES+ISSS)*, (pp. 69-78). CODES+ISSS.

Henkel, J., Ebi, T., Amrouch, H., & Khdr, H. (2013). Thermal management for dependable on-chip systems. In *Proceedings of ASP-DAC*, (pp. 113-118). ASP-DAC.

Horn, P. (2001). *Autonomic computing: IBM's perspective on the state of information technology (Tech. Report)*. IBM.

Joshi, R., Kanj, R., Adams, C., & Warnock, J. (2013). Making reliable memories in an unreliable world. In *Proceedings of 2013 IEEE International Reliability Physics Symposium* (pp. 3A.6.1-3A.6.5). IEEE.

Kamran, A., & Navabi, Z. (2013). Online periodic test mechanism for homogeneous many-core processors. In *Proceedings of 2013 IFIP/IEEE 21st International Conference on Very Large Scale Integration* (VLSI-SoC), (pp. 256-259). IEEE.

Karl, E., Sylvester, D., & Blaauw, D. (2005). Timing error correction techniques for voltage-scalable on-chip memories. In *Proceedings of IEEE International Symposium on Circuits and Systems*, (vol. 4, pp. 3563-3566). IEEE.

Keane, J., & Kim, C. (2011). Transistor aging. *IEEE Spectrum*. Retrieved Dec. 18, 2013, from http://spectrum.ieee.org/semiconductors/processors/transistor-aging

Kim, K. K., Wang, W., & Choi, K. (2010). On-Chip Aging Sensor Circuits for Reliable Nanometer MOSFET Digital Circuits. *IEEE Transactions on Circuits and Systems, 57*(10), 798–802. doi:10.1109/TCSII.2010.2067810

Knight, J. C. (2002). Dependability of embedded systems. In *Proceedings of the 24th International Conference on Software Engineering*, (pp. 685-686). IEEE.

Kobayashi, K., & Onodera, H. (2008). Best ways to use billions of devices on a chip - Error predictive, defect tolerant and error recovery designs. In *Proceedings of ASP-DAC*, (pp. 811-812). ASP-DAC.

Laddaga, R. (1997). *Self-adaptive software* (Tech. Rep. 98-12). DARPA BAA.

Lammers, D. (2010). The Era of Error-Tolerant Computing. *IEEE Spectrum*, *47*(11), 15. doi:10.1109/MSPEC.2010.5605876

Leem, L., Cho, H., Bau, J., Jacobson, Q., & Mitra, S. (2010). ERSA: Error Resilient System Architecture for Probabilistic Applications. In *Proceedings of the Conference on Design, Automation and Test in Europe* (pp. 1560-1565). Academic Press.

Lehtonen, T., Wolpert, D., Liljeberg, P., Plosila, J., & Ampadu, P. (2010). Self-Adaptive System for Addressing Permanent Errors in On-Chip Interconnects. *IEEE Transactions on Very Large Scale Integration Systems*, *18*(4), 527–540. doi:10.1109/TVLSI.2009.2013711

Memik, S.O., & Mukherjee, R., Ni, & Long. (2008). Optimizing Thermal Sensor Allocation for Microprocessors. *IEEE Transactions on Computer-Aided Design of Integrated Circuits and Systems*, *27*(3), 516–527. doi:10.1109/TCAD.2008.915538

Mushtaq, H., Al-Ars, Z., & Bertels, K. (2011). Survey of fault tolerance techniques for shared memory multicore/multiprocessor systems. In *Proceedings 2011 IEEE 6th Design and Test Workshop (IDT)* (pp. 12 -17). IEEE.

Nakura, T., Nose, K., & Mizuno, M. (2007). Fine-Grain Redundant Logic Using Defect-Prediction Flip-Flops. In *Proceedings of IEEE International Solid-State Circuits Conference*, (pp. 402-403, 611). IEEE.

Narayanan, V., & Xie, Y. (2006). Reliability concerns in embedded system designs. *Computer*, *38*(1), 118–120. doi:10.1109/MC.2006.31

Noguchi, J., Saito, T., Ohashi, N., Ashihara, H., Maruyama, H., & Kubo, M. … Hinode, K. (2001). Impact of low-k dielectrics and barrier metals on TDDB lifetime of Cu interconnects. In *Proceedings of 39th IEEE International Reliability Physics Symposium*, (pp. 355-359). IEEE.

Nose, & Sakurai. (2000). Optimization of VDD and VTH for low-power and high speed applications. In *Proceedings of the 2000 Asia and South Pacific Design Automation Conference* (pp. 469—474). IEEE.

Paul, B.C., Kang, Kufluoglu, H., Alam, M.A., & Roy, K. (2005). Impact of NBTI on the temporal performance degradation of digital circuits. *IEEE Electron Device Letters*, *26*(8), 560–562. doi:10.1109/LED.2005.852523

Pedram, M., & Nazarian, S. (2006). Thermal Modeling, Analysis, and Management in VLSI Circuits: Principles and Methods. *Proceedings of the IEEE*, *94*(8), 1487–1501. doi:10.1109/JPROC.2006.879797

Quach, N. (2000). High Availability and Reliability in the Itanium Processor. *IEEE Micro*, *20*(5), 61–69. doi:10.1109/40.877951

Radetzki, M., Feng, C. C., Zhao, X. Q., & Jantsch, A. (2013). Methods for fault tolerance in networks-on-chip. *ACM Computing Surveys*, *46*(1). doi:10.1145/2522968.2522976

Salehie, M., & Tahvildari, L. (2009). Self-adaptive software: Landscape and research challenges. *ACM Trans. Auton. Adapt. Syst.*, *4*(2).

Shang, L., Peh, L., Kumar, A., & Jha, N. K. (2004). Thermal Modeling, Characterization and Management of On-Chip Networks. In *Proceedings of the 37th International Symposium on Microarchitecture* (pp. 67-78). IEEE.

Simone, M., Lajolo, M., & Bertozzi, D. (2008). Variation tolerant NoC design by means of self-calibrating links. In Proceedings of Design, Automation and Test in Europe DATE '08, (pp. 1402--1407). DATE.

Sylvester, D., Blaauw, D., & Karl, E. (2006). ElastIC: An Adaptive Self-Healing Architecture for Unpredictable Silicon. *IEEE Design & Test of Computers*, *23*(6), 484–490. doi:10.1109/MDT.2006.145

Unsal, O. S., Tschanz, J. W., Bowman, K., De, V., Vera, X., Gonzalez, A., & Ergin, O. (2006). Impact of Parameter Variations on Circuits and Microarchitecture. *IEEE Micro*, *26*(6), 30–39. doi:10.1109/MM.2006.122

Wang, S., Hu, J., & Ziavras, G. S. (2008). Self-Adaptive Data Caches for Soft-Error Reliability. *IEEE Trans. on CAD of Integrated Circuits and Systems*, *27*(8), 1503–1507. doi:10.1109/TCAD.2008.925789

KEY TERMS AND DEFINITIONS

Aging: The performance degradation of semiconductor devices over time. Aging is a time-dependent degradation phenomenen. The speed of aging is influenced many factors, such as workload intensity, thermal stress and material.

Dependability: In general, dependability is "the ability to deliver service that can justifiably be trusted" (Avizienis et al., 2004). It has a broad spectrum of attributes- availability, reliability, safety, integrity, confidentiality and maintainability. The concerned scope of dependability is much beyond the conventional term of fault-tolerance (see its definition in the list).

Error Resilient Computing: Also called probablistic computing or stochastic pocessing.

Instead of automatically tracing and correcting an error once it occurs, this new paradigm allows a reasonable amount of error rate as long as it doesnot signficantly affect the quality of results (Lammers, 2010).

Fault Tolerance: In a strict taxonomical sense, fault tolerance is one of the means to achieve dependability, which is defined as "the means to avoid service failures in the presence of faults" (Avizienis et al., 2004). In practice, fault-tolerance is used in a much wider sense to cover other means of dependability- fault prevention, fault removal and fault forecasting.

Run-Time Management: In this chapter's context, it refers to management techniques (performance, power, dependability, etc.) when the system is running.

Self-Adaptive System: The original definition was given to self-adaptive software, which can be applied in general to the system: "self-adaptive software valuates its own behaviro when the evaluation indicates that it is not accmplishing what the sofware is intended to do, or when better functionality or performance is possible" (Laddaga, 1997).

SEU (Single Event Upset): A transient error caused by particle hits which are strong enough to flip the content of a state-saving element (Chandra & Aitken, 2008).

SoC (System-on-a-Chip): An integrated circuit (IC) that integrates all components of a computer or other electronic system into a single chip. System-on-Chips can be designed in various system architectures, containing different processing, communication and memory components.

Timing Error: The error caused by the late arrival of data signals compared to the clock signal, so that the data cannot be sampled correctly (Guang et al., 2012a).

Chapter 2
Security Issues in Tactical Software–Defined Radios:
Analysis of Attacks and Case Studies

Fabrício A. B. da Silva
Oswaldo Cruz Foundation (FIOCRUZ), Brazil

David F. C. Moura
Brazilian Army Technological Center, Brazil

Juraci F. Galdino
Brazilian Army Technological Center, Brazil & Military Institute of Engineering, Brazil

ABSTRACT

In recent years, the development of radio communication technology solutions has experienced a huge paradigm change – the Software-Defined Radio (SDR) technology upspring, in which previously hardware-based features became software-defined and users may also introduce new application waveforms on-the-fly. Given its growing importance for SDR application vendors and developers in different project domains, one of first steps in engineering a secure SDR system is the identification of classes of attacks on a SDR, along with the associated threats and vulnerabilities. Therefore, the identification of classes of attacks is necessary for the definition of realistic and relevant security requirements. One contribution of this chapter is to identify classes of attacks that Software Communications Architecture (SCA) compliant Software-Defined Radios (SDR) can suffer. It is noteworthy that, with the advancement of technology, new vulnerabilities emerge every day, and with them, new forms of threats and attacks on systems. The authors intend, however, to highlight in this chapter the classes of attacks that are more relevant for tactical software-defined radios, taking into account expected losses for legitimate radio network users. They also discuss, in this chapter, mitigation strategies for several identified attacks and how attack mitigation strategies can affect a SCA-compliant operating environment. Finally, the authors present several case studies, along with simulation results, illustrating the identified attack classes.

DOI: 10.4018/978-1-4666-6034-2.ch002

INTRODUCTION

In the past, military radio design was totally focused on dedicated electronic components. Afterwards, we have witnessed the appearance of software configurable radios (SCR), in which users have the opportunity to choose the most appropriate waveforms for different combat scenarios. In recent years, though, the development of radio communication technology solutions has experienced a huge paradigm change - the Software-Defined Radio (SDR) technology upspring, in which previously hardware-based features became software-defined and users may also introduce new application waveforms on-the-fly.

Such progress is due to several enhancements in different areas like embedded systems, analog-to-digital converters, digital transmission, digital signal processing, multi-band antennas, software architectures, and especially in novel General-Purpose Processors (GPP) evaluation capacity. Based on that, SDR foreshadows important consequences and advantages for the development of wireless solutions for military communications systems. Among the envisioned features, we can list interoperability, waveform portability, and the possibility to be updated with the most recent advances in radio communications without hardware replacement requirements. Moreover, the SDR is envisioned as the most appropriate platform for cognitive radio development.

At a glance, the high level functional model of a SDR consists of a front end RF subsystem which performs channel selection, down-conversion to baseband, and data forwarding onto a software-based processing unit, where the associated digital bitstream is submitted onto appropriate layers (e.g., data link, network, security modules) to perform suitable decoding tasks to extract the desired information. This process is reversed on the transmit side, where the input signal is coded and a modulated signal bearing the associated information suitable for transmission is created. This signal is then passed to the RF subsystem for insertion into the wireless channel.

Due to the multitude of concepts related to the described functional model, several efforts have been done towards the standardization of key elements within the SDR architecture, providing a common platform for the development of SDR sets. The standards supported may be proprietary or industry-developed through a consensus process – while the former approach brings product differentiation to manufacturers, the latter strategy commoditizes the technology, allowing support by third parties in creating the radio platform to achieve specific business objectives.

One of the most typical areas of standardization is the application framework, which provides a common software operation environment, with vendor-free interfaces to set up, configure, control and release application waveforms under operation on a SDR platform. As mentioned in (Silva, Moura & Galdino, 2012), examples of application frameworks relevant to SDR systems include the Open Mobile Alliance, the Service Availability Forum, and the Software Communications Architecture (SCA) supported by the Wireless Innovation Forum's SCA Working Group.

The SCA standard was originally proposed by the Joint Tactical Radio System program (JTRS) (SCA, 2001; SCA, 2006; SCA, 2012), which is a program for the development of military tactical radios sponsored by the US Department of Defense. The SCA / JTRS standard is becoming the de facto standard for the construction of tactical military radios. However, the interest in the SCA goes beyond the military domain, since this standard has inspired academic and commercial projects (Gonzalez,Carlos,Dietrich,Reed, 2009). Since October 2012, the Joint Tactical Networking Center (JTNC) absorbed JTRS mission and duties.

Given its growing importance for SDR application vendors and developers in different project domains, one of first steps in engineering a secure SDR system is the identification of classes of attacks on a SDR, along with the associated threats and vulnerabilities. It precedes the identification of security requirements and the development of security mechanisms (Myagmar,Lee,Yurcik,

2005). Therefore, the identification of classes of attacks is necessary for the definition of realistic and relevant security requirements.

In this chapter, we define an attack as a malicious action that aims to explore one or more vulnerabilities, subverting the system security policy. The adversary is a person or organization that performs the attack. Vulnerabilities are defects in the system (either in software or in hardware) relevant from the security viewpoint. A threat relates to a potential danger associated with a vulnerability, which can be exploited or not by an adversary. An intrusion represents a successful attack.

A systematic way of classifying intrusions is described in (Lindqvist and Jonsson, 1997). An intrusion is defined by the authors as a successful event from the adversary's point of view, and consists of 1) an attack, in which a vulnerability is exploited, and 2) a breach, which is the resulting violation of the security policy of the system. Essentially, the authors consider two dimensions when classifying intrusions, intrusion techniques and intrusion results.

Another type of analysis that can benefit from the knowledge about the classes of attacks that a tactical radio set can suffer is the evaluation of the SDR attack surface. A system's attack surface is the way in which the system will be successfully attacked (Manadhata and Wing, 2010), and can be used to evaluate the level of security provided by an existing SDR architecture. The attack surface is defined in terms of system resources: if a resource can be used to attack the system, this resource is part of the system's attack surface. By defining an attack surface metric, it is possible to determine if a version A of the system is more or less secure than a version B (Manadhata and Wing, 2010).

Several works available in the literature deal with SDR security aspects. In (Wireless Innovation Forum, 2010), the authors list classes of vulnerabilities, threats and attacks aimed at software-defined radios. It also proposes a Radio Platform Security Architecture, with focus on commercial and public safety radios. In (Myagmar et al., 2005),

it is presented a threat analysis aimed specifically at the GNU Radio platform, handled as a case study for the process of threat modeling based on data flow graphs. In (3GPP, 2001), classes of threats associated to 3G networks and security requirements related to these threats are listed. In (Murotake and Martin, 2004) vulnerabilities, threats and attacks on WiFi devices (IEEE 802.11) are described. In (SCA Security, 2001) the requirements associated with secure SCA military radios, as well as a high-level security architecture and the corresponding security mechanisms are presented. However, the classes of attacks and threats that inspired the requirements are not discussed. Beyond that, the requirements and mechanisms described in (SCA Security, 2001) were proposed more than 10 years ago. Therefore, several of the suggested security mechanisms have now known vulnerabilities.

Therefore, one objective of this chapter is to identify classes of attacks that Software Communications Architecture (SCA) compliant Software-Defined Radios (SDR) can suffer. Our emphasis is on attacks that target the radio set. Attacks on other radio network components, such as systems management software, download servers and other data sources that use the SDR as the physical layer, are not included in the scope of this paper. It is noteworthy that, with the advancement of technology, new vulnerabilities emerge every day, and with them new forms of threats and attacks on systems. We intend, however, to indicate the classes of attacks that are more relevant for tactical software-defined radios, taking into account expected losses for legitimate radio network users. We also present several case studies, along with simulation results, illustrating the identified attack classes. This chapter is a thoroughly revised and much improved version of a previous paper published at the International Journal of Embedded and Real-Time Communication Systems (Silva, Moura & Galdino, 2012).

The classification of attacks presented in this paper is an adaptation of the systematic classification presented by Lindqvist and Jonsson (1997).

In other words, classes of attacks are defined based on potential attack results on the radio set. This classification can also be associated with the adversary's objectives when planning an intrusion. Thus, the classification presented in this paper can be used to identify the resources that are part of a SDR attack surface, being the first step of the methodology of measuring the attack surface presented in (Manadhata and Wing, 2010).

This chapter is organized as follows: Section 2 contains a brief description of the Software Communications Architecture (SCA) standard. Section 3 provides definitions of attacks, vulnerabilities and threats, and identifies vulnerability classes and several classes of threats. The various classes of attacks on SCA-compliant radios are discussed in Section 4, as well as the assumptions and models associated with the definition of these classes. Section 5 presents mitigation strategies for several attacks described in the previous section. Section 6 discusses how attack mitigation strategies can affect a SCA-compliant operating environment. Section 7 presents several case studies and simulation results. Finally, Section 8 contains our concluding remarks.

SOFTWARE COMMUNICATIONS ARCHITECTURE (SCA)

The SCA standard (Gonzalez et al. 2009, SCA 2006, SCA 2001) is an open software infrastructure developed by the U.S. Department of Defense, through the JTRS program, to assist in the development of software-defined radios. The standard specifies mechanisms to create, deploy, manage and interconnect component-based radio applications in a distributed platform. The SCA standard has been proposed to enhance the interoperability of radio communications systems and reduce its implementation time and development costs.

The SCA was developed as a scalable architecture capable to support platforms with different capacities, ranging from fixed radio communication stations to portable devices. This scalability, together with the implementation of user-driven features, greatly enhances both the network interoperability and the portability of radio applications, which are desired features in military scenarios.

The SCA software structure defines a common operating environment (OE). The OE provides mechanisms for deploying applications on different platforms with different hardware components, device drivers or transport mechanisms, like Digital Signal Processors (DSP), Field Programmable Gate Arrays (FPGA), and General Purpose Processors (GPP). The operating environment also defines interfaces for managing and controlling applications and their components, and these interfaces are independent of particular implementations.

In the SCA context, a radio application is known as a waveform, which is defined as the result of a set of transformations performed in order to overcome disturbances caused by the radio channel conditions and possible enemy interference actions. Those transformations are performed on the transmitter and applied to the information that is transmitted through the air. Moreover, this definition also comprises the corresponding set of transformations performed on the receiver to convert the received signals back on the original information.

There is a multitude of typical waveforms (MIL-STD-188-110B 2000, NATO 2000), comprising physical, data-link, and network layers. Those waveforms describe data transmission security mechanisms, source coding (e.g., voice, image, and video compression), channel coding like FEC - Forward Error Correction and ARQ – Automatic Repeat Request, modulation and demodulation techniques, adaptive equalization, automatic gain control, synchronization, filtering, medium access control, among other features (Proakis 2001). Therefore, a SCA waveform is typically composed of interconnected components through well-defined interfaces. Every component has a well-defined functionality and can be reused in the development of other waveforms.

The OE can be divided into four layers of abstraction:

- Real-time operating system (RTOS);
- Middleware (CORBA);
- SCA Core Framework (CF);
- SCA Services.

The SCA requires a RTOS compatible with the Portable Operating System Interface (POSIX) standard. The SCA 2.2.2 standard defines a subset of POSIX called Application Environment Profile (AEP), which limits the operating system services available for radio applications (waveforms). Core Framework components, in turn, have full access to the operating system.

Concerning SCA Services, the SCA standard also defines the utilization of several CORBA services (logging, event and name services) as specified by the Object Management Group (OMG 2012).

The Core Framework (CF) defines a set of interfaces that rule the deployment and management of waveforms and their components (Gonzalez et al. 2009). The SCA Core Framework includes four main sets of interfaces:

- Base application interfaces.
- Base device interfaces.
- Framework control interfaces.
- Framework services interfaces.

From the security viewpoint, the *framework control* interfaces are particularly important. These interfaces define a set of components that provide management and control capabilities to the CF over the whole radio domain. For instance, the *Application Factory* interface is used to create instances of a specific waveform. After creating an application, the Application Factory returns an instance of the corresponding Application interface. The *Domain Manager* interface controls and maintains the overall state of the radio.

All the information concerning applications and platforms within the SCA is included in a set of XML files called the Domain Profile (Gonzalez et al. 2009). For instance, Software Package Descriptor (SPD) files describe software components and their implementation. The Device Package Descriptor (DPD) and the Device Configuration Descriptor (DCD) describe platform characteristics. Several other files are also defined. For a complete description of the Domain Profile see (SCA 2006, SCA 2001).

In addition to the OE, SCA standard version 2.2 (SCA 2001) includes a security supplement (SCA Security 2001) that defines the U.S. Government security requirements for JTRS radios, a security API and suggests a high level architecture for a secure SCA-compliant radio. The security architecture is composed of several separate partitions: the red partition, the black partition, and the cryptographic subsystem (see Figure 1). The red partition stores and processes sensitive information unencrypted or with a low level of protection, and black partition handles non-sensitive or encrypted information. Both partitions are connected by a cryptographic subsystem, which has a bypass mechanism for data that can be transferred unencrypted between black and red partitions.

A SCA-compliant radio set that implements the architecture depicted in Figure 1 is described in (Kurdziel, Beane, Fitton, 2005). The security supplement suggests a physical separation between partitions, i.e., it recommends separate processors for the red and black partitions, and a third one for the cryptographic subsystem. This physical partitioning of functions should also take into account compromising emanations, which involves both the distribution of power and electromagnetic emissions. Depending on the level of confidentiality of the information processed by the red partition, electromagnetic shielding, physical distancing, and electrical isolation of physical partitions are mandatory (Tempest 1982, Tempest 1995).

Nevertheless, it should be highlighted that the security supplement was not completely alien to

Figure 1. Basic JTRS security architecture (SCA Security 2001)

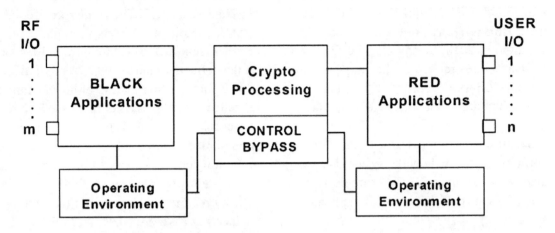

comments and criticisms (Bunnell and Trinidad 2004), indicating the reduced number of requirements that were covered by that document. The JTRS officially deprecate the Security Supplement after the release of the version 2.2.2 of the JTRS SCA (SCA 2006), in favor of a JTRS-specific security specification. This last document has not been released in public domain.

In the first quarter of 2012, the JTRS program released the SCA specification version 4.0 (SCA 2012). This new specification is based on the work developed by the SCA Next working group of the Wireless Innovation Forum (SCA Next 2011). One of the main requirements of this new specification was to maintain backward compatibility with the previous versions of the standard. Among the new features proposed by SCA v4 is the separation between Platform Independent Models (PIM) and technology specific Platform Specific Models (PSM). Different PSMs may represent radio sets with different capabilities. Another remarkable new feature is the utilization of transfer mechanisms other than CORBA, such as Data Distribution Service (DDS). SCA v 4.0 also defines two Application Environment Profiles: Full and Lightweight. The full AEP corresponds to the SCA 2.2.2 AEP. The lightweight AEP has reduced functionality and it is aimed at radio sets with smaller capacity.

ATTACKS, VULNERABILITIES AND THREATS

As defined in the Introduction, an attack is a malicious action that aims to explore one or more vulnerabilities, subverting the system security policy. In this paper, we identify the person or organization performing the attack as the adversary. A vulnerability is a defect in the system (either in software or in hardware) relevant for its security (Correia and Sousa 2010). The vulnerabilities can be classified as design, implementation, or operational, as follows.

- **Design Vulnerabilities** are introduced during system design. An example is to use weak encryption and signature mechanisms for secret data. Another example is a protocol design subject to replay attacks.
- **Implementation Vulnerabilities** are introduced during the manufacturing and delivery processes. This class of vulnerabilities can be divided into two subclasses: software implementation and hardware implementation vulnerabilities. An example of software implementation vulnerability is the lack of verification of the boundaries of a buffer, leading to buffer overflow vulnerabilities (Hsu, Guo,Chiueh, 2006).

An example of hardware implementation vulnerability is the introduction of hardware trojans (Karri,Rajendran,Rosefeld,Tehranipoor, 2010) during chip fabrication. Vulnerabilities related to the supply chain (Swanson,Bartol,Moorthy, 2010), e.g. device cloning (Gallo,Kawakami,Dahab, 2009), are also included in this subclass.

- **Operational Vulnerabilities** are vulnerabilities caused by how the system is operated or is configured. One example is the use of weak passwords in authentication systems.

Related to the concepts of attack and vulnerability, there is a third concept: the threat (Wireless Innovation Forum, 2010). The threat relates to a potential danger associated with a vulnerability, and it can be exploited or not by an adversary. Vulnerabilities create threats, which in turn can be used by attackers to exploit vulnerabilities of a system. Vulnerabilities may be associated with one or more threats. It is noteworthy that a threat does not always translate into an attack. From the adversary viewpoint, in some cases the expected cost of performing an attack related to a specific threat could be much higher than the expected reward (Gallo,Kawakami,Dahab, 2011).

For general-purpose systems, Microsoft has developed a methodology for identifying threats known as Stride (**S**poofing, **T**ampering, **R**epudiation, **I**nformation disclosure, **D**enial of service, **E**scalation of privilege) (Myagmar et al. 2005, Swiderski and Snyder 2004). A classification of threats aimed specifically at software-defined radios can be found in (Wireless Innovation Forum, 2010). However, the actual utilization of these classifications in real systems requires experts with extensive knowledge about possible attacks, which makes the application of these methodologies on complex systems inefficient and not scalable in practice (Dhillon 2011). Therefore, in this paper, we focus primarily on defining a classification of attacks that can be used later to perform an analysis of threats in a more scalable way.

In the next section, we define classes of attacks applicable to tactical SDRs, with special emphasis on SCA-compliant radio sets, through the analysis of attacks reported in the literature and depending on the defined scenario and adversary model. It is worth noting that a class of attacks might be related to several classes of threats, which in turn may be related to several classes of vulnerabilities and several individual vulnerabilities. The process of defining classes of attacks and threats related to a system precedes the definition of system security requirements, and it is the first step in engineering a secure SDR, as explained in the introduction (Myagmar et al. 2005).

CLASSES OF ATTACKS FOR GENERAL TACTICAL SDRS

In this section, we describe several classes of attacks that a tactical radio set can suffer. Our goal in this paper is to include the classes of attacks that we consider the most relevant, considering expected losses to legitimate users of the tactical radio network, and it is worth remembering that this classification is based on attack results. The scenario for defining attack classes is of a SCA-compliant radio set based on GPPs that implements waveforms that go up to the third OSI layer (network). The SCA-compliant radio set is part of a radio network consisting of a finite and predetermined number of radios. The data transmission between two devices on the radio network always uses some form of encryption. The adversary can act externally to the radio network, or can have direct access to the network by capturing one or more radio sets, with the loss being unnoticed during a finite amount of time. The adversary has the ability to monitor all frequency bands used by the radio network. The adversary is also capable to intervene in the process of manufacturing and delivery of a limited, but crucial, number of components that compose the radio set. As stated in the Introduction, the attacks considered in this paper target the radio set. Attacks aimed at other elements of the communication infrastructure (such as waveform download servers) are not included in the scope of this work.

Considering this scenario, the most important classes of attacks for general tactical SDRs identified in this work are:

- **Radio Control:** For this class of attacks, the adversary seeks to gain control (total or partial) of the SDR. Through radio control, for example, an adversary can force the device to behave in a Byzantine, seemingly random, way.
- **Personification:** For attacks belonging to this class, the goal of the adversary is to fool the radio set, or to introduce himself to the SDR as an entity belonging to the radio network or authorized to access it.
- **Unauthorized Data Modification:** This class includes attacks that alter the data transmitted by the radio set.
- **Unauthorized Accesses to Data:** The attacks of this class targets sensitive internal data and information, without modifying them.
- **Denial of Service:** In this case, the adversary objective is to make the SDR unavailable or non-operational.

Other classes of attacks are possible, but were considered not relevant given the proposed scenario. For example, in (Murotake and Martin 2004, 3GPP 2001), there were observed classes of attacks identified as "Unauthorized access to services" and "Rejection." The first case can be included in the class Personification, as explained in the following subsections. The second case concerns an *external user* that denies to have accessed to the network despite having done it. In the described scenario, the external user is necessarily an adversary. Therefore, any denial of the external user is not relevant and the described classes cover his /her possible actions. Table 1 shows the classes of attacks and related attack techniques and vulnerabilities. The associations among classes,

Table 1. Attack classes, associated attacks and vulnerabilities

Attack Classes	Attacks/Vulnerabilities
Radio Control	Software injection • Buffer overflow • Waveform download vulnerabilities • Start-up vulnerabilities
Personification	Replay • Protocol vulnerabilities Authentication break GPS Spoofing
Unauthorized data modification	Software injection Hardware injection • Implementation vulnerabilities • Hardware Trojans • Device cloning
Unauthorized access to data	Hardware injection Software injection Traffic analysis attacks Side-channel attacks Fault-based attacks Social Engineering attacks
Denial of Service	Hardware injection Software injection Jamming Flooding

attack techniques and vulnerabilities are described in detail in the following subsections.

Radio Control

In the *Radio Control* class of attacks, the adversary aims to gain control of all or part of the radio set. This goal can be achieved through the injection of spurious or malicious software that alters the proper functioning of a SDR partition in order to compromise the system security, e.g., by violating its security policy.

There are several ways of introducing malicious code on a SDR, exploiting several vulnerabilities, as exemplified in the following paragraphs. The code injection can be made through the RF interface or via physical access to the SDR, using, for example, a local interface. This type of attack can have different results, as to reduce the radio set functionality or even allowing the adversary to take complete control of it.

Buffer Overflow

In this attack, the adversary exploits buffer overflow vulnerabilities present in the system software to inject malicious code (Hsu et al. 2006; Riley,Jiang,Xu, 2010). A buffer overflow occurs when data stored in a buffer exceeds its capacity. There are two basic types of buffer overflow vulnerabilities: heap-based and stack-based (Correia and Sousa 2010). The heap-based buffer overflow occurs when the buffer is located in the heap, which is the area of memory where dynamic data are stored. The stack buffer overflow occurs when the buffer is located in the stack, which is the area of memory where local data and return addresses of a function are typically stored. For example, a stack overflow attack may be able to redirect the program flow to code under adversary's control, by rewriting the return address of a function. In a SCA-compliant SDR, buffer overflow vulnerabilities may be present in any component of a waveform, in any OSI model layer. In addition, vulnerabilities of this type can be present in the SCA Core Framework, CORBA middleware, operating system and hardware abstraction layer (Board Support Package-BSP + drivers).

Attacks that exploit buffer overflow vulnerabilities to inject malicious code are certainly among the most common and dangerous today (Riley et al., 2010). An example of this type of attack on a wireless system was identified in 2006 by researchers Dave Maynor and Jon Ellch (Murotake and Martin, 2009). This particular attack exploited vulnerabilities in the Atheros WiFi chipset device drivers used in Apple computers, and allowed the adversary to gain administrator access (root), even without the system being connected to a WiFi network. A similar attack was translated to the Broadcom WiFi chipset device drivers for Windows systems. These device drivers were used at that time by several manufacturers (Dell, HP, Gateway, among others).

WiFi interfaces are constantly scanning the spectrum to search for available networks, as soon as the system is turned on, regardless if the computer is connected or not to a network. The attack exploited a buffer overflow vulnerability by broadcasting a malformed SSID (network access identifier). The target machine received the malformed SSID, allowing the adversary to inject executable code at the device driver level and then getting root access, overlapping security mechanisms such as firewalls and virus control.

Waveform Downloading

Downloading waveforms through the air interface (RF) is one of the main features associated with SDRs in general and with SCA-compliant radios in particular. The waveform download process consists of three phases (Wireless Innovation Forum, 2002):

- **Pre-Download:** Includes service discovery, selection of security mechanisms, mutual authentication, user authorization, download acceptation;
- **Download:** Includes software physical transfer, integrity check of received code, potential retransmission request and storage in a safe place;
- **Post-Download:** Consists of installation, validation, non-repudiation and SDR reconfiguration.

However, this feature introduces several important security issues (Stango and Prasad, 2009). One of these security issues happens when malicious software is loaded instead of the original waveform. In this type of attack, the adversary may exploit vulnerabilities associated with the download protocol and the process of waveform instantiation on the SDR. Malicious software can be installed by waveform downloading if mechanisms to ensure code authenticity, integrity and versioning in the download and post-download phases are not available. Moreover, it may force a military network to operate with a non-optimal waveform – for instance, a waveform typically used for low-error rate channel conditions (e.g., a line-of-sight UHF low-mobility networks) may be downloaded for operation on a harsh environment (e.g., a VHF ground-to-air multipath propagation channel or a non-line of sight VHF urban scenario). This fact would severely degrade a previously agreed quality-of-service level, in spite of being a previously authorized waveform.

Radio Start-Up

The malicious software can also be embedded in non-volatile memory used for initialization (e.g. BIOS), during the radio set manufacturing process or later, through physical access to the SDR. This type of attack is associated with vulnerabilities in the system start-up (Gallery and Mitchell, 2006), which are generally related to the integrity verification of the start-up routines before execution.

For instance, malicious software may alter the transmission power level adopted by a radio set on a military network. Such modification does not keep the agreed communication to take place, but inserts vulnerabilities on allied communications, since it may allow an enemy force with electronic warfare capabilities to monitor, search, intercept and decode unauthorized data.

Personification

In the *Personification* class of attacks, the adversary presents himself as another entity, different from the original. For example, an adversary can impersonate a server, a network, a radio set or act as a man-in-the-middle. An attack of this class can have several goals, for example, to access or modify information in transit, to send outdated or invalid data to the SDR to reduce its functionality, or simply to allow the adversary to present himself as an authorized correspondent, gaining access or changing the SDR behavior.

This class of attacks can explore various types of vulnerabilities, often associated with protocols. For example, vulnerabilities in the data transmission protocol of a waveform may allow replay attacks (Wireless Innovation Forum, 2010). In the replay attack, the adversary captures a copy of the transmitted information and relays it later, and it can be exploited, for example, in any SCA-compliant data transmission waveform that does not implement sequence numbers, challenges or a freshness scheme (van Dijk, Rhodes, Sarmenta, Devadas, 2007). An example of a waveform that allows replay attacks is the IEEE 802.11-WEP (Berghel and Uecker, 2005). Replay attacks can also be used to distribute obsolete waveforms (possibly with known vulnerabilities), which were previously stored by an adversary, for several radio sets belonging to a radio network.

If the SDR deploys waveforms that require some form of authentication between devices, another type of attack can occur if the authentication protocol contains vulnerabilities. Examples of waveforms with authentication vulnerabilities are several implementations of IEEE 802.11 (WiFi) including WEP, WPA-PSK or EAP / LEAP (Berghel and Uecker 2005).

Several radio sets implement internal geolocators. Examples include military radios (SCA-compliant or not) and cognitive radios. One of the easiest ways to implement a geolocator is via a civilian GPS receiver, which can also be set by software in a SDR. GPS receivers may be subject to an impersonation attack known as GPS spoofing, which consists of sending wrong location signals to a receiver position (Humphreys, Ledvina, Psiaki, O'Hanlon, Kintner, 2008). These attacks are relatively simple to perform in civilian GPS receivers, since these receptors do not implement countermeasures against such attacks. In fact, impersonation attacks on civilian GPS receivers are performed even through commercial GPS signal simulators. However, this simple attack is also the easiest to detect. Attacks that are more sophisticated require the use of portable receiver-spoofer devices (Humphreys et al., 2008). A single device or a set of coordinated devices can do this latter attack.

Unauthorized Data Modification

In the *Unauthorized modification of data* class of attacks, the adversary aims to modify data that is stored or transmitted by the SDR. The unauthorized modification of data or software can impair the security or functionality of the radio set. This class of attacks includes the possibility of unauthorized modification of internal parameters of a SCA waveform. Therefore, it may alter different radio set features, like a protocol internal configuration setup, a control data flow, or even a user-driven message data that is under transmission.

As in previous cases, a number of vulnerabilities can be exploited to carry out an attack of this class. For example, an adversary can exploit buffer overflow vulnerabilities, via code injection attacks, to perform unauthorized data modification.

The unauthorized modification of data can also be performed by exploiting vulnerabilities in hardware. A processor is usually described in about 500 000 lines of VHDL code, which usually are not submitted to a formal verification process. Thus, one can expect that a processor may have multiple vulnerabilities exploitable by an adversary. Attacks can also make use of vulnerabilities introduced deliberately by an adversary. An example of vulnerability introduced by the adversary is the *hardware trojan* (Karri et al. 2010). A hardware trojan is a deliberate, malicious and difficult to detect hardware modification in an electronic device. The hardware trojan can modify the functionality of an integrated circuit and undermines the system reliability and security. For example, we may consider a cryptographic processor that normally sends encrypted data to an output. When the trojan is active, the encryption is disabled and the data is sent in clear, without the system operators' awareness (see Figure 2) (Karri et al., 2010). In this paper, we identify attacks that exploit vulnerabilities in hardware generally as *hardware injection*.

In addition to trojans, hardware vulnerabilities can be introduced through device cloning, which turns out to be another variant of hardware injection. In this case, we have a device (e.g. a board or an integrated circuit) that is physically the same as the original, but it contains malicious code or hardware and it does not have appropriate protection mechanisms to prevent the extraction of sensitive information (Gallo et al., 2009). Unlike a hardware trojan, which is introduced during the chip manufacturing process, device clones are introduced during the transportation of the chip or board from its factory to the SDR assembly location. Attacks such as device cloning, which exploit vulnerabilities in the supply chain, are also associated with a wider class of attacks known as Supply Chain Attacks. Supply chain attacks involve the manipulation of hardware, software

Figure 2. Hardware trojan example

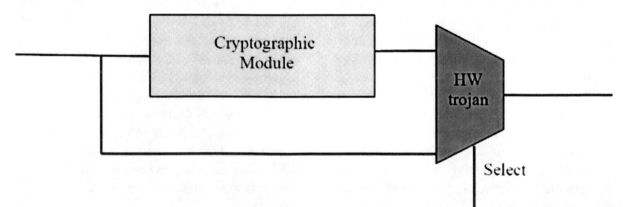

or services at any time during the life cycle of a device (Swanson et al., 2010). This type of attack is typically carried out or facilitated by individuals or organizations that have access to radio set components via commercial ties. The results of attacks on the supply chain are, in addition to modification of critical data, technology and sensitive data theft and deactivation of mission-critical features.

Unauthorized Access to Data

In the *unauthorized access to information* class of attacks, the objective is to access sensitive data or information stored in the SDR, without modifying them. Unauthorized access to sensitive information stored in the SDR can compromise the security and availability of the radio network, and this is a class of attacks usually difficult to detect. For example, attacks aimed at obtaining cryptographic keys and other critical security parameters. The access to cryptographic keys, in turn, could allow the adversary to access encrypted data transmitted between two radio sets, without regular users being aware of the eavesdropping. As described previously, it may affect different radio set features, like a protocol internal configuration setup, a control data flow, or even a user-driven message data that is under transmission.

Unauthorized access to information can be done by exploiting several vulnerabilities, such as those related to the injection of malicious code and the use of hardware trojans and device clones. For example, in SCA-compliant radios that have a single partition for storing sensitive data and non-confidential information, privileged execution of malicious code can allow the adversary to access all data stored in the nonvolatile memory of the radio set, including the cryptographic keys. Critical information can then be transmitted in clear using a data transmission waveform deployed in the SDR itself, or the adversary may use any other external interface available.

Another example is related to hardware injections in cryptographic processors. Malicious hardware injected in these processors can cause the bypass mechanism in the JTRS security architecture (SCA Security 2001) not to function properly, which can result in confidential information being transmitted in clear, like data protocol configuration parameters, which are vital to a proper communication protocol handshaking mechanism.

Another type of attack that belongs to this class is traffic analysis attacks. In such attack, the adversary aims to get mission-critical radio network information by observing traffic statistics (Fu, Graham, Bettati, Zhao, 2003; Wireless Innovation Forum, 2010). Critical information that can be obtained include senders and receivers identities, the establishment and termination of connections, consumption of bandwidth, burst traffic, signal strength, etc. This type of attack can be divided into two types: passive and active. In the passive traffic analysis, the adversary passively collects data and performs several analysis tasks on the collected data. In the active traffic analysis, the adversary uses active probes in the process of gathering information to obtain additional data not accessible by passive collection. In this case, the adversary seeks to analyze the behavior of radio network elements when subjected to a specific stimulus (Fu et al., 2003). Besides that, any modification on the transmission power level adopted by a military radio set is a potential tool of unauthorized access to any type of data, since it may allow an enemy force with electronic warfare capabilities to monitor, search, intercept and decode messages or even identify traffic patterns by mapping control and configuration data flow.

When the adversary has possession of an operational SDR, sensitive information (eg, cryptographic keys) can be obtained through side-channel attacks, as described in (Standaert, Malkin, Yung, 2009; Chevallier-Mames, Ciet, Joye, 2004). In this type of attack, the adversary collects and analyzes data related to several physical quantities, such as power consumption, processing time and electromagnetic emissions of the SDR internal circuitry, to gain access to sensitive information. As an illustration, there are several examples in the literature of how an attack of this type can be performed with the aim of breaking implementations of cryptographic algorithms (Kocher 1996; Kocher,Jaffe,Jun, 1999; Agrawal,Archambeault,Rao,Rohatgi, 2003; Bonneau and Mironov, 2006). For example, in (Agrawal et al., 2003) the authors focus on the analysis of electromagnetic signals emitted by the device under attack to break cryptographic algorithms and overcome countermeasures against other side-channel attacks, such as those based on power. The electromagnetic emissions of interest are originated from data processing operations, such as those observed in CMOS circuits. The authors describe successful attacks to block ciphers (DES) and public key algorithms (RSA) in chipcards, and to SSL accelerators using a single electromagnetic signal sensor (antenna). It is noteworthy that all the attacks described in (Agrawal et al., 2003) are non-intrusive, non-invasive and does not require precise positioning of the electromagnetic sensors.

Another encryption algorithm that is susceptible to side-channel attacks is the AES-Rijndael (Bonneau and Mironov, 2006; Bogdanov, 2008; Renauld,Standaert,Veyrat-Charvillon, 2009). For example, in (Renauld et al., 2009), it is shown that a class of side-channel attacks known as algebraic side-channel can be applied to the AES-Rijndael algorithm implemented in 8-bit controllers. The observation of a single trace (plaintext and corresponding ciphertext) may be sufficient to perform complete recovery of the encryption key in this context. The attack is directly applicable to certain masking schemes, i.e., schemes that make the power consumption of the processing device independent of the intermediate values processed by the cryptographic algorithm. Observe that the attack can be successful even if traces are not available. It is also interesting to note that the offline processing phase of the attack was limited, in that paper, to 3600 seconds.

Another type of attack that belongs to this class is fault-based attacks, which, as is the case of several side-channel attacks, requires ownership of an operational radio. In this type of attack, the adversary induces faults in hardware in order to gain access to sensitive information. An example of this type of attack is described in (Pellegrini,Bertacco,Austin, 2010). In this paper, the authors describe how to obtain a 1024-bit RSA private key in a Linux SPARC system through the insertion of processing errors by decreasing the supply voltage of the processor. By introducing faults in the processing of the fixed window exponentiation algorithm (FWE) used in OpenSSL-0.9.8i, the authors were able to get the 1024 bits private key after 100 hours of offline processing, in a Pentium 4 cluster with 81 nodes. It should be highlighted that this type of attack does not damage the target machine, leaving no signs that its security has been compromised, as is the case for side-channel attacks.

Social engineering attacks (Goodchild 2010; Jagatic,Johnson,Jakobsson,Menczer, 2007; Ferguson 2005), which consists on the manipulation of people to obtain confidential information, can also be used to obtain sensitive information from a radio set. These attacks involve the use of tricks to deceive one or more individuals within an organization, and often the adversary never meet personally with the misled individuals. Social engineering can be applied through various methods, e.g., personification of superiors or colleagues, phishing (Jagatic et al., 2007), etc. This attack involves individuals who have specific knowledge or have physical access to the radio device, and can be highly effective in military institutions (Ferguson, 2005). For example, the action of a single operator can enable an electromagnetic emission side-channel attack by placing a data collector device near the SDR, influenced by an adversary impersonating a superior officer. The operator would then return the data collector device to the adversary, whom can process the collected data offline.

Denial of Service

A *denial of service* attack aims to make the SDR unavailable or non-operational. This type of attack can exploit several vulnerabilities, for example: buffer overflow vulnerabilities, protocol vulnerabilities, hardware trojans, jamming and flooding. In the scope of this paper, jamming is defined as the deliberate transmission of radio signals that interfere with the radio communication between two devices by reducing the signal-to-noise ratio. Jamming attacks can be used in wireless data transmission to stop the flow of information between two communicating entities (Frankel, Eydt, Owens, Scarfone, 2007; Brown and Sethi, 2007). In turn, in the flooding attack an adversary sends a large number of messages, related to a particular waveform, to a radio device at a rate so high that the SDR cannot process all the messages in time (Frankel et al., 2007). This over-capacity processing can result in a partial or total denial of service.

As an illustration of possible attacks on military radio set, consider the frame structure for all MIL-STD-188-110 B waveforms, used to convey interoperability for data modems on several frequency bands. Based on that standard, an initial 287 symbol preamble is sent, being followed by a 72 frames sequence of alternating data and known symbols. Each data frame is made of a data block consisting of 256 data symbols, followed by a mini-probe sequence with 31 symbols of known data. After that 72 data frames sequence, a 72 symbol subset of the initial preamble is reinserted to facilitate late acquisition, Doppler shift removal, and synchronization adjustment. So, any change on the synchronization sequence or on the quantity of data frames make the military network unavailable or non-operational, since it keeps the receiver radio set from recognizing the adopted communication protocol.

A common denial of service attack occurs when the adversary sends a spurious signal that prevents the reception of the signal transmitted legitimately (Brown and Sethi, 2007). This brute force approach can be applied to any type of radio transmission. There are situations, in the case of data transmissions, in which such attacks are effective even when the adversary is transmitting low power spurious signals (Lin and Noubir, 2003).

Geolocators based on civilian GPS signals may also be subject to denial of service attacks (Brown and Sethi 2007). GPS signals are weak and easily masked by power emissions. In addition, GPS signals fade indoors, in dense urban areas or in steep terrain. The best strategy in this case is to implement, in the radio set, several different strategies to infer the geographic location beyond GPS.

MITIGATION MEASURES

In this section, we present several mitigation measures for the attacks described in the previous section. The list of mitigation measures presented in this section is illustrative and not exhaustive. It refers to several published works available in the literature.

Radio Control

Buffer Overflow

There are several ways to prevent and mitigate the effects of code injection attacks via buffer overflow. Examples are static code analysis or the implementation of dynamic protection mechanisms (Correia and Sousa 2010). Current operating systems (e.g. Windows, Linux) and C/C++ compilers already implement several mechanisms against buffer overflow attacks, such as canaries (Cowan et al. 1998), stack cookies (Howard and LeBlanc 2007), address space layout randomization (van de Ven 2005), non-executable stacks. As an example of a dynamic protection strategy based on an architectural approach, it was recently published (Riley et al. 2010) a promising proposal to prevent code injection attacks that exploit buffer overflow vulnerabilities. In this work, the authors proposed a virtualized Harvard architecture over the memory architectures of modern computers (von Neumann), including those without support for non-executable memory pages, as a way to prevent injection of malicious code. Harvard architecture is one in which data and code are entirely separate. Any code injection will be made on the data partition, so the malicious code cannot be executed. The authors implemented this proposal as a software patch for Linux, incurring a performance loss of about 10% to 20%. This approach, however, has limitations. It is not adequate to protect self-modifying programs (Giffin, Christodorescu,Kruger, 2005), and offers no protection when the attack executes code that was not injected by the adversary.

Another mechanism for dynamic protection of processes against code injection via buffer overflow vulnerabilities is described in (Abadi, Budiu, Erlingsson, Ligatti, 2009). This work describes a mitigation technique, the enforcement of Control Flow Integrity (CFI), that is effective against stack-based buffer overflow and heap-based "jump-to-libc" attacks. The CFI property states that software execution must follow a path of a Control Flow Graph (CFG) determined ahead of time. In this work, the construction of the CFG is done through a static binary analysis. The authors also presented fast, scalable implementations of CFI for Windows on the x86 architecture. All in all, It should be highlighted that choosing the most appropriate protection mechanism against code injection will depend on several factors: what are the processing elements available in the SDR, available memory, set of waveforms that will be deployed, etc.

Waveform Downloading

Stango and Prasad (2009) propose a mechanism based on policies to ensure the secure download of reconfiguration files, which includes waveforms, in a SDR. On the other hand,

(Brawerman,Blough,Bing, 2004) proposes a secure protocol for downloading reconfiguration files in a SDR. This paper proposes a lightweight version of SSL (LSSL) for the connection between the server and the radio set, and a protocol to download securely files to the SDR. An equivalent protocol is proposed in (Gallery and Mitchell 2006).

Radio Startup

When the SDR deploys several separate partitions, as in a MILS system (Alves-Foss, Harrison, Oman, Taylor, 2006), each partition may have a separate boot process. As a result, partitions different from the compromised one may not be affected. Still, such attack can make the SDR unusable or subject to external control, depending on the attacked partition. It is worth noting that the SCA v.2.2 security supplement defines safe boot requirements only for the cryptographic subsystem (SCA Security, 2001).

Unauthorized Data Modification

Hardware Injection

Several preventive measures are proposed in (Gallo et al., 2009). In this work, the authors introduce the concept of Cryptographic Identity (CID) as a unique identifier created at the physical inspection of the final hardware. The CID allows the establishment and verification of a root of trust for secure commodity hardware. The CID is statistically unique, copy and tamper protected and verifiable. The physical hardware inspection used to create the CID defines what the authors call a secure device epoch. The authors also define shared verification schemes to attest CID validity. Verifications can be made throughout the device lifetime in order to ensure that the hardware has not been altered between successive utilizations. One of the verification schemes proposed in (Gallo et al., 2009) was the Time-Based One Time

Verification Code (TOTV), and a variation of this verification scheme is used in the Brazilian electronic voting machines (Gallo et al., 2010). Even when the CID is not implemented, it is possible to detect hardware trojans through a number of countermeasures (Karri et al., 2010), for instance: to check the RTL code of the I/O unit for changes in the I/O protocol; to perform exhaustive memory testing; to analyze the side-channels; to check exhaustively for resource utilization changes; to communicate periodically with the device, even after its deployment; to skew clocks and observe the IC transient behavior; to scale dynamically the supply voltage while checking for transient characteristics; to do the concurrent detection for soft errors; to use path delay fingerprints (Jin and Makris, 2008). Regarding military systems, some agencies have restrained circuit designs to the factories that have passed certain certifications (Jin and Makris, 2008).

Unauthorized Access to Data

Side-Channel and Fault-Based Attacks

It is possible to deploy countermeasures to mitigate these attacks in the process of physical construction of the radio set. Several countermeasures are described in standards FIPS140-2, level 4 (FIPS, 2001) and FIPS 140-3, level 4 (FIPS, 2009). Some of the possible countermeasures are (Gallo et al., 2010; Agrawal et al., 2003; FIPS, 2001; FIPS, 2009): the implementation of constant-duration cryptographic operations to mitigate timing analysis attacks; to filter and to stabilize the power supply and to decouple all external communication connectors to mitigate attacks based on power variation; to involve circuit boards in inviolable coatings (e.g. resins); the use of electromagnetic shielding to reduce the power of compromising electromagnetic emanations; the frequent redefinition of cryptographic keys; the deletion of cryptographic keys and other critical data if the SDR casing is opened or tampered with.

Denial of Service

Jamming and Flooding

One way to make a radio network more robust to these attacks is the use of spread spectrum techniques, such as direct sequence spread-spectrum or frequency hopping (Strasser, Popper, Capkun, Cagalj, 2008). However, these techniques are only effective if the secret code shared between sender and receiver, which enables the spread spectrum, is not compromised. The use of cognitive radio techniques can also make this attack more difficult, since an adversary may have to transmit spurious signals simultaneously in various frequency bands, or to implement reliable techniques for detecting when the cognitive radio switches between different frequencies bands (Brown and Sethi 2007).

For denial of service attacks that involve the transmission of power signals that prevent the communication between radio sets, an interesting solution related to waveform implementation was proposed in (Strasser et al. 2008). The authors propose a non-coordinated frequency hopping mechanism, where radio sets do not need to share a secret key or code that allows for the frequency hop. The basic idea is that, after enough transmission attempts, the sender and the receiver will transmit and receive in the same channels in some intervals of time, even without a previous agreement on these intervals. The price to pay is a reduced transmission rate, but even this reduced rate of transmission is sufficient to perform a key sharing protocol that allows a more efficient subsequent communication. This mechanism can be very useful if the adversary succeeds in obtaining the secret code related to the frequency hop.

SCA IMPACTS

In this section, we provide a description of how attack mitigation strategies can influence the development of an SCA-compliant software infrastructure. In this section, we focus mainly on the components of the SCA operating environment. We start with an analysis of the basic JTRS security architecture (Figure 1) and related impacts on the location of SCA components and on overall radio security.

SCA Security Architecture

The basic JTRS security architecture described in (SCA Security 2001) and depicted in Figure 1 defines three independent partitions: red, black and the cryptographic subsystem. The red partition holds unencrypted (confidential) data, while the black partition only handles encrypted data. The RF module is connected to the black partition, while the red partition has access to the user I/O interface. Actually, the red-black paradigm was originated to address TEMPEST (TEMPEST 1982, TEMPEST 1985) separation of analog circuits, and this configuration influenced the organization of radio security partitions. It is noteworthy that the three partitions can be implemented either physically (e.g. using three processing elements physically separated) or logically (via separation kernels), as discussed in the next subsection (RTOS).

Both the red and black partitions contain a full SCA stack, and most SCA components are replicated in both partitions, with few exceptions. One major exception is the Domain Manager. In a SCA-compliant radio, the Domain Manager component controls the whole radio domain and there is one Domain Manager per radio set. There are advantages and disadvantages of allocating the radio control component either in the red or black partition (Davidson 2008). Usually, the Domain Manager is instantiated in the red partition and it uses the bypass mechanism to communicate with components in the black partition. If the black partition is compromised due, for instance, an attack that exploits buffer overflow vulnerabilities, the confidential data stored in the red partition is not exposed. Furthermore, due to the hardware implementation of cryptographic algorithms and other security features in the cryptographic subsystem, there is enough computing power in this architecture to encrypt/decrypt all data transmitted or received. Beyond that, the hardware components of the cryptographic subsystem provide a reliable root-of-trust for the radio set as a whole. A hardware root-of-trust is important because trusting all the radio software is an unreasonable option (Davidson 2008). However, the widespread utilization of the bypass mechanism in some implementations represents a major weakness of the red-black paradigm (Davidson 2008). Nevertheless, the basic security architecture may be effective to mitigate risks related to the attack classes *Unauthorized access to data, Unauthorized data modification, Radio Control*.

In order for a waveform to access the functionalities provided by the cryptographic subsystem, a set of APIs are implemented through either SCA Devices or SCA services. It is important to standardize this set of APIs to improve waveform portability. A first attempt in this direction was provided by SCA v2.2 Security Supplement. However, as mentioned previously, the supplement has been deprecated and updates of this document have not been made public.

This situation impulses the creation of the Security WG at the Wireless Innovation Forum, which in 2011 releases the International Radio Security Services (IRSS) API (Wireless Innovation Forum 2011). The working group focuses during the development of this API on the military and tactical domain. The API is intended to deployed on tactical radios implementing the JTRS SCA specification, though SCA is not a requirement for its use.

Following the aforementioned objective on waveform portability, the IRSS API intention is to maximize the portability degree between various radio platforms that provide the same API. This API enforces also the framework paradigm from the JTRS SCA standard, assuring, but not tied to it, allowing its deployment in non-SCA platforms.

Real-Time Operating Systems (RTOS)

The RTOS provides several services for the upper layers of the operating environment, including:

- Process/thread creation.
- Process/thread management.
- Inter-process communication.
- Timing services.
- Input/Output management.
- Scheduling management.

It is worth highlighting that process separation and data isolation are fundamental security features that should be supported by the operating system (SCA Security 2001). Beyond those features, the RTOS should also provide dynamic protection mechanisms against buffer overflow attacks. Examples of dynamic protection mechanisms currently available for Linux systems are Stack Smashing Protectors and Address Space Layout Randomization (Correia e Souza 2010). These measures increase the level of difficulty for an adversary to perform a successful intrusion. Even when an attack of the class Radio Control results in an intrusion, thanks to the isolation of partitions suggested in the JTRS security architecture, a possible impairment of the black partition by a buffer overflow based attack, e.g. via the RF interface, allows only partial control of SDR by the adversary. The security supplement recommends the physical isolation among partitions also because of concerns about spurious emanations (Tempest, 1982; Tempest, 1995). However, for SDRs in which requirements such as weight and battery duration are critical, the recent emergence of separation kernels with a high level of assurance, such as those classified as EAL 6+ (Common Criteria, 2008), makes the logical separation among partitions a viable alternative. The utilization of separation kernels also enables the implementation of Multilevel Secure (MLS) SDRs. An example of a MLS capable OS is the GEMSOS security kernel (GEMSOS, 2012). GEMSOS is rated Class A1: Verified Protection by the National Security Agency. Indeed, the problem with full MLS systems is that they must be rigorously analyzed for security before they can be certified. Every portion of the MLS system must be analyzed to ensure that it properly handles labeled data and that there is no possible violation of the security policy. Therefore, there is often too much to evaluate.

The concept of separation kernels is tightly related to MILS (Multiple Independent Levels of Security) capable systems (Alves-Foss et al. 2006; Beckwith, Vanfleet, MacLaren, 2004). The MILS architecture was developed to resolve the difficulty of certification of MLS systems, by separating out the security mechanisms and concerns into manageable components. MILS makes mathematical verification possible for the core systems software by reducing the security functionality to four key security policies (Beckwith et al., 2004):

- Information flow.
- Data Isolation.
- Periods processing.
- Damage limitation

A MILS system employs one or more separation mechanisms (e.g., Separation kernel, Partitioning Communication System, physical separation) to assure data and process separation. A MILS RTOS system is designed to minimize the size of the kernel in order to make it verifiable by formal analysis and proof-of-correctness methods. The price related to this approach is more context switch overhead. However, this cost has been made tolerable by hardware advances and careful design of the inter-partition communication services.

Concerning the flow of information inside the kernel, an information flow policy is required to avoid unauthorized bypass of security functions. The information flow policy should be enforced by the inter-process communication functionality inside the kernel. A particular threat arises from the possibility to forge IPC objects that are shared between processes (Kiszka and Wagner, 2007). In order to mitigate the risks associated with this threat, separated namespaces can be established to isolate adversaries' from legitimate registration requests (Kiszka and Wagner, 2003).

As pointed out in previous subsections, device drivers can be a major source of vulnerabilities inside an operating system kernel. The set of device drivers that should be available in a radio set depends on the underlying hardware, and in many cases, several custom drivers should be developed. Therefore, in order to minimize the number of vulnerabilities in a radio set, thorough analysis of device drivers is mandatory. For instance, in (Ball et al. 2006) the authors present the Static Driver Verifier (SDV) tool, which uses static code analysis to enhance both the observability and coverage of device driver testing.

In (Kiszka and Wagner, 2007), the authors introduced a class of attacks related specifically to real-time operating systems dubbed Denial of Determinism (DoD). This class groups attacks against hard real-time services, aiming at the denial to obtain those services at any time both correctly and timely. In (Kiszka and Wagner, 2007), several DoD attacks are described, e.g. timer storm attack, wakeup list attack, priority inheritance chain attack, etc. The foundation to mitigate the risks of DoD attacks is the Real-Time Domain and Type Enforcement (RT-DTE) model, which is an access control model applicable on hard-RT environment, as introduced in (Kiszka and Wagner, 2007). The main idea of the original Domain and Type Enforcement model is to group processes (subjects) to domains and objects to types and to define the access rights of subjects on objects or other subjects. This concept is extended in RT-DTE by additional, real-time related properties of domain and types. Normally, the RT-DTE needs to contain only a few domains and types compared to the large number available in non-real-time entities.

Vulnerabilities in the operating system relate to attacks of the classes Radio Control, Unauthorized Data Modification of Data, Unauthorized Access to Data, Denial of Service.

Middleware

The middleware security approach is to control message traffic among objects by restricting access to object references within the system (SCA Security 2001). For instance, each object, upon instantiation, registers with the ORB naming service by performing a *bind()* operation. This naming service may reside inside the SCA Domain Manager's process space. Therefore, only entities within the SCA Domain Manager process space shall be permitted to perform *resolve()* and *list()* operations, protecting object references from being discovered by unauthorized entities. One potential problem with this approach, related to the basic JTRS security architecture, is the proliferation of bypass channels to allow naming service access to components located both in red and black partitions.

SCA components can use CORBA sequences to send data between components. A CORBA sequence is a CORBA version of a variable length one-dimensional array. Like container classes (i.e. STL vector), it allows for an array of any data type including complex data types like structures. This particular feature help increase radio security by providing an error path if a malicious user inject bad data that causes the software to write data outside the sequence (Balister, Robert, Reed, 2006). Beyond that, the CORBA sequence provides bounds checking as well as memory management.

The SCA security supplement mandates implicit authorization for prevention of unauthorized access of CORBA objects. Implicit authorization means that handing over an object reference to a client authorizes the client to access the respective object. The problem with this approach is the risk of attacks where adversaries can fabricate object references for valid objects, sometimes demanding modest computational effort and time (Becker et al, 2007). In this case, the implementation of an additional logic of explicit access rights (e.g. through the CORBA Security Service specified by the OMG (OMG CORBA Security 2002)) may be a solution to avoid this particular attack. Other possible countermeasure, when eavesdropping in the transport connection is not possible, is the inclusion of an unpredictable bit sequence (used once) in each object key contained in object references (Becker et al, 2007).

The vulnerabilities associated to the middleware are mostly related to attacks of the class Unauthorized Access to Data.

SCA Core Framework

In (Bard and Kovarik 2007) several requirements related to the SCA core framework are listed. Some of the more important requirements are described below.

- **Application Interface:** In (Bard and Kovarik, 2007) are described several security requirements related to SCA applications. For instance, Application components' SPD implementation dependency *propertyref* elements shall indicate a dependency to a Red or Black device. Another relevant requirement is related to the disconnection of component's ports by the *releaseObject* operation. It shall only happen when authorized by an authentication service, which should be included as as one of the SCA services described in section 2. When the authorization is not granted, a Security_Alarm event should be logged.
- **ApplicationFactory Interface:** One important security requirement related to the Application Factory is the previous authorization of an authentication service when creating components through the create() operation (Bard and Kovarik, 2007).

- **DomainManager Interface:** Examples of security requirements related to the SCA Domain manager component is that Device SPD properties shall have an allocation property that indicates a red or black device. This information must be considered when installing or instantiating a new waveform. Another security requirement related to the DomainManager is free access across the red/black boundary of the cryptographic subsystem during power-up sequence (SCA Security, 2001). As mentioned before, all the required bypass channels may represent a source of vulnerabilities when the bypass mechanism is not carefully designed and verified.

Another possible source of vulnerabilities in a SCA-compliant radio set are SCA Devices. SCA devices should interact with OS device drivers, which may contain vulnerabilities as discussed in previous paragraphs. Beyond that, a SCA device should implement several concurrent state machines. The SCA v2.2.2 specification requires SCA Devices to implement behavior to support the administrative, usage, and operational states of the device (SCA, 2006). The SCA required state management behavior is a variation of the X.731 specification (X.731, 2012). If the implementation is not thoroughly checked, may let the system into a deadlock state under special circumstances. Adversaries with knowledge of internal implementation details can exploit these vulnerabilities.

R-check (Ezick and Springer 2011; Ezick, Springer, Litvinov, Wohlford, 2010) is a static source code analysis tools specially tailored to check SCA Core Framework requirements. Static analysis refers to the analysis performed by inspecting a program source or binary code without executing it. R-check has been used to analyze operating environments and production waveforms. Since some requirements are related to secure software development practices, such as avoiding memory leaks (Ezick and Springer 2011), this type of tool can also be used to minimize the number of vulnerabilities present in Core Framework implementations.

These security requirements relate to attacks of the classes Radio Control and Unauthorized Access to Data.

Standard SCA Services

The name service specified in the SCA specification (SCA 2001, SCA 2006) may be used by an adversary to discover system components' object references, if there is no mechanism to avoid unauthorized queries. As mentioned earlier, the SCA security supplement suggests the instantiation of the naming service inside the *DomainManager* process space. Therefore, only entities within the *DomainManager* process space shall be permitted to perform *resolve()* and list() operations, protecting object references from being discovered by unauthorized entities but introducing other potential vulnerabilities related to bypass channels. Another possibility is to include access control mechanisms in the process of object reference resolving, possibly using a security service like CORBAsec (OMG CORBA Security 2002). The need for an access control service that provides authentication and authorization has already been highlighted in previous subsections, and it should be included in the set of SCA services in a secure radio implementation.

CASE STUDIES

In this section, we describe several cases of vulnerabilities that a tactical radio is typically exposed to. Our goal in this section is to present case studies based on the taxonomy presented in this chapter. As a common feature amongst the following case studies, we describe military applications in which the radio sets are enabled with adaptive modulation features. Such applications have been widely

studied for use in mobile scenarios (Moura, Salles & Galdino 2009; Moura, Salles & Galdino 2010).

Adaptive modulation techniques require constant and fast-paced parameters changing pattern for features like application data traffic shaping, modulation and coding schemes, and frame time intervals. Based on that, the overall system performance may capitalize on proper radio environment conditions, due to higher spectral, energy, and/or cost efficiency.

Such techniques also present an embedded secrecy capacity (Amorim, Galdino 2010) bringing within a remarkable contribution to the transmission security and performance over tactical wireless communications links. Moreover, adaptive modulation techniques guarantee performance improvement (given by metrics in several network architecture layers, such as average spectral efficiency, packet loss rate, packet dropping probability, and packet error rate) over conventional modulation strategies in typical mobile radio communication scenarios (Moura, Salles & Galdino 2009; Moura, Salles & Galdino 2010).

Thus, we argue that the implementation of adaptive modulation techniques is suitable for tactical SDR and cognitive radios, since they present a software-based functional core embedded with security and performance enhancements usually required by tactical communications systems. Nevertheless, they also present several vulnerabilities and flaws that could be exploited by attackers, bringing unacceptable performance and/or security drawbacks for military operations (Silva, Moura & Galdino 2012; Moura, Silva & Galdino 2012).

Based on that, we state that this technique accomplishes a promising strategy in the context of strategic and tactical military communications, which justifies its choice as the physical layer technology in this chapter. Moreover, this assumption justifies the need for a deeper analysis over its potential sources of exposure to different classes of attacks.

Military Scenario Description

We present in Figure 3 an illustration of a point-to-point connection between a server and a subscriber, through a wireless link with two radio channels, representing forward and backward communication between stations. This setup is typical in several military communication scenarios, such as brigade, battalion and company tactical radio networks (Moura, Salles & Galdino 2009).

As a common scenario setup, we consider that the transmitter station has a MAC-layer finite-length queue (buffer), which operates in a first-in-first-out (FIFO) mode and feeds an adaptive modulation controller. This configuration is made in order to match transmission parameters (e.g., application data traffic pattern, modulation and coding schemes, and frame time intervals) to time-varying channel conditions, aiming to ensure a maximum performance and/or an agreed quality-of-service (QoS) level.

The adaptive modulation selector is implemented at the receiver to make an on-the-fly performance evaluation, giving the acquired results back to the transmitter through the feedback channel. We assume that the data link layer processing unit is a packet, which comprises multiple information bits. On the other hand, we define that the processing unit at the physical layer is a frame, consisting of multiple deterministically transmitted symbols, using a fixed frame rate pattern.

We assume that multiple transmission modes are available, with each mode representing a specific modulation format. Based on the channel state information (CSI) estimated at the receiver and accordingly transmitted through the feedback channel, the adaptive modulation selector determines the more adequate modulation scheme to be used by the transmitter at the following packet transmission (Moura, Salles & Galdino 2009).

There are several design strategies to perform modulation selection. According to some of the most suggested approaches, the adaptive modulation scheme is selected to guarantee a physical layer driven QoS level, since based on the instantaneous SNR on wireless channels characterized by flat fading effect. As presented in (Moura, Salles & Galdino 2010), the last approach is suboptimal when compared to a data link layer oriented QoS provision, in which the choice of modulation seeks to keep the total packet loss rate (evaluated as a physical-layer driven packet error rate and a data-link layer packet dropping rate) at an appropriate level. Such approach must guarantee spectral efficiency maximization, considering the required application-layer traffic model.

Given the receiver-driven adopted modulation selection scheme, CSI data with the proper modulation scheme is sent back to the transmitter through a feedback channel, for the adaptive modulation controller to update the transmission mode. It is worthwhile noting that coherent demodulation and maximum-likelihood (ML) decoding are employed at the receiver and that the decoded bit streams are mapped to packets, which are pushed upwards to layers above the physical layer (Pinto & Galdino 2009).

Figure 3. Typical point-to-point tactical wireless link

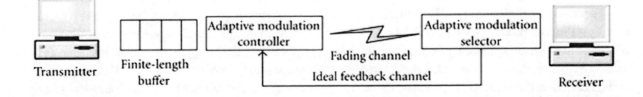

The objective of adaptive modulation is to maximize the achieved data rate by adjusting the modulation scheme usage to the present channel state, being subject to prescribed QoS constraints (e.g., a maximum physical-layer packet error rate). In order to achieve that, the entire signal-to-noise (SNR) ratio is partitioned into N+1 consecutive non-overlapping intervals, with boundary points denoted by (L_i), $0 \leq i \leq N+1$. As an example, this means that mode j is used when the instantaneous SNR (L) is such that $L_j \leq L \leq L_{j+1}$. Moreover, we state that no transmission occurs when $L \leq L_1$, which corresponds to mode 0.

Considering several practical operational assumptions, such as the user application's data traffic, the existence of flat fading effects over the wireless channel, a finite-length buffer at the data link layer, and the availability of a limited number of modulation schemes, there are important imperfections on the system performance. The most important drawbacks are the packet error rate (PER) at the physical layer, the packet dropping probability (PDP) to indicate the ratio of dropped packets at the data-link finite-length buffer, and the average spectral efficiency (ASE), which stands for the average number of transmitted packets at a single frame. We may infer that the overall system performance depends on parameters brought by several layers, namely the channel state, the queue occupancy discipline, and the traffic data rate that comes from the application layer.

Based on such assumptions, we consider in this chapter an adaptive modulation transmitter with five available modulation schemes (BPSK, QPSK, 8-QAM, 16-QAM, and 64-QAM) to perform data transmission, a fixed packet length of 1080 bytes, and a fixed frame transmission duration (time-unit) of 2 ms. Besides that, we assume a queue length of 50 packets, an application traffic pattern modeled by a Poisson process with an arrival rate of 2 packets/time-unit, and a Doppler frequency (which measures the wireless channel fading behavior) of 10 Hz. Based on the method proposed by (Moura, Salles & Galdino 2010), we can evaluate the best suited SNR intervals to maximize ASE given a prescribed maximum PER for different average signal-to-noise ratio (SNRav) scenarios.

Before describing more subtle system vulnerabilities, it is worthwhile noting this communication system is strongly subject to denial of service attacks. Namely, a jamming attack over any of the aforementioned channels (forward or reverse) interrupts the service, degrading overall performance. The following subsections present further classes of attacks and their consequences over adaptive modulation-enabled tactical communications systems.

Radio Control Attack

We evaluate a radio control attack, in which an adversary alters buffer occupancy by exploiting a MAC layer vulnerability, injecting malicious code and increasing the amount of stored data in a buffer, by exploiting heap- or stack-based vulnerabilities.

Here, system performance is deviated by an attacker that does not cause the buffer to exceed its capacity, which could be more easily perceived, but just inserts unexpected data packets into the finite length buffer described in Figure 3. Due to that, optimal parameters input are altered, changing the achieved value for the most suitable adaptive modulation threshold vector to maximize ASE or to minimize overall PLR, given a previously agreed signal-to-noise-ratio.

As a first case study, after identifying an available vulnerability, the attack strategy imposes an unwittingly 10-packets augmentation for a 30-packet-capacity buffer. Therefore, instead of storing 30 packets, the finite length buffer overflows with a 20-packet load.

Figures 4, 5, and 6 describe, respectively, the average spectral efficiency, the packet dropping probability, and the global packet loss rate (encompassing both physical- and data-link layer errors) for the currently planned and the under attack scenarios.

Figure 4. ASE in different scenarios

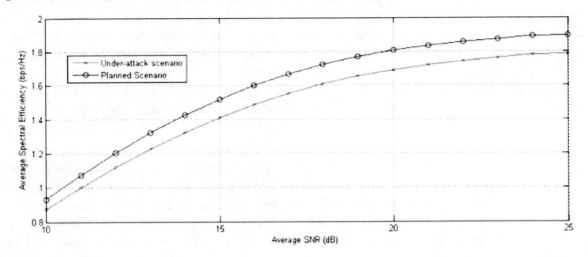

Figure 5. Packet dropping probability in different scenarios

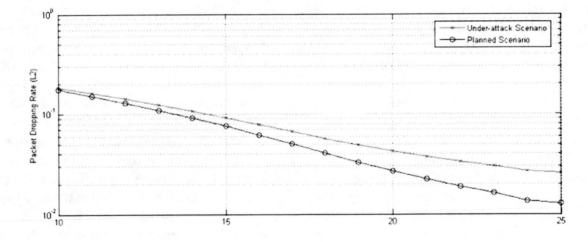

Figure 6. Global packet loss rate in different scenarios

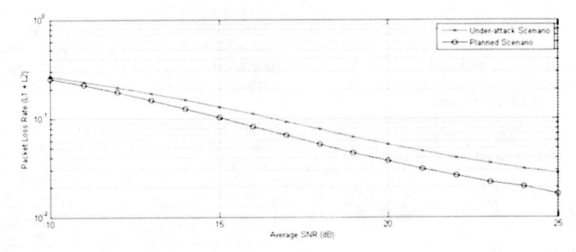

Based on that, we can state that the resulting breach exploited by the adversary intrusion brings an overall drawback in system performance, since it lowers ASE (e.g. as much as 0.1175 bps/Hz at 18 dB), while increases PDP (for instance, the dropping probability increases 50.6% at 25 dB) and, as a consequence, the overall PLR (e.g., the global losses in both physical- and data-link layer are 38.84% higher at 25 dB), even with a such slightly different adaptive modulation vector, which could easily pass unnoticed.

The reason for that difference relies on the SNR interval suitability for a non-present higher buffer length. Since there is a higher availability for buffer space, optimal throughput per frame transmission can be achieved with a less aggressive transmission strategy, which is based on the adoption of modulation schemes with lower spectral efficiency rates. Thus, data-link packet dropping rate is reduced, as well as the overall packet loss rate, without compromising overall performance.

However, in our scenario, the prescribed application data rate remains the same (it is worthwhile noting that the authorized user is unaware of the attack) and the attacker inserts an unexpected set of 10 packets into the L2 buffer. Therefore, the lower is the number of transmitted packets per time slot, as described in the previous paragraph, the higher is the average buffer occupancy. This setup not only increases the packet dropping probability and the overall packet loss rate as well, but also reduces overall system performance.

Thus, under such a threshold mismatch, the average queue occupancy in the link layer tends to increase, which reduces the overall spectral efficiency. Such a scenario can therefore dramatically increase the packet loss, bringing an unexpected spectral efficiency reduction.

Personification Attack/ Unauhorized Access of Data

In the second case study, we evaluate a personification attack, replaying a given modulation scheme instead of performing proper updates for the communication system operation. In this study, we consider that a transmitter adopts a modified adaptive modulation vector as a side effect, instead of using the proper SNR interval to communicate with an authorized receiver station.

For instance, we consider that the average SNR between the transmitter and the authorized receiver (SNRTR) is 25 dB, with a data-link queue length of 20 packets. However, instead of getting the optimal configuration, an unauthorized receiver replays the setup for a 10 dB average SNR between transmitter and receiver stations, under a data-link queue length of 30 packets.

The attacker may exploit a data transmission protocol vulnerability that allows replay attacks – for instance, relaying a copy of previously transmitted information onto a SCA-compliant waveform that does not implement challenge scheme to update a given adaptive modulation scheme vector, getting an authorized access to classified data. So, instead of using the evaluated adaptive modulation vector for an average SNR = 25 dB, the transmitter adopts the evaluated adaptive modulation vector for an average SNR = 10 dB, since the attacker is considered to be an authorized user and informs his average signal-to-noise ratio during the data transmission establishment phase.

Based on that, Table 2 describes the achieved performance at the two system configurations, before and after the breach exploitation, according to several performance metrics. The authorized user (AU) is evaluated under two scenarios, namely, the expected (i.e., before the breach) and the actually received (namely, after the breach) ones. The presented results indicate system performance degradation due to the personification attack, especially on data-link packet dropping events.

Table 2. System performance comparison between expected and actual user station experience

Metrics	Expected AU	Actual AU
ASE (bps/Hz)	1.7858	0.9326
PDP	0.0258	0.1754
PLR	0.0280	0.2550

Thus, among several options to harden tactical SDR security against personification attacks mixed with data access drawbacks, the implementation of sequence numbers, challenges, or freshness schemes present simple, but effective counter-measures to overcome such threats (Pellegrini, Bertacco, Austin 2010).

Unauthorized Data Modification

We also evaluate an unauthorized data modification attack, in which an adversary injects malicious code exploiting, for instance, a vulnerability at the SCA Data Component Descriptor (DCD) file. Here, we analyze the case where system performance is deviated due to an attacker that alters the proposed SNR intervals to maximize ASE for a given application. Among several options of possibly available vulnerabilities to achieve his goal, the adversary may exploit SDR adaptive modulation scheme reconfiguration during post-waveform download phase, by exploring errors either in code-authenticity security policy or in software-version control policy.

After identifying an available vulnerability, the imposed attack strategy described here provides an 1-dB augmentation for all elements into the SNR adaptive modulation interval – for instance, the evaluated SNR interval for an average SNR = 20 dB is L = [8.5971 10.9598 16.5522 18.2111 21.9357] dB, while the under-attack SNR interval is L = [9.5971 11.9598 17.5522 19.2111 22.9357] dB.

Based on that, Table 3 describes the achieved performance at the two system configurations for an average SNR of 20 dB, before and after the breach exploitation, according to several performance metrics. The authorized user (AU) is evaluated under two scenarios, namely, the expected and the actually received ones. The presented results indicate system performance degradation due to the unauthorized data modification, especially on data-link packet dropping events.

However, even with a such slightly different adaptive modulation vector, which could easily pass unnoticed, the resulting breach exploited by the adversary intrusion brings an overall drawback in system performance, since it lowers ASE, while increases PDP. It is worthwhile noting that the overall PLR is reduced, due to a physical-layer error rate reduction.

The reason for that performance is the SNR interval translation to higher levels. Due to that upper shift, it is less probable to adopt modulation schemes with higher spectral efficiency rates, lowering the average number of packets transmitted at a given time slot. Since the pre-scribed application data rate remains the same (it is worthwhile noting that the authorized user is unaware of the attack), the lower is the number of transmitted packets per time slot, the higher is the average buffer occupancy. Such an effect also enables the packet dropping probability increase, which also degrades overall system throughput (Moura, Salles & Galdino 2009), in spite of frame structure preservation.

Since the threshold increases, the physical layer QoS (namely, the bit error rate) is satisfied with an increasing slack, but the spectral efficiency tends to decrease, which increases the average queue occupancy in the link layer.

CONCLUSION

In this chapter, we identify several classes of attacks aimed at SCA-compliant radios after an extensive literature survey. We also identify several possible sources of vulnerabilities in a SCA-compliant operating environment, and present

Table 3. System performance comparison between expected and actual user station experience

Metrics	Expected AU	Actual AU
ASE (bps/Hz)	1.6894	1.6562
PDP	0.0424	0.0482
PLR	0.0548	0.0539

several case studies along with simulation results. Concerning attack classes, it is worth noting that the set of attacks and threads that are relevant to a radio set depend on its particular architecture, and therefore they are precisely defined during the execution of the radio-specific security engineering process.

Ideally, security engineering principles should be incorporated into the design of a system as soon as possible, that is, in the initial architecture specification. The sooner the security issues are addressed, the lower is the cost and the time needed to fix future security problems (Myagmar et al. 2005). The identification of classes of attacks and threats is part of the security modeling process, which is prior to defining the radio set security requirements and the design of the corresponding security architecture (Myagmar et al. 2005).

However, this is not a one-way process. The definition of security requirements and the identification of new threats, when designing the security architecture, may require the review and the improvement of the initial modeling. Thus, we consider that the attack classes identified in this study define a point of departure for the threat analysis and the requirements definition related to the security engineering process of a SCA-compliant radio.

REFERENCES

3GPP. (2001). *Security Threats and Requirements (Release 4), Technical Specification Group Services and System Aspects.*

Abadi, M., Budiu, M., Erlingsson, U., & Ligatti, J. (2009). Control-flow integrity principles, implementations, and applications. [TISSEC]. *ACM Transactions on Information and System Security, 13*(1). doi:10.1145/1609956.1609960

Agrawal, D., Archambeault, B., Rao, J. R., & Rohatgi, P. (2003). The EM Side—Channel(s). *Lecture Notes in Computer Science, 2523,* 29–45. doi:10.1007/3-540-36400-5_4

Alves-Foss, J., Harrison, W. S., Oman, P., & Taylor, C. (2006). The MILS Architecture for High Assurance Embedded Systems. *International Journal of Embedded Systems, 2*(3-4), 239–247. doi:10.1504/IJES.2006.014859

Amorim, A., & Galdino, J. F. (2010). I2TS02-Secrecy Rate of Adaptive Modulation Techniques in Flat-Fading Channels. *Latin America Transactions, IEEE, 8*(4), 332-339.

Ball, T., Bounimova, E., Cook, B., Levin, V., Lichtenberg, J., & McGarvey, C. … Ustuner, A. (2006). Thorough Static Analysis of Device Drivers. In *Proceedings of EuroSys'06*. Academic Press.

Ballister, P. J., Robert, M., & Reed, J. H. (2006). Impact of the use of CORBA for Inter-Component Communication in SCA Based Radio. In *Proceedings of the Software-Defined Radio Technical Conference*. Orlando, FL: Academic Press.

Bard, J., & Kovarik, V. J. Jr. (2007). *Software-Defined Radio – The Software Commmunications Architecture*. John Wiley and Sons. doi:10.1002/9780470865200

Becker, C., Staamann, S., & Salomon, R. (2007). Security Analysis of the Utilization of Corba Object References as Authorization Tokens. In *Proceedings of the 10th IEEE International Symposium on Object and Component-Oriented Real-Time Computing* (ISORC'07). IEEE.

Beckwith, R. W., Vanfleet, W. M., & MacLaren, L. (2004). High Assurance Security/Safety for Deeply Embedded, Real-time Systems. In *Proceedings of the Embedded Systems Conference 2004*. Academic Press.

Berghel, H., & Uecker, J. (2005). WiFi Attack Vectors. *Communications of the ACM, 48*(8), 21–28. doi:10.1145/1076211.1076229

Bogdanov, A. (2008). Multiple-Differential Side-Channel Collision Attacks on AES. In *Proceedings of the Workshop on Cryptographic Hardware and Embedded Systems 2008* (CHES 2008). CHES.

Bonneau, J., & Mironov, I. (2006). Cache-Collision Timing Attacks Against AES. In *Proceedings of the Workshop on Cryptographic Hardware and Embedded Systems 2006* (CHES 2006). CHES.

Brawerman, A., Blough, D., & Bing, B. (2004). Securing the download of radio configuration files for software-defined radio devices. In *Proceedings of the ACM International Workshop on Mobility Management and Wireless Access*. ACM.

Brown, T. X., & Sethi, A. (2007). Potential Cognitive Radio Denial of Service Attacks and Remedies. In *Proceedings of the international symposium on advanced radio technologies 2007* (ISART 2007). ISART.

Bunnell, R., & Trinidad, J. (2004). *The Challenge in Developing an SCA Compliant Security Architecture that Meets Government Security Certification Requirements*. Retrieved from http://www.omg.org/news/meetings/workshops/SBC_2004_Manual/06-2_Trinidad_etal_revised.pdf

Chevallier-Mames, B., Ciet, M., & Joye, M. (2004). Low-Cost Solutions for Preventing Simple Side-Channel Analysis: Side-Channel Atomicity. *IEEE Transactions on Computers, 53*(6), 760–768. doi:10.1109/TC.2004.13

Common Criteria Evaluation and Validation Scheme Validation Report. (2008). *Green Hills Software INTEGRITY-178B Separation Kernel*. Retrieved from http://www.commoncriteriaportal.org/files/epfiles/st_vid10119-vr.pdf

Correia, M. P., & Sousa, P. J. (2010). *Segurança no Software*. FCA Editora de Informática.

Cowan, C., Pu, C., Maier, D., Walpole, J., Bakke, P., & Beattie, S. … Hinton, H. (1998). StackGuard: Automatic adaptive detection and prevention of buffer-overflow attacks. In *Proceedings of the 7th USENIX Security Symposium*. San Antonio, TX: USENIX.

Davidson, J. A. (2008). On the Architecture of Secure Software-Defined Radios. In *Proceedings of the IEEE Military Communications Conference* (MILCOM 2008). IEEE.

Dhillon, D. (2011, July/August). Developer-Driven Threat Modeling: Lessons Learned in the Trenches. *IEEE Security and Privacy*.

Ezick, J., & Springer, J. (2011). The Benefits of Static Compliance Testing for SCA Next. In *Proceedings of the SDR'11 Technical Conference and Product Exposition, 2011*. SDR.

Ezick, J., Springer, J., Litvinov, V., & Wohlford, D. (2010. A Path Toward Cost-Effective SCA Compliance Testing. In *Proceedings of the SDR'10 Technical Conference and Product Exposition*. SDR.

Ferguson, A. J. (2005). Fostering e-mail security awareness: The West Point carronade. *EDUCAUSE Quarterly, 28*(1).

FIPS PUB 140-2. (2001). *Federal Information Processing Standards Publication 140-2 - Security Requirements for Cryptographic Modules*. Information Technology Laboratory, National Institute of Standards and Technology.

FIPS PUB 140-3. (2009). *Federal Information Processing Standards Publication 140-3 (Revised Draft 09/11/09) - Security Requirements for Cryptographic Modules*. Information Technology Laboratory, National Institute of Standards and Technology.

Fitton, J. J. (2002). Security Considerations for Software-Defined Radios. In *Proceedings of the SDR 02 Technical Conference and Product Exposition, 2002*. SDR.

Frankel, S., Eydt, B., Owens, L., & Scarfone, K. (2007). *Establishing Wireless Robust Security Networks: A Guide to IEEE 802.11i - Recommendations of the National Institute of Standards and Technology*. NIST Special Publication 800-97.

Fu, X., Graham, B., Bettati, R., & Zhao, W. (2003). Active Traffic Analysis Attacks and Countermeasures. In *Proceedings of the 2003 International Conference on Computer Networks and Mobile Computing* (ICCNMC'03). ICCNMC.

Gallery, E. M., & Mitchell, C. J. (2006). Trusted computing technologies and their use in the provision of high assurance SDR platform. In *Proceedings of the Software-Defined Radio Technical Conference*. Orlando, FL: Academic Press.

Gallo, R., Kawakami, H., & Dahab, R. (2009). On Device Establishment and Verification. In *Proceedings of the EuroPKI 2009*. EuroPKI.

Gallo, R., Kawakami, H., & Dahab, R. (2011). FORTUNA – A Probabilistic Framework for Early Design Stages of Hardware-Based Secure Systems. In *Proceedings of 5th International Conference on Network and System Security* (NSS 2011). NSS.

Gallo, R., Kawakami, H., Dahab, R., & Azevedo, R. Lima, S., & Araujo, G. (2010). T-DRE: A Hardware Trusted Computing Base for Direct Recording Electronic Vote Machines. In *Proceedings of the 26th Annual Computer Security Applications Conference* (ACSAC'10). ACSAC.

GEMSOS Security Kernel. (2012). Retrieved from http://www.aesec.com/

Giffin, J., Christodorescu, M., & Kruger, L. (2005). Strengthening Software Self-Checksumming via Self-Modifying Code. In *Proceedings of the 21st Annual Computer Security Application Conference* (ACSAC'05). ACSAC.

González, A., Carlos, R., Dietrich, C. B., & Reed, J. H. (2009). Understanding the software communications architecture. *IEEE Communications Magazine, 47*(9).

Goodchild, J. (2010). *Social Engineering: The Basics*. Retrieved from http://www.csoonline.com/article/514063/social-engineering-the-basics

Howard, M., & LeBlanc, D. (2007). *Writing Secure Code for Windows Vista*. Microsoft Press.

Hsu, F.-H., Guo, F., & Chiueh, T.-C. (2006). Scalable Network-based Buffer Overflow Attack Detection. In *Proceeding of ACM/IEEE Symposium on Architectures for Networking and Communications Systems 2006*. ACM/IEEE.

Humphreys, T. E., Ledvina, B. M., Psiaki, M. L., O'Hanlon, B. W., & Kintner, P. M., Jr. (2008). Assessing the Spoofing Threat: Development of a Portable GPS Civilian Spoofer. In *Proceedings of the 2008 ION GNSS Conference*. GNSS.

Jagatic, T., Johnson, N., Jakobsson, M., & Menczer, F. (2007). Social Phishing. *Communications of the ACM, 50*(10). doi:10.1145/1290958.1290968

Jin, Y., & Makris, Y. (2008). Hardware Trojan detection using path delay fingerprint. In *Proceedings of the 2008 IEEE International Workshop on Hardware-Oriented Security and Trust*. IEEE.

Karri, R., Rajendran, J., Rosefeld, K., & Tehranipoor, M. (2010). Trustworthy Hardware: Identifying and Classifying Hardware Trojans. *IEEE Computer, 43*(10), 39–46. doi:10.1109/MC.2010.299

Kiszka, J., & Wagner, B. (2003). Domain and Type Enforcement for Real-Time Operating Systems. In *Proceedings of the 9th IEEE International Conference On Emerging Technologies and Factory Automation*. IEEE.

Kiszka, J., & Wagner, B. (2007). Modelling Security Risks in Real-Time Operating Systems. In *Proceedings of the 5th IEEE International Conference on Industrial Informatics*. IEEE.

Kocher, P. (1996). Timing Attacks on Implementations of Diffie-Hellman, RSA, DSS and Other Systems. *Lecture Notes in Computer Science, 1109*, 104–113. doi:10.1007/3-540-68697-5_9

Kocher, P., Jaffe, J., & Jun, B. (1999). Differential Power Analysis: Leaking Secrets. In *Proceedings of Crypto '99 (LNCS)* (Vol. 1666, pp. 388–397). Berlin: Springer Verlag.

Kurdziel, M., Beane, J., & Fitton, J. J. (2005). An SCA security supplement compliant radio architecture. In *Proceedings of the Military Communications Conference.* IEEE.

Lin, G., & Noubir, G. (2003). Low Power DoS Attacks in Data Wireless LANs and Countermeasures. In *Proceedings of ACM MobiHoc.* ACM.

Lindqvist, U., & Jonsson, E. (1997). How to Systematically Classify Computer Security Intrusions. In *Proceedings of the 1997 IEEE Symposium on Security and Privacy.* IEEE.

Manadhata, P. K., & Wing, J. M. (2011). An Attack Surface Metric. *IEEE Transactions on Software Engineering, 37*(3), 371–386. doi:10.1109/TSE.2010.60

MIL-STD-188-110B. (2000). *Interoperability and Performance Standards for Data Modems.* United States Department of Defense Interface Standard.

Moura, D. F., Salles, R. M., & Galdino, J. F. (2010). Multimedia Traffic Robustness and Performance Evaluation on a Cross-Layer Design for Tactical Wireless Networks. In *Proceedings of the 9th International Information and Telecommunication Technologies Symposium* (I2TS'10). I2TS.

Moura, D. F. C., da Silva, F. A. B., & Galdino, J. F. (2012). Case Studies of Attacks over Adaptive Modulation Based Tactical Software Defined Radios. *Journal of Computer Networks and Communications.*

Moura, D. F. C., Salles, R. M., & Galdino, J. F. (2009). Generalized input deterministic service queue model: Analysis and performance issues for wireless tactical networks. *IEEE Communications Letters, 13*(12), 965–967. doi:10.1109/LCOMM.2009.12.091646

Murotake, D., & Martin, A. (2004). System Threat Analysis for High Assurance Software Radio. In *Proceedings of the SDR'04 Technical Conference.* SDR.

Murotake, D., & Martin, A. (2009). A High Assurance Wireless Computing System Architecture for Software-Defined Radios and Wireless Mobile Platforms. In *Proceedings of the SDR '09 Technical Conference and Product Exposition.* SDR.

Myagmar, S., Lee, A. J., & Yurcik, W. (2005). Threat Modeling as a Basis for Security Requirements. In *Proceedings of the Symposium on Requirements Engineering for Information Security (SREIS).* SREIS.

NATO. (2000). Profile for High Frequency (HF) Radio Data Communication Ed. 2 (STANAG 5066). North Atlantic Treaty Organization.

OMG CORBA Security Service v1.8. (2002). Retrieved from http://www.omg.org/spec/SEC/1.8/PDF/

OMG DDS Portal. (2012). Retrieved from http://portals.omg.org/dds/

OMG Web site. (2012). Retrieved from http://www.omg.org/

Pellegrini, A., Bertacco, V., & Austin, T. (2010). Fault-Based Attack of RSA Authentication. In Proceedings of Design, Automation and Test in Europe (DATE). DATE.

Pinto, E. L., & Galdino, J. F. (2009). Simple and robust analytically derived variable step-size least mean squares algorithm for channel estimation. *Communications, IET, 3*(12), 1832–1842. doi:10.1049/iet-com.2009.0038

Proakis, J. G. (2001). Digital Communications (4th Ed.). McGraw-Hill International Ed.s.

Renauld, M., Standaert, F. X., & Veyrat-Charvillon, N. (2009). Algebraic Side-Channel Attacks on the AES: Why Time also Matters in DPA. In *Proceedings of the Workshop on Cryptographic Hardware and Embedded Systems 2009* (CHES 2009). CHES.

Riley, R., Jiang, X., & Xu, D. (2010). An Architectural Approach to Preventing Code Injection Attacks. *IEEE Transactions on Dependable and Secure Computing, 7*(4), 351–365. doi:10.1109/TDSC.2010.1

SCA Next Specification v. 1.0. (2011). Retrieved from http://jtnc.mil/sca/Pages/sca1.aspx

SCA Security – Security Supplement to the Software Communications Architecture Specification. (2001). Retrieved from http://jtnc.mil/sca/Pages/sca1.aspx

SCA - Software Communications Architecture Specification, v.2.2. (2001). Retrieved from http://jtnc.mil/sca/Pages/sca1.aspx

SCA - Software Communications Architecture Specification, v.2.2.2. (2006). Retrieved from http://jtnc.mil/sca/Pages/sca1.aspx

SCA - Software Communications Architecture Specification, v.4.0. (2012). Retrieved from http://jtnc.mil/sca/Pages/sca1.aspx

Silva, F. A. B., Moura, D. F., & Galdino, J. F. (2012). Classes of attacks for tactical software defined radios. [IJERTCS]. *International Journal of Embedded and Real-Time Communication Systems, 3*(4), 57–82. doi:10.4018/jertcs.2012100104

Standaert, F.-X., Malkin, T. G., & Yung, M. (2009). A Unified Framework for the Analysis of Side-Channel Key Recovery Attacks. *Lecture Notes in Computer Science, 5479*, 443–461. doi:10.1007/978-3-642-01001-9_26

Stango, A., & Prasad, N. R. (2009). Policy-Based Approach for Secure Radio Software Download. In *Proceedings of the SDR '09 Technical Conference and Product Exposition*. SDR.

Strasser, M., Popper, C., Capkun, S., & Cagalj, M. (2008). Jamming-resistant Key Establishment using Uncoordinated Frequency Hopping. In *Proceeding of the 2008 IEEE Symposium on Security and Privacy*. IEEE.

Swanson, M., Bartol, N., & Moorthy, R. (2010). *Piloting Supply Chain Risk Management Practices for Federal Information Systems*. Draft NIST IR 7622.

Swiderski, F., & Snyder, W. (2004). *Thread Modeling*. Microsoft Press.

TEMPEST. (1982). *Tempest fundamentals, NSA-82-89, NACSIM 5000, National Security Agency*. Retrieved from http://cryptome.org/jya/nacsim-5000/nacsim-5000.htm

TEMPEST. (1995). *NSTISSAM TEMPEST/2-95, Red/Black Installation Guidance*. Retrieved from http://cryptome.org/jya/tempest-2-95.htm

van de Ven, A. (2005, July). Limiting buffer overflows with ExecShield. *Red Hat Magazine*.

van Dijk, M., Rhodes, J., Sarmenta, L. F. G., & Devadas, S. (2007). Offline Untrusted Storage with Immediate Detection of Forking and Replay Attacks. In *Proceedings of the The Second ACM Workshop on Scalable Trusted Computing* (STC'07). STC.

Wireless Innovation Forum. (2002). *Requirements for Radio Software Download for RF Reconfiguration, Approved Document SDRF-02-S-007-V1.0.0, November 2002*. Author.

Wireless Innovation Forum. (2010). *Securing Software Reconfigurable Communications Devices, Approved Document WINNF-08-P-0013, Version 1.0.0, July 2010*. Author.

Wireless Innovation Forum. (2011). *International Radio Security Services API Task Group. IRSS API Specification WINNF-09-S-0011-V1.0.0*. Author.

X.731 - ITU X.731 ISO/IEC10164-2 State Management. (2012). Retrieved from http://www.itu.int/rec/T-REC-X.731-199201-I/en

KEY TERMS AND DEFINITIONS

Platform Independent Model (PIM): A model of a subsystem that contains no information specific to a specific platform, or the underlying technology used to realize it.

Platform Specific Model (PSM): The PSM is a model of a subsystem that incorporates technology specific information.

SCA Core Framework (CF): The SCA CF provides the essential ("core") set of open software interfaces and profiles that provide for the deployment, management, interconnection, and intercommunication of software application components in an embedded, distributed-computing communication system.

SCA Component: A SCA component is an autonomous unit whose manifestation is replaceable within its environment. A component exposes a set of ports that define the component specification in terms of provided and required interfaces.

SCA Device: A SCA device corresponds to an implementation of the Device interface. A SCA device is an abstraction of a hardware device that defines the capabilities, attributes, and interfaces for that device

SCA Port: A SCA Port is an implementation of the Port interface that identifies a source (Provides Port) or a sink (Uses Port) for data and/or commands.

Waveform: A waveform is the set of transformations applied to information that is transmitted over the air and the corresponding set of transformations to convert received signals back to their information content.

Chapter 3

Interference and Spoofing:
New Challenges for Satellite Navigation Receivers

Fabio Dovis
Politecnico di Torino, Italy

Nicola Linty
Politecnico di Torino, Italy

Luciano Musumeci
Politecnico di Torino, Italy

Marco Pini
Istituto Superiore Mario Boella, Italy

ABSTRACT

This chapter deals with one of the major concerns for reliable use of Global Navigation Satellite Systems (GNSS), providing a description of intentional and unintentional threats, such as interference, jamming, and spoofing. Despite the fact that these phenomena have been studied since the early stages of Global Positioning System (GPS), they were mainly addressed for military applications of GNSS. However, a wide range of recent civil applications related to user safety or featuring financial implications would be deeply affected by interfering or spoofing signals intentionally created. For such a reason, added value processing algorithms are being studied and designed in order to embed in the receiver an interference reporting capability so that they can monitor and possibly mitigate interference events.

INTRODUCTION

Global Navigation Satellite Systems (GNSS) are communication infrastructures enabling a generic user to compute position, velocity and time at its current location anywhere on the Earth, processing the signals transmitted from a constellation of satellites and performing a trilateration with respect to the satellites taken as reference points. One of the main characteristics of GNSS signals is the low signal power level reaching a receiving antenna on the ground. Despite of the weakness of the signals, the spread spectrum nature of the transmission allows navigation receivers to recover timing information and to compute the user's position by exploiting the gain obtained at the output of the correlation block (Enge & Misra, 2006).

Among all the different error sources that can potentially corrupt satellite navigation waveforms, such as errors introduced by ionospheric and tropospheric propagation, obsolete satellites ephemeris, errors due to satellites clocks, Radio Frequency

DOI: 10.4018/978-1-4666-6034-2.ch003

Interference (RFI) is particularly harmful since, in some cases, it cannot be mitigated by the correlation process. In fact, even if the correlation process is theoretically able to mitigate the presence of nuisances in the bandwidth of interest, a real limitation can be the finite dynamic range of the receiver front-end (Motella, Savasta, Margaria, & Dovis, 2009).

The presence of undesired RFI and other channel impairments can result in degraded navigation accuracy or, in severe cases, in a complete loss of signal tracking. It is also to be considered that the number of electromagnetic sources that are candidate to become unintentional interferers for GNSS signals is large. There is in fact an increasing number of systems that we depend on in daily life that rely on the transmission of Radio Frequency (RF) energy in the L-band or close to it (Kaplan & Hegarty, 2005). Depending on the spectral and power features of the interfering sources they can affect the performance of GNSS receivers (CEPT, 2002).

Furthermore, GNSS threats include intentional attacks aiming at decepting the target receiver. Recalling that GNSS bandwidths are protected and that signal radiation on frequency bands allocated to radio navigation is not legal, the transmission of false GNSS-like signals, usually known as RF spoofing, may become quite dangerous also for civil use of GNSS, as the number of applications increases.

An example of application that can be severely damaged by jamming or spoofing attacks are the new Pay As You Drive (PAYD) systems, in which costs of motor insurance depend on the type of vehicle used and are measured against time, distance and place. PAYD insurance usually involves an on-board unit equipped with a GNSS receiver able to collect and transmit data related to the vehicle's use to an operational center: the device can measures how safely the vehicle is being driven in terms of speed, type of road, driving time and distance. GNSS is a key component in this segment and also it enables additional services, like

the last estimated position in case of a distress call after an accident (European emergency call) or for tracking a stolen vehicle. This application, taken as example, demonstrates that RF spoofing is a potential risk leading to frauds towards insurances. The driver might want to take advantages providing false positions to the service provider (e.g. pretending he is driving on a different road, parking in a different park slot, etc.). In this case, we refer to self-spoofing (or limpet spoofing), since the target receiver is owned by the attacker. Although RF spoofing in the road sector seems just a research curiosity, the problem can be real in the future years, since the market for insurance telematics is now approaching its maturity phase. In 2010, the telematics-enabled services market was worth in Europe an estimated of approximately 5 b€, or 4% of the whole European motor insurance market, estimating 2 million users worldwide (PTOLEMUS Consulting Group, 2012).

The scope of this chapter is to provide an updated overview of both the traditional and the most advanced techniques currently under investigation for unintentional interference, jamming and spoofing detection, in order to implement user alerts and, possibly, to mitigate their effects on the receiver performance.

In the first part of the Chapter the different threats are considered and described in details, providing a general classification. Then, an overview of techniques for detection and mitigation of interference and jamming is presented, along with some mathematical details and examples, showing their features and performance. To conclude a brief note on the complexity tradeoff is given and the future research directions are outlined.

BACKGROUND: THREATS CLASSIFICATION

GNSS receivers' interference can be divided in unintentional and intentional. Intentional interference includes jamming, meaconing and different levels of spoofing. This classification is accepted by most of the scientists working on GNSS, although it is not exhaustive. A detailed classification of interfering signals and an overview of electronics devices and other existing communication systems that can represent a serious threat for GNSSs is presented in Savasta (2010).

Unintentional Interference

All communication systems with carrier frequencies close to the GNSS band are a potential source of interference for a receiver. It is likely that some out-of-band energy from signal frequencies located near the GNSS bands could interfere, due to harmonics or power leakages that may be negligible for a communication system but threatening for a navigation receiver that has to deal with very low-power signals. Thus, interfering sources can be divided between those falling in the GNSS frequency bands (in-band RFI) or those far from them (out-of-band RFI). Since there are almost no in-band emissions in L1 frequency, interfering signals come from out-of-band or spurious emissions and are usually located in small portions of the frequency band. Depending on the nature of the interfering signal, the effect on the receiver can range from a worsening of the position accuracy to the appearance of biases in the measurements, and in some cases to a loss of tracking.

Several real cases have been experienced in the past due to different systems, such as analog and digital TV channels, other Very High Frequency (VHF) channels the harmonics of which fall in the L1/E1 band, portable electronic devices, or ultra-wide band systems (Savasta, 2010; Wildemeersch, Rabbachin, Cano, & Fortuny, 2010). Particular focus was given around 2011 and 2012 to the potential interference generated by 4G Long Term Evolution satellite broadband communication network, developed by LightSquared. The radio band in which the system was supposed to operate is close to the L1 band and it was proved by a technical working group that LightSquared transmissions could produce adverse effects on GPS receivers (Federal Communication Commission, 2011). In February 2012, following extensive testing and analysis, the US Federal Communications Commission announced it would not allow LightSquared's terrestrial operations and planned to withdraw LightSquared's authorizations.

Jamming

Jamming refers to intentional transmission of RF energy to hinder the navigation service, by masking GNSS signals with noise (Scott, 2004). The objective of jammers is to cause the receiver to lose tracking and to impede signal re-acquisition.

It represents a growing threat for many GNSS-based applications. Systems involving safety and liability-critical operations (e.g. safe navigation in ports, system for smart parking and tolling, GNSS-based synchronization of power networks) might be heavily impaired by jamming attacks. The level of threat associated to jamming cannot be disregarded, considering that portable jammers are available on line and can be purchased at a very low cost. Although the use of jammers is not legal, the interest of individuals willing to break the law may result in fraudulent actions towards GNSS-enabled systems. Mitch et al. 2011 tried to characterize commercial jammers, demonstrating that they can affect GPS receivers' functionality even if located up to 9 km away; others assessed the effects of jammers on GPS and Galileo receivers (Borio, O'Driscoll, & Fortuny, 2012). A famous incident occurred at Newark Airport (NJ, USA) in January 2010, where one of the ground based augmentation system receivers was occasionally jammed by a portable device installed on board of a truck. In some cases jamming is a preliminary step to spoofing, forcing the target receiver to lose the tracking of the real signal in order to re-acquire a stronger spoofing signal.

Meaconing

In Wesson, Rothlisberger, and Humphreys (2011) the meaconing attack is defined as the reception and playback of an entire block of RF spectrum containing an ensemble of GNSS signals to confuse victim receivers. It is worth noticing that constituent GNSS signals from different satellites are typically not separated during record and playback. In addition, the meaconing principle can be extended if the received signal is recorded and replayed in a remote location (e.g. by receiving signal, transmitting it by means of a modem and replay it in a remote location).

Meaconing can be seen as an intermediate spoofing attack, without artificial relative offsets on the counterfeit signal components from different satellites. However, differently from spoofing attacks, a meaconing attack does not attempt to control the tracking loops of the target receiver. As a consequence the victim receiver, and thus the Position Velocity Time (PVT) solution, is not forced astray. A meaconing attack does not necessarily produce erroneous positions; rather, the accuracy of the user positions is degraded or at most the victim receiver can lose the tracking of some satellites in view, without being able to re-acquire them. Generally, a meaconing attack produces effects if the counterfeit signal power is higher than the real signal power. Furthermore, it introduces a positive delay on all the received signals, such that the meaconed signal arrives at the target receiver later if compared to the authentic signal: this kind of attack can be harmful in case of GNSS receivers are used to provide precise timing reference to some applications.

A meaconing attack can be easily generated with two antennas and a Low Noise Amplifier (LNA). The receiving antenna is connected to the LNA and the output signal is then broadcasted in the target area, without any additional signal processing. An example of meaconing attack can be found in Akos (2012), where the author utilizes a simple repeater for live testing. Other examples of modified meaconing attacks are also known in literature, including the use of a modem for re-transmitting signals from a remote antenna or introducing a variable-delay in order to steer a GNSS timing receiver without being detected.

RF Spoofing

Scott (2013) defines the spoofing as a process whereby someone tries to control reported position out of a device. Spoofing forces the reporting of incorrect PVT and in its most general form, spoofing can also involve cyber-attacks such as malicious software, falsified maps, man-in-the-middle attacks. Instead, RF spoofing is the transmission of false GNSS signals that force the victim to compute erroneous positions.

RF spoofing is more malicious than jamming, because the victim receiver is fooled without any notice. Smart spoofers track the location of the target receiver and use this information, along with a GNSS signal generator, to create strong signals that initially match the actual weaker signals in Time Of Arrival, until loop capture is assured, and then lead the target astray.

At the time of writing, the analysis of spoofing and the design of effective countermeasures for civilian systems is relatively new in the field of civil satellite navigation, but is already a hot research topic. Safe navigation in ports or constrained waters, fishery and road tolling are just few examples of applications that would be deeply affected by spoofers in order to elude public authorities or service providers.

Although the vulnerability of GNSS-based civilian infrastructures is understood, few recognize that severe attacks can be carried out with self-made spoofing devices, composed by a software receiver and trivial RF front- ends, as anticipated some years ago by Scott (2003) and

Nicola, Musumeci, Pini, Fantino, and Mulassano (2010). It is important to understand how a spoofer operates, in order to design and to apply countermeasures that might be undertaken against an attack. Baldini and Hofher (2008) provide a first comprehensive classification of civilian spoofers, grouped in simplistic, intermediate and sophisticated, depending on their complexity and on the level of robustness required to the associated anti-spoofing techniques. Such a classification is recalled below.

Simplistic Spoofing Attack

The simplest spoofer is a GNSS signal generator connected to a transmitting antenna. These spoofers can be easily detectable, because most of the generators available on the market are not able to synchronize the generated signals with the constellation in view. Although a receiver could be fooled by a GNSS signal generator (particularly if the target receiver is first jammed and forced to reacquire), the counterfeit signal generated in this way typically looks to a receiver like noise, rather than a usable signal. In addition, simplistic spoofing attacks come with a cost, as they require a hardware GNSS signal generator, which is generally an expensive, not portable, device.

Intermediate Spoofing Attack

A more malicious attack can be accomplished combining a GNSS receiver with a transmitting RF front-end. Intermediate spoofers are able to frequency synchronize and code-phase align the real and counterfeit signals, mainly if the victim position is known. The role of the receiver in the processing chain is fundamental; when it tracks the Signal In Space (SIS), it perfectly knows both the Doppler shift and the spreading code delay. The local code and local carrier are synchronized

to the incoming signal and are used to maintain the lock of the tracking loops. These local replicas can be used to generate the counterfeit signal.

Some years ago, a few research works (Humphreys, Ledvina, Psiaki, O'Hanlon, & Kinter, 2008; Nicola et al. 2010) demonstrated that an intermediate spoofer can be built in lab, using a software receiver. Recently, effective attacks have been demonstrated in different scenarios (see for example the experiment carried out to bring a ship off course (Divis, 2013)). The high level of flexibility of Software Defined Radio (SDR) receivers facilitates the development of PC-based receivers, where all the functionalities are performed on conventional general purpose processors. SDR receivers can be appropriately modified and converted into spoofers, thus reverting the receiving chain, just adding some offsets to each satellite signal and irradiating a modified version of the received signal in the air.

This type of spoofers is able to vary the signal strength of the constituent signals so that it appears at the target antenna with the same relative strength of the authentic signals. Intermediate spoofing attacks can deceive the target receiver, because the receiver itself is not able to distinguish between the counterfeit and the genuine signal. In addition they accurately reproduce the code phase and frequency and are able to predict the navigation data bits and furtively align the correlation peak to the genuine one. After the alignment, the counterfeit signal power is gradually increased until it begins to control the tracking loops of the target receiver. As demonstrated by Nicola et al. (2010) this is possible, although not trivial.

Intermediate spoofing configuration has one transmitting antenna and can be built with software parts freely available, using RF components that may be purchased for a few hundred dollars. This type of spoofers is moderately complex but requires a deep knowledge of GNSS signal processing. It might become a serious threat in the next years, considering the growing use of SDR receivers and the increasing availability of open-source software on the Web.

Sophisticated Spoofing Attack

A sophisticated GNSS spoofing attack consists of multiple coordinated and synchronized spoofing devices. In practice, these spoofing devices act as a beamforming antenna array, simulating the different angles of arrival for different satellites and thereby defeating the receiver's angle of arrival defense. This can be accomplished either by keeping each spoofer fixed and transmitting the signals of all satellites with appropriately calculated delays, or by having each spoofer transmitting the signal of exactly one satellite and mechanically moving the spoofers around the target receiver.

Creating sophisticated spoofers based on GNSS receivers is theoretically possible but technically cumbersome. In fact the true satellite constellation has to be perfectly simulated, taking into account not only the local code and local carrier synchronization, as well as the signal strength and the number of SIS, but also the SIS angle of arrival. This type of attack is the most complex to implement and to deploy, the most expensive (both in hardware costs and in developer efforts) and the hardest to defend against.

Several attacks can be designed modifying or combining spoofing and meaconing. Such hybrid attacks are currently considered of the most dangerous type: as an example a modified version of the classical meaconing has been presented in Wesson, Rothlisberger and Humphreys (2012). The scope is to take control of the delay introduced by the meaconer and fool any implemented countermeasure (e.g. based on the monitoring of the clock drift). Wesson et al. (2012) state that high performance digital signal processing hardware permits to drive the delay introduced by the meaconer to values on the order of nanoseconds. In the limit, if the delay approaches zero the meaconed and the authentic signals result code-phase aligned. They conclude that such an alignment enables a seamless lift-off of the target receiver's tracking loops, following which a meaconer can increase the delay at a rate that is consistent with the target receiver clock drift and gradually impose a significant timing delay.

INTERFERENCE MONITORING, DETECTION AND MITIGATION TECHNIQUES

Nowadays, with the lack of frequencies allocation, the need to find solutions to cope with the interference issues is becoming of paramount importance in the framework of GNSSs. Recent works tried to extend quality metrics related to the prediction of the interference effect, such as spectral separation coefficients, to include in the assessment not only the spectral characteristics but also the receiver set-up and architecture. Such works introduced the concept of Interference Error Envelope, which allows for a prediction of the interfering effects on a GNSS receiver and may be useful during the receivers design phase (Motella, Savasta, Margaria, & Dovis, 2011).

As far as the detection and mitigation techniques are concerned they can be classified according to the point in which they are applied along the processing chain of a GNSS receiver, as depicted in Figure 1:

- Antenna techniques, that can be applied only in static environments and require a particular hardware configuration, such as antenna arrays;
- Automatic Gain Control (AGC)/Analog to Digital Converter (ADC) monitoring techniques, based on the AGC behavior observation;
- Raw observables monitoring techniques, processing the signal at the front-end output;
- Post-correlation techniques, based on the analysis of the shape of the correlation function, in most cases exploiting a multi-correlator receiver.

Figure 1. Classical GNSS receiver architecture

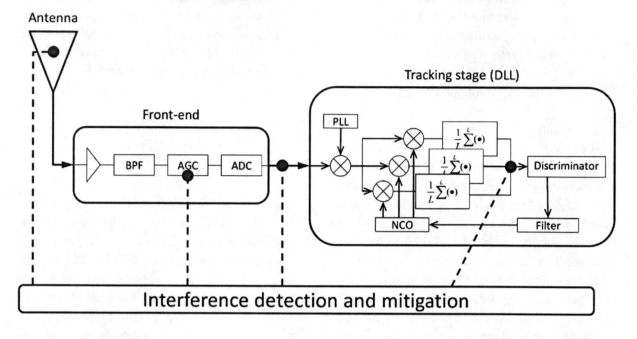

AGC Monitoring

The AGC is an important device used in the GNSS receiver chain and required anytime a multi-bit quantization ADC is present. It is, in fact, an adaptive variable gain amplifier, whose main role is to minimize the quantization losses and to exploit the maximum dynamic range of the ADC.

In an interference free environment, the useful GNSS SIS power at the receiver is below the noise floor; therefore the AGC is driven by the environment and thermal noise rather than by the useful signal power. When interference is present the input power level increases, leading to the saturation of the ADC and causing quantization losses. For this reason the AGC has to provide a different gain in order to adapt the incoming power level to the ADC input range.

Figure 2 shows the behavior of the AGC gain in the absence and in the presence of interference: when interference is present, the input signal power level changes and its variance increases. Therefore also the AGC gain variation is higher. The gain provided by the AGC gives information on the composite RF interference level entering the receiver and suggests a meter for the Jammer-to-Noise power ratio (J/N) computation.

Taking a different standpoint to analyze this effect, the degradation of the AGC behavior caused by the interference signals can be seen as a mean to detect the presence of the threat. More details about this concept can be found in Bastide, Akos, Macabiau, and Roturier (2003). However, this approach requires the access to the hardware of the AGC device to observe the trend of the control feedback command that is not the case when using common analog front-ends.

Raw Observable Level

A different set of techniques for detection and possibly mitigation of the interference is obtained processing the digital raw sample of the SIS at the output of the front-end stage, immediately after the ADC. The most traditional methods provide detection or mitigation looking at the features of the incoming digitized signal either in the frequency domain, based on Fourier transform

Figure 2. AGC behavior in the presence and in the absence of interference

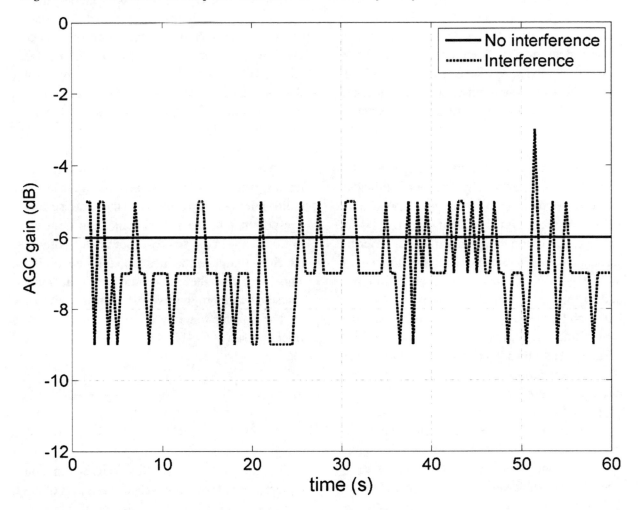

analysis or notch filtering, or in the time domain, as the simple blanking technique. However, in the last years researchers of the GNSS community have focused their attention on a new family of algorithms based on the use of advanced signal processing techniques which manipulate the signal at the ADC output in order to provide a representation in a different domain where interference components can be better identified and separated from the useful GNSS signal components. Such techniques will be referred to in the remainder of the Chapter as Transformed Domain (TD) techniques. A different approach, recently proposed aims at assessing some statistical property of the signal samples in the probabilistic domain, such as

the Goodness of Fit (GoF) technique, that will be described of an example of this class or methods.

It is important to highlight that while some of them only perform interference detection, raising an alert to the user and not foreseeing any solution, some others, such as the transformed domain techniques, are also able to mitigate the effect of signal impairments.

Frequency Domain: Discrete Fourier Transform and Notch Filtering

A first group of techniques includes detection algorithms based on spectral estimation of the incoming signal, obtained by applying signal processing techniques such as the Discrete Fourier Transform. They are typically performed comparing the spectrum of the received signal with a theoretical threshold determined according to a statistical model representing the received signal. In this framework, mitigation techniques are performed applying notch filtering or frequency domain adaptive filtering, which remove the interference frequency components.

In particular, notch filtering has been proved to be an efficient mitigation algorithm for Continuous Wave Interference (CWI), i.e. pure sinusoids or very narrowband signals, which appear as a spike in the spectral domain. This kind of interfering signals can be generated by Ultra High Frequency and VHF TV, VHF Omnidirectional Radio-range and Instrument Landing System (ILS) spurious harmonics, caused by power amplifiers working in non-linearity region or by oscillators present in many electronics devices (Borio et al., 2012). Notch filters are usually characterized by a frequency response which is null in correspondence of the CW carrier frequency, thus providing attenuation of the interfering signal and preserving as much as possible the useful GNSS signal spectral components. The most common class of notch filters, which has been already proposed for CW countermeasure in the past, is represented by Infinite Impulse Response filters with constrained poles and zeroes. For these notch filters the zeros are constrained on the unit circle and the poles lie on the same radial line of the zeros. In Borio,

Camoriano, and Lo Presti (2008), the design of a two-pole notch filter integrated with an adaptive unit is presented, and the determination of the CW frequency component perturbing the received GNSS signal is based on the removal of the constraint on the location of the filter zeros whose amplitude is adjusted by an adaptive unit. Through this algorithm, the notch filter is able to detect the presence of the interfering signal and to decide whether to use its filtered output or input signal. In presence of multiple sinusoids, a multi-pole notch filter, based on the use of several two-pole notch filters in cascade, can be used. In this scenario the first two-pole notch filter in the chain mitigates the most powerful disturbing signal, whereas the others remove the residual sinusoids with progressively decreasing power.

The transfer function of the two-pole notch filter is given by:

$$H(z) = \frac{1 - 2\,\mathrm{Re}\left(z_0\right)z^{-1} + \left|z_0\right|^2 z^{-2}}{1 - 2k_a\,\mathrm{Re}\left(z_0\right)z^{-1} + k_a^2\left|z_0\right|^2 z^{-2}}$$

The numerator of the filter transfer function represents the Moving Average (MA) part of the two-pole notch filter, the structure of which is depicted in Figure 3. Here z_0 represent the zero placed in correspondence of the interfering frequency:

Figure 3. Notch filters structure

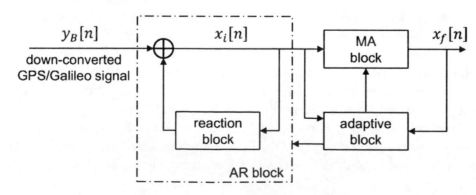

$$z_0 = \beta \exp\left\{ j2\pi f_i \right\}$$

Since the interfering signal is unknown, an adaptive block is used for providing the zero estimation which is then fed to the MA and Autoregressive (AR) blocks. In order to compensate for the effect introduced by the MA part, an AR is added, the transfer function of which is represented by the denominator in the expression of $H(z)$, where the parameter $0 < k_\alpha < 1$, known as pole contraction factor, determines the width of the notch filter. The more k_α is close to the unity the more the notch is narrow, which in turns means a reduction of the distortion on the useful GNSS signal. However k_α cannot be chosen arbitrarily close to unity for stability reasons and thus a compromise has to be adopted.

The core of notch filter structure is represented by the presence of the adaptive block, which is in charge of estimating the interference frequency and tracking its variation over the time. The presence of the adaptive block makes the entire notch filter suitable also for suppressing the harmful interference produce by jammers. Such devices, which can be easily found on the Web, transmit strong chirp signals sweeping several MHz in few µs, thus appearing in the spectrum as wide-band interference. More details on the use of such two-pole notch filter for jamming suppression can be found in Borio et al. (2012).

The adaptive algorithm, proposed in Camoriano et al (2008) is based on an iterative normalized Least Mean Square (LMS) which minimizes the following cost function:

$$f_C[n] = \mathrm{E}\left\{ \left| x_f[n] \right|^2 \right\}$$

where is $x_f[n]$ is the output of the filter.

The minimization is performed with respect to the complex parameter , using the iterative rule:

$$z_0[n+1] = z_0[n] - \mu[n] \cdot g\left(f_C[n] \right)$$

where $g\left(f_C[n] \right)$ is the stochastic gradient of the cost function $f_C[n]$ and $\mu[n]$ is the algorithm step, which is set to:

$$\mu[n] = \frac{\delta}{\mathrm{E}_{x_i[n]}},$$

with $\mathrm{E}_{x_i[n]}$ being an estimate of:

$$\mathrm{E}\left\{ \left| x_i[n] \right|^2 \right\},$$

which is in turn the power of the AR block output $x_i[n]$. δ is the un-normalized LMS algorithm step that controls the convergence properties of the algorithm.

In Troglia Gamba, Falletti, Rovelli, and Tuozzi (2012) it is shown how the position of the z_0 with respect the unit circle is impacting on the distortion of the signal at the notch filter output. Here a different adaptive algorithm, consisting in forcing the zero of the filter to move on the unit circle, is proposed. Furthermore, in order to improve the convergence speed of the adaptive algorithm, a run-time change of the pole contraction factor and of the LMS step is performed. In the absence of interferences the notch width is wide and the LMS step is large. When the interference appears, the notch becomes narrower, the convergence step smaller and the zeros is forced to move on the unit circle, according to:

$$z_0^F = \frac{z_0}{|z_0|}$$

where z_0 is the zero produced by the adaptive block and z_0^F is the zero employed in the filter transfer function.

Time Domain: Pulse Blanking

Although the notch filter represents an effective countermeasure when dealing with CW interferers, it does not represent the best solution when coping with interference in the aviation scenarios. In this case the interference environment for the an board GNSS receiver is quite harsh, as described in De Angelis, Fantacci, Menci, and Rinaldi (2005). Many Aeronautical Radio Navigation Services systems based on pulsed signal broadcasting, such as the Distance Measuring Equipment (DME) or the Tactical Air Navigation (TACAN), transmit on the same frequency range of GPS L5 and Galileo E5 signals. In such a scenario, the interference affecting the on-board GNSS receiver is represented by the composite strong pulsed signals transmitted from all the DME/TACAN ground stations in line of sight. The implementation of notch filters for suppressing multiple narrow-band interference spread all over the GNSS useful signal spectrum would become extremely complicated, as mentioned in Gao (2007). In case of pulsed interference, pulse blanking is the most common mitigation technique implemented in on board receivers. It blanks the samples of the signal whose amplitude exceeds a certain threshold level. The implementation of the blanker is rather simple; however it requires to manage an excellent synchronization of the activation/de-activation of the blanker with the time interval in which the pulsed interference appears. Otherwise, either a late activation of the blanker might leave interfered samples in the signal digital stream or a late deactivation might unnecessarily remove a portion of useful signal.

Such delays, might be due to implementation issues of the blanker, and are named reaction time and recovery time, respectively. They both depend by the behavior of the AGC.

The reaction time is defined as the time interval between the instant in which a strong interference appears at the input of the AGC and the instant when the AGC starts to compress the signal in order to match the dynamics of the ADC, as previously described. When the interference ends, since the AGC does not react instantaneously to the variation of the input signal power, there is a drop of its provided gain. Thus the AGC recovery time is the time interval between the end of the interference and the time instant where the AGC reaches again its nominal gain value. These two parameters have to be really small in order to follow the dynamics in time of the input signal.

Moreover, pulse blanking circuit performance can be negatively influenced by the impact of pulsed signals on other components in the receiver front-end. Very strong pulses or very strong received power due to the combination of multiple pulses can cause the saturation of GNSS receivers' active components, such as amplifiers, which may require a recovery time to go back to a normal state when the interference ends. In Hegarty, Van Dierendonck, Bobyn, Tran, and Grabowski (2000) it is mentioned that for a particular commercial receiver, an interference pulse signal with peak power 15 dB above the thermal noises is sufficient to saturate the last amplification stage within the receiver front-end. Under this interference environment conditions, pulse blanking may perform signal suppression even during the off state of the pulse for a time period similar to the recovery time needed by the amplifiers to resume normal operation. For a commercial receiver, typical recovery times for amplification stages are about 40 ns/dB of input level beyond the saturation point (Hegarty et al., 2000). In general pulsed interference signals impact on receiver front-end component might be different, depending especially on the pulse peak power level and on the pulse duration.

Moreover, the envelope of the pulsed interference generated by the DMEs is Gaussian shaped; thus, not all the samples belonging to the pulsed interference are blanked, leaving a residual contribution to the noise floor due to the tails of the Gaussian shape, whose amplitude is below the threshold.

In Bastide, Chatre, Macabiau, and Roturier (2004), a simulation and a theoretical derivation of the signal to noise density ratio degradation due to the DME and TACAN signals is presented. The study carried out in this paper has shown that, by simulation, in a hard pulsed interference environment which can be experienced by an on board receiver, flying over the European hotspot (Frankfurt), where the interference duty cycle can reach 50% value, the experienced degradation for a receiver with a digital pulse blanking implemented can reach 10 dB, against the 18 dB observed during a flight trial carried out by the German Aerospace Center over the same location (Denks, Steingaß, Hornbostel, & Chopard, 2009). Theoretically, the expected degradation for a receiver with a digital pulse blanker can be estimated as

$$\left(\frac{C}{N_0}\right)_{\text{eff}} = \frac{1 - BDC}{1 + R_I}$$

where *BDC* is the total effective blanker duty cycle of all pulses strong enough to activate the blanker while R_I is the aggregate ratio of total below-threshold pulsed RFI power density to receiver thermal noise density. Such amount of residual interfering power can be assessed to be:

$$R_I = \left(\frac{1}{N_0 \cdot BW}\right) \cdot \sum_{i=1}^{N} P_i \cdot dc_i$$

where N is the total number of low-level RFI sources, P_i is the peak received power of the *i-th* RFI signal source, *BW* is the pre-correlator IF bandwidth and dc_i is the duty cycle of the *i-th*

low-level signal source exclusive of pulse collision (RTCA DO-292, 2004; Erlandson, Kim, Hegarty, & Van Dierendonck, 2004).

Due to this strong dependence of the pulse blanker on the receiver architecture, it becomes extremely important to find solutions "receiver architecture independent."

Probabilistic Domain: Goodness of Fit

The GNSS digital signal at the output of the ADC can be modeled as

$$y[n] = s_{IF}[n] + w[n]$$

where $s_{IF}[n]$ is a sequence of samples of the SIS and $w[n]$ is a realization of a zero-mean white discrete-time additive Gaussian noise. In the presence of interference or in general of signal nuisances, the signal model above becomes

$$x[n] = y[n] + v \cdot q[n]$$

where $q[n]$ is the interference signal at Intermediate Frequency (IF) and v is a generic amplitude factor.

In statistical techniques (Motella, Pini, & Lo Presti, 2012), interference detection consists in the inference of the presence of $q[n]$, based on the measurement of N samples of $x[n]$. This is a classical problem in detection theory and is widely used in disciplines like economics and biostatistics, even though rarely exploited in GNSS. It can be formulated in terms of some hypothesis test, by represented the measured data set by a random process. In the example introduced above, the random process:

$$X[n] = Y[n] + V \cdot Q[n]$$

is considered: V is equal to zero in the absence of interference, and known or deterministic (random variable with a given Probability Density Function (PDF)) in presence of interference.

The Chi-square test on GoF can be applied by formulating the hypothesis testing problem as:

$$\begin{cases} H_0 \text{ (RFI absent)}: V = 0: p_X\left(x\right) = p_Y\left(x\right) \\ H_1 \text{ (RFI present)}: V = 1: p_X\left(x\right) \neq p_Y\left(x\right) \end{cases}$$

where $p_X\left(x\right)$ and $p_Y\left(x\right)$ are first order PDFs of a stationary random process. It is noted that the knowledge of the process distribution when there are no interfering signals (H_0) is the only requirement posed by Chi-square GoF; no other information on the interference characteristics H_1 are required. The method works according to the following steps:

1. The discrete version of the PDF of $X\left[n\right]$ is evaluated when the H_0 hypothesis is verified and the reference discrete histogram ($E = \left\{E_1, E_2, ..., E_k\right\}$) representing $p_X\left(x\right)$ is obtained.
2. A set of measurements of the signal are taken, and the observed histogram ($O = \left\{O_1, O_2, ..., O_k\right\}$) representing $p_Y\left(x\right)$ is built accordingly.
3. The test statistic is evaluated:

$$T_X\left(\mathbf{x}_m\right) = \sum_{i=1}^{k} \frac{\left(O_i - E_i\right)^2}{E_i}$$

When the two histograms coincide, no nuisances are present and $T_X\left(\mathbf{x}_m\right) = 0$; the higher is the value of $T_X\left(\mathbf{x}_m\right)$, the larger is the difference between the two histograms.

4. Since $T_X\left(\mathbf{x}_m\right)$ can be seen as an instance of the Chi-square distributed random variable $T_X\left(\mathbf{x}\right)$, a p-value can be evaluated as:

$$p_m = \Pr\left\{T_X\left(\mathbf{x}\right) > T_X\left(\mathbf{x}_m\right)\right\}$$

It is observed that $p_m \approx 1$ means that the histograms are almost identical and that no interference is presence; vice versa with $p_m \approx 0$ the two distributions are different.

5. The decision is taken by fixing a threshold p_α, known as level of significance:

$$\begin{cases} p_m > p_\alpha : H_0 \text{ is accepted} \\ p_m < p_\alpha : H_0 \text{ is rejected} \end{cases}$$

Figure 4 shows and example of result of the Chi-square GoF test on real GNSS signals. In particular, IF samples are taken at the output of a RF front-end sampling GPS L1 signals, as described in Balaei, Motella, and Dempster (2008); it can be proved that this type of interference strongly degrades the signal processing and induces errors in the estimated position. Chi-square GoF test is able to reveal the presence of interference: while when no interference is present, the p-value remains constant and close to 1, satisfying H_0 (continuous line), when CW interference is injected the p-value tends to zero (marked line).

Transformed Domain Techniques

The potential advantage of the TD techniques is the possibility to clearly detect an interference waveform affecting the received signal, which presence could be masked both in the time domain and in the frequency domain.

Figure 5 provides a block scheme of the logical steps implemented in such techniques. The received digital signal is at first processed through the use of mathematical transformations in order to

Figure 4. p-value of the Chi-square GoF test, applied to real GPS signals in the absence and in the presence of CW interfering signal

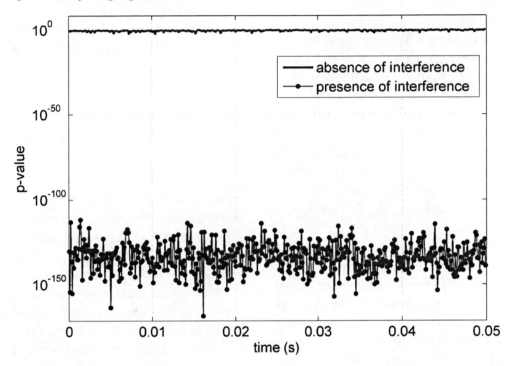

Figure 5. Detection and mitigation algorithms based on the representation of the received signal in a different domain

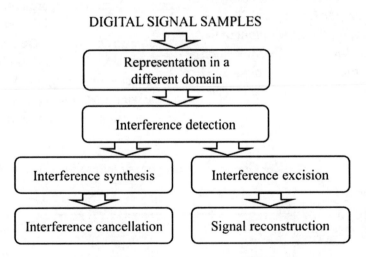

obtain a new representation of the signal itself in a new domain. Then, an a-priori criterion, chosen according to a statistical analysis of the received signal, is adopted to detect interference components in the new transformed domain. Finally, interference mitigation is performed, either by a simple excision of the components belonging to the interfering signals, or, in some cases, the spurious coefficients can be used to perform a synthetic reconstruction of the interference signal to be subtracted from the composite received signal.

It has to be noted that much emphasis has been granted to these techniques, due to the increased available computational capabilities; at present, the major efforts in the development and analysis of innovative interference mitigation techniques are following this direction, which seems to be promising.

The study on this field has been focused on testing different kind of TD techniques, in order to identify the best transformed domain in which the interfering components can be more clearly seen and split from the useful components. In many cases, an interfering signal may appear for a limited time and present a very variable behavior in frequency. In such cases, by means of a Time-Frequency (TF) representation of the received signal, the presence of an interference signal is limited to a region of the TF plan and it is possible to better monitor the interference contribution (Sun & Jan, 2011).

Borio, Lo Presti, Savasta, and Camoriano (2008) describe an interference mitigation technique based on the TF representation approach. The TF representation of the GNSS received signal is obtained by performing an Orthogonal-like Gabor expansion on the samples at the output of the ADC. The Gabor expansion is the decomposition of a signal into a set of time-shifted and modulated versions of an elementary window function, and is defined as

$$s(t) = \sum_{m=-\infty}^{\infty} \sum_{n=-\infty}^{\infty} C_{m,n} h_{m,n}(t)$$

where $C_{m,n}$ are the so called Gabor coefficients,

$$h_{m,n}(t) = h(t - mT) e^{jn\Omega t},$$

and T and Ω represent time and frequency sampling intervals respectively. The synthesis function $h(t)$ is subject to a unit energy constraint. More mathematical details on the Orthogonal-like Gabor expansion can be found in Savasta (2010). An advantage of this detection and mitigation method is that it does not require any hypothesis on the nature of the interference signals.

Figure 6 shows the steps performed in this mitigation algorithm. After the conversion of the signal to its analytical representation, in order to avoid interference cross terms between negative and positive signal frequencies in the TF domain, the expansion in terms of discrete Gabor coefficients is performed. The following step is the application of a pre-determined mask on the previously obtained TF representation, in order to isolate the coefficients related to the interference components. This mask is obtained with a Gabor expansion on the expected GNSS received signal that would be present in an interference-free environment. The mask represents a kind of threshold

Figure 6. Steps of the orthogonal-like Gabor expansion based mitigation algorithm

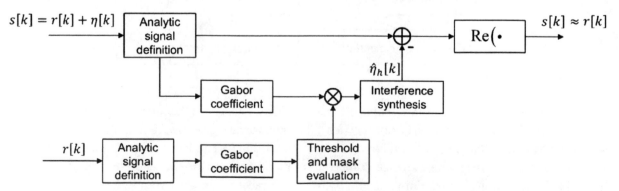

used to detect the portion of the TF plan where the interference is present. Its determination is based on a false interference detection probability, under the assumption of a free interference received signal distribution. Then the interference coefficients obtained are processed by performing an inverse transformation, in order to obtain an interference signal synthesis, which is subtracted from the initial signal (interference excision). In Borio et al. (2008), the performance of this TF plan interference exicision algorithm has been assessed looking at the ambiguity function. The considered interference scenario consists in a GNSS signal corrupted by a chirp interferer such that the J/N is equal to 10 dB. Considering this scenario, the search space obtained without performing any TF excision appears noisy, instead, the search space obtained after the TF excision is less noisy and the correlation peak clearly emerges.

Another class of mitigation algorithms aiming at obtaining a representation of the digitized received signal in a different domain exploits the time-scale transformation, which can be obtained by means of techniques based on the wavelet transform. The proposed wavelet based mitigation algorithm is based on the Wavelet Packet Decomposition (WPD), where the discrete-time signal is decomposed with respect to a set of functions derived by a wavelet basis. The WPD is implemented by means of a uniform digital filter bank. Each stage of the uniform filters bank is composed by a filtering process that produces a decomposition of the signal in high frequency components and low frequency components, followed by a down-sampling operation.

The output of each branch of the filter provides a set of coefficients representing the projection of the signal on a WPD waveform, and thus it brings the information of the power contribution localized in a determined frequency portion of the spectrum.

In presence of narrowband interference it is possible to iterate the WPD until a best matching of the interfering source with the wavelet decomposition is found. An example of the decomposition of the received signal in the different wavelet

Figure 7. Spectral representations of the wavelet sub-bands for a signal affected by a narrowband interference

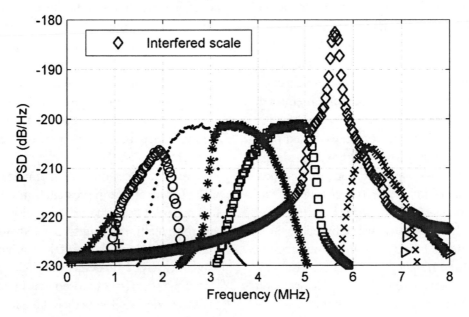

packets is provided in Figure 7. This figure shows the decomposition of the GNSS received signal interfered by a narrowband interference after 3 stages of WPD. Increasing the number of decomposition stages leads to the generation of narrower wavelet filters allowing the decomposition of the incoming signal in several sub-bands with higher resolution. After 3 decomposition stages, 8 wavelet packets are obtained and only one of them is containing completely all the frequency region affected by the interference.

It has been shown that this wavelet based method is particularly suitable for detecting and mitigating narrowband interference signals creating "narrowband" notch filters (Dovis & Musumeci, 2011). As well known, a CWI appears in the spectrum of the received signal as a spike; in this case the best option for mitigation is to operate an interference excision using a notch filter. If the interfering signal has a narrow-band spectrum, notch filtering could not remove all the interference frequency components, and the sub-band decomposition may outperform other mitigation techniques.

Statistical observation of the different wavelet coefficients is the basis for the detection of anomalies and the presence of interference. The basic idea to provide detection and mitigation in the wavelet domain is that it is possible to define a threshold in each block of a wavelet decomposition, in order to isolate the presence of an interfering signal block by block. Once the interference components in the time-scale domain are identified, two possible methods can be considered for the interference excision: the former is based on a synthetic reconstruction of the interfering signal, by means of an anti-transformation process operated on the interference coefficients, which can be then subtracted from the composite received signal, as in the Gabor expansion based method; whilst the latter is based on a direct suppression process of the interference coefficients in the time-scale domain followed by an anti-transformation operation for the signal reconstruction

In Dovis and Musumeci (2011), the performance of this algorithm has been observed at acquisition and tracking stage, looking at the acquisition metric, which represents the ratio between the highest correlation peak and the mean value of correlation floor and Delay Locked Loop jitter respectively, by means of the N-GENE GNSS fully software receiver (Fantino et al., 2010). The interference scenario consists in a GNSS signal affected by a 300 kHz of narrowband interference of -130 dBW. The considered nominal C/N_0 in absence of interference is equal to 45 dB-Hz. Without applying the wavelet interference excision the receiver is not able to acquire the received signal. After the interference excision the correlation peak clearly emerges from the noise floor with an acquisition metric equal to 13.6 dB against the previous 6 dB observed without applying the excision. Improvements on the quality of the received signals have been assessed; the C/N_0 observed after the interference excision is 43 dB-Hz, against 39 dB-Hz experienced without using the wavelet method.

Even if based on a wavelet decomposition, a different approach has been proposed by Paonni et al. (2010) for the mitigation. In this latter case the mitigation is based on the zero-forcing of the coefficients containing the interference components. The last solution is more suitable for those scenarios where the interference properties change in time, such as the pulsed interference addressed by the paper. In Dovis, Musumeci, and Samson (2012), the authors demonstrate how the wavelet based algorithm can represent a high performing solution even in the presence of heavy pulsed interference.

Even more complex mathematical tools, always based on the representation of the signal in a different domain, can be exploited to detect and mitigate interfering signals. Szumski (2011) analyzes the performance of the Karhunen-Loève Transform (KLT) to detect weak RF signals. This technique was first proposed by Maccone (2010) to detect very weak signals hidden in noise in the framework of the Search for Extra Terrestrial Intelligence program. However this technique can be extended to the GNSS scenario and offers several advantages with respect to the Fast Fourier Transform based approach. First the ability to work either with narrowband and wideband signals; second the flexibility in the choice of the basis functions; then, the ability to merge a deterministic and a stochastic analysis of the signal and finally the capability to detect much weaker signals than other methods.

The signal is decomposed in a vectorial space using eigenfunctions, which can have, in principle, any form, provided that they are ortho-normal, and therefore they can better adapt to the signal being processed and increase the detection performance. The KLT decomposition of a general time dependent function is

$$X(t) = \sum_{n=1}^{\infty} Z_n \phi_n(t)$$

where Z_n are random scalar variables and $\phi_n(t)$ are the eigenfunctions. In particular, the decomposition starts with the estimation of the linear auto-correlation function of the signal; afterwards, the Toeplitz matrix is computed and a set of eigenfunctions with their corresponding eigenvalues are extracted by solving linear equations. It can be proved that a signal containing only noise is characterized by KLT eigenvalues equal to one (Maccone, 2010). Therefore, the presence of higher eigenvalues is a proof of the presence of deterministic signals buried in noise. Indeed, the greater benefit of the KLT transform is the capability to successfully detect not only CWI but also narrowband, wideband and chirp interferers, which are usually arduous to handle.

Nevertheless, the biggest drawback of the KLT technique is the complexity and computational burden required to extract a very large number of eigenvalues and eigenfunctions. A possible improvement is given by the Bordered Autocorrelation Method, which in principle allows to reduce the complexity, despite limiting the detection performance (Maccone, 2010).

Another technique for interference suppression, similar to the KLT method and based on the subspace filtering idea, has been recently presented in Kurz et al. (2012). Here the use of antenna array together with a digital embedded spatial filtering technique which acts as a digital beam-forming operation is described. Given an antenna array with M sensors elements, it is possible to define the digital signal at the input of the interference mitigation block at epoch k in a matrix form:

$$\mathbf{X}[k] = \mathbf{S}[k] + \mathbf{Z}[k] + \mathbf{N}[k]$$

where

$$\mathbf{X}[k] = \begin{bmatrix} \mathbf{x}[(k-1)N+1], \dots, \mathbf{x}[(k-1)N+n], \\ \dots, \mathbf{x}[(k-1)N+N] \end{bmatrix},$$

as well as $\mathbf{S}[k]$, $\mathbf{Z}[k]$ and $\mathbf{N}[k]$ are $M \times N$ complex matrices containing respectively the composite received signal, the useful GNSS received signal component, the interference component and the noise component coming from the M different front-end connected to each of the sensors present in the antenna array.

The spatial covariance matrix of the received signal considering the k-th period can be given by as

$$\mathbf{R}_{xx}[k] = \mathrm{E}\left[\mathbf{x}[(k-1)N+n]\mathbf{x}^H[(k-1)N+n]\right]$$

and due to the un-correlation between useful GNSS signal, interference and noise components, the spatial covariance matrix becomes:

$$\mathbf{R}_{xx}[k] = \mathbf{R}_{ss}[k] + \mathbf{R}_{zz}[k] + \mathbf{R}_{nn}[k]$$

Since the power of the GNSS signal is completely buried in the noise floor, and it is extremely smaller compared to the interference power, the spatial covariance matrix can be approximated as follows:

$$\mathbf{R}_{xx}\left[k\right] \approx \mathbf{R}_{zz}\left[k\right] + \mathbf{R}_{nn}\left[k\right]$$

Thus, the eigen-decomposition of the spatial covariance matrix becomes:

$$\mathbf{R}_{xx}\left[k\right] \approx \left[\mathbf{U}_I\,\mathbf{U}_N\right]\left(\begin{bmatrix} \Lambda_I & 0 \\ 0 & 0 \end{bmatrix} + \sigma_n^2\mathbf{I}_M\right)\begin{bmatrix} \mathbf{U}_I^H \\ \mathbf{U}_N^H \end{bmatrix}$$

where the columns of the unitary matrix $\mathbf{U}_I \in \mathbb{C}^{M\times I}$ span the interference subspace, the columns of the unitary matrix $\mathbf{U}_N \in \mathbb{C}^{M\times(M-I)}$ span the noise subspace, and Λ_I denotes a diagonal matrix which contains the non-zero eigenvalues $\lambda_1,\dots\lambda_i,\dots\lambda_I$ with respect to the interference subspace in the noise free case. For all the eigenvalues $\lambda_i \gg \sigma_n^2$ a pre-whitening matrix to suppress interference in $\mathbf{X}\left[k\right]$ can be derived as:

$$\mathbf{R}_{xx}^{-1/2}[k] \approx \frac{1}{\sqrt{\sigma_n^2}}\mathbf{U}_N\mathbf{U}_N^H = \frac{1}{\sqrt{\sigma_n^2}}\mathbf{P}_I^\perp[k]$$

where $\mathbf{P}_I^\perp[k]$ is the projector onto the interference free sub-space for the k-th period.

Thus, interference suppression can be achieved applying the projector matrix to the received digital signal as follows:

$$\tilde{\mathbf{X}}[k] = \mathbf{P}_I^\perp[k]\mathbf{X}[k]$$

The projector matrix $\mathbf{P}_I^\perp[k]$ can be derived from an eigen-decomposition of an estimate of the pre-correlation spatial covariance matrix of the k-th period.

Post-Correlation Techniques

Several research works focused the attention also on post-correlation techniques, i.e. processing the outputs of the correlators employed by the tracking stage of the receiver. The advantage of basing the interference counteracts on the observation at this stage of the receiver is that in some cases, even if a disturbance might be present (and detected) at the pre-correlation stage, it could be filtered out by the correlation process and therefore become harmless. As mentioned in Macabiau, Julien, and Chartre (2001), if a CW carrier frequency perfectly matches with the strongest line of a certain pseudo-random noise code spectrum, it likely produces post-correlation signal degradation. On the other hand, if the CW carrier falls in-between two code lines, it will be filtered out. In this case, two monitoring algorithms working at the pre and post-correlation stages might produce different results (Troglia Gamba, Motella, & Pini, 2013). Another benefit of post-correlation techniques is their low computational burden, that makes them good candidates for the implementation in software or real time receivers.

A first set of techniques is based on the use of multiple correlators. In fact, providing the correlation information at different relative delays, they allow for a monitoring of the shape of the correlation function. An example of this shape-monitoring technique was first introduced by Fantino, Molino, Mulassano, Nicola, and Rao (2009) in order to detect multipath through the analysis of the distortions of the correlation peak, by means of a statistical analysis of the values at the output of the multiple correlators, in terms of mean value and variance of the correlators. This approach can be extended to the interference detection using proper metrics.

A similar approach is based on the harmonic analysis of the correlation function. In the case in which CW interferences are present in the incoming signal, a sinusoidal component, result of cross-correlation between the interfering signal and the local code, can be extracted from the multiple correlators values performing a spectral analysis with parametric techniques, allowing both interference detection and mitigation (Macabiau et al., 2001). In Falletti, Fantino, Linty, Parizzi, and Torchi (2012) this method is enhanced by releasing the dependence on the analysis of the correlation function as output of a certain tracking channel and by using a variable local frequency. In addition, a proper statistical analysis of the metric and of the detection threshold is proposed.

Other kinds of statistical methods can be applied at the post-correlation level, to detect signal distortions. The Chi-square GoF Test and the Sign Test, are exploited and applied to the correlator outputs in Pini, Motella, and Troglia Gamba (2013). Both of them use sets of Early (E) and Late (L) correlations to build the test statistic, which is used to discriminate two hypotheses: the correlation is distorted or not, i.e.:

1. E and L correlations have the same statistics (the correlation is not distorted);
2. E and L correlations have different statistics (the correlation is distorted).

When no disturbances affect the signals and the correlation shape is not distorted, the distribution of the measured data on the E and L slopes of the correlation is similar. In this case, the test statistic assumes small values and the probability that the two correlators have the same statistical characteristics tends to one. On the contrary, in a critical scenario where spoofing or interference distort the correlation, the distribution of the E and L correlators change significantly. In this case, the test statistic assumes higher values and can be used to flag the anomaly to the receiver control logic that might exclude the corresponding channel from the navigation solution.

The Sign Test is a non-parametric check based on the comparison between the *p*-value and a threshold. The algorithm is based on the fact that, if there are no disturbances, the correlation is an even function. In nominal conditions, a pair of E and L multi-correlators equally spaced from the Prompt (P) can be modeled as normally distributed random variables with the same mean:

$$\mu_E = \mu_L = \mu$$

Furthermore, if the E-L spacing is larger than chip, it is possible to show that E and L are independent and $D = E - L$ results to be a normally distributed random variable with zero mean, $\mu_D = 0$. The sign test is then used to test the null hypothesis

$$H_0 : \mu_D = 0 :$$ the correlation function is not distorted

In case an impairment affects the correlation function (due to interference, spoofing, or multipath) and it is mixed with the SIS, the condition H_0 is not verified any more and H_0 has to be rejected.

As an example, Figure 8 shows the result of the sign test applied to Very Early (VE)-Very Late (VL) correlators for a GPS L1 signal affected by a pulsed CWI. The test decision assumes positive values only in correspondence of the second and fourth segments where the interference is present, demonstrating again the effectiveness of the algorithm. The fact that not all the *p*-value points within the affected sections exceed the threshold is due to the variable impact of the CWI on the relative phase between the CW and the SIS carrier.

In Pini et al. (2013) the same algorithm is considered, but taking into account some additional and new aspects. Both the GoF and the Sign Test are applied to the correlator output to monitor the correlation shape and the two methods are also employed with a bench of correlators pairs. In this way, the method is not limited to the detection of the disturbance, but it also enables the estimation of its temporal evolution.

Figure 8. Results of the sign test applied to Very Early – Very Late correlators in a CW interfered scenario

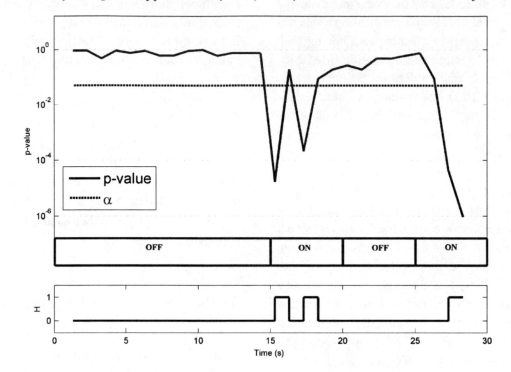

In particular, despite the fact that both the techniques are valid for signal quality monitoring and that both can reveal the temporal evolution of the counterfeit signal, if applied to a bench of correlators, it is shown that the GoF test outperforms the Sign Test.

SPOOFING DETECTION

As previously remarked, the design of countermeasures to spoofing is becoming of relevant importance, due to the implications that a malicious attack can have on the applications based on the position estimated by the GNSS receiver. Moreover, low-cost and low-complexity anti-spoofing techniques would be desirable in order to be easily implemented also in mass-market receivers, where cost and computational burden are the main drivers. It has to be considered that spoofing detection is then the main objective, in order to raise a warning to the user, and it can be accomplished through monitoring and cross-

checking of observables, intermediate measurements and position solutions. Nevertheless, all the spoofing attacks are in general hard to mitigate, once the victim receiver tracks counterfeit signals, and their detailed description is out of the scope of this chapter.

Simplistic attacks are detectable through the monitoring of the navigation solution. In fact, simplistic spoofing signals are not synchronized with the constellation in view, therefore this type of attack produces "jumps" in the computed positions. Intermediate and sophisticated spoofing attacks can be detected with ad hoc algorithms. In particular, the observation of the navigation message and of the bit transitions can provide hints for the presence of a fake signal. In fact, the generation of the navigation message by the spoofer (as replica of the demodulated message) introduces a delay that can be easily monitored; alternatively, in case of prediction of the data bits, the jamming can be detected at the change of the navigation message (i.e. at the ephemeris update) or when a new satellite comes into view.

Several strategies are based on the fact that a spoofing signal appearing in the scenario is a replica of the authentic GNSS signal (called vestigial signal) that "looks like" multipath but is characterized by different power, different delay and different phase, delay and amplitude variations in time (Humphreys et al., 2008). If the receiver is able to split the two components, the anomalies can be detected. As an example, if the relative delay increases but the amplitude does not, it could be an unlikely behavior of a multipath reflection. Several different techniques are present in literature: Humphreys et al. (2008) describe a method based on the monitoring of the shape and of the distortion of the correlation function: after the receiver is locked on the malicious signal, a vestigial peak can be observed along with the signal generated by the jammer in the correlation function domain. Some recent works exploited the capability of some receiver architectures, denoted Coupled Amplitude Delay Locked Loops and originally designed for estimating and separately tracking multipath components of the SIS, to estimate and track both the vestigial signal and the jamming signal (Dovis, Chen, Cavaleri, Khurram, & Pini, 2011).

A similar architecture, denoted vestigial signal defense is the basis for receivers able to identify a spoofing fake signal, as described in Wesson, Shepard, Bhatti, and Humphreys (2011). Different metrics, based on the outputs of a multi-correlator receiver, previously proposed and widely used for multipath detection, such as the Delta Metric, the Ratio Test or the E-L Phase Metric, have been analyzed and extended to process spoofed signals.

Wen, Huang, Dyer, Archinal, and Fagan (2005) describe several methods based on the monitoring of different observables in a GNSS receiver, such as the maximum signal power, its changing rate, the relative power between signals coming from the same satellite but in different bands, code and phase range rates, Doppler shifts and differences between different bands and in general jump detections introduced by the jammer switching on. For instance, the technique denoted L1-L2 range

differences is based on phase and code range differences between L1 and L2 signals caused by the ionosphere propagation; since the spoofed signal is propagating in the lower layer of atmosphere, it behaves differently from the authentic signal, and the spoofing can be detected.

Troglia Gamba et al. (2013) describe a statistical method based on the hypothesis testing to detect the presence of the spoofing signals in the correlation function in the case of an intermediate spoofing attack. It is shown that the sign test effectively detects correlation distortions using VE and VL correlations, while with the E-L pairs there is a significant chance of misdetection. In addition the low complexity associated to the test seams an advantage for real implementations.

Recently a technique for autonomous spoofing and meaconing detection, based comparison and statistically testing of the measured Direction Of Arrival (DOA) against the expected DOAs has been presented in Meurer, Konovaltsev, Cuntz, and Hättich (2012). In absence of spoofing, the expected and estimated DOAs should be consistent and Neyman-Pearson criterion is used to detect these biases, by setting a threshold on the sum of squares of errors test metric, defined according to some desired false alarm rate. In Konovaltsev, Cuntz, Haettich, and Meurer (2013) the estimated DOAs are used not only to determine the spoofing source direction, but also to mitigate its effect by exploiting digital beamforming methods, i.e. placing spatial zeros in the antenna reception pattern.

A complex method that processes the beat carrier phase measurements from a single moving antenna, in order to detect spoofing has been developed by Psiaki, Powell, and O'Hanlon (2013). A specially equipped GNSS receiver can detect sophisticated spoofing attacks, which cannot be detected using Receiver Autonomous Integrity Monitoring techniques. The spoofing detection algorithm correlates high-pass filtered versions of a motion component with high-pass filtered versions of carrier phase variations. Correlation patterns produced by true signals and spoofed signals exhibits differences, which are used to develop a hypothesis test, in order to detect a spoofing attack. The methods is claimed to yield false alarm probabilities and missed detection probabilities on the order of 10^{-5} or lower, when working with typical numbers of GPS signals available at typical patch-antenna signal strengths.

COMPLEXITY TRADEOFF

Given the scenario of all the possible threats and related suitable countermeasures, one may wonder which is the impact of the different techniques on the receiver architecture. In particular a non-negligible complexity burden has to be afforded in order to make a receiver robust to external attacks.

Depending on the user requirements there are, of course, several implementation options. However, several factors have to be considered and traded-off.

First, interference mitigation techniques processing the signal at the ADC output need a good representation of the received signal. Thus, wide-band RF filters, higher sampling frequencies and a larger number of quantization bits are required. At the same time, all the techniques based on the representation of the signal in a different domain and, in general, aiming at elaborating the signal in a certain way, require often complex mathematical operations, which increase the computational requirements. Spoofing countermeasures can also take advantage of a larger number of quantization bits, especially in the case in which they aim at distinguishing the malicious signals from multipath.

The complexity of a spoofing/interference monitoring unit may be challenging and the requirements for the computational capabilities and memory size make, at present, these techniques more suitable for high-end receivers (where generally more resources, in terms of memory, computational capabilities and speed are available and the cost of the device is negligible with respect to its requirements) rather than for mass-market units.

CONCLUSION

In this chapter, an overview of some traditional and recent techniques for spoofing detection and interference mitigation has been provided. It has been outlined that interference and spoofing are challenging arguments. Especially with the advent of new GNSS systems like Galileo and the modernized GPS a wide range of civil applications related to user's safety or featuring financial implications, which would be deeply affected by intentional interfering or spoofing signals, will be developed in the next years. For such a reason, it is urgent to keep investigating effective interference detection and mitigation algorithms and techniques in order to improve accuracy and robustness of a GNSS receiver and to assure the reliability of the estimated position and time solution also in mainstream receivers, in order not to threat the plethora of applications for which reliability is essential.

All the interference mitigation techniques presented in the chapter are expected to give good detection results; nevertheless the performance of each method strongly depends on the resources available, on the operational scenario and on the characteristics of the interference. The decision on the algorithms that can be elected as the best interference mitigation technique and recommended for implementation strongly depends on the likely scenario in which the receiver is expected to operate as well as on the computational resources available. As an example, raw osbservable techniques are more suitable for ground based monitoring stations, where the computational burden is not a strict requirement, while AGC/ADC algorithms are often used in military receivers. Antenna techniques are certainly not feasible on partable low-cost devices. Finally post-correlation techinques offer a good trade-off between performance and complexity.

While detection may be rather easy for jammers and simplistic spoofers, it becomes quite challenging in the presence of spoofers of intermediate complexity, which can be designed and implemented at an affordable cost and a limited technological effort. Future works on this topic are still needed, in order to properly prevent a malicious user to spoof GNSS signals.

REFERENCES

Akos, D. (2012). *GNSS RFI/Spoofing: Detection, Localization, & Mitigation*. Paper presented at Stanford's 2012 PNT Challenges and Opportunities Symposium. Stanford, CA.

Balaei, A., Motella, B., & Dempster, A. (2008). A Preventive Approach to Mitigating CW Interference in GPS Receivers. *GPS Solutions, 12*(3), 199–209. doi:10.1007/s10291-007-0082-8

Baldini, G., & Hofher, J. (2008). *IPSC Projects based on Satellite Navigation Systems. E. C. Institute for the Protection and Security of the Citizens, 1st MENTORE event*. Italy: Ispra.

Bastide, F., Akos, D., Macabiau, C., & Roturier, B. (2003). Automatic Gain Control (AGC) as an Interference Assessment Tool. In *Proceedings of the 16th International Technical Meeting of the Satellite Division of The Institute of Navigation (ION GPS/GNSS 2003)* (pp. 2042-2053). Portland, OR: ION GPS/GNSS.

Bastide, F., Chartre, E., Macabiau, C., & Roturier, B. (2004). GPS L5 And GALILEO E5a/E5b Signal-to-Noise Density Ratio Degradation Due to DME/TACAN Signals: Simulations and Theoretical Derivation. In *Proceedings of the 2004 National Technical Meeting of The Institute of Navigation* (pp. 1049-1062). San Diego, CA: Academic Press.

Borio, D., Camoriano, L., & Lo Presti, L. (2008). Two-Pole and Multi-Pole Notch Filters: A Computationally Effective Solution for GNSS Interference Detection and Mitigation. *IEEE Systems Journal, 2*, 38–47. doi:10.1109/JSYST.2007.914780

Borio, D., Lo Presti, L., Savasta, S., & Camoriano, L. (2008). Time-frequency Excision for GNSS Application. *IEEE Systems Journal, 2*, 27–37. doi:10.1109/JSYST.2007.914914

Borio, D., O'Driscoll, C., & Fortuny, J. (2012). GNSS Jammers: Effects and Countermeasures. In *Proceedings of the 6th ESA Workshop on Satellite Navigation Technologies and European Workshop on GNSS Signals and Signal Processing (NAVITEC 2012)* (pp. 1-7). Noordwijk, The Netherlands: NAVITEC.

De Angelis, M., Fantacci, R., Menci, S., & Rinaldi, C. (2005). Analysis of Air Traffic Control Systems Interference Impact On Galileo Aeronautics Receivers. In *Proceedings of the 2005 National Technical Meeting of The Institute of Navigation* (pp. 346-357). San Diego, CA: Academic Press.

Denks, H., Steingaß, A., Hornbostel, A., & Chopard, V. (2009). GNSS Receiver Testing by Hardware Simulation with Measured Interference Data from Flight Trials. In *Proceedings of the 22nd International Technical Meeting of The Satellite Division of the Institute of Navigation (ION GNSS 2009)* (pp. 1-10). Savanna, GA: ION GNSS.

Divis, D. A. (2013). GPS Spoofing Experiment Knocks Ship off Course – University of Texas at Austin team repeats spoofing demonstration with a superyacht. *InsideGNSS news*. Retrieved December 5, 2013, from http://www.insidegnss.com/node/3659.

Dovis, F., & Musumeci, L. (2011). Use of Wavelet Transform for Interference Mitigation. In *Proceedings of the 2011 International Conference on Localization and GNSS (ICL-GNSS)* (pp. 116-121). Tampere, Finland: GNSS.

Dovis, F., Musumeci, L., & Samson, J. (2012). Performance Comparison of Transformed Domain Techniques for Pulsed Interference Mitigation. In *Proceedings of the 25th International Technical Meeting of the Satellite Division of The Institute of Navigation (ION GNSS 2012)* (pp. 3530-3541). Nashville, TN: GNSS.

Dovis, F., Chen, X., Cavaleri, A., Khurram, A., & Pini, M. (2011). Detection of Spoofing Threats by Means of Signal Parameters Estimation. In *Proceedings of the 24th International Technical Meeting of The Satellite Division of the Institute of Navigation (ION GNSS 2011)* (pp. 416-421). Portland, OR: GNSS.

Enge, P., & Misra, P. (2006). *Global Positioning System: Signal Measurements and Performance*. Ganga-Jamuna Press.

Erlandson, R. J., Kim, T., Hegarty, C., & Van Dierendonck, A. J. (2004). Pulsed RFI Effects on Aviation Operations Using GPS L5. In *Proceedings of the 2004 National Technical Meeting of The Institute of Navigation* (pp. 1063–1076). San Diego, CA: Academic Press.

European Radiocommunication Committee within the European Conference of Postal and Telecommunication Administrations (CEPT). (2002). *The European Table of Frequency Allocations and Utilizations covering the frequency range 9–275 GHz*. Lisboa, Portugal: Author.

Falletti, E., Fantino, M., Linty, N., Parizzi, F., & Torchi, A. (2012). *Italian Patent No. TO2012A000408*.

Fantino, M., Molino, A., & Nicola, M. (2010). N-Gene: a Complete GPS and Galileo Software Suite for Precise Navigation. In *Proceedings of the 2010 International Technical Meeting of The Institute of Navigation* (pp. 1075-1081). San Diego, CA: Academic Press.

Fantino, M., Molino, A., Mulassano, P., Nicola, M., & Rao, M. (2009). Signal Quality Monitoring: Correlation mask based on Ratio Test metrics for multipath detection. In *Proceedings of International Global Navigation Satellite Systems Society (IGNSS) Symposium 2009*. Surfers Paradise, Australia: IGNSS.

Federal Communication Commission. (2011). *Light-Squared Technical Working Group final report*. Washington, DC: Academic Press.

Gao, G. X. (2007). DME/TACAN Interference and its Mitigation in L5/E5 Bands. In *Proceedings of the 20th International Technical Meeting of the Satellite Division of The Institute of Navigation (ION GNSS 2007)* (pp. 1191-1200). Fort Worth, TX: ION GNSS.

Hegarty, C., Van Dierendonck, A. J., Bobyn, D., Tran, M., & Grabowski, J. (2000). Suppression of Pulsed Interference through Blanking. In *Proceedings of the IAIN World Congress and the 56th Annual Meeting of The Institute of Navigation* (pp. 399-408). San Diego, CA: IAIN.

Humphreys, T., Ledvina, B., Psiaki, M., O'Hanlon, B., & Kinter, P. (2008). Assessing the Spoofing Threat: Development of a Portable GPS Civilian Spoofer. In *Proceedings of 21th ION GNSS International Technical Meeting of Satellite Division (ION GNSS 2008)* (pp. 2314-2325). Savannah, GA: GNSS.

Kaplan, E., & Hegarty, C. (2005). *Understanding GPS Principles and Applications* (2nd ed.). Artech House.

Konovaltsev, A., & Cuntz, M. Haettich, C., & Meurer, M. (2013). Autonomous Spoofing Detection and Mitigation in a GNSS Receiver with an Adaptive Antenna Array. In *Proceedings of the 26th International Technical Meeting of The Satellite Division of the Institute of Navigation (ION GNSS 2013)* (pp. 2937-2948). Nashville, TN: ION GNSS.

Kurz, L., Tasdemir, E., Bornkessel, D., Noll, T. G., Kappen, G., Antreich, F., et al. (2012). An Architecture for an Embedded Antenna-array Digital GNSS Receiver using Subspace-based Methods for Spatial Filtering. In *Proceedings of the 6th ESA Workshop on Satellite Navigation Technologies and European Workshop on GNSS Signals and Signal Processing (NAVITEC 2012)* (pp. 1-8). Noordwijk, The Netherlands: GNSS.

Macabiau, C., Julien, O., & Chartre, E. (2001). Use of Multicorrelator Techniques for Interference Detection. In *Proceedings of the 2001 National Technical Meeting of The Institute of Navigation (ION GNSS 2013)* (pp. 353-363). Long Beach, CA: GNSS.

Maccone, C. (2010). The KLT (Karhunen-Loève Transform) to extend SETI Searches to Broadband and Extremely Feeble Signals. *Acta Astronautica, 67*(11-12), 1427–1439. doi:10.1016/j.actaastro.2010.05.002

Meurer, M., Konovaltsev, A., Cuntz, M., & Hättich, C. (2012). Robust Joint Multi-Antenna Spoofing Detection and Attitude Estimation using Direction Assisted Multiple Hypotheses RAIM. In *Proceedings of the 25th International Technical Meeting of The Satellite Division of the Institute of Navigation (ION GNSS 2012)* (pp. 3007-3016), Nashville, TN: GNSS.

Mitch, R. H., Dougherty, R. C., Psiaki, M., Powell, S., O'Hanlon, B., Bhatti, J., & Humphreys, T. (2011). Signal Characteristics of Civil GPS Jammers. In *Proceedings of the 24th International Technical Meeting of The Satellite Division of the Institute of Navigation (ION GNSS 2011)* (pp. 1907-1919). Portland, OR: GNSS.

Motella, B., Pini, M., & Lo Presti, L. (2012). GNSS Interference Detector Based On Chi-square Goodness-of-fit Test. In *Proceedings of the 6th ESA Workshop on Satellite Navigation Technologies and European Workshop on GNSS Signals and Signal Processing (NAVITEC 2012)* (pp. 1-6). Noordwijk, The Netherlands: GNSS.

Motella, B., Savasta, S., Margaria, D., & Dovis, F. (2011). A Method for Assessing the Interference Impact on GNSS Receivers. *IEEE Transactions on Aerospace and Electronic Systems, 47*(2), 1416–1432. doi:10.1109/TAES.2011.5751267

Motella, B., Savasta, S., Margaria, D., & Dovis, F. (2009). Assessing GPS Robustness in Presence of Communication Signals. In *Proceedings of the International Workshop on Synergies in Communications and Localization, IEEE International Conference on Communications* (pp. 1-5). Dresden, Germany: IEEE.

Nicola, M., Musumeci, L., Pini, M., Fantino, M., & Mulassano, P. (2010). Design of a GNSS Spoofing Device Based on a GPS/Galileo Software Receiver for the Development of Robust Countermeasures. In *Proceedings of the European Navigation Conference (ENC 2010)*. Braunschweig, Germany: ENC.

Paonni, M., Jang, J. G., Eissfeller, B., Wallnert, S., Rodriguez, J. A., Samson, J., et al. (2010). Innovative Interference Mitigation Approaches, Analytical Analysis, Implementation and Validation. In *Proceedings of the 5th ESA Workshop on Satellite Navigation Technologies and European Workshop on GNSS Signals and Signal Processing (NAVITEC)* (pp. 1-8). Noordwijk, The Netherlands: GNSS.

Pini, M., Motella, B., & Troglia Gamba, M. (2013). Detection of Correlation Distortions Through Application of Statistical Methods. In *Proceedings of the 26th International Technical Meeting of The Satellite Division of the Institute of Navigation (ION GNSS 2013)* (pp. 3279-3289). Nashville, TN: GNSS.

Psiaki, M. L., Powell, S. P., & O'Hanlon, B. W. (2013). GNSS Spoofing Detection using High-Frequency Antenna Motion and Carrier-Phase Data. In *Proceedings of the 26th International Technical Meeting of The Satellite Division of the Institute of Navigation (ION GNSS 2013)* (pp. 2949-2991). Nashville, TN: GNSS.

PTOLEMUS Consulting Group. (2012). *Global Insurance Telematics Study 2012*. Author.

RTCA DO-292. (2004). *Assessment of Radio Frequency Interference Relevant to the GNSS L5/E5A Frequency Band*.

Savasta, S. (2010). *GNSS Localization Techniques in Interfered Environment*. (Unpublished Ph.D. Dissertation). Politecnico di Torino, Italy.

Scott, L. (2003). Anti-spoofing and Authenticated Signal Architectures for Civil Navigation Systems. In *Proceedings of the 16th International Technical Meeting of the Satellite Division of The Institute of Navigation* (pp. 1543-1552). Portland, OR.

Scott, L. (2004). *GPS and GNSS RFI and Jamming Concerns IV. Navtech Tutorial 410D*. Long Beach, CA: ION GNSS.

Scott, L. (2013). Spoofing – Upping the Anti. *InsideGNSS, 4*.

Sun, C.-C., & Jan, S.-S. (2011). GNSS Interference Detection and Excision Using Time-Frequency Representation. In *Proceedings of the 2011 International Technical Meeting of The Institute of Navigation* (pp. 365-373). San Diego, CA: Academic Press.

Szumski, A. (2011). Finding the Interference, Karhunen-Loève Transform as an Instrument to Detect Weak RF Signals. *InsideGNSS, 3*, 56-64.

Troglia Gamba, M., Falletti, E., Rovelli, D., & Tuozzi, A. (2012). FPGA Implementation Issues of a Two-pole Adaptive Notch Filter for GPS/Galileo Receivers. In *Proceedings of the 25th International Technical Meeting of The Satellite Division of the Institute of Navigation (ION GNSS 2012)* (pp. 3549-3557). Nashville, TN: GNSS.

Troglia Gamba, M., Motella, B., & Pini, M. (2013). Statistical Test Applied to Detect Distortions of GNSS Signals. In *Proceedings of the 2013 International Conference on Localization and GNSS (ICL-GNSS)* (pp. 1-6). Torino, Italy: GNSS.

Wen, H., Huang, P. Y., Dyer, J., Archinal, A., & Fagan, J. (2005). Countermeasures for GPS Signal Spoofing. In *Proceedings of the 18th International Technical Meeting of the Satellite Division of The Institute of Navigation (ION GNSS 2005)* (pp. 1285-1290). Long Beach, CA: GNSS.

Wesson, K., Rothlisberger, M., & Humphreys, T. (2011). A Proposed Navigation Message Authentication Implementation for Civil GPS Anti-spoofing. In *Proceedings of the 24th International Technical Meeting of The Satellite Division of the Institute of Navigation (ION GNSS 2011)* (pp. 3129-3140). Portland, OR: GNSS.

Wesson, K., Rothlisberger, M., & Humphreys, T. (2012). Practical Cryptographic Civil GPS Signal Authentication. *Navigation. Journal of The Institute of Navigation, 59*(3), 177–193. doi:10.1002/navi.14

Wesson, K. D., Shepard, D. P., Bhatti, J. A., & Humphreys, T. E. (2011). An Evaluation of the Vestigial Signal Defense for Civil GPS Anti-Spoofing. In *Proceedings of the 24th International Technical Meeting of The Satellite Division of the Institute of Navigation (ION GNSS 2013)* (pp. 2646-2656). Portland, OR: GNSS.

Wildemeersch, M., Rabbachin, A., Cano, E., & Fortuny, J. (2010). Interference assessment of DVB-T within the GPS L1 and Galileo E1 band. In *Proceedings of the 5th ESA Workshop on Satellite Navigation Technologies and European Workshop on GNSS Signals and Signal Processing (NAVITEC 2010)* (pp. 1-8). Noordwijk, The Nederlands: GNSS.

ADDITIONAL READING

Bastide, F., Chartre, E., & Macabiau, C. (2001). GPS Interference Detection and Identification Using Multicorrelator Receivers. *Proceedings of the 14th International Technical Meeting of the Satellite Division of The Institute of Navigation (ION GPS 2001)* (pp. 872-881). Salt Lake City, UT.

Betz, J. W. (2000). Effect of Narrowband Interference on GPS Code Tracking Accuracy. *Proceedings of the 2000 National Technical Meeting of The Institute of Navigation* (pp. 16-27), Anaheim, CA.

Borio, D. (2010). GNSS Acquisition in the Presence of Continuous Wave Interference. *Aerospace and Electronic Systems. IEEE Transactions on, 46*(1), 47–60.

Borio, D., Camoriano, L., Lo Presti, L., & Mulassano, P. (2006). Analysis of the One-Pole Notch Filter for Interference Mitigation: Wiener Solution and Loss Estimations. *Proceedings of the 19th International Technical Meeting of The Satellite Division of the Institute of Navigation (ION GNSS 2006)* (pp. 1849-1860). Fort Worth, TX.

Borio, D., Lo Presti, L., & Mulassano, P. (2006). Spectral separation coefficients for digital GNSS receivers. *Proceedings of 14th European Signal Processing Conference (EUSIPCO)*.

Borre, K. (2007). *A software-defined GPS and Galileo receiver: a single-frequency approach.* Springer.

Boulton, P., Borsato, R., Butler, B., & Judge, K. (2011). GPS Interference Testing, Lab, Live and LightSquared. *InsideGNSS, 4*, 32-45.

Chen, X., Dovis, F., & Pini, M. (2010). An Innovative Multipath Mitigation Method Using Coupled Amplitude Delay Locked Loops in GNSS Receivers. *Proceedings of IEEE/ION PLANS 2010* (pp. 1118-1126). Indian Wells, CA.

Dovis, F., Musumeci, L., Linty, N., & Pini, M. (2012). Recent Trends in Interference Mitigation and Spoofing Detection. [IJERTCS]. *International Journal of Embedded and Real-Time Communication Systems, 3*(3), 1–17. doi:10.4018/jertcs.2012070101

Elhabiby, M., El-Ghazouly, A., & El-Sheimy, N. (2001). A new wavelet-based multipath mitigation technique. *Proceedings of the 21st International Technical Meeting of the Satellite Division of The Institute of Navigation (ION GNSS 2008)* (pp. 625-631), Savannah, GA.

Fantino, M., Molino, A., & Nicola, M. (2009). N–Gene GNSS receiver: Benefits of software radio in navigation. *Proceedings of the European Navigation Conference-Global Navigation Satellite Systems (ENCGNSS)*, Naples, Italy.

Konovaltsev, A., De Lorenzo, D., Hornbostel, A., & Enge, P. (2009). Mitigation of Continuous and Pulsed Radio Interference with GNSS Antenna Arrays. *Proceedings of the 21st International Technical Meeting of the Satellite Division of The Institute of Navigation (ION GNSS 2008)* (pp. 2786-2795), Savannah, GA.

Motella, B., Pini, M., & Dovis, F. (2008). Investigation on the effect of strong out-of-band signals on global navigation satellite systems receivers. *GPS Solutions, 12*(2), 77–86. doi:10.1007/s10291-007-0085-5

Musumeci, L., & Dovis, F. (2013). Performance assessment of wavelet based techniques in mitigating narrow-band interference. *Proceedings of the 2013 International Conference on Localization and GNSS (ICL-GNSS)* (pp. 1-6). Torino, Italy.

Musumeci, L., & Dovis, F. (2012). A comparison of transformed-domain techniques for pulsed interference removal on GNSS signals. *Proceedings of the 2012 International Conference on Localization and GNSS (ICL-GNSS)* (pp. 1-6). Tampere, Finland.

Musumeci, L., Samson, J., & Dovis, F. (2012). Experimental assessment of Distance Measuring Equipment and Tactical Air Navigation interference on GPS L5 and Galileo E5a frequency bands. *Proceedings of the 6th ESA Workshop on Satellite Navigation Technologies and European Workshop on GNSS Signals and Signal Processing (NAVITEC)* (pp. 1-8). Noordwijk, The Nederlands.

Nielsen, J., Dehghanian, V., & Dawar, N. (2013) GNSS Spoofing Detection Based on Particle Filtering *Proceedings of the 26th International Technical Meeting of The Satellite Division of the Institute of Navigation (ION GNSS 2013)* (pp. 2997-3005). Nashville, TN.

O'Driscoll, C., Rao, M., Borio, D., Cano, E., Fortuny, J., Bastide, F., & Hayes, D. (2012). Compatibility analysis between lightsquared signals and L1/E1 GNSS reception. *Proceedings of the 2012 IEEE/ION Position Location and Navigation Symposium (PLANS)* (pp. 447-454). Myrtle Beach, South Carolina.

Pozzobon, O. (2011). Keeping the spoofs out: Signal authentication services for future GNSS. *Inside GNSS, 6*(3), 48-55.

Raimondi, M., Julien, O., Macabiau, C., & Bastide, F. (2001). Mitigating pulsed interference using frequency domain adaptive filtering. *Proceedings of the 19th International Technical Meeting of the Satellite Division of The Institute of Navigation (ION GNSS 2006)* (pp. 2251-2260). Fort Worth, TX.

Savasta, S., Lo Presti, L., & Rao, M. (2013). Interference Mitigation in GNSS Receivers by a Time-Frequency Approach. *IEEE Transactions on Aerospace and Electronic Systems, 49*(1), 415–438. doi:10.1109/TAES.2013.6404112

Savasta, S., Motella, B., Dovis, F., Lesca, R., & Margaria, D. (2008). On the interference mitigation based on ADC parameters tuning. *Proceedings of the 2008 IEEE/ION Position Location and Navigation Symposium (PLANS)* (pp. 689-695). Monterey, CA.

Omura, J. K., & Scholtz, R. A. (1994). *Spread spectrum communications handbook* (Vol. 2). New York: McGraw-Hill.

Soubielle, J., Vigneau, W., Samson, J., Banos, D. J., & Musumeci, L. (2010). Description of an Interference Test Facility (ITF) to assess GNSS receivers performance in presence of interference. *Proceedings of the 5th ESA Workshop on Satellite Navigation Technologies and European Workshop on GNSS Signals and Signal Processing (NAVITEC 2010)* (pp.1-7). Noordwijk, The Nederlands.

Wesson, K., Shepard, D., & Humphreys, T. (2012). Straight talk on anti-spoofing. *GPS World*, *23*(1), 32–39.

KEY TERMS AND DEFINITIONS

GNSS: GNSSs are communication infrastructures, enabling a generic user to compute position, velocity and time at its current location anywhere on the Earth, processing the signals transmitted from a constellation of satellites and performing a trilateration with respect to the satellites taken as reference points.

GPS: GPS is the first fully operational and currently most used GNSS, owned and operated by the U.S. Government and free of charge for civilian users.

Interference Detection: Interference detection is the process by means of which a certain interfering component is declared present or absent in a received composite signal.

Interference Mitigation: Interference mitigation is the process whereby a detected interference is suppressed from the received composite signal or its effect is reduced.

Interference: The interference is the disturbance caused to GNSS signal processing by any communication systems with carrier frequencies or harmonics affecting the GNSS band.

Jamming: Jamming is the intentional transmission of RF energy in GNSS bands to hinder the navigation service, obtained by masking GNSS signals with noise.

Spoofing: Spoofing is a process whereby someone tries to control reported position out of a device, in order to convince the user that he is somewhere he is not.

Section 2
Innovations for Wireless Networks and Positioning

Chapter 4
A Multi–Hop Software Update Method for Resource Constrained Wireless Sensor Networks

Teemu Laukkarinen
Tampere University of Technology, Finland

Jukka Suhonen
Tampere University of Technology, Finland

Lasse Määttä
Tampere University of Technology, Finland

Marko Hännikäinen
Tampere University of Technology, Finland

ABSTRACT

Wireless Sensor Networks (WSNs) require automated over the air software updates for fixing errors or adding new features. Reprogramming nodes manually is often impractical or even impossible. Current update methods require an external memory, additional computation, and/or external WSN transport protocol. In this chapter, the authors propose Program Image Dissemination Protocol (PIDP) for WSNs. Combining PIDP with an application description language provides a complete method for WSN firmware management. PIDP is reliable, lightweight, and supports multi-hopping. PIDP does not require external memory, is independent of the WSN implementation, transfers firmware reliably, and reprograms the whole program memory. In addition, PIDP allows several levels of WSN node heterogeneity. PIDP was implemented on an 8-bit node platform with a 2.4 GHz radio. Implementation requires 22 bytes of data memory and less than 7 kilobytes of program memory. PIDP updates 178 nodes within 5 hours. One update consumes under 1% of the energy of two AA batteries.

INTRODUCTION

A Wireless Sensor Network (WSN) consists of autonomous sensor nodes (Akyildiz, Weilian, Sankarasubramaniam, & Cayirci, 2002). The goal of sensor node hardware development is to create tiny battery-powered low-cost disposable nodes.

Increasing the performance or memory capacity increases the physical size, energy consumption and manufacturing costs. Thus, nodes are limited in computation, storage, communication and energy resources. These limitations must be addressed when designing and implementing protocols in WSNs.

DOI: 10.4018/978-1-4666-6034-2.ch004

It is not always possible to physically access the nodes in the field once they are deployed. Yet, adding new features, applications and program error fixes necessitates updating the program image that contains the software and protocols running on a node. The solution is a WSN reprogramming protocol, which is used to inject new software into a WSN.

Five general challenges affecting reprogramming in WSNs can be identified (Wang, Zhu, & Cheng, 2006). First, large program images must be transferred reliably through an error prone medium. Thus, the receiver should be able to detect errors and request the corrupted segments again. Second, processing speed and memory capacity in nodes set limits to the time and space complexity of designed protocols. Third, battery powered WSN nodes inherently require the reprogramming protocols to be energy efficient. Fourth, the reprogramming protocol must be scalable enough to handle WSNs that consist of hundreds or thousands of nodes deployed in varying densities. And fifth, the operating system, which is used in nodes, can set limits on the program image format and the reprogramming protocol.

Several protocols (Wang, Zhu, & Cheng, 2006) have been proposed for reprogramming a WSN. A common approach is to equip each node with external memory storage where the new program image is stored. Once the image has been received and verified, a dedicated image transfer program copies the new program image over the old image. This approach allows uninterrupted operation as the new image is transferred in the background. However, the additional memory increases hardware price and takes place on the circuit board, therefore necessitating expensive or energy consuming platforms that prohibit the vision of long term, disposable nodes. Furthermore, many protocols (Hui & Culler, 2004; Levis, Patel, Culler, & Shenker, 2004; Levis & Culler, 2002) support a particular operating system only. Triggering method is often required to start the update after successful image transfer. This can be difficult to achieve reliably, if the update breaks the backwards compatibility of the protocol stack (Langendoen, Baggio, & Visser, 2006; Brown & Sreenan, 2013). If the network splits or one node is temporarily out of range, when the update triggering message is sent, the network can end up situation, where parts of it cannot be connected anymore. Unfortunately, this problem is not often discussed in the research or is left under little attention.

In this paper we present the design, implementation and experimental results of a Program Image Dissemination Protocol (PIDP) for autonomous adhoc multihop WSNs. PIDP consists of firmware version handshakes between nodes, periodic firmware version advertisements and a reliable program image transfer, as shown in Figure 1. Firmware version advertisements are used between neighboring nodes to advertise and compare firmware versions and check for compatibility. The reliable image transfer is used to transfer program images between nodes and to rewrite the program memory. A small bootloader program locates and executes the loaded program image. PIDP is lightweight, energy efficient, reliable and, unlike other reprogramming protocols, does not require external memory for temporary storage of program images. A PIDP update in one part of the WSN does not disturb the whole network, thus, allowing a continuous operation of the non-affected nodes. Furthermore, PIDP is not restricted to a particular operating system or WSN protocol. PIDP provides heterogeneity support for different hardware sensors and for different application logics when combined with Process Description Language (PDL) presented in (Laukkarinen, Suhonen, & Hännikäinen, 2013).

PIDP was evaluated using the TUTWSN prototype (Kuorilehto, Kohvakka, Suhonen, Hämäläinen, Hännikäinen, & Hämäläinen, 2007). TUTWSN is a state of the art adhoc multihop WSN technology for resource-constrained WSNs developed by Department of Computer Systems at Tampere University of Technology. TUTWSN

Figure 1. The logical structure and the memory layout of PIDP and the WSN stack. PIDP is a separate protocol stack. Firmware version advertisements and handshaking co-operate with the WSN stack to disseminate version information and to begin reliable program image transfer.

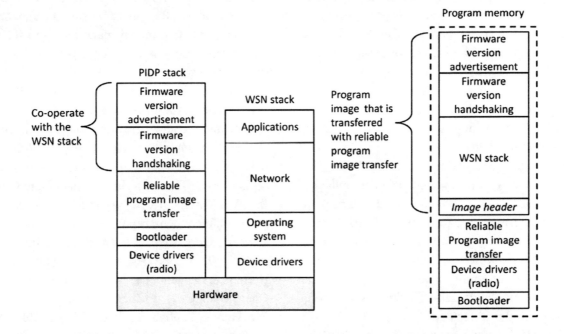

features an energy efficient medium access control (MAC), which uses time-division multiple access (TDMA), a cost-aware routing protocol (Suhonen, Kuorilehto, Hännikäinen, & Hämäläinen, 2006) and multiple custom designed hardware platforms. The operating principle of the MAC layer of TUT-WSN is similar to the beacon enabled clustered mode in IEEE 802.15.4 (IEEE Standards, 2008). Therefore, a similar implementation can be applied to ZigBee (ZigBee Alliance, 2010).

The paper is organized as follows. First, related work is covered. Second, the design of PIDP is presented. Third, implementation is shown. Fourth, evaluation, performance measurements are given. Finally, the paper is concluded.

Portions of this paper have been published in (Määttä, Suhonen, Laukkarinen, Hämäläinen, & Hännikäinen, 2010) ©2010 IEEE. Reused with permission.

RELATED WORK

Two major approaches are distinguished for re-programming a WSN: a full update and an incremental update. The full update transfers a whole program image. The incremental update transfers either a segment of the image or instructions to modify existing image. The image is transferred either using the sensor network protocol stack or a separate parallel maintenance network (Brown & Sreenan, 2013). Currently, incremental updates have gained more attention than full updates due to the smaller transferred data amount.

A number of reprogramming protocols for WSN are built on the TinyOS (Hill, Szewczyk, Woo, Hollar, Culler, & Pister, 2000) operating system. TinyOS does not support loadable modules. Thus, a program image must be loaded as a single binary image.

XNP (Crossbow Technologies, 2003) is one of the first reprogramming services for TinyOS and the MICA2 platform. It features a single-hop reprogramming scheme where the program image is sent as unicast to a particular node or broadcasted to a group of nodes. The single-hop nature limits the scalability of XNP and it only serves as an alternative to manual wired reprogramming.

The successor to XNP is Deluge (Hui & Culler, 2004). Deluge is an epidemic multihop protocol that allows nodes to store several different program images in an external EEPROM memory. One of these images can act as the so called Golden image, which is used as a backup image if the main program image is corrupted. The 2.0 version (Hui, 2005) also adds support for resuming incomplete program image downloads and additional program image verification. MOAP (Stathopoulos, Heidemann, Estrin, & SENSING, 2003) is similar to Deluge.

The Maté virtual machine (Levis & Culler, 2002), which is built upon TinyOS, bypasses the lack of loadable modules by presenting a high-level virtual machine instruction set. Maté bytecode programs are smaller than full program images, which lowers the energy cost of disseminating them. The downside is that interpreting the bytecode creates energy overhead. If new software is disseminated only seldom, the energy consumption of the code interpretation is dominant.

Unlike the TinyOS-based approaches, individual applications and services can be loaded individually in the Contiki operating system (Dunkels, Gronvall, & Voigt, 2004; Dunkels, Finne, Eriksson, & Voigt, 2006). Like Maté, this saves energy as only parts of the whole image need to be disseminated. This dynamic loading only applies to the applications, while the operating system and the protocol stack can only be updated with a separate special image transfer program.

The requirement for external memory storage is common to all these reprogramming protocols, as they use transport layer dissemination protocols to transfer program images. These dissemination protocols are stored within the main program image, which cannot be overwritten as long as it is being executed. This can cause problems e.g. in (Langendoen, Baggio, & Visser, 2006), where unreliable MAC protocol made Deluge useless. As a result, nodes were updated by hand on the deployment site.

In incremental update approaches, a delta file is generated that describes the differences between the old and the new firmware images. The delta file contains instructions to modify and relocate parts of the existing code and the new inserted parts as well. The delta file is disseminated with the WSN protocol or a separate dissemination protocol. Incremental approaches are proposed in several papers, such as in (Mukhtar, Kim, Kim, & Joo, 2009), a processor specific approach in (Reijers & Langendoen, 2003), R2 in (Dong, Liu, Chen, Bu, Huang, & Zhao, 2013), R3 in (Dong, Mo, Huang, Yunhao, & Chen, 2013) and Hermes in (Pant & Bagchi, 2009). Incremental update efficiency degrades if the firmware image changes significantly. Also, parsing the new image according to the delta file requires execution time, program memory and data memory, which are all constrained resources on WSNs. As a result, incremental approaches fit well for situations, where small updates are done often, e.g. when a WSN is developed. When the WSN is deployed and the updates become rarer and often larger, the incremental approaches add unnecessary overhead compared to the full updates. However, incremental updates do not necessarily require external flash, but then they do require a reliable transport protocol layer.

Reliable transport protocols for code dissemination have been presented e.g. in (Stathopoulos, Heidemann, Estrin, & SENSING, 2003) and (Miller & Poellabauer, 2008). Rateless codes can be used to improve dissemination. The rateless code allows reproducing missing packets, provides compression, and/or increases reliability, since the disseminated firmware packets are encoded with rateless codes. A modification of Deluge with rateless codes is presented in (Hagedorn, Starobinski, & Trachtenberg, 2008). Synapse++ is another rateless code proposal that works separately of the WSN implementation (Rossi, Bui, Zanca, Stabellini, Crepaldi, & Zorzi, 2012). As with incremental updates, using rateless codes add execution, program memory, and data memory overheads compared to a direct full update method.

As opposed to Contiki, Maté or difference models, PIDP transfers complete program images. In our experience, the ability to update individual applications is seldom needed as programming error fixes and new features often affect multiple modules of the program image. In addition, loading individual applications requires either support

from the operating system or a separate mechanism for handling runtime relocation of modules. PIDP requires no such support and is operating system independent. PIDP does not require an external reliable transport protocol and it can be used to update completely different image to the network.

PIDP DESIGN

PIDP design consists of firmware version handshaking, periodic firmware version advertisements, and reliable image transfer. Figure 2 presents the PIDP design in action. Following sections present the design in detail. PIDP minimizes

Figure 2. Node B has a newer version. Node A and Node B execute firmware handshaking on WSN association. Then they start program image update and move to the PIDP reliable image transfer. Meanwhile Node C and D continue normal WSN operation. Eventually, Node A is updated and nodes reboot. Node B associates with Node C, starts the update, and continues disseminating image further.

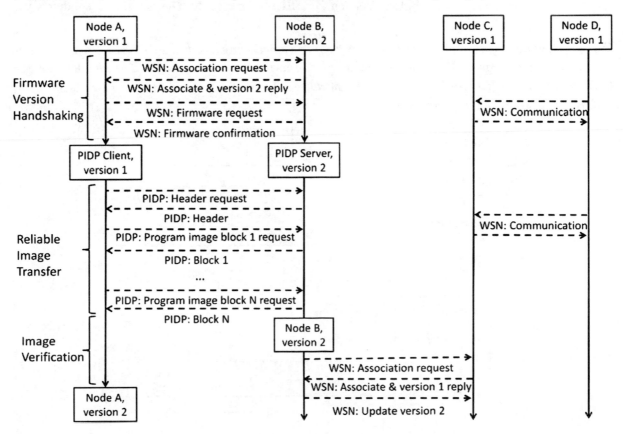

communication and memory overhead, therefore allowing very resource-constrained implementations. Also, PIDP design includes new firmware injection, security, operating system support, and support for heterogeneous networks.

Firmware Version Handshaking

Program image transfer begins automatically when a node detects that one of its neighbors has a new version of a compatible program, as shown in Figure 2 between the nodes A and B. The node with a lower version number sends a *firmware request* and the other node responds with a *firmware confirmation* at the WSN protocol level. After this, the nodes jump to the reliable image transfer of the PIDP, which is independent of the WSN stack.

PIDP assumes that a node performs a handshake with its neighbors after powering up to find new routes. This is the case in most of the sender decided WSN protocols, such as ZigBee. Version

information is exchanged in PIDP when a node exchanges routing information or synchronizes with its neighbors, which adds a small overhead. Either node participating in the handshaking can start the update operation. Both nodes reboot after the update and perform handshaking with their neighbors. This guarantees that program images will propagate epidemically in the WSN.

It is important to note that version information is exchanged on a hop-by-hop basis between neighbors without flooding the version information further into the network. If a network contains multiple nodes with incompatible program images, it may limit the propagation of program images.

Periodic Firmware Version Advertisements

Nodes periodically advertise their program image version on an advertisement channel, as shown in Figure 3. After each advertisement the source listens for a reply. The parameter T_a adjusts the

Figure 3. Example of periodic advertisements and their functioning as a fail-safe. Node A transmits advertisements with interval T_a. Node B is updating its firmware with Node C at the data channel of Node C. Node B encounters a problem at t_{error}, scans the advertisement channel and begins a new transmission with Node A using the data channel of Node A.

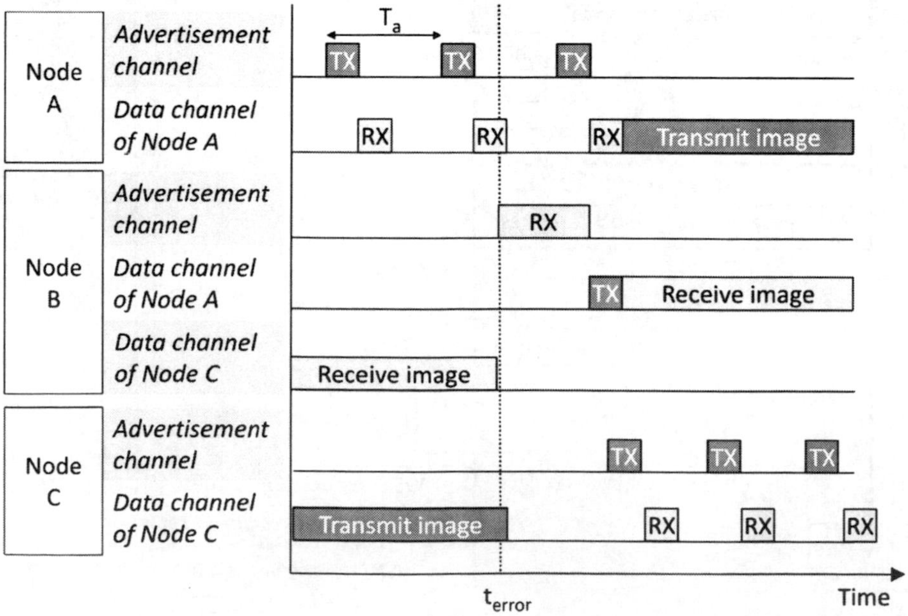

interval between these periodic advertisements. The main purpose of the periodic advertisements is to act as a failsafe. If a node encounters a problem while reprogramming or the image transfer is disrupted, the node may listen for the periodic advertisements to find a new source for image transfer as shown in Figure 3. In addition, periodic advertisements allow nodes to perform image acquisition even if their protocol stacks might otherwise be incompatible. As periodic advertisements are only transmitted seldom and nodes do not listen for them during normal operation, they use very little energy. WSNs that use synchronized MAC protocols, such as IEEE 802.15.4, can embed the version information in synchronization beacons, which nodes transmit periodically.

Reliable Image Transfer

Information about the program image is stored in a header, which consist of hardware identifier, firmware version, message authentication code, and valid and dissemination bits. Program images are identified by a combination of the hardware platform identification number and the firmware version number. Platform identification numbers are used to limit the transfer of program images between incompatible sensor nodes. Furthermore, the header contains a valid bit that indicates whether or not the program image has been successfully validated and can be safely executed. The dissemination bit decides if the node will disseminate the image to the network.

Unlike other reprogramming proposals, the image transfer in PIDP operates independently of the main WSN stack. This allows the image transfer protocol to achieve a better energy-efficiency, as minimizing the number of protocols layers used in the transfer also minimizes the amount of overhead in the transmission of program images. Furthermore, simple independent stack can be tested thoroughly and possible problems of unreliable transfer protocols cannot prevent program image update. As the WSN protocol stack is part of the updated image, the possible problems on WSN protocols can be fixed with PIDP.

The image transfer protocol follows the general client-server architecture as seen in Figure 2. The receiver of the program image acts as a PIDP client, while the sender acts as a PIDP server. Communication between a PIDP client and a PIDP server is performed at a channel selected by the PIDP server, which is transmitted within the firmware advertisements. PIDP servers may choose to use a single network-wide dedicated channel for the image transfers or they may use an appropriate channel selection algorithm to choose channels that are not being used. Choosing different channels is preferred, as this lowers the chance of collisions with nearby image transfers.

The PIDP client begins by requesting the header of the program image as presented in Figure 2. Once the header is received the PIDP client marks the current header invalid and requests the contents of the program image in blocks. After each block the PIDP client immediately writes the data to the program memory, thus invalidating the previous program image. After the whole image is received the PIDP client validates the program image by calculating the message authentication code and comparing it to the one in the header. If the validation calculation is correct, the header is marked valid. Otherwise, the header remains marked invalid.

After the transfer the PIDP client and the PIDP server reboot and return to the normal WSN operation. The PIDP client uses the PIDP bootloader program to check that the header is valid and begins executing code from the beginning of the program image. If the header is not valid then the PIDP client begins to scan the advertisement channel for firmware advertisements and re-executes the image transfer.

Program Image Injection

Three alternative ways exist for new program image injection with PIDP. First, a new node with a new image may be brought to the coverage area of the WSN. The new image is then disseminated to the network automatically by PIDP. Second, a *PIDP cloner device* may be used to transfer program image to one node in the network, which then starts advertising the new version. Third, the new image can be uploaded to a server, which delivers it to the gateways. The gateways advertise the new image to the network and PIDP will first update the nearest nodes using the gateways as relays.

The PIDP cloner device is a specially programmed node that does not act as a part of the WSN. It only advertises the new image on a special cloning channel. The nodes do not normally listen for this channel. When the node is rebooted while a button is pressed, the node will listen for the cloning channel for a period. If there is a PIDP cloner device nearby and the hardware platform identifier match, the node will start the image

transfer. If the dissemination bit is set, the newly programmed node will then start disseminating the image further.

The server injection is presented in Figure 4. The image is first compiled from the C code to a HEX file and then modified with a script to a XML file, and the XML file is finally uploaded to the database. The server indicates to the gateways that there is a new image to advertise. When a node notices the new image from the advertisements, the gateway requests the new image piece by piece from the server and relays it to the node.

Security

Three major security questions concern program image updating (Deng, Han, & Mishra, 2006). First, the new image must be from a reliable source. Second, the new image must be valid. Third, the image must be transferred securely to preserve intellectual property. PIDP accepts only images with correct message authentication code. This ensures that unknown source cannot

Figure 4. Program image injection starts with compilation of the image from a source code to a hex file, then formatting it to an xml file, and uploading it to the database. The server will retrieve image information and relay it to the gateways. The gateways will advertise the image to the WSN and relay the image, when a node requests the new image.

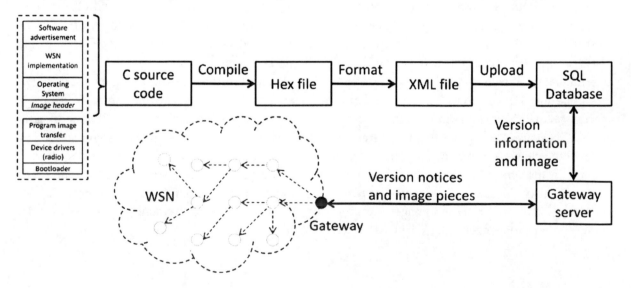

inject a new image to the network and hijack the network. The message authentication code is calculated with a one-way function that uses a secret key, the program image, and a magic number as parameters. As the program image is used in calculation of the message authentication code, the image validity is secured at the same time. The secret key can be used to encrypt and decrypt the program image packets with AES algorithm after the handshaking to prevent stealing the program image with sniffing.

Operating System Support

WSN operating systems have two approaches for updating. The whole image including the application and the operating system are disseminated (TinyOS and Deluge), or only the applications are disseminated to the network (Contiki). PIDP supports both ways as presented in Figure 5. The operating system can be a part of the whole program image as in Figure 5a. This is similar to Deluge and TinyOS. The operating system can be left out of the program image, as in Figure 5b, but the applications are treated as one image. Injecting one new application requires re-injecting all the existing applications.

PIDP allows as many image version headers as there are room in the version advertisement packet. Thus, program image can be split in several parts as presented in Figure 5c. These parts can be separately updated. The selected program image part is indicated in the handshaking between the PIDP client and server. Then, the PIDP will update only the selected part. This can be used to inject new applications to the network without injecting the remaining ones again and the operating system can be updated separately. However, each image requires a new header. The header overhead would increase and the amount of applications would be limited. Furthermore, the applications should always fit inside a certain space and some applications would waste the program memory. Every program image requires known entry functions, which will add some complexity in the development.

Figure 5. a) A typical use of PIDP, where operating system and WSN stack form the program image. b) If the operating system is reliable and will not require new features, it can be left out of the program image. c) With small modifications, PIDP can update the program image in two or more parts. This allows granular updating, but image headers increase overhead.

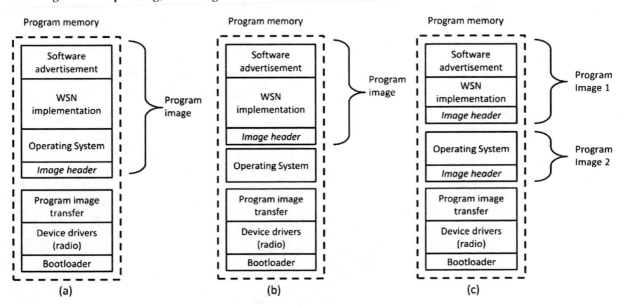

Multiple program image support is not currently incorporated to the PIDP design nor implemented. Instead, we use hardware independent application description language PDL (Laukkarinen, Suhonen, & Hännikäinen, 2013) to create application logic, and to update applications over the WSN protocol stack without need for actual reprogramming. PDL is a similar approach to a virtual machine, but it describes only the application logic that is smaller in size than virtual machine programs.

PIDP does not restrict the operating system from using its own protocols to update applications. For example, PIDP can update Contiki and Contiki can use its own protocols to handle applications. The image validation should then only cover the area, which is not modified by the program code dissemination of Contiki.

Heterogeneous Node Support

WSNs are seldom homogenous; nodes have different sensors, different roles, and different applications. PIDP separates heterogeneity only between hardware devices. If the hardware is not same, the image transfer is not started. To overcome this limitation, we have developed an auto-configurator. The node is configured during the building process to its configuration: the connected sensors, desired roles, and required protocol configurations of the node are set to the EEPROM of the node. A program image is used, which contains all the necessary code for these configurable parts. Node selects used role and sensors at the startup according to the configuration. This allows us to use one single program image for the whole WSN with various node configurations.

Figure 6. PIDP combined with PDL for a complete solution with wider heterogeneity support

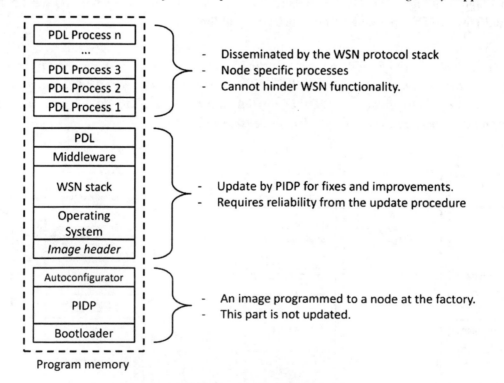

Combining PIDP with PDL (Laukkarinen, Suhonen, & Hännikäinen, 2013) improves heterogeneity support. Each node can have several different application logics, which are distributed over the WSN protocol stack. This combination is presented in Figure 6 and it provides following benefits: first, PIDP ensures that the network does not cease or split if the update procedure fails. Second, PIDP ensures that all nodes are eventually updated. Third, PIDP allows updates on WSN OS and protocol stack without of risk losing the compatibility between the nodes. Finally, PDL allows different applications on each node without any reprogramming. Application changes are byte code dissemination task that works separately of the actual program image. As a result, PIDP does not need complex code relocation operations or update triggering. If the WSN protocol stack and OS are mature and stable, and the program image updates are infrequent, the small overhead PIDP with separate application logic updates of PDL is a versatile and efficient approach to update deployed WSNs.

The design of PIDP allows factory programming, where only the bootloader and PIDP are programmed on each node MCU at the factory. Upon the first time use of a node, the auto-configurator is used to program the sensor and network configuration to the node and the up-to-date protocol stack is autonomously programmed by PIDP when the node is installed. From there on, PDL can be used to reconfigure applications without any reprogramming overheads.

IMPLEMENTATION

PIDP and TUTWSN protocol stack are implemented using the C programming language and the Microchip MPLAB C compiler (Microchip Technology, 2009).

TUTWSN Protocol Stack

The TUTWSN MAC protocol forms a clustered tree topology (Kuorilehto, Kohvakka, Suhonen, Hämäläinen, Hännikäinen, & Hämäläinen, 2007). Each cluster contains a cluster head, a *headnode*, and several cluster members. Cluster members can be leaf nodes or headnodes of other clusters forming a tree of clusters. Each cluster within interference range operates on a separate *cluster channel* that is used for intra-cluster communications. Nodes share a common *network channel* that is used by the headnodes to advertise their clusters. Nodes scan the network channel at least once every hour to find new clusters. In addition, network scans occur when nodes lose their route to the network gateway.

The headnodes maintain a data exchange schedule. Time is divided into fixed length access cycles. Each access cycle begins with a superframe, which contains slots for data transfers at the cluster channel, and ends in an idle time. The length of the access cycle is set to two seconds. Cluster advertisements are sent on the network channel in the beginning of each superframe. Three additional cluster advertisements are also sent during the idle time with approximately 500 millisecond intervals between them.

A TUTWSN sensor node includes an 8 bit Microchip PIC18LF8722 microcontroller (Microchip Technology 2008) with 128 kilobytes of program memory and 3936 bytes of data memory. The microcontroller has an internal 1024 byte EEPROM memory. A Nordic Semiconductors nRF24L01 (Nordic Semiconductors, 2007) is used as the radio, which has a payload size of 32 bytes and a configured transmission rate of 1 megabit per second. The radio does support carrier sensing. It operates on the 2.4 gigahertz band and offers 126 channels and transmission powers of -18 dBm...0 dBm. TUTWSN node has a simple user interface, which consists of a push button and two light emitting diodes. TUTWSN sensor nodes can be equipped with multiple sensors, such as accelerometers, temperature sensors, and humidity sensors. Two 1.5 volt LR6-sized batteries are used as the power source. A TUTWSN sensor node circuit board is shown in Figure 9.

PIDP Implementation

Firmware version handshaking was embedded to the MAC layer of TUTWSN. Thus, nodes exchange version information when they perform association with each other. This allows rapid firmware dissemination within a TUTWSN cluster tree.

The periodic firmware advertisements are transmitted in the TUTWSN network channel, which allows nodes to receive advertisements while they are performing normal neighbor discovery. An advertisement is sent on each access cycle during the idle time. Thus, the interval T_a between the advertisements matches the length of the access cycle. In addition, advertisements on the TUTWSN network channel allow the program image to propagate between different cluster trees, but this method of dissemination is limited by the low frequency of network scans. The cluster channel is used for program image transfer to minimize collisions between concurrent program image transfers.

The bootloader and the program image transfer protocol are stored in a reserved segment in the beginning of the program memory. They are followed with the program image header and the main program image. The message authentication codes are implemented by using a modified 4 byte RC4 code similar to the code described in (Zhang, Yu, Huang, & Yang, 2008).

The program image transfer protocol uses a packet size of 32 bytes. Each packet has a 6 byte header followed by a payload with a length of 26 bytes.

Reliable image transfer is located on a memory section that cannot be updated with PIDP. It includes only the necessary modules to perform the program image transfer and the program memory rewrite. Modules are a radio driver, PIDP server, PIDP client, program memory writer, program image verification, and bootloader. This memory section has to be kept as small as possible since it reduces amount of available memory for the WSN implementation.

The reliable image transfer and the main program are never executed concurrently. Thus, the image transfer can utilize data memory segments that are normally reserved for the main program. Overlaying the data memory significantly reduces data memory requirements of the image transfer protocol. Despite the overlaying, a small amount of dedicated memory is needed for passing version information between the main program and the image transfer.

EVALUATION

Evaluation of PIDP was performed by analyzing the memory consumption, propagation speed and energy consumption impact.

Memory Consumption

Memory consumption was analyzed from the compiled program images for the TUTWSN platform.

Table 1. Memory consumption of PIDP with TUTWSN integration in bytes

Component	Data Memory (B)	Program Memory (B)
Reliable image transfer	793 (overlayed)	3578
Version handshaking	15	1386
Firmware advertisements	7	1425
Total	22 + (793)	6389

From the results in Table 1 we can see that the memory consumption of the PIDP protocol is split in two parts. The first part contains the image transfer while the second part is stored within the program image and contains the necessary support for accessing the image transfer protocol and the implementation of the firmware advertisement scheme.

Although PIDP requires 815 bytes of data memory in total, the absolute increase in the data memory requirements stays at 22 bytes. The image transfer overlays data memory with the WSN stack.

Propagation Time

In order to give a reference point for the measured propagation times, the program image transmission and verification times between two nodes were first measured. Transferring a 123 kilobyte image between two nodes in optimal conditions was 51

seconds on average, thus achieving a transfer rate of 2.4 kilobytes per second. The program image verification time was a constant 23 seconds. Thus, the minimum time for updating a sensor node with this particular program image was 74 seconds.

Program image propagation experiments were performed in a typical office environment with various interference sources such as several WLAN routers operating on the same frequency band. The first experiment included one gateway and 25 sensor nodes. The nodes were placed on a table in one group. Size of the table was less than one square meter. The purpose of this experiment was to see how PIDP performed in a situation where every node had multiple neighbors in close proximity and the amount of network activity was high. Update speed is presented in Figure 7. PIDP successfully reprogrammed the nodes in 12 minutes. 8 concurrent image transfers were observed during this period at the time of 500

Figure 7. The updating speed graph of PIDP on the 25 node experiment. The extra three updates were result of failed updates, which caused re-update. Ideal T_a presents how the network would be updated, if a node could start updating another one immediately after receiving the new image. Approximated T_a of the experiment indicates that one node disseminates 35 seconds after the update.

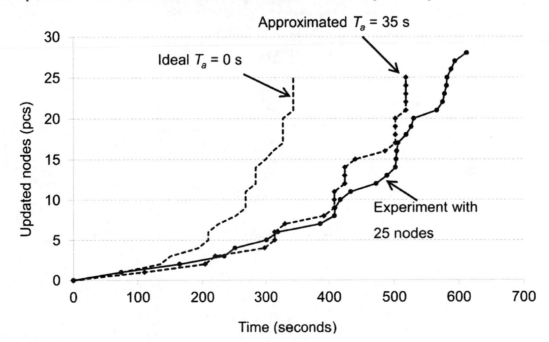

Figure 8. a) Tampere University of Technology campus, the coverage of the campus WSN, and the new program image injection point of the experiment in Computer Science building. b) A graph presenting the TUT Campus WSN experiment progress in percentage of updated nodes. The dissemination was stalled, because two neighboring nodes belonged to different clusters and had good routes to different gateways. Therefore, they did not associate until one hour periodic scan was due. Labels indicate the time when the update was completed for that building.

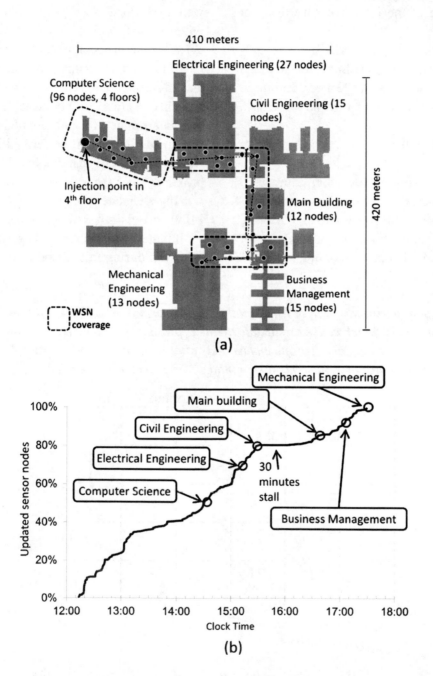

and 600 seconds from the start. 28 updates were performed, which indicates that 3 updates failed and 3 nodes had to be updated again. Reason for these failures is unknown. T_a was approximated based on the measurements. It took 35 seconds from a node to be capable to disseminate received image.

For the second experiment, the performance of PIDP was measured using the Tampere University of Technology campus WSN. This campus network has 178 sensor nodes and 13 gateways distributed in six buildings around the university campus. Figure 8a presents the campus and the coverage area of the campus WSN. The Computer Science building has sensor nodes in four floors while the others have nodes in only one floor. Distance between nodes ranges from 5 meters to 20 meters. The campus WSN is used as an application platform for students to implement their own applications on a WSN course. Measurement data of the campus WSN is provided for property maintenance. Figure 9 presents a humidity, luminance, and temperature measurement node at the campus of the Civil Engineering building. In addition, carbon dioxide and passive infrared based human activity are measured in the campus WSN. Students attending to the course may carry a node with them, which is tracked by the campus WSN.

The new program image was injected into the WSN by updating a single node on the 4th floor of the Computer Science building. Each node sent a report after a successful update procedure.

The results of the second experiment show that PIDP successfully propagated the program image through the campus WSN in five hours, as shown in Figure 8b. A delay was experienced between a pair of nodes located between the Civil Engineering building and the Main building.

Figure 9. A TUTWSN node platform, TUTWSN node in an enclosure, and two nodes installed to the Civil Engineering building of Tampere University of Technology campus in the TUT campus WSN

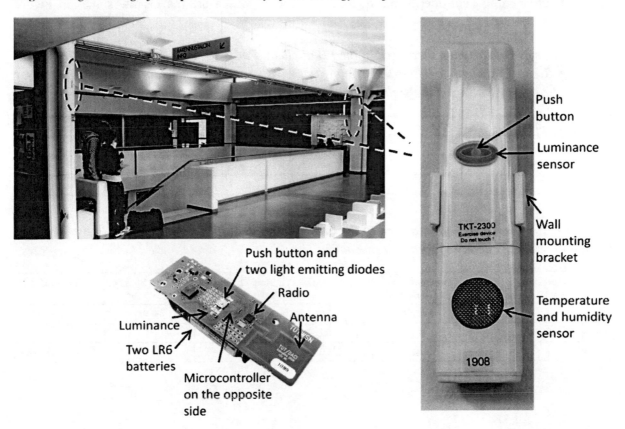

Once the image had spread to the Main building, it continued to propagate to the rest of the WSN. As a result, program image had to travel over 50 hops to achieve the last node in the Mechanical Engineering building (700 meters long path and average hop distance of 12.5 meters equals 56 hops).

The dissemination speed during the second test was mostly limited by the long interval between network scans. The new image propagated quickly within individual clusters e.g. inside one building, but spread slowly from one cluster to another. As most of the nodes had good routes to the nearest network gateway, they had no reason to perform additional network scans. This limitation can be avoided by disseminating the program image through the network gateways.

The transmission time between nodes varied from one minute to four minutes. This was caused by the differences in link reliability between different nodes. Due to the hardware restrictions, PIDP client chooses first received advertiser to be a PIDP server without considering link quality.

This can lead to a situation where image transfer has to be attempted several times between nodes that are too far apart or suffer from low link reliability due to interference.

Energy Consumption Impact

PIDP energy consumption impact was measured on the TUTWSN hardware platform for the PIDP client and the PIDP server. Measurements were conducted with a stable power source and series resistor of known value. The voltage over the series resistor was measured with an oscilloscope to determine drawn root-mean-square current. Duration and current of the image transfer and verification operations were measured. Slight differences to the measurements in propagation time section are result of different updated image. Figure 10a and Figure 10b present screen captures of the oscilloscope, where the series resistor voltage is drawn over the time.

The PIDP client sends requests to the PIDP server, which are the narrow spikes in Figure

Table 2. Energy consumption measured of the PIDP client. Radio RX denotes for radio listening and receiving. Radio TX denotes for radio transmitting. MCU energy consumption during the image transfer includes the energy consumed in the program memory writing.

Operation	Time Consumed (s)	MCU (mJ)	Radio RX (mJ)	Radio TX (mJ)	Total (mJ)
Image transfer	49.62	711.99	436.14	48.46	1196.59
Image verification	24.44	272.44	0	0	272.44
Total:	74.06	984.43	436.14	48.46	1469.03

Table 3. Energy consumption measured of the PIDP server. Radio RX denotes for radio listening and receiving. Radio TX denotes for radio transmitting.

Operation	Time Consumed (s)	MCU (mJ)	Radio RX (mJ)	Radio TX (mJ)	Total (mJ)
Image transfer	49.62	553.03	932.92	64.85	1550.80
Image verification	0	0	0	0	0
Total:	49.62	553.03	932.92	64.85	1550.80

Figure 10. Screen captures of the oscilloscope drawing the series resistor voltage during the energy consumption measurements for the PIDP client and server. The scale is horizontally 10 ms/div and vertically 200 mV/div. a) The PIDP client sends a request packet and listens for the image packet. 2-3 image packets are received before writing to the program memory. b) The PIDP server listens for requests of the PIDP client, prepares an image packet and sends it. Note: a) is not in synchronization with b).

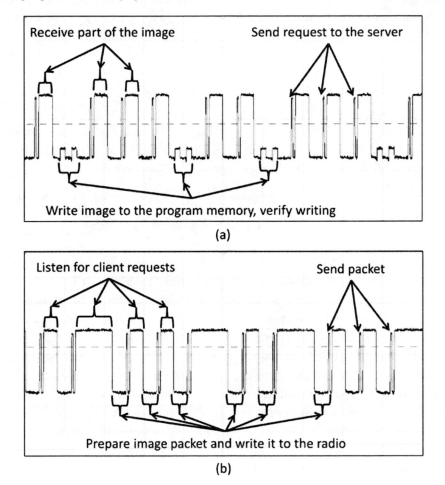

(a)

(b)

10a. Then the PIDP client listens for the image packets, which are the wide spikes in the capture. One packet has 26 B of payload and the program memory writing has to be done in 64 B blocks. Therefore, two to three packets are received before writing. The success of writing a block of the program image is always verified immediately. The whole image is verified after the image has been completely transferred. The PIDP server listens for requests of the PIDP client (wide spikes in Figure 10b), prepares a packet, writes it to the radio and sends it to the PIDP client (narrow spikes).

The TUTWSN platform was run on 4 MHz clock frequency and 2.5 V supply voltage. 4 MHz is the highest possible clock frequency with the specified supply voltage. The highest possible clock frequency is used to achieve the fastest possible dissemination time and the shortest affection time to the normal operation. The radio used a random channel from the 2.4 GHz ISM band and the highest possible transmission power of 0 dBm. The MCU is continuously active during the update operation. On the PIDP server, the radio is active for 35 second, which is 67.3% of the image transfer time and it is receiving 93.5% of that time. On the PIDP client, the radio is active for 17 seconds, which is 32.7% of *the image transfer time* and it is receiving 90% of that time. These values were obtained from the oscilloscope.

Energy consumption results of the PIDP client are presented in Table 2. Typical lithium AA batteries have approximately 20000 J of usable energy. Thus, one update consumes under 0.1‰ of the available energy of the PIDP client. If the expected node lifetime is two years, updating a node once a day would reduce lifetime approximately 10%. Impact is irrelevant on moderate amounts of updating. However, PIDP is not suitable for continuous application dissemination. This is due to the whole program image updating. If the image is sliced to smaller pieces as presented in Operating System Support section, the energy consumption impact will be reduced.

PIDP server energy consumption is presented in Table 3. The image transfer consumes more energy with the PIDP server, since it has to listen for longer periods. However, the PIDP server does not need to write or verify the image after the transfer and the total amount of consumed energy is similar to the PIDP client. In normal operation, one node acts once as a PIDP client and zero to multiple times as a PIDP server. Therefore, it is difficult to determine actual energy consumption impact of server duty in a network. In our experiments, one node acted as a PIDP server zero to three times. Thus, energy consumption impact varies between 1500 mJ – 6000 mJ, which is under 1‰ of the available energy.

PIDP is an energy efficient method to disseminate new program image to the network. The energy consumption impact increases significantly only if the network is updated often, e.g. once a day. Furthermore, the energy consumption impact is divided evenly across the whole network, since the image disseminates one hop a time and most of the nodes act once as the PIDP client and similar amounts as the PIDP server.

The energy efficiency could be improved by decreasing the radio transmission power or the radio listening time. If the radio transmission power is decreased that the total transmission energy consumption is half of the current consumption, it would reduce energy consumption 2% in total. Thus, the transmission power is not significant energy consumer. In optimum case, radio listening time would be the same as the transmission time. Therefore, reducing the radio listening time would reduce maximum of 26% of the PIDP client and 56% of the PIDP server energy consumption. Reducing the radio listening time is one major task in future work.

COMPARISON

Feature comparison of known multi-hop reprogramming methods for WSNs is presented in Table 4. PIDP manages to function without operating system. Also, it can function with any OS and update the OS as well. PIDP does not require external flash or reliable transport protocol, since it has own transport protocol. The preparation and loading overhead of the firmware image on PIDP are minimal. Finally, PIDP usually updates the whole program image, but it can be modified to update it in parts.

CONCLUSION

This paper presents a lightweight, reliable and energy efficient program image dissemination protocol for WSNs. Unlike other dissemination protocols, PIDP does not require external memory storage, is independent of the WSN stack, offers a low overhead protocol for transferring program images, and can reprogram the whole WSN stack. PIDP is implemented using low-power WSN prototype nodes and tested in actual real-world conditions. The experimental results show that PIDP can reprogram 178 nodes in 5 hours and

Table 4. Feature comparison of known WSN reprogramming approaches

Protocol	Requires OS	Supports Multiple OS	Requires Ext. Flash	Requires Transport Protocol	Image Preparation Overhead	Update Scope
PIDP	No	Yes	No	No	Small: verification	Whole image or parts
Deluge	TinyOS	No	Yes	Yes	Small: verification	Whole image
Contiki	Yes	No	No	Yes	Medium: dynamic loading	Application dissemination
Maté	TinyOS	No	No	Yes	Virtual machine	Application dissemination
MOAP	TinyOS	No	Yes	Yes	Small	Whole image
R3	TinyOS	?	Yes	Yes	High: construct a new image	Whole image through a delta file
SYNAPSE++	No	Yes	Yes	No	High: decode and construct	Whole image using rateless codes

requires less than 7 kilobytes of ROM and 22 bytes of RAM and that it is possible to create a dissemination protocol that does not require external memory and yet achieves the epidemic dissemination capabilities of traditional dissemination protocols with low energy consumption.

Future work on PIDP will include new methods for inter-cluster advertisements, reliability improvements, energy consumption minimizing, and use of it for application dissemination with operating systems.

The new inter-cluster advertisement methods will speed up the propagation of the program image. This would remove stalls as seen in Figure 8b, where network was partitioned and both partitions considered their network situation satisfactory. In addition to multiple injection points, this can be solved with application level software advertisements, where nodes are informed in the application level that there might be newer program image available. Then the nodes could seek more eagerly for the new image.

To improve reliability, the PIDP protocol must select the transfer channel from non-interfering channels. Also, the PIDP client should start the image transfer with the best possible neighbor. These

are difficult tasks to do and require novel designs and implementations to fit the PIDP design.

Energy consumption can be reduced significantly by reducing the radio listening time. This requires strictly synchronized protocol. The PIDP client and server could negotiate a timetable in every transmission for the next packet. This introduces research problems of what to do after unsuccessful transmissions and how to fit such a complex protocol on a restricted space.

REFERENCES

Akyildiz, I., Weilian, S., Sankarasubramaniam, Y., & Cayirci, E. (2002, August). A survey on sensor networks. *Communications Magazine, IEEE, 40*(8), 102–114. doi:10.1109/MCOM.2002.1024422

Alliance, Z. (2010, December). *ZigBee Specification*. Retrieved December 16, 2010, from http://www.zigbee.org/Standards/ZigBeeSmartEnergy/Specification.aspx

Brown, S., & Sreenan, C. J. (2013). Software Updating in Wireless Sensor Networks: A Survey adn Lacunae. *Journal of Sensor and Actuator Networks, 2*, 717–760. doi:10.3390/jsan2040717

Crossbow Technologies, I. (2003, March). *Mote In-Network Programming User Reference.* Retrieved 11 13, 2009, from http://www.tinyos.net/tinyos-1.x/doc/Xnp.pdf

Deng, J., Han, R., & Mishra, S. (2006). Secure code distribution in dynamically programmable wireless sensor networks. In *Proceedings of Information Processing in Sensor Networks* (pp. 292–300). IPSN. doi:10.1145/1127777.1127822

Dong, W., Liu, Y., Chen, C., Bu, J., Huang, C., & Zhao, Z. (2013). R2: Incremental Reprogramming Using Relocatable Code in Networked Embedded Systems. *IEEE Transactions on Computers, 62*(9), 1837–1849. doi:10.1109/TC.2012.161

Dong, W., Mo, B., Huang, C., Yunhao, L., & Chen, C. (2013). R3: Optimizing relocatable code for efficient reprogramming in networked embedded systems. In *Proceedings of INFOCOM, 2013 Proceedings IEEE* (pp. 315-319). IEEE.

Dunkels, A., Finne, N., Eriksson, J., & Voigt, T. (2006). Run-Time Dynamic Linking for Reprogramming Wireless sensor Networks. In *Proceedings of the Fourth ACM Conference on Embedded Networked Sensor Systems (SenSys 2006)*. ACM.

Dunkels, A., Gronvall, B., & Voigt, T. (2004). Contiki - a lightweight and flexible operating system for tiny networked sensors. In *Proceedings of Local Computer Networks, 2004: 29th Annual IEEE International Conference on* (pp. 455-462). IEEE.

Hagedorn, A., Starobinski, D., & Trachtenberg, A. (2008). Rateless Deluge: Over-the-Air Programming of Wireless Sensor Networks Using Random Linear Codes. *International Conference on Information Processing in Sensor Networks, IPSN '08* (pp. 457-466). St. Louis, MO: IPSN.

Hill, J., Szewczyk, R., Woo, A., Hollar, S., Culler, D., & Pister, K. (2000). System architecture directions for networked sensors. *SIGPLAN Not., 35*(11), 93–104. doi:10.1145/356989.356998

Hui, J. W. (2005, July 28). *Deluge 2.0 - TinyOS Network Programming.* Retrieved November 13, 2009, from http://www.cs.berkeley.edu/~jwhui/deluge/deluge-manual.pdf

Hui, J. W., & Culler, D. (2004). The dynamic behavior of a data dissemination protocol for network programming at scale. In *Proceedings of the 2nd international conference on Embedded networked sensor systems* (pp. 81-94). New York: Academic Press.

Kulkarni, S., & Wang, L. (2005). MNP: Multihop Network Reprogramming Service for Sensor Networks. [IEEE.]. *Proceedings of Distributed Computing Systems, 2005*, 7–16.

Kuorilehto, M., Kohvakka, M., Suhonen, J., Hämäläinen, P., Hännikäinen, M., & Hämäläinen, T. D. (2007). Ultra-Low Energy Wireless Sensor Networks. In *Practice: Theory, Realization and Deployment*. John Wiley.

Langendoen, K., Baggio, A., & Visser, O. (2006). Murphy loves potatoes: experiences from a pilot sensor network deployment in precision agriculture. In *Proceedings of Parallel and Distributed Processing Symposium*. IPDPS.

Laukkarinen, T., Suhonen, J., & Hännikäinen, M. (2013). An embedded cloud design for internet-of-things. *International Journal of Distributed Sensor Networks, 13*.

Levis, P., & Culler, D. (2002, December). Maté: A Tiny Virtual Machine for Sensor Networks. *SIGOPS Oper. Syst. Rev., 36*(5), 85–95. doi:10.1145/635508.605407

Levis, P., Patel, N., Culler, D., & Shenker, S. (2004). Trickle: a self-regulating algorithm for code propagation and maintenance in wireless sensor networks. In *Proceedings of the 1st conference on Symposium on Networked Systems Design and Implementation*. USENIX Association.

Microchip Technology, I. (2008, February 10). *PIC18F8722 Product Page*. Retrieved December 7, 2009, from http://www.microchip.com/

Microchip Technology, I. (2009). *MPLAB C Compiler for PIC18 MCUs*. Retrieved December 13, 2009, from http://www.microchip.com/

Miller, C., & Poellabauer, C. (2008). PALER: A Reliable Transport Protocol for Code Distribution in Large Sensor Networks. In *Proceedings of Sensor, Mesh and Ad Hoc Communications and Networks, 2008: SECON '08. 5th Annual IEEE Communications Society Conference on* (pp. 206-214). IEEE.

Mukhtar, H., Kim, B. W., Kim, B. S., & Joo, S.-S. (2009). An efficient remote code update mechanism for Wireless Sensor Networks. In *Proceedings of Military Communications Conference*. IEEE.

Määttä, L., Suhonen, J., Laukkarinen, T., Hämäläinen, T., & Hännikäinen, M. (2010). Program image dissemination protocol for low-energy multihop wireless sensor networks. In Proceedings of System on Chip (SoC), 2010 International Symposium on (pp. 133-138). Tampere: IEEE.

Nordic Semiconductors. (2007, July). *nRF24L01 Product Specification*. Retrieved December 7, 2009, from http://www.nordicsemi.com/

Pant, R. K., & Bagchi, S. (2009). Hermes: Fast and Energy Efficient Incremental Code Updates for Wireless Sensor Networks. [Rio de Janeiro: IEEE.]. *Proceedings - IEEE INFOCOM, 2009,* 639–647.

Reijers, N., & Langendoen, K. (2003). Efficient code distribution in wireless sensor networks. In *Proceedings of the 2nd ACM international conference on Wireless sensor networks and applications (WSNA '03)* (pp. 60-67). New York: ACM.

Rossi, M., Bui, N., Zanca, G., Stabellini, L., Crepaldi, R., & Zorzi, M. (2012). SYNAPSE++: Code Dissemination in Wireless Sensor Networks Using Fountain Codes. *IEEE Transactions on Mobile Computing, 9*(12), 1749–1765. doi:10.1109/TMC.2010.109

Standards, I. E. E. E. (2008, August 31). *Part 15.4: Wireless Medium Access Control (MAC) and Physical Layer (PHY) Specifications for Low-Rate Wireless Personal Area Networks (WPANs)*. Retrieved December 16, 2010, from http://standards.ieee.org/getieee802/download/802.15.4a-2007.pdf

Stathopoulos, T., Heidemann, J., Estrin, D., & Sensing, C. U. (2003). *A remote code update mechanism for wireless sensor networks*. Citeseer.

Suhonen, J., Kuorilehto, M., Hännikäinen, M., & Hämäläinen, T. (2006). Cost-Aware Dynamic Routing Protocol for Wireless Sensor Networks - Design and Prototype Experiments. In *Proceedings of Personal, Indoor and Mobile Radio Communications, 2006 IEEE 17th International Symposium on* (pp. 1-5). IEEE.

Wang, Q., Zhu, Y., & Cheng, L. (2006, June). Reprogramming wireless sensor networks: Challenges and approaches. *Network, IEEE, 20*(3), 48–55. doi:10.1109/MNET.2006.1637932

Zhang, C., Yu, Q., Huang, X., & Yang, C. (2008). An RC4-Based Lightweight Security Protocol for Resource-constrained Communications. In *Proceedings of Computational Science and Engineering Workshops* (pp. 133–140). IEEE. doi:10.1109/CSEW.2008.28

KEY TERMS AND DEFINITIONS

Bootloader: A part of firmware that loads the actual program image into execution on embedded devices.

Dissemination Protocol: A method to disseminate data over a sensor network.

Embedded Networked System: A system consisting of two or more embedded devices, which communicate over a medium, such as radio or Internet.

Firmware: Software that runs on an embedded device.

Program Image: The part of the firmware that is updated in reprogramming. Often used as a synonym for firmware.

Reprogramming Protocol: A method to disseminate and program new program image to embedded device.

Sensor Node: An embedded device in a sensor network that consist of a processor, sensors/actuators, communication medium, and energy source.

Wireless Sensor Network: A wireless network of embedded device that autonomously communicate and deliver measurement data. Often resource constrained in terms of restricted computation, communication, and energy.

Chapter 5
Asymmetric Link Routing in Location-Aware Mobile Ad-Hoc Networks

Pramita Mitra
Ford Research and Innovation Center, USA

Christian Poellabauer
University of Notre Dame, USA

ABSTRACT

Recent experimental research has revealed that the link conditions in realistic wireless networks vary significantly from the ideal disk model, and a substantial percentage of links are asymmetric. Many existing geographic routing protocols fail to consider asymmetric links during neighbor discovery and thus discount a significant number of potentially stable routes with good one-way reliability. This chapter provides a detailed overview of a number of location-aware routing protocols that explicitly use asymmetric links in routing to obtain efficient and shorter (low latency) routes. An asymmetric link routing protocol, called Asymmetric Geographic Forwarding (A-GF) is discussed in detail. A-GF discovers asymmetric links in the network, evaluates them for stability (e.g., based on mobility), and uses them to improve the routing efficiency.

INTRODUCTION

Routing in mobile ad-hoc and sensor networks typically assumes that wireless links are bidirectional, i.e., wireless devices have identical transmission ranges. However, some recent empirical studies (Ganesan, Estrin, Woo, Culler, Krishnamachari & Wicker, 2002; Woo, Tong & Culler, 2003; Zamalloa & Krishnamachari, 2007; Zhao & Govidan, 2003) show that approximately 5-15% of the links

in a low-power wireless network are asymmetric, and an increasing distance between nodes also increases the likelihood of an asymmetric link. This trend is further exacerbated by the increasing use of power management techniques that may cause the nodes in a network to operate at different transmission ranges. Link asymmetry is also caused by node mobility, heterogeneous radio technologies, and irregularities in radio ranges and path and packet loss patterns. Based on these

DOI: 10.4018/978-1-4666-6034-2.ch005

observations, it is expected that link asymmetry will become more common in future wireless ad-hoc and sensor networks.

With increasing use of Global Positioning System (GPS) and many other (possibly less accurate but more resource-efficient) localization schemes, Geographic Forwarding (GF) is becoming an attractive choice for widely scalable routing in wireless ad-hoc and sensor networks. GF incurs very low overhead since no prior route discovery is required before forwarding the data packets. Many existing GF protocols are designed under the assumption of symmetric wireless links. That is, whenever a node receives a beacon packet from another node, it considers that node as its neighbor as it assumes the link is bi-directionally reachable. Such an assumption may not be realistic for practical wireless ad-hoc and sensor networks, since wireless links are often asymmetric. In Figure 1, node A discovers its neighbors B, C, and D by receiving beacons from them, which means that node A is within the wireless transmission ranges of all three nodes. Both node C and node D are also within the wireless transmission range of node A, so both the links A↔C and A↔D are symmetric. However, node B is outside node A's wireless transmission range and therefore, there is an asymmetric link B→A between these two

nodes. Note that while all wireless links are to some extent asymmetric (i.e., the signal strength being stronger in one direction than the other but the wireless link still being connected in both directions), A-GF focuses on maximally asymmetric links (perfect connectivity one direction, zero in the other), as shown in Figure 1.

Some geographic routing protocols (Couto, Aguayo, Bicket & Morris, 2003; Zhou, He, Krishnamurthy & Stankovic, 2004) detect asymmetric links and explicitly ignore them while making routing decisions. Eliminating asymmetric links discounts a substantial number of potentially stable routes with good one-way reliability and may lead to (1) routes that are longer than necessary and (2) no routes at all if there is not at least one fully symmetric path between the sender and receiver (Marina & Das, 2002; Sang, Arora & Zhang, 2007). Therefore, it is difficult to achieve high network connectivity, high data transmission rates, and low transmission latencies if the network has many asymmetric links. Furthermore, some recent empirical studies (Chen, Hao, Zhang, Chan & Ananda, 2009; Ganesan, Estrin, Woo, Culler, Krishnamachari & Wicker, 2002) show that asymmetric links tend to span longer distances than symmetric links. As a result, inclusion of asymmetric links in routing decisions can further improve the network performance. On the other hand, in a highly mobile setting, communications among mobile nodes may only be temporary and short-lived. For example, while the wireless connection between two cars driving in the same direction on a highway may last for a long time, the connection between cars driving in opposite directions will break very quickly. Therefore, it may be necessary to not only identify the presence of an asymmetric link, but to also measure or predict the time-varying quality or stability of such a link.

Through a survey of the key research approaches to asymmetric link routing in location-aware mobile ad-hoc networks, this chapter provides a detailed overview of the routing protocols that explicitly focus on exploiting asymmetric link for

Figure 1. Concept of asymmetric links

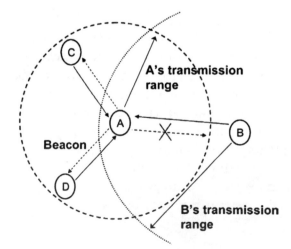

improving routing efficiency. This chapter also discusses in detail one such protocol called *Asymmetric Geographic Forwarding (A-GF)* (Mitra & Poellabauer, 2011), which discovers and exploits asymmetric links in a network, i.e., instead of eliminating these links, A-GF uses them for routing to reduce the routing hop count (and latency) and to increase the routing reliability when there are no symmetric paths available. A-GF also proactively monitors the changes in conditions of discovered links and ranks neighbors based on perceived link stability. During routing, a node considers both this ranking and the progress a neighbor can make towards the sink location in terms of geographical distance (thereby considering both reliability and performance). The remainder of this chapter is organized as follows: Section 2 describes a study of the origins and characteristics of link asymmetry in wireless ad-hoc environments. Section 3 provides a detailed overview of a number of routing protocols that explicitly discover and use asymmetric links in routing. Section 4 provides the details of A-GF. Finally, Section 5 concludes the chapter.

CHARACTERISTICS OF ASYMMETRIC LINKS IN WIRELESS NETWORKS

In conventional routing protocols, whenever a node receives a beacon packet from another node, it considers that node as its neighbor and the link as bi-directional. Such an assumption may not be realistic for practical wireless sensor networks, since wireless links are often asymmetric. The varying transmission range of wireless devices is the primary reason for link asymmetry and can be attributed to these following three factors:

1. The increasing heterogeneity in mobile ad-hoc networks, e.g., networks consisting of different types of wireless devices, where these devices have different radio capabilities.

2. Transmission ranges are also affected by the transmit power used by a wireless device, which is dynamically varied by resource management protocols in low-power networks.

3. Irregularities in path and packet loss patterns and varying noise floors of wireless devices, caused by distance, interferences and obstacles, and hardware variability, respectively, lead to varying transmission ranges (Cerpa, Wong, Kuang, Miodrag & Estrin, 2005; Reijers, Halkes & Langendoen, 2004; Srinivasan, Dutta, Tavakoli & Levis, 2006).

This section presents an empirical study of how some of these factors contribute to link asymmetry in wireless ad-hoc and sensor networks.

Heterogeneous Radios

Simulations (performed in the JiST/SWANS[1] simulator) are used to study the impact of heterogeneous radio technologies on link asymmetry (Mitra & Poellabauer, 2011). The mobile nodes in the network have heterogeneous radio capabilities as a result of varying transmit powers, radio reception sensitivities, and radio reception thresholds of the nodes. 4 different types of radios, each with different transmission range, were used in the simulations. The number of nodes was varied from 25 to 200 in steps of 25. The total area of the network is scaled in proportion so that the node density remains unchanged.

Figure 2 shows the percentage of symmetric and asymmetric links in the network with varying number of nodes. The percentage of asymmetric links in the network increases with increasing numbers of nodes with heterogeneous radio capabilities. For example, in a mobile network of size 75, the number of asymmetric links is 11.33% of the total number of links in the network. The number of asymmetric links increases to 24.79% of the total number of links when there are 200 nodes with heterogeneous radio capabilities in

Figure 2. Percentage of symmetric and asymmetric links vs. network size

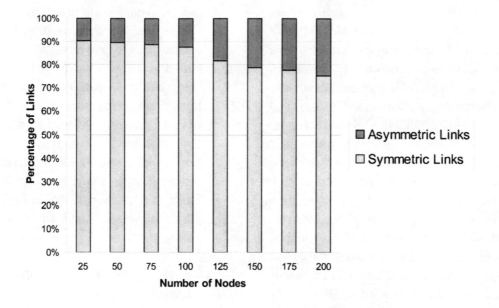

the network. Nodes with heterogeneous radio capabilities have different transmission ranges, which results in non-uniform wireless connectivity over both directions of a wireless link, thus causing asymmetric links in the network. Note that existing symmetric routing protocols, e.g., Greedy Perimeter Stateless Routing (GPSR) (Karp & Kung, 2000), do not consider asymmetric links, and therefore, discover only 75-90% of the total number of links in the network, which in turn may limit the data transmission success of the symmetric routing protocols.

Varying Transmission Power and Path Loss

Varying transmission power and path loss are two major reasons of varying transmission ranges leading to link asymmetry. Outdoors field tests are conducted to study the impact of transmit power and path loss variance on link asymmetry (Mitra & Poellabauer, 2011). The testbed consists of 2 Crossbow Stargate[2] wireless devices, placed 60 centimeters above the ground. The Stargate devices have an ARM processor running Linux, and

each device is equipped with an 802.11b compact flash card, based on the PrismII chipset. The tests are conducted in an open yard at the University of Notre Dame (Figure 3) during class hours. In Figure 3, the left-most dot represents the location of the static receiver device and the other dots represent the locations of the mobile sender device. The sender device was initially placed at a distance of 20 meters from the receiver device and then the distance was varied up to 140 meters along a straight line (in steps of 20 meters). At each step, the mobile sender device used 5 different transmit power levels, and broadcasted 200 beacon packets at the rate of 10 packets/second to the receiver device.

The field tests results are used to study the effects of varying transmit power and distance (path loss falls off by square of distance) on the successful packet transmission rate of the sender device. Figure 4 shows the corresponding successful packet transmission rate of the sender device with varying transmit power and distance. The figure shows that the successful packet transmission rate drops with increasing distance and decreasing transmit power levels. Table 1 provides the

Figure 3. Outdoors field tests with two Crossbow Stargate devices

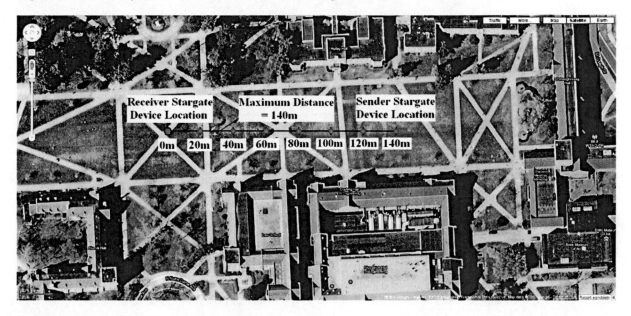

Figure 4. Successful packet transmission rate with varying distance and transmit power

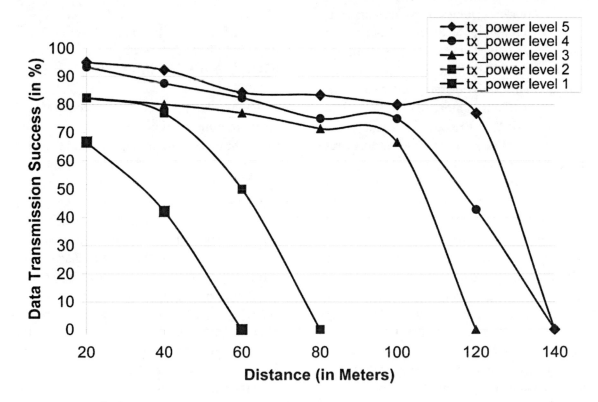

Table 1. Measured RSSI and Noise Floor values with varying distances and transmit powers

Transmit Power	Metric	Distance					
		20m	40m	60m	80m	100m	120m
Level 5	RSSI	45.63	40.87	40.73	40.63	36.32	33.48
	Noise	30.8	31.36	32.2	32.23	33.18	34.35
Level 4	RSSI	44.00	43.31	37.11	37.38	34.12	32.27
	Noise	30.69	31.14	31.3	31.53	31.69	32.35
Level 3	RSSI	43.66	42.43	39.44	34.42	33.56	30.17
	Noise	29.89	29.92	30.97	32.13	32.4	34.21
Level 2	RSSI	41.41	39.04	37.38	31.75	30.20	29.93
	Noise	29.86	29.92	30.6	30.67	30.78	32.5
Level 1	RSSI	37.78	35.44	33.86	32.67	32.13	32.08
	Noise	28.93	30.19	30.79	31.25	32.18	32.2

Received Signal Strength Indication (RSSI) and noise floor values measured at the receiver device for the corresponding data transmission from the sender device. It can be observed from Table 1 that (1) the measured RSSI values drop with increasing distance, corroborating the previous observation that the successful packet transmission rate drops with increasing distance, and (2) when the successful packet transmission rate drops to zero, the measured noise floor is more than or very close to the measured RSSI value, i.e., no meaningful data reception was possible because the noise was stronger than the received signal.

The above outdoors experiment shows that the successful packet transmission rate drops with increasing distance and decreasing transmit power levels. In other words, the data transmission quality (a good measure of connectivity) over a certain wireless link decreases with increasing distance and decreasing transmit power. In mobile ad-hoc networks, the transmit powers of the nodes may change over time when the nodes' resource managers attempt to preserve energy. In mobile ad-hoc networks, the distances between the nodes will also change frequently. As a result, there will be changes in wireless connectivity over a certain direction (e.g., when one node changes

it's transmit power) or both directions (e.g., when a node moves) of a wireless link, thereby leading to asymmetric links.

ASYMMETRIC LINK ROUTING IN LOCATION-AWARE MOBILE ENVIRONMENTS

Geographic routing protocols are becoming highly popular routing solutions for mobile ad-hoc networks. The underlying principle used in these protocols involves selecting the optimal next routing hop from the set of neighbors of the forwarding node – usually the neighbor that is geographically closest to the destination is selected. Since the forwarding decision is based entirely on local knowledge, it obviates the need to discover and maintain routes for each destination. By virtue of these characteristics, geographic routing protocols are highly scalable and particularly robust to frequent changes in the network topology. Furthermore, since the forwarding decision is made on the fly, each node always selects the optimal next hop based on the most current topology. However, historically, many existing geographic routing protocols have failed to consider asymmetric links during neighbor discovery and thus discount a significant number of potentially stable routes with good one-way reliability. On the other hand, many recent research efforts have focused on explicitly finding and using asymmetric links for improved routing efficiency. To that end, this section analyses and characterizes a few key asymmetric link routing protocols along three axes: (1) asymmetric link discovery, (2) asymmetric link monitoring and management, and (3) asymmetric link routing and acknowledgment.

Asymmetric Link Discovery

The most common approach used to detect asymmetric links is *exchanging neighbor lists* as part of the Hello messages that are periodically exchanged among one-hop neighbors (Marina & Das, 2002; Zhou, He, Krishnamurthy & Stankovic, 2006). By performing a simple look-up for itself in the received neighbor list, each node can deduce whether the wireless link connecting them is symmetric (i.e., the node is included in the neighbor list) or asymmetric (i.e., the node is absent in the neighbor list). Chen, Hao, Zhang, Chan and Ananda (2009) use unicast beacon packets for the detection of asymmetric links. When a node receives a MAC layer notification of failed delivery of these unicast beacon packets, the neighbor locally broadcasts a special beacon packet to announce the unidirectional link. While exchanging neighbor lists is definitely the easiest and most effective solution to asymmetric link discovery, this approach could result in huge bandwidth overheads in dense networks where each node has a fairly large amount of neighbors.

On the other hand, some routing protocols *exchange transmission range* to discover asymmetric links in the network. COMPOW (Narayanaswamy, Kawadia, Sreenivas, & Kumar, 2002) calculates a common transmission range that will be used by all the nodes in a heterogeneous environment. This common transmission range is used to reduce the interference, prune asymmetric links and ensure symmetric connectivity between nodes. In the EUDA (Ko, Lee, & Lee, 2004) routing protocol, parameters such as transmission range, noise level and minimum signal-to-noise ratio (SNR) are included in the Hello messages. When a node B receives a Hello message from a node A, it compares its highest transmission range to estimate the distance between itself and node A. If the value of the estimated distance from node B to A is larger than the transmission range of node B, node B considers the radio link to A as asymmetric. The TRIF (Le, Sinha, & Xuan, 2010) protocol, which is used jointly with RREQ/RREP-based routing protocols, assumes that the transmission range is adjustable. Hence, each successive RREQ is sent with decremented transmission range level. The source node adds in the header of the RREQ, the transmission range level used when sending this request. If the power level used to send the RREQ is higher than the power level available at the receiver node, then the receiving node concludes that it has received this request via an asymmetric link. The main drawback of EUDA and TRIF is that the estimation of the distance based on SNR is not a good metric to evaluate proximity. Both protocols are also limited in their assumption of varying transmission range to be the only source of link asymmetry.

Common neighbor-based feedback mechanisms are also widely used for asymmetric link discovery and notification. Sang, Arora and Zhang (2007) proposed a novel neighbor discovery technique based on a new one-way link metric to identify high-reliability forward asymmetric links and present a local procedure for their estimation. They employ the help of mutual witness neighbor nodes for asymmetric link discovery. To that end, when a node realizes that it has two neighbors connected by an asymmetric link it volunteers itself to relay the link discovery and maintenance information to both neighbors. Such mechanism can result in duplicated control messages when more than one neighbor volunteers themselves. Some suppression techniques can be used to reduce such duplications but the performance of the protocol will rely on the efficiency of those techniques. In Discover and Exploit Asymmetric Links (DEAL) protocol proposed by Chen, Hao, Zhang, Chan, and Ananda (2009), data link layer beacon messages are used for finding asymmetric links. DEAL uses a feedback scheme called Source-Specified Relay (SSR). SSR uses local knowledge at link layer to find the relay nodes for asymmetric link feedback over the poor direction of asymmetric links. In both these protocols the common neighbor changes very frequently with changes in network topology in mobile environments, which may lead to many failed transmissions and large network traffic overheads.

All of these routing protocols employ advanced techniques to discover asymmetric links in the network. However, a few of them treat these asymmetric links as an anomaly and neglect them while making routing decisions. Such schemes scale well when the number of asymmetric links in the network is small. However, when the number of asymmetric links is large and the network still remains a connected graph, the routing protocols that do not utilize asymmetric links perceive the network as an unconnected graph. As a result, the scalability and performance of such schemes decrease drastically with the increase in the number of asymmetric links in the network.

Asymmetric Link Monitoring And Management

Due to node mobility or dynamic adjustments in transmission ranges, symmetric links can become asymmetric links, or asymmetric links can turn into broken links. Therefore, it is also important to monitor the condition of the discovered links and evaluate the reliability of each link, in order to ensure successful data transmission.

Du, Shi and Sha (2005) proposed the Neighborhood Link Quality Service (NLQS) protocol and the Link Relay Service (LRS), to monitor the *timeliness link quality* with neighbors and

build a relay framework to alleviate effects of link asymmetry. They leverage the Window Mean Exponentially Weighted Moving Average Estimator (WMEWMA) to estimate the Packet Reception Rate (PRR) between neighbors. Each node periodically broadcasts control packets, which contain node ID and packet ID. Other nodes overhear the channel for the control packets to measure and estimate the link quality from their neighbors. WMEWMA uses a time window to observe the received packets and adjusts the estimation result using latest average value of the measured PRR. When nodes have estimated PRR for their inbound neighbors (i.e., nodes that you can hear from) they will send the estimated results back to those inbound neighbors. When their inbound neighbors get the estimated results from their outbound neighbors (i.e., nodes that can hear from you), they can identify those outbound neighbors and build the outbound neighbors table with the estimated results. As we have discussed before, due to the link asymmetry, nodes may not directly receive packets from their outbound neighbors. They will get the estimated PRR to their outbound neighbors from LRS over a short unicast routing path.

Gondalia and Kathiriya (2005) proposed Probabilistic Geographic Routing Protocol (PEOGR) that *combines link estimation with asymmetric link discovery* at the beginning of the deployment. The discovery time is decided at the deployment. The longer the discovery period is, the better the initial link estimation is. During the discovery phase each node sends "Hello" messages that contain the geographic location of the node, its NLV (New Link Value), and a list of its neighbors with the associated NLVs. At the end of the discovery phase, every node has a list of all the neighbors it can hear from with the associated NLV and evaluation. Lets assume that the size of the neighbor table of node i is N. At the end of the discovery phase, node i picks N of its neighbors with the highest NLVs. After the initial discovery phase, the nodes enter the maintenance phase during which each node continues to send "Hello" messages as before. The frequency of the beaconing can be decreased in the maintenance phase to reduce the energy consumption and the protocol overhead. In order to keep the neighbor table up-to-date, the nodes refresh their neighbor table every T seconds. When the NLV of a neighbor decreases over time, it gets replaced by a new neighbor with higher NLV.

Given the high cost associated with transmitting periodic "Hello" messages (e.g., energy and bandwidth consumption, packet collision and retransmission, etc.), it makes sense to adapt the frequency of beacon updates to the node mobility and the traffic conditions within the network. To that end, if certain nodes are frequently changing their mobility characteristics (speed, direction of movement, etc.), it makes sense to frequently broadcast their updated locations. Furthermore, if only a small percentage of the nodes are involved in forwarding packets, it is unnecessary for nodes which are located far away from the forwarding path to engage in frequent periodic beaconing because these updates are not useful for forwarding the current traffic. Heissenbuttel, Braun, Walchli and Bernoulli (2007) have proved that that the outdated entries in the neighbor table are the major source of performance degradation in highly mobile environments. They proposed three simple optimization techniques for *adapting beacon interval to node mobility and network traffic load*, namely distance-based beaconing, speed-based beaconing and reactive beaconing. In the distance-based beaconing, a node transmits a beacon when it has moved a given distance d. The node removes an outdated neighbor if the node has not received any beacon from the neighbor while the node has moved more than k-times the distance d, or after a maximum timeout period has elapsed. This approach therefore is adaptive to the node mobility, e.g., a faster moving node sends beacons more frequently. However, this approach has two problems, i.e., (1) a slow node may have many outdated neighbors in its neighbor list since

the neighbor timeout interval at the slow node is longer, and (2) when a fast moving node passes by a slow node, the fast node may not detect the slow node due the infrequent beaconing of the slow node. In speed-based beaconing, the beacon interval is determined from a predefined range with the exact value being inversely proportional to its speed. The neighbor timeout interval of a node is a multiple k of its beacon interval. Nodes piggyback their neighbor timeout interval in the beacons. A neighbor receiving the beacon compares the received timeout interval with its own timeout interval and selects the smaller one as the timeout interval for this neighbor. In this way, a slow node can have short timeout interval for its fast moving neighbor and therefore reduce the number of updated neighbors in its neighbor list. However, the speed-based beaconing still suffers from the problem that a fast node may not detect the slow nodes. In reactive beaconing, the beacon generation is triggered by data packet transmissions. When a node has a packet to transmit, the node first broadcasts a beacon request packet. The neighbors overhearing the request packet respond with beacons. Thus, the node can build an accurate local topology before the data transmission. However, this process is initiated prior to each data transmission, which can lead to high number of beacon broadcasts and increased data transmission latency, particularly when the traffic load in the network is high. Chen, Kanhere and Hassan (2013) proposed a novel beaconing strategy for called *Adaptive Position Updates* (APU). APU incorporates two rules for triggering the beacon update process. The first rule, referred as Mobility Prediction (MP), uses a simple mobility prediction scheme to estimate when the location communicated in the previous beacon will become inaccurate - the next beacon is broadcasted only if the predicted error in the location estimate is greater than a certain threshold, thus tuning the update frequency to node mobility. The second rule, referred as On-Demand Learning (ODL), aims at improving the accuracy of the knowledge of local topology along the routing paths. ODL uses an on-demand learning strategy, whereby a node broadcasts beacons when it overhears the transmission of a data packet from a new neighbor in its vicinity. This ensures that nodes involved in forwarding data packets maintain a more up-to-date view of the local topology. On the contrary, nodes that are not in the vicinity of the forwarding path are unaffected by this rule and do not broadcast beacons very frequently.

Asymmetric Link Routing and Acknowledgment

Asymmetric link routing protocols in location-aware networks employ a variety of metrics to find the best next hop node. On the other hand, data acknowledgment from the sink to the source node along the reverse routing path is not trivial because of the unidirectional nature of asymmetric links. Romdhani, Barthel, and Valois (2012) proposed a *contention-based asymmetric routing and acknowledgment scheme* in which during routing each node along the routing path is ranked based on their distance to the sink node – the node closest to the sink node has the smallest rank. When a sensor node has data to send to the sink node, it broadcasts the data message in its neighborhood. In the header of this data message, the sender node, adds its ID, its rank and its neighborhood table. Each sender node starts a timer, called timeout_relayed, during which it verifies if its message is relayed. If the timer expired and the sender node is not informed that its message was relayed by another node, it reattempts a second time to send its data message to the sink node. When receiving the data message broadcast from a source node, each of the receiving neighbors will first verify if it is closer to the sink node by comparing its rank with that of the sender node. If it is farther than the sender from the sink, then that node is not a candidate to relay that message and it will drop the received message. On the other hand, if it is a candidate to relay this message, the node

computes a timer called timeout_to_relay and enters in a contention phase. The objective of this contention is to favor the node closest to the sink node – to that end, if a candidate node detects that the message was forwarded by another node, the contention phase is ended. If the timer expired and no other node has forwarded the message, then this node proceeds to forward the data message. It first verifies whether the link between itself and the sender node is symmetric – relaying the data message will serve as an implicit acknowledgment message to the source node connected with a symmetric link. Else, if the candidate node deduces that the link between the sender node and itself is an asymmetric link, it attempts to find in its neighbor table a common neighbor with the sender node. If such a node exists, the candidate node forwards the data message and sends an explicit acknowledgment message to the common neighbor. The latter forwards this acknowledgment until it reaches the source node. On the other hand, if no such common node exists, the candidate node attempts to find in its neighbor table whether any of its neighbors, called Inter, satisfies the two conditions: (1) one of the neighbors heard by the source node can receive the message sent by this Inter node, and (2) the Inter node has a symmetric link with the candidate node. If such an Inter node is found, the candidate node forwards the data message and sends an explicit acknowledgment message to the Inter node, which will then forward that acknowledgment ultimately to be received by the source node.

Gade (2013) proposed a piezoelectric-based energy-harvesting sensor network in which variable transmission power levels caused by non-uniform energy-harvesting rates often lead to asymmetric links. They proposed a *scheme that combines weighted and lazy acknowledgment mechanisms*, in order to efficiently route sensor data messages to a sink over asymmetric links. To that end, this protocol uses multipath routing in order to improve message delivery in the presence of time-varying asymmetric links. As a result, the sink may receive more than one data messages from the source. Upon receiving the data message, the sink checks whether it already has received the message – it accepts up to three duplicated data messages received along three different routing paths and replies with acknowledgment messages along those paths to the data source. Whenever the sink replies an Ack, it piggybacks an incremental factor (initially set by 1) into the Ack, called Weighted Ack (WACK). Each node along the reverse routing path maintains a ratio of number of delivered messages to the sink (d) to the number of forwarded messages (f), DF = d/f. During routing when a node forwards a data message, it increments the number of forwarded messages. On the other hand, during the acknowledgment phase when the node receives an Ack back from the sink through multi-hop relaying, it increments the number of delivered messages. Note that the DF indicates a historical statistics of routing. If a node has higher DF, it was frequently involved in successfully delivering data messages to the sink. The proposed protocol multiplicatively adjusts the incremental factor to clearly see the effect of the WACK on the performance. Whenever the sink repetitively replies an Ack for the same data message, it reduces the incremental factor, i.e., 1, 0.6, and 0.4. Higher weight is assigned to the Ack for data message arriving earlier because the route could be either the shortest path or less congested path. When a node receives the Ack, it adds the piggybacked incremental factor to the current number of delivered messages, thus resulting in increasing the DF with different rates. Furthermore, whenever a node receives a data message, it executes an asymmetric link aware back-off mechanism in order to avoid possible data contention and collision -- a node with higher DF has lower back-off period. Additionally, a node operating with higher transmission range and thus most likely with higher number of neighbors has lower back-off period because of its potential for reducing the transmission latency.

Node mobility affects the lifetime of asymmetric links and hence some new routing protocols (Morgenthaler, Braun, Zhao, Staub & Anwander, 2012; Zhao, Rosario, Braun, Cerqueira, Xu & Huang, 2013; Arockia & Anbumani, 2013) use *mobility prediction schemes* to evaluate asymmetric links during routing. Morgenthaler, Braun, Zhao, Staub and Anwander (2012) proposed an Unmanned Aerial Vehicle Network (UAVNet), which is a mobile ad-hoc network where the mobility of UAVs is not random. To that end, the movements of UAVs are coordinated and they follow certain steering rules. As a consequence of these non-random mobility characteristics, UAVNets perform special movement behaviors. They employ a mobility prediction algorithm that estimates of the validity time of a link between two connected UAVs, and this estimation is used in the routing decision. To that end, a Link Validity Estimation (LIVE) protocol runs at each node to estimate the validity time (TLV) of each link with its 1-hop neighbors. This value is used to decide how long the unicast transmission will last. When this link validity timer expires, the sender starts another broadcast process to find a better forwarding node. Arockia and Anbumani (2013) argued that the performance of location-aware routing protocols in presence of mobility-induced location errors is not well understood. Greedy forwarding mode always forwards a packet to the neighbor that is located closest to the destination. However, due to node mobility the selected node may not exist within the radio range while it is listed as a neighbor. This situation is defined as lost link (LLNK) problem. A linear neighbor location prediction scheme is introduced as a solution to the LLNK problem. To avoid the bad next hop node selection that may result in LLNK problems, the current locations of neighbor nodes are estimated at the moment of routing. The proposed linear location prediction scheme is simple and it performs well when the beacon interval is reasonably small. The latest location and transmission range of each node are incorporated in the prediction scheme to account for asymmetric links. They assume each node knows (or estimates) it's approximate radio range and would not forward a packet to a neighbor node that is currently located outside of its range based on the estimated position.

ASYMMETRIC GEOGRAPHIC FORWARDING (A-GF)

This section describes the A-GF protocol in details. We assume that each node knows its own geographic location and that every node in the network can be fully mobile. GF's Perimeter Face Routing technique to handle networks voids or dead ends is left unchanged.

Detection of Asymmetric Links

In GF, each node broadcasts periodic Hello or beacon messages to its one-hop neighbors, containing the node's *node ID* and *location*. The exchange of periodic Hello messages helps to (1) discover new nodes that moved in the one-hop neighborhood; (2) update the neighbor list with the latest locations of known neighbors, and (3) identify and remove old neighbors that are have failed or moved out of range from the neighbor list.

A-GF modifies the Hello message approach of GF to discover asymmetric links in the network. Toward that end, in A-GF, each node augments the Hello messages with the list of neighbors it received beacons from previously. In Figure 5(a), when node B gets a periodic Hello packet from node A, B checks if it is recorded in the neighbor list in the received packet. If B is listed in A's neighbor list, then both A and B can hear each other and the link A↔B is symmetric. If B is not listed in A's neighbor list, B knows that the link A → B is asymmetric. We define A as the *up-link node* and B as the *down-link* node. In general, an asymmetric link is usable in the direction from the up-link node to the down-link node.

Notifying the Up-Link Node

At this point, only B is aware of the asymmetric link A→B, but A would not be able to exploit it. Sending a direct (single hop) report from B to A is not possible; therefore, A-GF finds a multi-hop *tunneling path* from the down-link node to the up-link node to notify the up-link node of the existence of the outgoing asymmetric link. It is reasonable to expect that the tunneling path will be built around the asymmetric link and thus would be short in length.

In Figure 5(b), the down-link node B routes a small control message, called *Asymmetric Notification (AN)*, towards the location of the

Figure 5. Asymmetric link discovery in A-GF

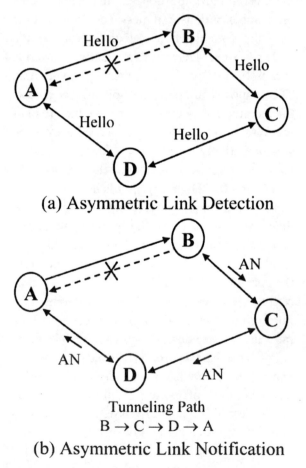

(a) Asymmetric Link Detection

Tunneling Path
B → C → D → A

(b) Asymmetric Link Notification

up-link node A. In each step, the AN message is forwarded to the neighbor that minimizes the geographic distance to node A. If no neighbor is closer to the up-link node than the down-link node, then face routing helps to recover from that situation and find a path to another node, where greedy forwarding can be resumed (Bose, Morin, Stojmenovic & Urrutia, 1999). In Figure 5(b), the AN message is routed to node A along the tunneling path B→C→D→A. When node A receives the AN message, it realizes the existence of the asymmetric link A→B, and puts node B in its neighbor table while marking the link as asymmetric.

Proactive Link Condition Monitoring

Due to frequent changes in the topology of mobile ad-hoc/sensor networks, (1) symmetric links can become asymmetric links, and (2) asymmetric links can turn into broken links, as discussed below.

Case 1: A symmetric link connecting node A and node B (A↔B) turns into an asymmetric link if node B moves out of node A's wireless transmission range, but node A is still reachable from node B (Figure 6). In this case, the asymmetric path containing the link B→A is still usable in the forward direction.
Case 2: An asymmetric link connecting node A and node B in the forward direction (A→B) turns into a broken link if node B moves out

of node A's wireless transmission range. Node A was already out of node B's wireless transmission range due to link asymmetry as shown in Figure 6.

Therefore, it is important to monitor the condition of the discovered links to account for these changes, and thereby ensure successful data transmission. Toward that end, A-GF *proactively monitors* the changes in conditions of discovered links and ranks neighbors based on perceived link stability. This proactive strategy, called *Proactive Link Condition Monitoring (PLCM)*, is discussed in detail below.

Propagation of Implicit Hello Messages

Neighbors connected via symmetric links receive periodic Hello messages from each other, which allow them to refresh their neighbor tables with the most recent locations of their known neighbors and to identify and remove "old" neighbors, i.e., neighbors that have failed or moved out of range. In case of an asymmetric link, the down-link node receives the periodic one-hop Hello messages from its up-link node. However, it is not possible for the up-link node to receive direct Hello messages from the down-link node because the asymmetric link is not usable from the down-link node to the up-link node. On the other hand, the down-link node could send periodic multi-hop Hello messages to the up-link node, along the same tunneling

Figure 6. As node B moves away from node A, the symmetric link A B in (a) turns into an asymmetric link A B in (b), and then into a broken link in (c)

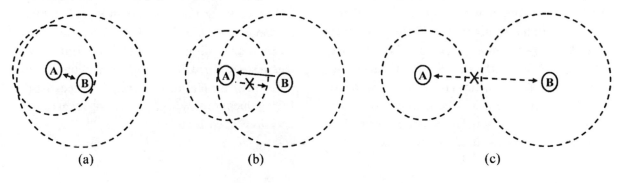

(a)   (b)   (c)

path that was used for routing the AN message in the asymmetric link notification phase (Section 4.2). However, sending periodic Hello messages over multiple hops for each asymmetric link in the network would not be a suitable solution for bandwidth constraint ad-hoc networks.

A-GF uses a bandwidth-efficient technique for the propagation of indirect Hello messages from the down-link node to the up-link node. During the periodic exchange of periodic one-hop Hello messages, an *Annotation* is inserted in the header of Hello messages sent by the down-link node to its neighbors. The Annotation contains the latest location of the down-link node and a confirmation that the down-link node received the latest periodic one-hop Hello message from the up-link node. The down-link node requests one of its one-hop neighbors to forward the Annotation to the up-link node. The neighbor which minimizes the geographic distance to the up-link node is requested for forwarding the Annotation. The requested neighbor inserts the Annotation in its next one-hop Hello message and repeats the same algorithm, and ultimately the Annotation is received by the up-link node. Thus the Annotation serves as an implicit multi-hop Hello message from the down-link node to the up-link node. The A-GF algorithm consumes very little extra bandwidth for propagating the Annotation from down-link node to up-link node because no actual messages are sent but only the size of the Hello messages increases a little due to the embedded Annotation. In this scheme, the maximum latency of the Annotation is bound by the upper limit of $n \times t_{HI}$, where n is the number of nodes forwarding the Annotation and t_{HI} is the periodic Hello message interval in the network. Note that during the forwarding of the Annotation from the down-link node to the up-link node, if no neighbor is closer to the up-link node, then a neighbor is randomly selected with the hope that a node would be found soon that would be closer to the up-link node. Such an assumption is reasonable because the up-link node is a neighbor of the down-link node and in a connected network the implicit path used for forwarding the Annotation from the down-link node to the up-link node should be short.

In Figure 7(a), the down-link node B puts an Annotation for node A and an *Annotation Forwarding Request (ANF-REQ)* for neighbor C in its periodic Hello messages. Similarly, node C inserts the Annotation in its next periodic Hello messages, with an ANF-REQ for neighbor D (Figure 7(b)). Finally, node A receives the Annotation from node D (Figure 7(c)), and updates the LCR value for the down-link node B in its neighbor table.

Monitoring of Link Conditions

A-GF uses the one-hop Hello messages to proactively monitor the changes in conditions of discovered symmetric links. Similarly, the implicit multi-hop Annotations are used by the up-link node, associated with an asymmetric link, to monitor the changes in condition of the down-link node. If subsequent Hello messages (or Annotations in case of asymmetric links) are received periodically from a neighbor node, the associated link is perceived to be stable. On the other hand, if a node misses subsequent Hello messages (or Annotations in case of asymmetric links) from a neighbor, the associated link is perceived to be less stable. Each node rates its one-hop neighbors based on this perceived link stability. This rating is called *Link Condition Rating (LCR)*. Note that this rating is personalized that means each node uses a self-policing evaluation of other nodes in its one-hop neighborhood, rather than a global evaluation for each node in the network. Thus it is possible for different nodes to have different LCR values for the same neighbor. We adopt a personalized evaluation scheme because (1) in a decentralized ad-hoc network, there is no centralized resource to maintain a global evaluation, and (2) a global evaluation is not required because the LCR values are only used for making local routing decisions at each node.

Figure 7. Annotation method for propagation of indirect Hello messages

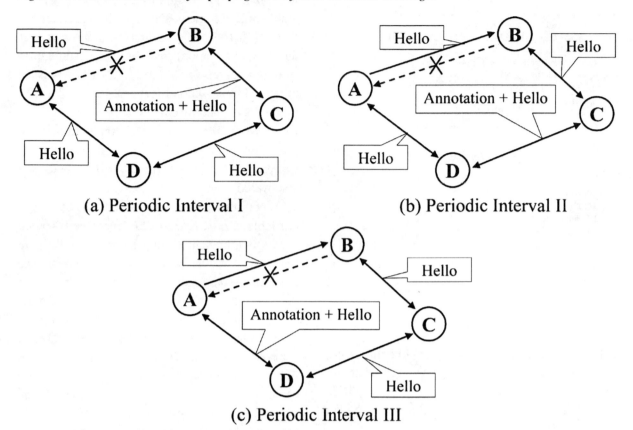

(a) Periodic Interval I

(b) Periodic Interval II

(c) Periodic Interval III

Each node runs a background daemon that periodically updates the LCR values in the neighbor table. Algorithm 1 illustrates the *update_LCR_value* method. If subsequent Hello messages (or Annotations in case of asymmetric links) are received periodically from a neighbor node, its LCR value is incremented by a step size Δ_{lcr}. On the other hand, if a node misses subsequent Hello messages (or Annotations) from a neighbor, the LCR value is decremented by the same step size (i.e., Δ_{lcr}). The maximum possible LCR value for a neighbor is *MAX_LCR*. A neighbor entry is deleted from the neighbor table when its LCR value falls to or below zero. For a lossy connection, the LCR value is aggressively decreased so that it quickly falls down to zero for broken links. On the other hand, the LCR value is aggressively increased for a good connection and quickly reaches to MAX_LCR for stable links.

The timeouts are different for symmetric and asymmetric links – toward that end, the timeout for any type of link (i.e., symmetric or asymmetric) is equal to the interval at which an update message is expected to be received from the node on the other end of the link. However, the interval at which direct one-hop Hello messages are received over a symmetric link and the interval at which implicit Hello messages in form of Annotation are received over an implicit multi-hop path that abstracts an asymmetric link, are not equal. In fact, the implicit Hello messages in form of Annotation take much more time to travel the implicit multi-hop path than the direct one-hop Hello messages. Therefore, it is important to have different timeouts for symmetric and asymmetric links, in order to ensure fair rating of the link conditions. The timeout for a symmetric link is equal to $(t_{HI} + t_{GP})$ where t_{HI} is the periodic Hello message interval and t_{GP} is a

Algorithm 1. Update_LCR_value

```
Require: Neighbor Entry Variable ne
Ensure: the LCR value of the neighbor entry is made
up-to-date
    1. if (is_new_neighbor (ne)) then
    2. ne.lcr = MIN_LCR
    3. add_new_neighbor (ne)
    4. return
    5. else
    6. if (is_timedout_update_interval (ne)) then

    7. ne.lcr = ne.lcr - Δ_lcr
    8. elseif (ne.lcr < MAX_LCR) then

    9. ne.lcr = ne.lcr + Δ_lcr
    10. endif
    11. if (ne.lcr ≤ 0) then
    12. delete_neighbor (ne)
    13. initiate_link_recovery
    14. endif
    15. endif
    16. if (ne != NULL) then
    17. reset_timeout_neighbor (ne)
18. endif
```

certain grace period. Similarly, the timeout for an asymmetric link is equal to $(n \times t_{HI} + t_{GP})$ where n is the number of nodes that forwarded the implicit multi-hop Annotation from the down-link node to the up-link node.

A typical entry in the neighbor table maintained at each node has the format shown in Table 2. Each entry consists of five fields: (1) the node ID of the neighbor, (2) the latest location of the neighbor in terms of (x, y) coordinates, (3) the type of the link, i.e., symmetric or asymmetric, (4) the timestamp value stating when this entry was last updated, and (5) the LCR value assigned to the neighbor. Let us assume that the example shown in Table 2 is an entry in the neighbor table maintained at node A. The first field of the entry tells that the node ID of the neighbor associated with this entry is B. The second field tells that latest location of B

is (5, 10). The third and fourth fields tell us that neighbor B is connected to A with an asymmetric link A→B (i.e., A is the up-link node and B is the down-link node) and the latest Annotation from B was received at time 13:23:40 (expressed in hour-minute-second format). Lastly, the fifth field tells us that the current LCR value for node B is 3 which means at least 3 Annotations from down-link node B were received since this entry was created (i.e., the asymmetric link A→B was discovered).

When a neighbor is deleted from the neighbor table, a *Link Failure (LF)* message is routed to that neighbor, notifying it that the link is no longer usable. On receiving an LF message, the reaction by the receiver depends on the type of the associated link. If the link was originally an asymmetric link, then the neighbor sending the LF message is deleted from the neighbor table of the receiver of the LF message. However, if the link was originally a symmetric link then there can be on of the following two situations: (1) the receiver of the LF message is still receiving Hello messages from the sender of the LF message. In that case, the previously symmetric link has now turned into an asymmetric link in the direction from the sender of the LF message to the receiver of the LF message. Therefore, the receiver of the LF message responds by sending an AN message to notify the other node of the newly formed asymmetric link. (2) The receiver of the LF message is missing Hello messages from the sender of the LF message. In that case, the symmetric link is now broken, and the receiver of the LF message responds by sending a new LF message to sender of the LF message so that both parties are removed from each other's neighbor table. Figure 6 (which

Table 2. An example of neighbor table entry

Node ID	Location	Type of Link	Timestamp	LCR value
B	(5, 10)	Asymmetric	13:23:40	3

was previously discussed in Section 4.3) shows an example of how the link conditions change with node mobility and how the nodes respond to it, as follows:

In Figure 6(b), node B misses subsequent Hello messages from node A, as node B has moved out of node A's wireless transmission range. Therefore, node B sends an LF message to node A, which means that the symmetric link A↔B is no longer usable in the forward direction, (i.e., A→B). However, when node A receives this Hello message, it finds out that it is still receiving Hello messages from node B, which means that this link is still usable in the reverse direction, i.e, B→A. In other words, the symmetric link A↔B has turned into an asymmetric link B→A. In that case, node A sends an AN message to notify node B of the newly formed asymmetric link.

However, if the LF message from a down-link node is received at an up-link node, it means that the associated asymmetric link is no longer usable, and the neighbor entry is deleted from the neighbor table. In Figure 6(c), the down-link node B misses subsequent Hello messages from the up-link node A, so node B sends an LF message to node A. When node A receives the LF message, it realizes the asymmetric link A →B is no longer usable, and it deletes node B from its neighbor table. Thus, A-GF is able to detect the changes in condition of already discovered links, which helps the nodes to be aware of the latest topology changes in the neighborhood and ultimately results in increased reliability of data transmission.

Elimination of Transient Asymmetric Links

In A-GF, each node knows its own location through some positioning mechanism, and each node periodically advertises its location and one-hop neighbor table using Hello messages. While periodic Hello messages can be used to discover asymmetric links at the routing layer (similar techniques could be employed at the link layer), the knowledge of geographic locations of a node's one-hop neighbors can also assist in careful planning of bandwidth usage for such asymmetric link discovery and maintenance mechanisms. In A-GF, once an asymmetric link has been detected at a down-link node, the up-link node notification method is initiated so that the up-link node is informed about the existence of the asymmetric link. As described earlier in Section 4.2, a control packet called Asymmetric Notification (AN) is sent from the down-link node to the up-link node for asymmetric link notification. However, it may not be beneficial to initiate such notifications when the asymmetric link is transient, i.e., the cost of discovering, announcing, and exploiting an asymmetric link may be greater than the benefit of the asymmetric link (i.e., the potential availability of a shorter path).

As a consequence, A-GF initiates the up-link notification (Annotation), only after an initial *probation phase* to ensure that the discovered asymmetric link will be stable long enough to justify the cost of additional control packets. Algorithm 2 illustrates the *observe_new_neighbor* method that implements the probation phase. When an asymmetric link is detected for the first time at a down-link node, the up-link node associated with the asymmetric link enters a probation phase. During the probation phase, the changes in the distance between the up-link node and the down-link node are monitored. The duration of the probation phase is set equal to:

$$\frac{DIST(uplink\_neighbor\_location, my\_location)}{2 \times uplink\_neighbor\_velocity},$$

i.e., until the current distance between the up-link node and the down-link node is at least half of the initial distance. The *Pythagorean Theorem with parallel meridians* is used to calculate the distance between the up-link node at (Φ_{uplink}, λ_{uplink})

and the down-link node at ($\Phi_{downlink}$, $\lambda_{downlink}$) where the (Φ, λ) pairs indicate the location of the node in latitude and longitude. The formula being used to calculate the distance is as follows:

$$DIST(uplink\_node, downlink\_node) =$$
$$R\sqrt{(\Phi_{uplink} - \Phi_{donwlink})^2 + (\lambda_{uplink} - \lambda_{donwlink})^2}$$

R is the radius of the Earth in the above formula.

At the end of the probation phase, the up-link node is notified about the asymmetric link only if it passes a test that includes the following three conditions:

1. More than one direct one-hop Hello messages from the up-link node were received at the down-link node during the probation phase;
2. The distance between the up-link node and the down-link node is either constant or decreasing over the probation phase;
3. The link is still asymmetric.

An asymmetric link has to meet all of these three conditions to pass this test. These three conditions, if all met; ensure that the asymmetric link will be stable for long enough to justify the cost of additional control packets cost for link notification. For example, condition 1 ensures that the up-link node is still in the neighborhood and the wireless connectivity from the up-link node to the down-link node is good. Condition 2 ensures that the up-link node and the down-link node are not moving away from each other, so the connection will most likely be available for some time. Condition 3 ensures that the link has not turned into a symmetric link, which is quite possible if conditions 1 and 2 are met and both nodes are within each other's transmission range. However, if one or more of these three conditions are not met, then the asymmetric link is not a stable one. In that case, the up-link node is not notified about the asymmetric link, and the correspond-

ing cost of the asymmetric link notification and maintenance is avoided.

The algorithm for updating LCR value is updated to include the initial probation phase. When an asymmetric link is detected for the first time at a down-link node, the up-link node associated with the asymmetric link is entered into a probation phase. If the asymmetric link successfully passed the test at the end of probation phase, then the associated neighbor is entered into the neighbor table and the asymmetric link notification method is initiated. On the other hand, if the asymmetric link failed the test at the end of probation phase, then the associated neighbor is removed from the probation list. The LCR value is decremented either when a Hello message is lost or the distance between the up-link and the down-link node has increased during the time elapsed between the previous and new Hello messages were received. Similarly, the LCR value is incremented when either a new Hello message is received or the distance between the up-link and the down-link node has decreased during the time elapsed between two subsequent Hello messages.

Forwarding of Data Packets

When a node makes a data forwarding decision, it considers (1) the LCR rank of the neighbor and (2) the geographical progress that neighbor makes towards the destination location (thereby considering both reliability and performance). Toward that end, when a node forwards a data packet M towards a sink node S, it only considers next-hop neighbors with LCR values of at least *MAX_LCR/2*. The forwarding node calculates a *Next Hop Quality (NHQ)* value for each of its neighbors N_i. *NHQ (N_i)* consists of a weighted linear combination of the two above-mentioned parameters. The data packet will be forwarded to the neighbor with the highest NHQ value.

$$NHQ(N_i) = (1 - \alpha) \times (DIST(S, M)$$
$$-DIST(S, N_i)) + \alpha \times LCR(N_i), i \in \{1, 2 \ldots n\}$$

The weight value (i.e., α) is the network designer's choice, i.e., if α is set to zero then A-GF works as the standard GF protocol. However, A-GF works as a proactive link condition based routing protocol if α is set to 1. Simulation results revealed that the protocol yields the desired performance (i.e., a combination of link reliability and geographic progress towards the sink location) when the value of α is set in the range 0.4-0.56.

Evaluation

In order to evaluate the performance of A-GF, simulations performed in the JiST/SWANS simulator are used to study large and complex network topologies. JiST (Java in Simulation Time) is a discrete event simulator designed to run over a standard Java Virtual Machine. SWANS (Scalable Wireless Ad-Hoc Network Simulator) is built on top of the JiST platform to provide the tools needed to construct a wireless mobile ad-hoc network.

Setup

The nodes in the simulated network have varying transmission ranges, as a result of differing transmit powers, radio reception sensitivities, and radio reception thresholds of the nodes. The default values of transmit power, radio reception sensitivity, and radio reception threshold are 16 dBm, -91 dBm, and -81 dBm, respectively. The radio transmission ranges of the nodes are chosen in the range *tx_radius* ± *((tx_radius * x) / 100)*, where tx_radius is the default transmission range, and x is varied from 5 to 30 in steps of 5. It is reasonable to assume that the number of asymmetric links in the network will go up with an increasing variance of transmission ranges. There are 200 nodes in the network and they are placed randomly in a 1000

meters × 1000 meters field. Node mobility follows the Random Waypoint model, with a maximum velocity of 5 m/s. MAX_LCR is chosen as 4 and α is chosen as 0.4.

Twenty runs of experiments were conducted for each metric, and the final results were taken as the average of the twenty experiments. In each simulation, n/2 CBR traffic flows are generated where n is the number of nodes in the network, and each flow sends 100 packets of 1000 bytes each. Each simulation runs for 1000 seconds.

Simulation Results

This section evaluates and compares the performance of A-GF to standard GPSR as well as *Symmetric Geographic Forwarding (S-GF)* (Zhou, He, Krishnamurthy, & Stankovic, 2004). S-GF allows a node to add the IDs of all its neighbors it has discovered into the periodic Hello messages. When a node receives a Hello message, it registers the sender as its neighbor in its local neighbor table, and then checks whether its own ID is in the Hello message. If the receiver finds its own ID in the neighbor list in the Hello message, then it marks the communication link connecting it to the sender as symmetric. Otherwise, it marks the wireless link between them as asymmetric. Whenever a node needs to forward a packet, it selects only those neighboring nodes connected via symmetric links. Thus, S-GF uses a similar approach as in A-GF to discover the asymmetric links in the network. However, S-GF ignores asymmetric links whereas A-GF exploits them while making routing decisions, and this results in an increase in data transmission success and a decrease in latency (or route length), as shown in the simulation results.

Figure 8 shows the *data transmission success* for these three protocols. The data transmission success is defined as the ratio of the number of data packets received at the destinations to the number of data packets sent by the source node. Figure 8 shows that A-GF performs as well as S-GF and

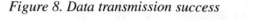

Figure 8. Data transmission success

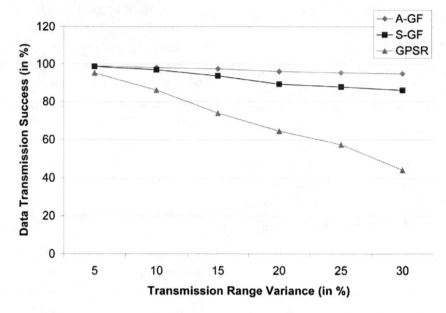

GPSR when the nodes in the network have only 5% variance in their transmission ranges. However, as the variance of transmission ranges of nodes in the network increases (i.e., number of asymmetric links increases), the data transmission success of both S-GF and GPSR decreases. Both A-GF and S-GF always result in better performance than GPSR because GPSR ignores asymmetric links and consequently suffers from low network connectivity and data transmission failures. S-GF discovers the asymmetric links in the network and avoids them during routing, thereby reducing the transmission failures incurred by GPSR in an asymmetric network. A-GF not only discovers the asymmetric links in the network, but also exploits the asymmetric links in routing, thereby offering better performance than S-GF.

Figure 9 shows the *average route length* for these three protocols, which is the average of total number of hops traveled by all data packets in the network. Figure 9 shows that when the variance of transmission range of the nodes in the network increases (i.e., number of asymmetric links increases), the average route lengths for all three protocols go up. However, the increase in the

average route length in A-GF is less substantial than in S-GF and GPSR, because A-GF explicitly exploits asymmetric links for routing. As indicated earlier in Section 1, asymmetric links often tend to span longer distances than symmetric links. Therefore, using asymmetric links in routing decisions often results in shorter routes, whereas ignoring them results in routes longer than necessary (as in S-GF and GPSR).

Figures 10, 11 and 12 compare the overhead of the A-GF, S-GF, and GPSR protocols. The total overhead in any routing protocol in wireless ad-hoc networks consists of the following main components: (1) control traffic generated by the protocol, (2) overheads from data traffic that is forwarded over sub-optimal routes, and data transmission failures due to node movement or failure, and (3) link maintenance overheads expressed in form of the average size of Hello messages. Figure 10 shows the *control packet overhead* in all three protocols. The control packet overhead is defined as the total number of control packets in the network. Figure 10 shows that the control packet overhead in all three protocols increases when the transmission range variance of

Figure 9. Average route length

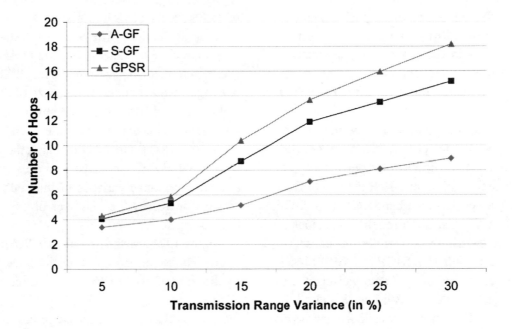

Figure 10. Control packet overhead

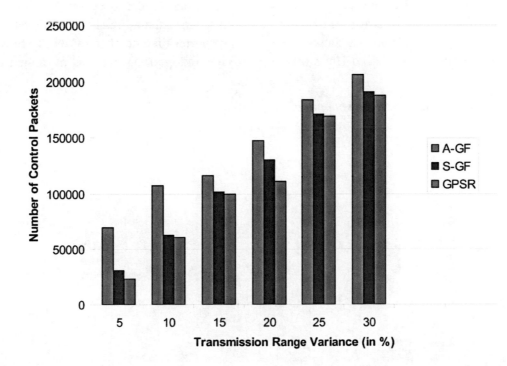

the nodes in the network increases (i.e., number of asymmetric links increases). However, A-GF always results in higher control packet overhead than S-GF and GPSR, because A-GF adds more control packets in the network for asymmetric link discovery and maintenance.

Figure 11 shows the *total data bandwidth consumption* of these three protocols. The total data bandwidth consumption is defined as the total number of bytes per second in the network that are needed to successfully deliver all data packets to their destinations. Figure 11 shows that A-GF consumes less data bandwidth than S-GF or GPSR. A-GF needs fewer retransmissions of data packets to reliably deliver the data packets to all destinations, whereas S-GF and GPSR result in higher retransmissions of data packets due to transmission failures over asymmetric links. Furthermore, A-GF eliminates transient asymmetric links, thereby increasing data transmission success and further reducing retransmissions.

Figure 12 shows the *link maintenance overhead* of these three protocols. The link maintenance overhead is defined as the average size of the Hello messages in bytes. Figure 12 shows that GPSR always results in fixed sized Hello mes-

sages, and that is because in GPSR the one-hop Hello messages always contain a fixed number of fields that include the node ID and current location of the node. The S-GF Hello messages have the neighbor table inserted in their header, and therefore, are bigger in size than GPSR Hello messages. Note that the average size of the S-GF Hello messages does not vary much over varying transmission ranges, because S-GF ignores asymmetric links. On the other hand, the average size of A-GF Hello message increases when the transmission range variance of the nodes in the network increases (i.e., number of asymmetric links increases).

Figure 13 shows the *asymmetric link utility* of A-GF. The asymmetric link utility is defined as the lifetime of asymmetric links vs. the number of times these links were used by A-GF in making routing decisions. The average lifetime of an asymmetric link in the simulations was 17 seconds, with the least and most available links being active for 1 second and 25 seconds, respectively. Figure 13 shows that A-GF chooses the asymmetric links with longer lifetime than the asymmetric links with shorter lifetime, thereby successfully identifying and ignoring transient asymmetric links

Figure 11. Total data bandwidth consumption

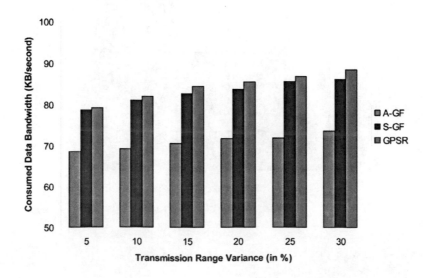

Figure 12. Link maintenance overhead

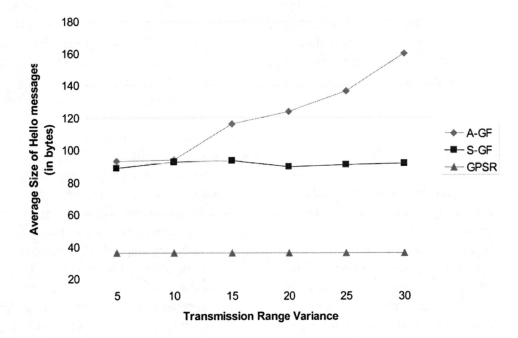

Figure 13. Asymmetric link utility

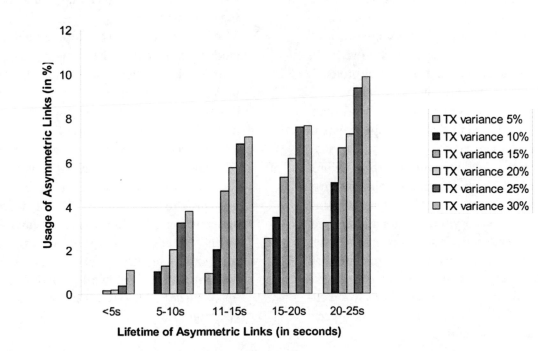

for efficient routing. Figure 13 also shows that the overall usage of asymmetric links in routing goes up when the variance of transmission range of the mobile nodes increases; this is because increasing variability in the radio capabilities of the mobile nodes results in an increase in the total number of asymmetric links in the network.

CONCLUSION AND FUTURE RESEARCH DIRECTIONS

This chapter provides an overview of a number of location-aware routing protocols that explicitly use asymmetric links in routing to obtain efficient and shorter routes. To that end, the asymmetric link protocols have been analyzed and characterized in details along three axes, namely (1) asymmetric link discovery, (2) asymmetric link monitoring and management, and (3) asymmetric link routing and acknowledgment. This chapter also discusses A-GF, a location-aware routing protocol that actively exploits asymmetric links to increase the reliability and performance of ad-hoc routing. Simulation results indicate that A-GF succeeds in reducing latencies and increasing successful route discoveries, while keeping the overheads small. The detailed study of these routing protocols clearly indicates that utilizing asymmetric links is overall beneficial, even when occasional asymmetric links fail due to dynamic adjustments in transmission range and node mobility. The study also reveals that while the different routing protocols discussed in this chapter employ a variety of advanced techniques to address various issues associated with link asymmetry, none of them yet provides a comprehensive solution that addresses all major issues associated with asymmetric link routing in highly mobile environments. To that end, future research efforts should focus on addressing all of the following three crucial aspects of asymmetric link routing:

- **Bandwidth-Efficient Beaconing:** Given that periodic beaconing is the most widely used mechanism for asymmetric link discovery; future routing protocols should make an effort to minimize the bandwidth and energy consumption by adopting efficient beaconing strategies, e.g., adapting beacon interval to node mobility and network traffic load (Heissenbuttel, Braun, Walchli & Bernoulli, 2007).

- **Link Condition Monitoring:** Due to node mobility or dynamic adjustments in transmission ranges, symmetric links can become asymmetric links, or asymmetric links can turn into broken links. Therefore, it is highly important to monitor the condition of the discovered links (both symmetric and asymmetric) and evaluate them for reliability, in order to improve data transmission success. The A-GF protocol discussed in this chapter presents a simple mechanism to monitor the newly discovered links for stability and eliminate the transient ones for further improving routing efficiency.

- **Mobility-Aware Next Hop Selection:** Node mobility affects the lifetime of asymmetric links. To that end, sudden changes in the mobility characteristics of the down-link node could cause it to move out of the transmission range of up-link node which, on the other hand, would not be aware of this change until several multi-hop updates from the down-link node are missed. This may lead to high number of failed data transmission in highly dynamic environments. Future routing protocols should use mobility prediction schemes to predict the availability of a down-link node before making a decision to select that node as the next hop forwarder.

REFERENCES

Arockia, D., & Anbumani, P. (2013). GRGPS – Geographic Routing based on Greedy Perimeter Stateless Position for Multi- Hop Mobile Ad-hoc Networks. *International Journal of Emerging Trends & Technology in Computer Science, 2*(2).

Bose, P., Morin, P., Stojmenovic, I., & Urrutia, J. (1999). *Routing with Guaranteed Delivery in Ad-Hoc Wireless Networks.* Paper presented at the 3rd International Workshop on Discrete Algorithms and Methods for Mobile Computing and Communications. New York, NY.

Cerpa, A., Wong, J. L., Kuang, L., Miodrag, P., & Estrin, D. (2005). *Statistical Model of Lossy Links in Wireless Sensor Networks.* Paper presented at the 4th International Symposium on Information Processing in Sensor Networks. Los Angeles, CA.

Chen, B. B., Hao, S., Zhang, M., Chan, M. C., & Ananda, A. I. (2009). *DEAL: Discover and Exploit Asymmetric Links in Dense Wireless Sensor Networks.* Paper presented at the 6th Annual IEEE Communications Society Conference on Sensor, Mesh and Ad Hoc Communications and Networks (SECON). Rome, Italy.

Chen, Q., Kanhere, S., & Hassan, M. (2013). Adaptive Position Update for Geographic Routing in Mobile Ad Hoc Networks. *IEEE Transactions on Mobile Computing, 12*(3).

Couto, D., Aguayo, D., Bicket, J., & Morris, R. (2003). *A High-Throughput Path Metric for Multi-Hop Wireless Routing.* Paper presented at the 9th Annual International Conference on Mobile Computing and Networking. San Diego, CA.

Duros, E., & Dabbous, W. (1996). *Handling of Unidirectional Links with OSPF.* Retrieved from http://tools.ietf.org/html/draft-ietf-ospf-unidirectional-link-00

Gade, T. (2013). *Acknowledgment Strategies for Efficient Asymmetric Routing in Energy Harvesting Wireless Sensor Networks.* (MS Thesis). Texas Tech University.

Ganesan, D., Estrin, D., Woo, A., Culler, D., Krishnamachari, B., & Wicker, S. (2002). *Complex Behavior at Scale: An Experimental Study of Low-Power Wireless Sensor Networks* (Technical Report No. CSD-TR 02-0013). University of California at Los Angeles.

Gondalia, A. K., & Kathiriya, D. R. (2012). Performance Tuning for Geographic Routing in Wireless Networks. *International Journal of Management, IT and Engineering, 2*(5).

Heissenbuttel, M., Braun, T., Walchli, M., & Bernoulli, T. (2007). Evaluating the Limitations and Alternatives in Beaconing. *Elsevier Ad Hoc Networks, 5*(5).

Johnson, D. B., & Maltz, D. A. (1996). Dynamic Source Routing in Ad Hoc Wireless Networks. *Mobile Computing, 5*, 153–181. doi:10.1007/978-0-585-29603-6_5

Karp, B., & Kung, H. T. (2000). *Greedy Perimeter Stateless Routing for Wireless Networks.* Paper presented at the 6th Annual International Conference on Mobile Computing and Networking (MobiCom). Boston, MA.

Kim, D., Toh, C. K., & Chou, Y. (2000). *RODA: A New Dynamic Routing Protocol Using Dual Paths to Support Asymmetric Links in Mobile Ad Hoc Networks.* Paper presented at the 9th International Conference on Computer Communications and Networks (ICCCN). Las Vegas, NV.

Kim, Y. J., Govidan, R., Karp, B., & Shenker, S. (2004). *Practical and Robust Geographic Routing in Wireless Networks.* Paper presented at the 2nd International Conference on Embedded Networked Sensor Systems (SenSys). Baltimore, MD.

Ko, Y., Lee, S., & Lee, J. (2004). *Ad-hoc Routing with Early Unidirectionality Detection and Avoidance*. Paper presented at the International Conference on Personal Wireless Communications (PWC). Delft, The Netherlands.

Le, T., Sinha, P., & Xuan, D. (2010). Turning Heterogeneity into an Advantage in Wireless Ad-Hoc Network Routing. *Elsevier Ad Hoc Networks*, *8*(1), 108–118. doi:10.1016/j.adhoc.2009.06.001

Liu, R. P., Rosberg, Z., Collings, I. B., Wilson, C., Dong, A., & Jha, S. (2008). *Overcoming Radio Link Asymmetry in Wireless Sensor Networks*. Paper presented at the International Symposium on Personal, Indoor and Mobile Radio Communications. Cannes, France.

MacDonald, J. T., & Roberson, D. A. (2007). *Spectrum Occupancy Estimation in Wireless Channels with Asymmetric Transmitter Powers*. Paper presented at the 2nd International Conference on Cognitive Radio Oriented Wireless Networks and Communications (CrownCom). Orlando, FL.

Marina, M. K., & Das, S. K. (2002). *Routing Performance in the Presence of Unidirectional Links in Multihop Wireless Networks*. Paper presented at the 3rd International Symposium on Mobile Ad Hoc Networking and Computing. Lausanne, Switzerland.

Misra, R., & Mandal, C. R. (2005). *Performance Comparison of AODV/DSR On-Demand Routing Protocols for Ad-Hoc Networks in Constrained Situation*. Paper presented at the IEEE International Conference on Personal Wireless Communications (ICPWC). New Delhi, India.

Mitra, P., & Poellabauer, P. (2011). Asymmetric Geographic Forwarding: Exploiting Link Asymmetry in Location Aware Routing. *International Journal of Embedded and Real-Time Communication Systems*, *2*(4). doi:10.4018/jertcs.2011100104

Morgenthaler, S., Braun, T., Zhao, Z., Staub, T., & Anwander, M. (2012). *UAVNet: A Mobile Wireless Mesh Network Using Unmanned Aerial Vehicles*. Paper presented at the 3rd International Workshop on Wireless Networking for Unmanned Autonomous Vehicles. Anaheim, CA.

Narayanaswamy, S., Kawadia, V., Sreenivas, R. S., & Kumar, P. R. (2002). *Power Control in Ad-Hoc Networks: Theory, Architecture, Algorithm and Implementation of the COMPOW Protocol*. Paper presented at the European Wireless Conference. Florence, Italy.

Reijers, N., Halkes, G., & Langendoen, K. (2004). *Link Layer Measurements in Sensor Networks*. Paper presented at the 1st IEEE International Conference on Mobile Ad-hoc and Sensor Systems (MASS). Fort Lauderdale, FL.

Romdhani, B., Barthel, D., & Valois, B. (2012). *Exploiting Asymmetric Links in a Convergecast Routing Protocol*. Paper presented at the 11th International Conference on Ad Hoc Networks and Wireless (ADHOC-NOW). Belgrade, Serbia.

Roosta, T., Menzo, M., & Sastry, S. (2005). *Probabilistic Geographic Routing in Ad Hoc and Sensor Networks*. Paper presented at Wireless Networks and Emerging Technologies (WNET). Banff, Canada.

Sang, L., Arora, A., & Zhang, H. (2007). *On Exploiting Asymmetric Wireless Links via One-way Estimation*. Paper presented at the 8th International Symposium on Mobile Ad Hoc Networking and Computing. Montreal, Canada.

Son, D., Helmy, A., & Krishnamachari, B. (2004). The Effect of Mobility-induced Location Errors on Geographic Routing in Ad Hoc Networks: Analysis and Improvement using Mobility Prediction. *IEEE Transactions on Mobile Computing*, 233–245. doi:10.1109/TMC.2004.28

Srinivasan, K., Dutta, P., Tavakoli, A., & Levis, P. (2006). *Understanding the Causes of Packet Delivery Success and Failure in Dense Wireless Sensor Networks.* Paper presented at the 4th International Conference on Embedded Networked Sensor Systems. Boulder, CO.

Srinivasan, K., & Levis, P. (2006). *RSSI is Under Appreciated.* Paper presented at the 3rd Workshop on Embedded Networked Sensors. Cambridge, MA.

Wang, G., Ji, Y., & Turgut, D. (2004). *A Routing Protocol for Power Constrained Networks with Asymmetric Links.* Paper presented at the 1st International Workshop on Performance Evaluation of Wireless Ad Hoc, Sensor, and Ubiquitous Networks (PE-WASUN). Venice, Italy.

Woo, A., Tong, T., & Culler, D. (2003). *Taming the Underlying Challenges of Reliable Multi-hop Routing in Wireless Networks.* Paper presented at the 1st International Conference on Embedded Networked Sensor Systems. Los Angeles, CA.

Zamalloa, M. Z., & Krishnamachari, B. (2007). An Analysis of Unreliability and Asymmetry in Low-Power Wireless Links. *ACM Transactions on Sensor Networks (TOSN), 3*(2).

Zhao, Y. J., & Govidan, R. (2003). *Understanding Packet Delivery Performance in Dense Wireless Sensor Network.* Paper presented at the 1st International Conference on Embedded Networked Sensor Systems. Los Angeles, CA.

Zhao, Z., Rosario, D., Braun, T., Cerqueira, E., Xu, H., & Huang, L. (2013). *Topology and Link Quality-aware Geographical Opportunistic Routing in Wireless Ad-hoc Networks.* Paper presented at the 9th International Wireless Communications and Mobile Computing Conference. Sardinia, Italy.

Zhou, G., He, T., Krishnamurthy, S., & Stankovic, J. A. (2004). *Impact of Radio Irregularity on Wireless Sensor Networks.* Paper presented at the 2nd International Conference on Mobile Systems, Applications, and Services. Boston, MA.

Zhou, G., He, T., Krishnamurthy, S., & Stankovic, J. A. (2006). Models and Solutions for Radio Irregularity in Wireless Sensor Networks. *ACM Transaction on Sensor Networks, 2,* 221–262. doi:10.1145/1149283.1149287

KEY TERMS AND DEFINITIONS

Down-Link Node: A node having an incoming asymmetric link is called a down-link node. It can only receive packets on this asymmetric link, but cannot use it to send packets in the opposite direction.

Hello/Beacon Messages: The control messages that are periodically exchanged between neighbor nodes to discover and maintain knowledge of local topology. In location-aware routing, these messages often include the latest location of a node, transmission range, list of neighbors, etc.

Link Asymmetry: This is caused by discrepancy in radio transmission ranges of two neighboring nodes. For two neighboring nodes A and B, if node A is within the transmission range of node B, but not vice versa, then there is a link asymmetry between nodes A and B.

Location-Aware Routing: A routing principle that relies on geographic location. It is mainly proposed for wireless networks and based on the idea that the source sends a message to the geographic location of the destination(s) over multiple hops, instead of using the network address and without any prior path discovery. This routing paradigm requires that each node can determine its own location and that the source is aware of the location of the destination.

Mobile Ad-Hoc Networks: A decentralized wireless network formed by autonomous and highly mobile nodes that are willing to forward data on behalf of other nodes. The determination of which nodes forward data is made dynamically based on the network connectivity.

Neighbor Discovery: A mechanism that involves exchange of control messages among mobile nodes within each other's' transmission ranges. In location-aware routing, nodes need to possess knowledge about their neighborhood, and hence the local topology, in order to efficiently route packets to the destination(s).

Up-Link Node: A node that has an outgoing asymmetric link is called the up-link node. It can use the asymmetric link to send packets, but cannot receive packets on that link.

ENDNOTES

[1] http://jist.ece.cornell.edu

[2] http://platformx.sourceforge.net/Links/resource.html

Chapter 6
Implementation Strategies for High-Performance Multiuser MIMO Precoders

Maitane Barrenechea
University of Mondragón, Spain

Andreas Burg
École Polytechnique Fédérale de Lausanne (EPFL), Switzerland

Mikel Mendicute
University of Mondragón, Spain

John Thompson
University of Edinburgh, UK

ABSTRACT

The multiuser MIMO environment enables the communication between a base-station and multiple users with several antennas. In such a scenario, the use of precoding techniques is required in order to detect the signal at the users' terminals without any cooperation between them. This contribution presents various designs and hardware implementations of a high-capacity precoder based on vector perturbation. To this aim, three tree-search techniques and their associated user-ordering schemes are investigated in this chapter: the well-known K-Best precoder, the fixed-complexity Fixed Sphere Encoder (FSE), and the variable complexity Single Best-Node Expansion (SBE). All of the aforementioned techniques aim at finding the most suitable perturbation vector within an infinite lattice without the high computational complexity of an exhaustive search.

INTRODUCTION

The demand for high-speed communications required by cutting-edge applications has put a strain on the already saturated wireless spectrum. The incorporation of antenna arrays at both ends of the communication link has provided improved spectral efficiency and link reliability to the inherently complex wireless environment, thus allowing for

DOI: 10.4018/978-1-4666-6034-2.ch006

the thriving of high data-rate applications without the cost of extra bandwidth consumption. As a consequence to this, multiple-input multiple-output (MIMO) systems have become the key technology for wideband communication standards both in single-user and multi-user setups.

The main difficulty in single-user MIMO systems stems from the signal detection stage at the receiver, whereas multi-user downlink systems struggle with the challenge of enabling non-cooperative signal acquisition at the user terminals. In this respect, precoding techniques perform a pre-equalization stage at the base station so that the signal at each receiver can be interpreted independently and without the knowledge of the overall channel state. The non-linear vector precoding (VP) technique has been proven to enable non-cooperative signal acquisition in the multi-user broadcast channel with a feasible complexity (Hochwald, Peel, Swindlehurst, 2005). The performance advantage with respect to the more straightforward linear precoding algorithms is the result of an added perturbation vector which enhances the properties of the precoded signal. Nevertheless, the computation of the perturbation signal entails a search for the closest point in an infinite lattice, which is known to be in the class of non-deterministic polynomial-time hard (NP-hard) problems.

This chapter addresses the difficulties that stem from the perturbation process in VP systems from a hardware implementation perspective. This study is focused on tree-search techniques that, by means of a strategic node pruning policy, reduce the complexity derived from an exhaustive search and yield a close-to-optimum error-rate performance. In the last section of the chapter, the incorporation of alternative norms in the distance computation process will be analyzed, which can greatly reduce the computational cost of all the reviewed tree-search schemes.

BACKGROUND

With the advent of new communication technologies, the interest in MIMO has recently evolved towards the development of multi-user schemes which consider more complex albeit realistic scenarios with multiple terminals sharing the time, space, bandwidth and power resources available in a wireless network. Consequently, a great part of the latest research on innovative wireless multi-antenna technologies has been focused on multi-user MIMO (MU-MIMO) environments.

A multi-antenna and multi-user system provides a set of advantages over point-to-point MIMO transmissions. One of the main features of MU-MIMO is its greater immunity to propagation shortcomings derived from antenna correlation. Being the antennas hosted at scattered users, the correlation coefficients are inherently low, which allows to overcome the usual problems related to channel rank loss. Another interesting property of MU-MIMO is that direct line of sight propagation, which greatly degrades the quality of the communication link in single-user MIMO systems with spatial multiplexing, does not pose a problem in a multi-user setup. Furthermore, MU-MIMO enables obtaining a spatial multiplexing gain at the base station without the requirement of multi-antenna receivers. This allows for the implementation of small, low-cost and low-power terminal devices as the computational load is transferred to the base station (Gesbert, Kountouris, Heath & Chae, 2007).

Nevertheless, the multi-user setup also poses a set of problems that do not exist in the single-user model. For example, the lack of interaction between the users forces the base station to acquire instantaneous knowledge of the channel in order to allow for independent detection of each user's information stream at the receivers. Additionally, the independence between the receive antennas may also incur in an outage situation if the sub-channel directed to a single-antenna user undergoes severe fading. Such a situation in MIMO systems can be overcome with simple diversity techniques.

Generally speaking, the multi-user MIMO environment is composed of two channels that communicate the base station with the user terminals: the multiple access channel (MAC), also known as the uplink channel, covers the communication from the terminals to the base station, whereas the broadcast channel (BC), or downlink channel, carries the transmissions that stem from the base station and end at the user terminals, as is shown in Figure 1. Hitherto, MU-MIMO techniques for the uplink channel have been widely studied as the detection problem in such systems is equivalent to that of a MIMO channel with multi-user detection (Verdú, 1998). However, the study of the downlink channel entails a greater complexity, as it will be shown throughout this chapter.

The interest in developing efficient algorithms for the broadcast channel in MIMO systems has arisen in recent years. Special focus has been drawn to this problematic scenario, where the impossibility of cooperation between the terminals prevents the use of well-known detection techniques to retrieve the users' information at the receiving end of the communication. However, by performing a signal shaping stage at the base station prior to transmission, it is possible for the terminals to detect their information stream without any interaction between them. This technique is referred to as precoding and it can follow a linear (Joham, Utschick & Nossek, 2005; Peel, Hochwald & Swindlehurst, 2005) or non-linear (Hochwald, Peel & Swindlehurst, 2005; Schmidt, Joham & Utschick, 2008) scheme.

The linear precoding techniques perform inter-user interference cancelation by premultiplying the signal aimed at the terminals by a precoding matrix. The latter can be designed so as to eliminate all the interference among the users' streams. This technique is commonly referred to as zero-forcing (ZF) precoding (Peel, Hochwald & Swindlehurst, 2005) and utilizes the inverse of the channel matrix as the transmission filter matrix. If the channel is ill-conditioned, the power of the unscaled precoded signal is increased significantly, which results in a poor signal to noise ratio (SNR) at the receivers. One way to overcome this noise boost is to allow for some interference among the users by regularizing the channel inverse. This approach, also known as regularized precoding, outperforms the previous technique especially at low SNR (Peel, Hochwald & Swindlehurst, 2005). It is also possible to design the precoding matrix following a Wiener filter approach in such a way that the mean square error (MSE) is minimized. Having a lower MSE, this approach outperforms the previous ones in terms of bit error rate (BER) performance (Joham, Utschick & Nossek, 2005).

Figure 1. Multi-user MIMO broadcast channel with M transmit antennas and N single-antenna user terminals

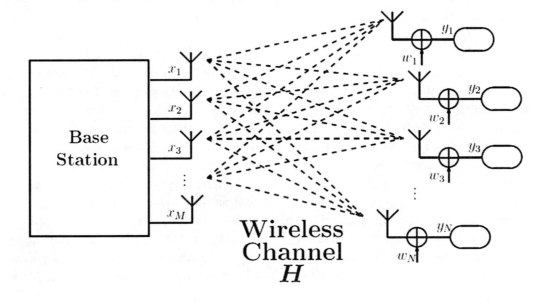

Costa introduced the concept of writing on dirty paper (Costa, 1983), which established the base for the non-linear precoding approaches that perform best in multiuser scenarios. Despite its relevance in the theoretical field due to its capacity-achieving property, the implementation of dirty paper coding (DPC) still remains an open issue. Other non-linear transmission schemes try to reach the capacity bound set by DPC by using a simplified design. This is the case of Tomlinson-Harashima precoding (THP) (Fischer, Windpassinger, Lampe & Huber, 2002; Harashima & Miyakawa, 1972; Tomlinson, 1971), which was initially designed to mitigate the effect of intersymbol interference (ISI). In THP, the interference between the streams is cancelled in a sequential fashion by means of a feedback filter. Nevertheless, the most important feature of this precoding approach is the insertion of a modulo operation, at both the transmitter and the receiver, to reduce the unscaled transmit power. A more efficient non-linear technique which also makes use of the modulo channel is VP (Hochwald, Peel & Swindlehurst, 2005). The modulo operation at the transmitter is replaced by the addition of a perturbing signal, which is optimized directly, therefore avoiding the successive interference cancelation performed in THP.

Despite its enhanced performance, the main issue with VP is the computation of the perturbing signal, which implies the search of the closest point in an infinite lattice. Several techniques have been proposed so far to deal with this matter. One of the most popular is the sphere encoder (SE) (Hochwald, Peel & Swindlehurst, 2005, Schmidt, Joham & Utschick, 2008), which performs a sphere search to obtain the optimum perturbation signal. In Windpassinger, Fischer and Huber (2004) a suboptimal method for the computation of the perturbation signal is proposed, where the Babai's approximate closest point solution is used along with the Lenstra-Lenstra-Lovász's (LLL) lattice reduction algorithm (Lenstra, Lenstra & Lovász, 1982). Even though the performance of this technique is very close to that of the optimal

sphere encoder, the hardware implementation of the required LLL preprocessing stage is a complex matter. In Habendorf and Fettweis (2007) and Zhang and Kim (2005) the use of the K-Best algorithm (Anderson & Mohan, 1984) is proposed to simplify the search for the perturbation vector. Nevertheless, the ordering stage that is required in each one of the tree levels hinders the implementation of this preprocessing scheme. A fixed-complexity search technique aimed at VP systems was presented in Barrenechea, Mendicute, Del Ser and Thompson (2009). The algorithm, which was named the fixed sphere encoder (FSE), yielded performance results close to the optimum while its architecture was specially designed for an efficient hardware implementation (Barrenechea, Barbero, Mendicute & Thompson, 2012).

So far, the amount of published work in the field of implementation for multiuser MIMO algorithms in the broadcast channel has been limited. The implementation of the first very-large-scale integration (VLSI) lattice reduction-aided VP system on an application-specific integrated circuit (ASIC) is described in Burg, Seethaler and Matz (2007). This VP approach avoids the process of searching for the optimum perturbation vector, which leads to a suboptimum BER performance. Another noteworthy design is the one developed in Bhagawat, Wang, Uppal, Choi, Xiong, Yeary and Harris (2008), where the first hardware implementation on a Xilinx Virtex II field-programmable gate array (FPGA) of a simple variant of DPC is analyzed. The data processing speed of the algorithm is of just 51 Mbps for a considerable device occupation. In Barrenechea, Mendicute and Arruti (2013) the implementation of a fully-pipelined vector precoder on a FPGA target device was carried out. For this implementation, fixed-complexity techniques such as the FSE and the K-Best were presented, rising the achievable throughput up to 5.6 Gbps. On the other hand, the hardware implementation of a variable-complexity and low-cost algorithm was performed in Barrenechea, Burg and Mendicute

(2012). In this case, a 65 nm technology ASIC was considered as the target device.

TREE-SEARCH TECHNIQUES FOR VECTOR PRECODING

System Model and Notation

In the remainder of the paper boldface letters will denote matrix-vector quantities. The transpose, expectation, conjugate transpose and matrix inversion operations will be represented by $(\cdot)^T$, $E[\cdot]$, $(\cdot)^H$ and $(\cdot)^{-1}$, respectively. The i-norm of a vector will be denoted by $\|\cdot\|_i$, the $N \times N$ identity matrix will be represented by $\boldsymbol{I}_N$ and $|\cdot|$ will represent the absolute value of a scalar variable. Real and imaginary parts of a complex variable will be derived by using the $\mathbb{R}(\cdot)$ and $\mathbb{I}(\cdot)$ operators, respectively. The flooring operation will be denoted by $\lfloor \cdot \rfloor$. Finally, the set of integers and complex numbers of dimension N will be denoted by $\mathbb{Z}^N$ and $\mathbb{C}^N$, respectively.

As for the system model, consider a MIMO broadcast channel with M antennas at the transmitter and N single-antenna users, denoted as $M \times N$. We assume that the channel between the base station and the N users is represented by a complex matrix $\boldsymbol{H} \in \mathbb{C}^{N \times M}$, whose element $h_{n,m}$ represents the channel gain between transmit antenna m and user n. The entries of the channel matrix satisfy $E\left[\left|h_{n,m}\right|^2\right] = 1$. The received data vector can be written by using the equation:

$$y = Hx + w,$$

where

$$\boldsymbol{y} = \left[y_1, \ldots, y_N\right]^T$$

represents the data received at the N users and the transmitted data vector is contained in:

$$\boldsymbol{x} = \left[x_1, \ldots, x_M\right]^T.$$

The additive white Gaussian noise (AWGN) vector added to the signal at the user terminals is **w**:

$$w = \left[w_1, \ldots, w_N\right]^T$$

with covariance matrix:

$$E\left[ww^H\right] = \sigma^2 \boldsymbol{I}_N.$$

The transmitted power at the base station E_{TR} is constrained to $E_{TR} = M$.

In a VP system, the data vector to be transmitted:

$$s = \left[s_1, \ldots, s_N\right]^T$$

is perturbed by a complex signal:

$$a \in \tau \mathbb{Z}^N + j\tau \mathbb{Z}^N,$$

where τ is the modulo constant (Figure 2). This value has to be chosen large enough so that unambiguous decoding can be performed. Nevertheless, small values of τ yield denser perturbation lattices and hence the ability of the precoding scheme to reduce the norm of the transmitted signal is enhanced. It is suggested in most of the literature on VP that τ is chosen as:

$$\tau = 2d_{max} + \Delta,$$

Figure 2. Block diagram of a VP system with M *transmit antennas and* N *single-antenna user terminal*

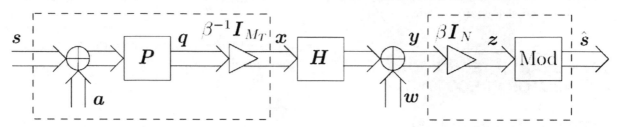

where d_{max} is the absolute value of the constellation symbol with the largest magnitude and " is the minimum spacing between constellation points. Note that the proper selection of the modulo constant allows for the extensions of the modulo alphabet in the complex plane to be centered around $\mathbb{CZ}$, where:

$$\mathbb{CZ} = \mathbb{Z} + j\mathbb{Z}$$

represents the set of Gaussian integers (Dietrich, 2008).

The complex components of the perturbation vector a belong to an infinite lattice. However, for the sake of simplicity, only those candidate points that lie within a boundary around the origin of the lattice are considered for hardware implementation. This selection can be made by either selecting those candidate points whose absolute values are below a certain threshold, or by reducing the number of feasible integers to a reasonable value. It has been observed that there is a higher concentration of perturbation elements within the area:

$$\mathfrak{U} \triangleq \left\{ d : \mathbb{R}(d) < 2\tau, \mathbb{I}(d) < 2\tau \right\}$$

centered at the origin of the lattice. Therefore, the constrained lattice $\mathfrak{U}$ composed of $B = 25$ elements will be utilized in the precoding scheme.

After the perturbation process, the precoding matrix P shapes the signal to be transmitted and a scaling factor β^{-1} is applied prior to transmission to comply with the transmit power constraints. At the user terminals, the received signal is scaled

by Δ again to meet the modulo operation requirements. This non-linear operation at the receivers is essential if the effects of the perturbing signal a are to be reversed.

The modulo operator can be equivalently described as the addition of integer multiples of the modulo constant, in such a way that the input signal is mapped into the fundamental Voronoi region of the lattice $\tau\mathbb{CZ}$. The fundamental Voronoi region of this particular lattice is a square of side τ centered at the origin, namely:

$$\mathcal{V}_M \triangleq \left\{ x + jy \mid x, y \in \left[-\tau/2, \tau/2 \right) \right\}.$$

The symbols in this equivalent region require a smaller transmit power, and hence their transmission generates a better detection SNR. The modulo operator works independently on the real and imaginary components of the data signal as shown by the following equation:

$$\mathrm{Mod}(d) = d - \left\lfloor \frac{\mathbb{R}(d)}{\tau} + \frac{1}{2} \right\rfloor \tau - j \left\lfloor \frac{\mathbb{I}(d)}{\tau} + \frac{1}{2} \right\rfloor \tau.$$

Note that:

$$\mathrm{Mod}(d + f) = \mathrm{Mod}(d)$$

holds if $f \in \tau\mathbb{CZ}$ and $\mathrm{Mod}(d) = d$ is satisfied if $d \in \mathcal{V}_M$.

The VP approach that achieves the best performance in terms of BER is the Wiener filter VP (WF-VP), which jointly optimizes the perturbing signal, the precoding matrix and the power scaling factor to reach the minimum mean square error (MMSE) solution (Schmidt, Joham & Utschick, 2008). The WF-VP model is supported by the following equations:

$$\left(HH^H + \xi I_N \right)^{-1} = U^H U,$$

$$a = \operatorname*{argmin}_{\hat{a} \in \tau \mathbb{Z}^N + j\tau \mathbb{Z}^N} \left(\left\| U \left(s + a \right) \right\|_2^2 \right), \qquad (1)$$

and

$$x = \beta^{-1} H^H \left(HH^H + \xi I_N \right)^{-1} \left(s + a \right),$$

where $\xi = N\sigma^2 / E_{TR}$ represents the inverse of the SNR and the triangular matrix U is obtained through the Cholesky decomposition of $\left(HH^H + \xi I_N \right)^{-1}$. From Equation (1) one can notice that the computation of the perturbing signal a entails a search for the closest point in a lattice. In the following sections several techniques to efficiently obtain the perturbation signal will be analyzed.

The Sphere Encoder

The SE algorithm, which represents the adaptation of the sphere decoding algorithm in Damen, El Gamal and Caire (2003) and Viterbo and Boutros (1999) to precoding systems, restricts the search for the perturbation vector to a set of points that lie within a hypersphere of radius R centered around a reference signal. By avoiding the search over the whole lattice, the complexity of the perturbation process is greatly reduced. The computation of the perturbation signal in Equation (1) can be rewritten to include the sphere constraint R as:

$$a = \operatorname*{argmin}_{\substack{\hat{a} \in \tau \mathbb{Z}^N + j\tau \mathbb{Z}^N \\ \left| U(s+a) \right|_2^2 < R}} \left(\left| U \left(s + a \right) \right|_2^2 \right). \qquad (2)$$

Given that U is a triangular matrix, the computation of the squared Euclidean norm can be performed recursively following a tree search fashion. This way, the calculation of the distances in Equation (2) (or equivalently Equation (1) for systems with no sphere constraint) is divided into N stages, where the sphere constraint is updated at every level. This process is represented by the following equations:

$$\left| a_i + z_i \right|^2 \le \frac{T_i}{u_{i,i}^2}, \qquad (3)$$

$$z_i = s_i + \sum_{j=i+1}^{N} \frac{u_{i,j}}{u_{i,i}} \left(a_j + s_j \right)$$

and

$$T_i = R^2 - \sum_{j=i+1}^{N} u_{j,j}^2 \left| a_j + z_j \right|^2.$$

The solution to Equation (2) can be obtained recursively, starting from level $i = N$ and traversing the tree backwards until $i = 1$. By accumulating the N distance increments, the Euclidean distance of a certain tree branch can be computed. At each level, B child nodes originate from each parent node of the tree. However, the sphere constraint ensures that only those child nodes that fulfil the constraint in Equation (3) are considered as feasible solutions. Once a node has been pruned out of the tree, all its descendants are discarded as well. Every time a leaf node is reached ($i = 1$), the path leading to that node is stored as a candidate solution and the search radius R is updated with the new Euclidean distance. The search process is then resumed with the updated sphere constraint until a leaf node is reached and there are no more branches that fulfil the sphere constraint. To determine the order in which the child nodes are to be visited, the Schnorr-Euchner enumeration method (Schnorr & Euchner, 1994) is recommended as it sorts out the lattice points in accordance with their distance to z_i, as opposed to the Fincke-Pohst enumeration method (Fincke & Pohst, 1985) where the lattice points are ordered arbitrarily (Figure 3).

Due to its optimum performance, the SE has been regarded as the best option for high-performance VP systems. However, the implementation of a sphere encoder entails two main drawbacks: On one hand, its variable complexity and dependence on the channel conditions make it unsuitable for communication systems where the data is expected to be processed at a fixed rate. On the other hand, the sequential nature of the algorithm derives in a suboptimum resource usage from a hardware architectural point of view. With the aim of overcoming the difficulties of a SE hardware implementation, fixed complexity tree-search-based algorithms such as the K-Best or the FSE have been proposed to solve the vector precoding problem.

Tree Traversal and User Ordering

The sequential processing and variable complexity of the SE stem from the sphere constraint that is checked at various stages in the algorithm. Even if this restricted search provides the algorithm with the ability to find the optimum solution to Equation (1), it is required that the constraint is removed or relaxed if a fixed-complexity approach is to be followed. For a breadth-first unconstrained tree search, the Euclidean distance computation in Equation (1) can be divided into N stages, as stated by the following equations:

$$D_i = u_{i,i}^2 \left| a_i + z_i \right|^2 + \sum_{j=i+1}^{N} u_{j,j}^2 \left| a_j + z_j \right|^2 = d_i + D_{i+1}$$

(4)

and

$$z_i = s_i + \sum_{j=i+1}^{N} \frac{u_{i,j}}{u_{i,i}} \left(a_j + s_j \right),$$

where the partial Euclidean distance (PED) of level i is denoted as d_i, while the accumulated Euclidean distance (AED) down to level i is represented by D_i.

Figure 3. Complex lattice enumeration techniques

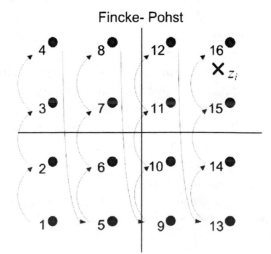

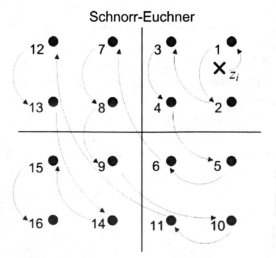

Fixed-complexity tree search algorithms usually require of a matrix ordering preprocessing stage to improve their performance. The ordering strategy determines the order in which the users' streams are visited during the tree search. Matrix ordering techniques are not exclusive of fixed-complexity tree traversal algorithms and they have been widely studied for their application to the SE. However, for such algorithm, the matrix ordering has no effect whatsoever on its final error-rate performance, but can be used to reduce the complexity of the search process.

For a certain matrix ordering strategy $\mathfrak{O}$, the computation of the perturbation signal in Equation (1) can be rewritten as

$$a = \Pi_{\mathfrak{O}}^{T} \operatorname*{argmin}_{\hat{a} \in \tau \mathbb{Z}^{N} + j\tau \mathbb{Z}^{N}} \left(\left\| U \left(\Pi_{\mathfrak{O}} s + a \right) \right\|_{2}^{2} \right),$$

with

$$\Pi_{\mathfrak{O}} \left(HH^{H} + \xi I_{N} \right)^{-1} \Pi_{\mathfrak{O}}^{T} = U^{H} U,$$

where the ordering matrix is defined as

$$\Pi_{\mathfrak{O}} = \sum_{i=1}^{N} e_{i} e_{\mathfrak{O}_{i}}^{T},$$

where e_{i} is the ith column of the identity matrix and $\mathfrak{O}_{i}$ represents the ith element in the ordering tuple determined by $\mathfrak{O}$.

The K-Best Encoder

As one can guess from its name, the K-Best precoder selects the K best tree branches at each level of the tree search regardless of the sphere constraint. Therefore, at each stage ($i = N,...,1$) of the K-Best tree-search algorithm, an ordering procedure has to be performed on the eligible KB candidate branches based on their AEDs

down to level i. After the sorting procedure, the K nodes with the minimum accumulated distances are passed on to the next level of the tree. Once the final stage of the tree has been reached, the path with the minimum Euclidean distance is selected as the K-Best solution. By following this tree-traversal procedure, the total amount of computed AEDs equals K. However, the amount of calculated distance increments required to reach the K-Best solution is remarkably higher than KN and strongly depends on the sorting approach implemented at each stage. If a fully-pipelined high-throughput architecture is to be used, the computation of a high amount of PEDs or AEDs derives in a high resource and multipliers usage.

The potential for parallel implementation of the algorithm comes at the price of a complex sorting procedure required at each stage of the search tree. This process, which performs the selection of the K candidate nodes, represents a bottleneck in K-Best-based VP systems. A full sorting procedure which computes all KB AEDs and subsequently sorts them is a simple method to carry out the ordering at the N stages of the tree but its complexity becomes prohibitive when medium-to-high values of K are used. The candidate sorting process can be simplified if the merging of K ordered lists is used (Wenk, Zellweger, Burg, Felber & Fichtner, 2006) along with the Schnorr-Euchner enumeration. Nevertheless, the complicated trigonometric operations required to order the values of the complex lattice as stated by the enumeration procedure has led to the dominance of equivalent real-valued models. However, direct operation on the complex plane is preferred from an implementation point of view as the length of the search-tree is doubled when real-value decomposition (RVD) is utilized. By using a sorted list merging approach, the complexity of the ordering strategy is reduced as only K^{2} AEDs need to be computed and processed. In Mondal, Ali and Salama (2008) a novel ordering strategy is proposed. The so-called winner-path extension (WPE) algorithm selects the K candidate branches by means of the computation of just $2K - 1$ AEDs, as opposed to the KB and K^2 branch metric computations required by the previous methods. As is the case with the list merging approach, the implementation of a Schnorr-Euchner enumerator is required. If the complex lattice model is to be used, the implementation of a complex-plane enumerator would be needed. A K-level enumerator for the complex plane has been recently presented in Barrenechea, Barbero, Jiménez, Arruti and Mendicute (2011) which can be utilized along with the WPE or list merging algorithms to carry out the ordering procedure.

As has been previously stated, fixed-complexity tree search strategies require of the rearrangement in the columns of the precoding matrix in order to achieve better performance results. The quality of the pruning process in a K-Best tree search can be enhanced by placing the most reliable streams at the top of the tree, so that a better solution can be found in the subsequent node selection stages. This ordering approach has been reported to achieve good error-rate performance results when applied to precoding scenarios. More specifically, the sorted QR matrix ordering ($\mathfrak{O}_{SQR}$) presented in Wübben, Böhnke, Rinas, Kühn and Kammeyer (2001) has been used along with the K-Best fixed-complexity tree-search scheme in Habendorf and Fettweis (2007).

The Fixed Sphere Encoder

An efficient search algorithm for the perturbation signal is one of the greatest challenges when implementing a VP system. Implementation-wise, the search engine is preferred to be parallelizable so that all the resources in the target device are efficiently utilized. A fully-pipelined architecture is also of great benefit as it accelerates the execution speed of the algorithm. Furthermore, it is an advantageous property that the complexity of the

algorithm is fixed so that the transmit symbols can be encoded at a fixed rate. The FSE algorithm gathers all of the aforementioned properties, which makes it a great candidate for performing the search of the perturbation signal in a VP scheme.

FSE Tree-Search Architecture

The proposed fixed complexity algorithm is based on the fixed sphere decoder (FSD) (Barbero, 2006), which was designed to overcome the two main drawbacks of the sphere decoder (SD) (Damen, El Gamal & Caire, 2003; Viterbo & Boutros, 1999) detection scheme in single-user MIMO systems: variable complexity and sequential search.

The main feature of the FSE is that, instead of constraining the search to those nodes whose AEDs are within a certain distance from the reference signal z_i, the search is performed with no regard to the AEDs. The tree search is defined instead by a tree configuration vector $\boldsymbol{n} = \left[n_1, \ldots, n_N\right]$, which determines the number of child nodes to be considered at each level (n_i). Therefore, only n_i PEDs are computed per parent node and the Schnorr-Euchner enumeration (Schnorr & Euchner, 1994) is used in order to select the n_i nodes to be processed, being the total number of AEDs $n_T = \prod_{i=1}^{N} n_i$. When the bottom of the tree is reached, that is $i = 1$, the path whose AED is the smallest is selected as the solution. A representation of an FSE tree search is depicted in Figure 4 for a 4×4 system with a constrained lattice of four elements and $\boldsymbol{n} = \left[1, 1, 2, 4\right]$.

As is the case with the K-Best encoder, a fixed amount of AEDs are computed during the FSE search process. Nevertheless, the intricate node ordering procedure required by the K-Best algorithm is not necessary when performing an FSE search as all the child nodes that need to be sorted out stem from the same parent node. Therefore, a

Figure 4. FSE tree-search in a 4×4 system with a constrained lattice of four elements and $\mathrm{n} = \left[1, 1, 2, 4\right]$

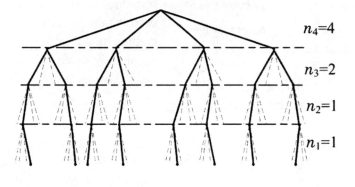

$n_4 = 4$

$n_3 = 2$

$n_2 = 1$

$n_1 = 1$

Schnorr-Euchner enumerator is sufficient in order to select the candidate child nodes.

The tree configuration vector of the FSE provides a flexible error-rate performance and complexity trade-off. Heuristically, one can argue that the higher the number of calculated paths is, the better the performance of the overall system will be, and viceversa. This fact is generally true, as an extended set of candidate vectors increases the probability of finding the optimum solution vector. Nevertheless, it should be noted that, for a given non-prime value of n_T, several tree configuration vectors can be defined, where each one will provide a certain error-rate performance. In the original description of the algorithm, namely the FSD technique for signal detection, the use of the $\boldsymbol{n} = \left[1, 1, \ldots, 1, n_T\right]$ tree configuration vector was recommended in consideration of the specific features of the FSD ordering strategy. Nevertheless, due to the key differences between MIMO detection and multi-user precoding scenarios, the guidelines for the design of the optimum tree configuration vector presented in Barbero and Thompson (2006) are not valid for the current application of the sort-free fixed-complexity tree search.

However, there are several lessons to be learned from the tree-search structure of the FSD. For example, by distributing the values within the tree configuration vector such that $n_1 < n_2 < \ldots < n_N$, a broader range of eligible values can be achieved at the top of the tree, which enables the selection of more suitable perturbation values at the rest of the levels of the tree. Additionally, n_1 should be set to 1 as setting a higher value for this element would not provide any error-rate improvement if the Schnorr-Euchner enumeration is followed.

We shall refer to the set of $\mathcal{W}$ factoring prime integers of n_T as:

$$\mathfrak{W} \triangleq \left\{\eta_1, \eta_2, \ldots, \eta_W\right\}$$

with $\prod_{j=1}^{w} \eta_j = n_T$. For a given value of n_T, the set of eligible tree configuration vectors can be represented as

$$\mathcal{N}_{n_T} = \left\{\left[1, \ldots, 1, n_T\right],\right.$$

$$\left[1, \ldots, \eta_1, n_T / \eta_1\right],$$

$$\left[1, \ldots, \eta_2, n_T / \eta_2\right], \ldots,$$

$$\left[1, \ldots \eta_1, \eta_2, n_T / (\eta_1 \eta_2)\right], \ldots,$$

$$\left.\left[1, \ldots \eta_1, \eta_2, \ldots, \eta_W\right]\right\}$$

Even if all of the tree-configuration vectors in the set $\mathcal{N}_{n_T}$ consider the same amount of traversed branches through the search-tree, the error-rate performance and computational effort required for each one of them is different. The BER performance curves of FSE-based VP systems in the 6×6 and 8×8 antenna setups considering the tree configuration vector candidate sets $\mathcal{N}_{12}$ and $\mathcal{N}_{24}$ are shown in Figure 5 and Figure 6, respectively. The data depicted in these figures show that the tree configuration vectors of the type $\boldsymbol{n} = \left[1,1,\ldots,n_T\right]$ achieve the worst performance among the eligible vectors in the set $\mathcal{N}_{n_T}$ for both antenna setups. Moreover, those tree structures with more dispersedly distributed values of n_i achieve the best performance, being the performance gap with respect to other tree configuration vectors with higher values of n_i more noticeable in the 8 antenna case.

Apart from the impact on the error-rate performance of the FSE, the implementation of a certain tree configuration vector also affects the computational complexity of the tree traversal. As already stated, all the candidate tree configuration vectors in $\mathcal{N}_{n_T}$ yield the same amount of total computed branches or AEDs. Nevertheless, this does not imply that the amount of computed PEDs is equal for all the considered configuration vectors. Therefore, the selection of the tree configuration vector will determine the computational complexity of the FSE, as most of its computational load is due to PED calculations. The amount of required PED calculations (C_{PED}) for a given FSE tree structure is given by the following formula:

$$C_{PED} = \sum_{i=1}^{N}\prod_{j=1}^{N} n_i \qquad (5)$$

Figure 5. *BER performance of the FSE with different tree configuration vectors in a* 6×6 *antenna setup*

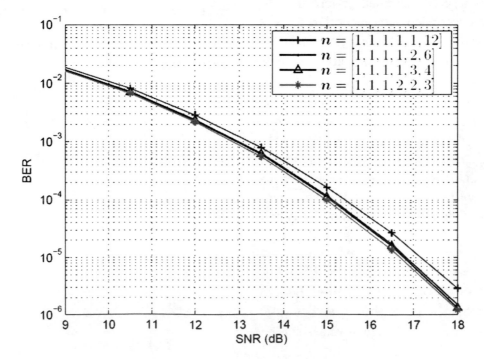

Figure 6. BER performance of the FSE with different tree configuration vectors in an 8×8 *antenna setup*

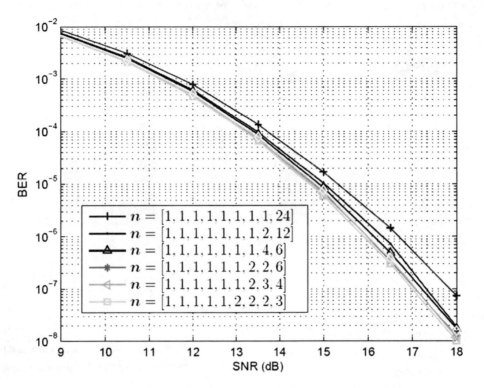

Additionally to the Euclidean distance computations, the number of intermediate point calculations should also be taken into account when assessing the computational complexity of a certain tree-search structure, as the high volume of complex-valued multiplications involved in their computation represents an important part of the computational load of the FSE tree traversal. Therefore, the amount of intermediate points to be computed (C_{IP}) can be calculated as:

$$C_{IP} = \sum_{i=1}^{N-1} \prod_{j=i+1}^{N} n_i \qquad (6)$$

Clearly from Equations (5) and (6), those tree configuration vectors with reduced values of n_i at each level will render the smallest computational complexity among the candidate vectors in $\mathcal{N}_{n_T}$. To better support this argument, the number

of PED computations and z_i calculations have been computed for all the tree configuration vectors with $n_T = 24$ in a $N = 8$ user system. Additionally, the total number of operations has been calculated for each one of these cases, where for the sake of simplicity, all the arithmetic operations (multiplication, addition and subtraction) have been considered to have the same weight on the final operation count. The results of this study are depicted in Figure 7. As already anticipated, the provided data reflect the dependency between the computational complexity of the FSE tree traversal and its tree-search structure. More specifically, those tree-search architectures with a greater amount of expanded nodes at the top levels require a higher number of operations than those with a more distributed node expansion scheme.

However, when it comes to hardware implementation, the multiplication count of a certain design is a factor of mayor importance as the em-

Figure 7. Number of PED computations (C_{PED}) and z_i calculations (C_{IP}) required for several tree configuration vectors with $n_T = 24$. Additionally, the total amount of operations (right axis) is depicted for each one of the FSE tree-search structures

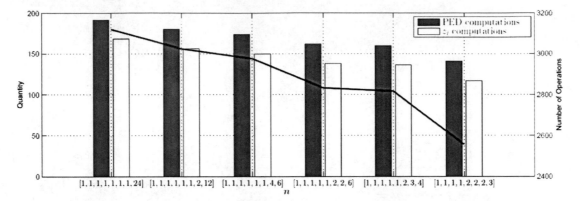

bedded multipliers are a scarce resource in FPGA devices and occupy a considerable device area in ASIC devices. In this case too, the structure of the tree search will determine the amount of required multiplication units, as is shown in Figure 8. The results shown in the aforementioned figure have been obtained by considering that 3 multipliers are required for the product of two complex variables. This has been achieved by rearranging the terms involved in the computation of the complex multiplication from the most straightforward approach structured as

$$(a + jb)(c + jd) = (ac - bd) + j(ad + cb),$$

where 4 multiplication operations are performed, to a more computationally-efficient method, namely:

$$(a + jb)(c + dj) =$$
$$[a(c - d) + d(a - b)]$$
$$+j[b(c + d) + d(a - b)]$$
,

Figure 8. Amount of required multipliers for different tree configuration vectors in an FSE tree search

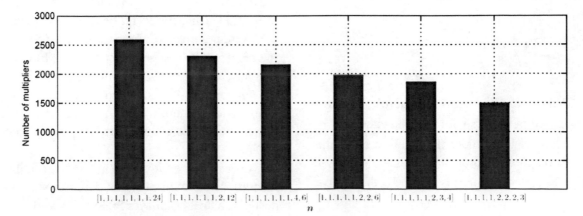

that reduces the number of required multipliers due to the repeated factor $d(a-b)$ (Barbero, 2006, Barrenechea, Barbero, Mendicute & Thompson, 2012). The data shown in Figure 8 reflect the diverse computational complexity of the tree configuration vectors in the set $\mathcal{N}_{24}$. The tree configuration vectors that expand more nodes at the higher levels of the tree search require more intermediate point computations and PED calculations, as already seen in Figure 7, which ultimately results in a high volume of allocated multipliers. Hence, by selecting the most disperse tree configuration vector within the candidate set $\mathcal{N}_{24}$, the amount of required embedded multipliers can be reduced in a 42% when compared to the originally proposed $\boldsymbol{n} = \left[1, 1, \ldots, n_T\right]$ tree structure.

In the light of the error-rate performance and computational complexity results, it is concluded that the optimum tree configuration vector for the precoding scenario is obtained by factorizing the amount of desired computed branches n_T and arranging the resulting values such that $n_1 \leq n_2 \leq \ldots n_N$. Hence, if the tree traversal is performed starting from $i = N$ and working backwards until $i = 1$, the tree configuration vector is set as

$$n_i = \begin{cases} \max \omega^{(i)} & for\ N - i < \mathcal{W} \\ 1 & otherwise \end{cases}$$

where $\omega^{(i)}$ represents the set of factoring prime integers that have not yet been selected at level i.

The performance of the algorithm in terms of BER is highly related to the selection of the tree configuration vector and the matrix ordering that determines the position of each user's symbols in the search tree. In the following section, different user ordering strategies will be discussed.

Matrix Preprocessing for the FSE

The close-to-optimum performance of the original FSD algorithm proposed for the single-user MIMO detection scenario is due to a unique combination of an unconventional ordering strategy and an optimized tree-search structure. The key factor of the FSD ordering approach is that, in order to minimize the probability of a wrong decision, the weakest stream must be placed at the top of the tree, where all possible nodes are considered for expansion. However, this approach is not applicable in precoding scenarios as the set of candidate nodes at the top level of the tree does no longer have a finite number of elements. This fact, along with the inherent differences between decoding and precoding scenarios, motivates the implementation of a different ordering strategy for the FSE algorithm.

A more suitable ordering strategy can be used for the FSE by exploiting the similarities between VP and THP schemes. After all, the VP system model is transformed into the THP precoding scheme if an FSE tree traversal with $\boldsymbol{n} = \left[1, 1, \ldots, 1\right]$ is used to obtain the precoding vector. Hence, the THP solution is always contained within the set of candidate branches considered by the FSE regardless of the tree configuration vector in use. In a K-Best system however, such a statement cannot be made as the THP solution vector may be dropped out during any of the sorting stages of the algorithm. The following section summarizes the main features of the preferred ordering for THP systems.

THP can be regarded as the precoding counterpart of decision-feedback equalization (DFE), where a successive interference cancelation procedure is performed on the received signal vector. With the aim of minimizing the error propagation through each detection step, the spatial streams in a DFE approach are ordered following a best-first rule (Kusume, Joham, Utschick, & Bauch, 2007). This ordering strategy states that the strongest data stream is detected first while the worst stream is left for the final detection stage. Nevertheless, the picture is different in precoding scenarios. In a THP system for example, the first data stream is transmitted unaltered (only the channel shaping matrix is applied to this signal), whereas the last precoded signal has to avoid interfering all the previous data streams. Hence, the task of precoding the last user is much more arduous. Consequently, following a best-last rule instead and precoding the best stream last is a sensitive decision, as this data stream has less degrees of freedom. This reversed ordering approach represents the optimum ordering strategy for THP systems ($\mathfrak{D}_{BL}$) (Joham, Brehmer & Utschick, 2004), and hence, it will be used along with the FSE tree-search to enhance its error-rate performance.

Once the strategy of the ordering procedure has been defined, we shall focus on the criteria used to select the best/worst stream at each iteration. Unlike in point-to-point MIMO detection, where the ordering was determined based on the SNR associated with each transmitted stream, the MSE of the perturbed symbols in a block transmission will be the parameter used to establish the optimum user permutation strategy for VP. Note that, the perturbation vector in Equation (1) is computed so as to minimize the MSE for a given block of data symbols. Consequently, the best user stream will be the one that contributes the least to the MSE, or equivalently the one with a smallest diagonal value in the:

$$\Psi = \left(HH^H + \xi I_N \right)^{-1} \text{ matrix.}$$

Table 1 shows the pseudocode for the joint user order determination and triangular matrix computation. The reverse direction in which the matrix preprocessing (from $i = 1$ to N) and the tree traversal (from $i = N$ down to $i = 1$ since U is upper triangular) are carried out is noticeable from the algorithm displayed in this table. The

Table 1. Computation of the upper-triangular matrix U with best-last ($\mathfrak{D}_{BL}$) ordering

$$\Psi = \left(HH^H + \xi I_N \right)^{-1}$$
$$\Pi = I_N$$
for $i = 1, \ldots, N$
$$\quad q = \operatorname*{argmin}_{\hat{q} \in \{i, \ldots, N\}} \Psi(\hat{q}, \hat{q})$$
$$\quad \Xi = \tilde{I}_N^{(i,q)}$$
$$\quad \Pi = \Xi \Pi$$
$$\quad \Psi = \Xi \Psi \Xi^T$$
$$\quad D(i,i) = \Psi(i,i)$$
$$\quad \Psi(i:N,i) = \Psi(i:N,i) / \Psi(i,i)$$
$$\quad \Psi(i+1:N, i+1:N) =$$
$$\quad \Psi(i+1:N, i+1:N) -$$
$$\quad \Psi(i+1:N, i) \Psi(i+1:N, i)^H D(i,i)$$
end
$$L = lower\ triangular\ part\ of\ \Psi$$
$$U = D^{1/2} L^H$$

calculation of the triangular matrix in Equation (1) is performed following the computationally-efficient Cholesky factorization with symmetric permutation method presented in Kusume, Joham, Utschick and Bauch (2005, 2007), whose complexity is similar to that of a sorted QR decomposition.

The performance of the proposed FSE tree search is depicted in Figure 9 for a 4×4 antenna system and a tree configuration vector $n = \begin{bmatrix} 1, 1, 2, 5 \end{bmatrix}$. The triangular matrix used for the fixed-complexity tree traversal has been rearranged following the aforementioned $\mathfrak{D}_{SQR}$ and $\mathfrak{D}_{BL}$ ordering strategies. The BER performance curve of the unordered FSE tree search has additionally been included for completion. The benefits of performing an appropriate preprocessing stage on the triangular matrix are clearly visible from the data displayed in this figure. As one can notice, a performance gain of 5 dB at a BER of 10^{-5} can be attained by simply rearranging

Figure 9. BER performance of the proposed FSE algorithm with $\mathbf{n} = \lfloor 1,1,2,5 \rfloor$ and ordering strategies $\mathfrak{O}_{SQR}$ and $\mathfrak{O}_{BL}$. Additionally, the unordered case is included for completion

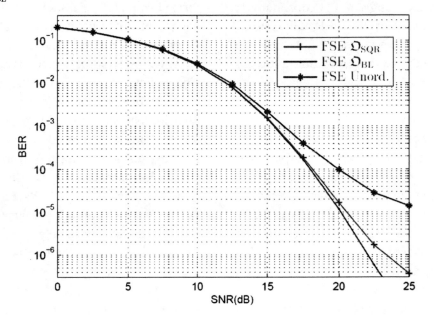

the order in which the users are processed within the search tree. The data displayed in this figure also show that a similar BER performance is obtained by the FSE at the low-SNR range with any of the considered ordering strategies. Nevertheless, the FSE with the $\mathfrak{O}_{SQR}$ ordering suffers considerable error-rate performance degradation in the high-SNR regime.

An alternative precoding version of the FSD has been recently published in Mohaisen and Chang (2011). The presented algorithm, also referred to as FSE, deals with the parallel computation of the branches in a fixed-complexity tree search. Nevertheless, the differences with respect to the algorithm presented in this chapter are numerous. First, no guidelines for the configuration of the fixed-complexity tree search are provided. Additionally, the approach in Mohaisen et al. (2011) features a RVD model, with the consequent double length of the tree search, and does not consider the use of a matrix preprocessing method to enhance the error-rate performance of the system. Due to

these facts, the approach in Mohaisen et al. (2011) achieves a considerably worse BER performance than the FSE scheme reviewed in this chapter.

The Variable-Complexity SBE Algorithm

The so-far addressed K-Best and FSE bounded breadth-first techniques represent a low-complexity alternative to the widely-used SE algorithm. The advantages of a non-iterative tree traversal, such as the fixed-complexity of the tree search and the high data processing throughput, have been presented and analyzed in previous sections of this chapter. Nevertheless, the branch pruning strategy in bounded breadth-first schemes also presents other shortcomings.

Given that the selection of the nodes at each stage is performed based on the PEDs or AEDs up to that level and not the Euclidean distance associated with the entire branch, the validity of the pruning is not certain until the bottom of the tree is reached. This speculative pruning strategy results in a great amount of unnecessary distance computations and excessive power consumption. Consequently, many of the PED calculations performed during a bounded breadth-first tree search, along with the corresponding allocated hardware resources and power usage, can be spared by traversing the tree in both forward and backward directions.

The depth-first model of an originally breadth-first algorithm for the computation of the perturbation vector in VP systems will be presented in the following sections. The tree-search structure of the sequential best-node expansion (SBE) scheme is inspired by the so-called conditioned ordered successive interference cancelation (COSIC) detector introduced in Hess, Wenk, Burg, Luethi, Studer, Felber & Fichtner (2007) (Wenk, 2010). The aforementioned tree-search approach also features a novel user ordering strategy that provides a good trade-off between error-rate performance and complexity of the algorithm in terms of amount of visited nodes (Barrenechea, Burg & Mendicute, 2012).

This novel scheme features two main concepts: a tree traversal architecture where only the most promising nodes are expanded and an innovative matrix preprocessing strategy. The former is based on the COSIC tree search described for signal detection in MIMO scenarios in Hess et al. (2007) and Wenk (2010), while the latter has been specially designed to yield a good trade-off between performance and run-time of the SBE algorithm.

SBE Tree-Search Structure

The tree structure of the SBE is a depth-first implementation of an originally breadth-first algorithm, which enables the usage of a sphere constraint for additional pruning. This way, the Euclidean distance related to the first computed branch is considered as the initial radius R, which is updated every time a leaf node with a smaller AED is reached. The main motivation for the

change of tree traversal strategy is the reduction of the unnecessary distance computations performed in the fully-parallel scheme. As a consequence of adopting a sequential scheme, the complexity of the tree search is no longer fixed. However, by avoiding a speculative node-expansion policy, an overall smaller computational effort is required when compared to parallel branch-processing approaches. The tree-search architecture under consideration uses different node expansion policies throughout the tree traversal (Barrenechea, Burg & Mendicute, 2012):

On-Demand Sibling Node Expansion: This node expansion approach is only performed at the root level of the tree, namely $i = N$. In the initial step of the algorithm, the best node according to the Schnorr-Euchner enumeration is selected and passed on to the next level. When the algorithm returns to the root of the tree, the next most favorable node is expanded only if its PED is smaller than R. Note that the ordered child node sequence at this level can be obtained without performing any PED calculations and subsequent sorting procedures.

Single Node Expansion: A single child node is expanded per parent node in levels $i < N$, being a radius check performed at each one of these levels. This way, if a node with $D_i > R$ is reached, all its descendants are discarded and the algorithm returns to the root level. Otherwise, the tree traversal is resumed. When a leaf node with $D_1 < R$ is found, the branch leading to that node is stored as a candidate solution and the search radius is updated accordingly.

The sequential nature and the radius reduction procedure allow for an efficient pruning of undesired branches. Additionally, the single node expansion policy provides a more regular data path than that of the SE and removes the need for tracking candidate sibling nodes throughout the tree traversal.

Following the inherent sequential nature of the SBE algorithm, the hardware implementation of the tree search can be carried out by concatenating N single-branch distance processing units (DPUs) and iterating through the architecture as necessary. This implementation approach derives in a modest hardware usage and a more efficient pruning, but also yields a reduced and variable throughput. Nevertheless, the former disadvantage can be alleviated by running several DPU instances in parallel, hence processing various vector symbols at the same time. The data processing throughput of the SBE algorithm can be enhanced by relaxing the premise of a strictly sequential tree traversal. This way, the architecture of the SBE tree can be parallelized to some degree by employing several single-branch DPUs to simultaneously process various branches within the same search tree. This approach results in a higher data processing throughput derived from an earlier shrinking of the sphere radius, but also involves a less computationally-efficient tree search. Several tree-search structures for the SBE are investigated in Barrenechea, Burg and Mendicute (2012), where a 65 nm technology ASIC was considered as the target device.

Matrix Preprocessing Algorithm for the SBE

The rearrangement of the user streams in the search tree has a great impact on the amount of nodes that are visited during an iterative tree traversal. In systems where there is some sort of limitation on the number of nodes to be processed, such as the K-Best or the FSE, the procedure of user ordering solely affects the error-rate performance of the algorithm. Due to the special pruning strategy performed in the SBE algorithm, the rearrangement of the user streams will impact both its performance and run-time.

The optimum ordering approach for SBE-based tree-search schemes ($\mathfrak{O}_{SBE}$) was presented in Barrenechea et al. (2012). The aforementioned ordering strategy consists of following a best-last approach when determining the user order in levels $i < N - 1$. Therefore, the ordering proce-

dure starts by selecting the best data stream at the bottom level of the tree, namely $i = 1$, and proceeds equally through the rest of the levels until $i = N - 2$. Once the second level of the tree is reached ($i = N - 1$), the worst stream is selected instead.

The pseudocode for the joint user order determination and triangular matrix computation of the $\mathfrak{O}_{SBE}$ strategy is shown in Table 2.

Comparative BER Performance under an Overall Run-Time Constraint

The iterative tree-search algorithms discussed in this chapter require a variable amount of cycles to find the perturbation vector. This variable nature usually constitutes a problem in practical systems, where the user data streams are supposed to be processed at a fixed rate. It is therefore important to assess the error-rate performance of the tree-search algorithms under an overall run-time constraint. This way, a fair comparison of the error-rate performance of different tree-search techniques is possible.

The data in Figure 10 reflect the error-rate performance of an 8×8 VP system where an overall run-time constraint of $\varpi = 32$ evaluated nodes has been imposed on the analyzed tree-search algorithms. Given the fixed-complexity of the K-Best and FSE approaches, the design parameters that yield a node expansion count of ϖ have been selected. Consequently, values of $K = 4$ and $n_T = 4$ have been considered for the K-Best and FSE tree-search approaches, respectively. Note that the optimum tree configuration vector for FSE that yields $n_T = 4$ in an 8-user system entails the computation of 30 PEDs. This is the closest amount of PED computations to $\varpi = 32$ that can be achieved with the FSE, as dictated by Equation (5).

As is shown in Figure 10, the performance of the run-time-constrained algorithms is similar, being the K-Best the one that performs slightly

Table 2. Computation of the upper-triangular matrix **U** *with (* $\mathfrak{O}_{SBE}$ *) ordering*

$$\Psi = \left(HH^H + \xi I_N \right)^{-1}$$

$$\Pi = I_N$$

for $i = 1,\dots,N$

if $i = N - 1$

$$q = \operatorname*{argmax}_{\hat{q} \in \{i,\dots,N\}} \Psi(\hat{q},\hat{q})$$

else

$$q = \operatorname*{argmin}_{\hat{q} \in \{i,\dots,N\}} \Psi(\hat{q},\hat{q})$$

end

$$\Xi = \tilde{I}_N^{(i,q)}$$

$$\Pi = \Xi\Pi$$

$$\Psi = \Xi\Psi\Xi^T$$

$$D(i,i) = \Psi(i,i)$$

$$\Psi(i:N,i) = \Psi(i:N,i)\,/\,\Psi(i,i)$$

$$\Psi(i+1:N,i+1:N) =$$

$$\Psi(i+1:N,i+1:N) -$$

$$\Psi(i+1:N,i)\,\Psi(i+1:N,i)^H\,D(i,i)$$

end

$$L = lower\ triangular\ part\ of\ \Psi$$

$$U = D^{1/2}L^H$$

better in the low and mid-SNR ranges. Nevertheless, the K-Best shows its characteristic error-rate performance degradation in the high-SNR regime. Despite their suboptimum performance, the additional pruning strategy carried out in the fixed-complexity and SBE-based systems favors the expansion of the most promising nodes as opposed to the SE, which considers all eligible nodes at every level. As a consequence to this, a slightly higher BER performance degradation is introduced by the SE when imposing an overall run-time constraint (Barrenechea, 2012).

Implementation of an Approximate Norm for Hardware Resource Reduction

A significant portion of the hardware resources in the implementation of any tree-search algorithm is dedicated to computing the ℓ^2 norms required by the cost function in Equation (1). Additionally, the long delays associated with squaring operations required to compute the PEDs account for a significant portion of the latency of the fixed-complexity tree-search architectures. It is

Figure 10. BER performance of various tree-search approaches with an overall run-time constraint of $\varpi = 32$ *evaluated nodes for an* 8×8 *antenna setup*

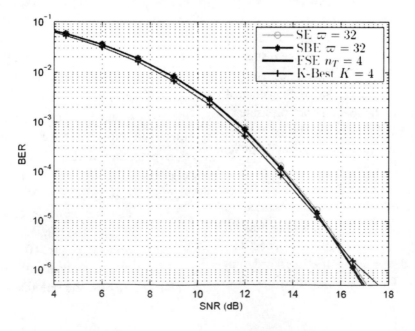

possible to overcome these problems by using an alternative norm that prevents the use of the computationally expensive squaring operations.

The application of the modified-norm algorithm (MNA) (Burg, Wenk, Zellweger, Wegmueller, Felber, & Fichtner, 2004) entails two main benefits: on one hand, a simplified distance computation scheme that immediately reduces silicon area and delay of the arithmetic units can be performed, and on the other hand, a smaller dynamic range of the PEDs is achieved. The key point of the MNA is to compute the square root of the accumulated and partial distance increments, namely $E_i = \sqrt{D_i}$ and $e_i = \sqrt{d_i}$, respectively. Hence, the accumulation of the distance increments in this equivalent model gives:

$$E_i = \sqrt{E_{i+1}^2 + e_i^2}$$

An approximate norm can now be applied to get rid of the computationally-expensive squaring and square root operations, such that

$$E_i \approx f\left(\left|E_{i+1}\right|, \left|e_i\right|\right)$$

This way, the distance accumulation and computation in Equation (4) can be reformulated as

$$E_i = E_{i+1} + e_i,$$

with

$$e_i = u_{i,i}\left(\left|\mathbb{R}\left(a_i + z_i\right)\right| + \left|\mathbb{I}\left(a_i + z_i\right)\right|\right)$$

for the $\tilde{\ell}^1$-norm variant of the algorithm. The norm approximation can also be performed following the $\tilde{\ell}^\infty$-norm simplified model, in which case the following expressions should be considered:

$$E_i = \max\left(E_{i+1}, e_i\right),$$

with

$$e_i = u_{i,i}\left[\max\left(\left|\mathbb{R}\left(a_i + z_i\right)\right|, \left|\mathbb{I}\left(a_i + z_i\right)\right|\right)\right].$$

The error-rate performance degradation introduced when approximating the optimum ℓ^2-norm varies depending on the tree-search technique being analyzed and the norm used in the approximation. Due to the better approximation of the optimum norm performed by the $\tilde{\ell}^1$-norm, a slightly better error-rate performance can be attained by the aforementioned norm variant (Barrenechea, Mendicute, Arruti, 2013).

As for the achievement in complexity reduction, a considerable decrease in the hardware resources has been documented for the K-Best and FSE tree-search architectures in Barrenechea et al. (2013). More specifically, the impact of adopting a norm approximation scheme on a 4×4 vector precoder implemented on a Xilinx Virtex VI FPGA (XC6VHX250T-3) is documented in the aforementioned study. A 27.5% reduction in the required DSP48e1 blocks is achieved for the FSE tree-search with $n_T = 10$, while a 22.5% decrease is attained for the K-best with $K = 7$. Note that this considerable resource saving is achieved at the cost of a minor error-rate performance degradation (about 0.22 dB in the high-SNR regime).

FUTURE RESEARCH DIRECTIONS

Three are the main directions this research work can follow in the future:

- Assessment of the performance of the reviewed tree-search algorithms, in a more realistic scenario. The adverse conditions to be considered include imperfections in the channel state information or interference generated from users in an adjacent cell. Moreover, it would be of great interest to analyze the performance of the aforementioned tree-search algorithms with real measured channel coefficients instead of the widely-used Rayleigh fading model.

- Since the number of antennas used in base stations is growing, it would be interesting to extend the algorithm and results provided in this research to scenarios with a large amount of antennas. Even if linear precoding schemes are normally used in such scenarios, VP-based techniques could reduce the required number of transmission chains while achieving a similar performance.

- Considering the capacity of the latest families of programmable logic devices, such as the Xilinx 7 family of FPGAs, it would be of great interest to implement all the aforementioned precoders on such target devices. This would show the current silicon cost of these algorithms and would allow a complete and fair comparison among all the architectures.

CONCLUSION

This chapter addresses the problem of an efficient computation of the perturbation vector in VP systems. Several tree-search techniques have been proposed in the literature to solve the selection of the perturbation vector, being the SE one of the most widely used tree-search techniques due to its optimum error-rate performance. The good performance of the algorithm is a consequence of the identification and management of the admissible set of nodes at each stage of the tree search, which ultimately leads to a variable complexity

of the algorithm and rather intricate hardware architecture of the tree traversal.

Due to the simplicity of its architecture and the possibility of parallel processing, the fixed-complexity K-Best algorithm has been regarded as a prominent candidate for an efficient and high-speed hardware implementation of vector precoders. Despite its non-iterative nature, which greatly simplifies the tree traversal, the sorting stages required to select the candidate branches contribute to the high computational complexity of the algorithm. With the aim of overcoming the main shortcomings of the K-Best precoder, a sort-free fixed-complexity algorithm has been proposed in the literature. The FSE scheme traverses the tree in a non-iterative fashion, where the selection of the nodes at each level is dictated by a tree configuration vector. This design parameter offers a flexible trade-off between error-rate performance and complexity. Furthermore, the optimum combination of matrix preprocessing strategy and structure of the FSE tree search that results in a minimum computational complexity and best error-rate performance has been reviewed.

This chapter has also dealt with the design of sequential and low-complexity tree-search algorithms for VP. The main motivation for the change of tree traversal strategy is the implementation of a sphere constraint that allows for additional pruning. This way, a considerable reduction of the unnecessary distance computations performed in the fully-parallel scheme can be achieved.

The SBE approach features, on one hand, a simplified tree-search architecture that only considers the most favorable nodes for expansion and, on the other hand, a specially tailored ordering strategy that provides a good trade-off between performance and computational complexity. Provided BER results have shown that the proposed approach achieves an error-rate performance that is close to the optimum set by the SE.

REFERENCES

Anderson, J. B., & Mohan, S. (1984). Sequential coding algorithms: A survey and cost analysis. *IEEE Transactions on Communications, 32*(2), 169–176. doi:10.1109/TCOM.1984.1096023

Barbero, L. (2006). *Rapid prototyping of a fixed-complexity sphere decoder and its application to iterative decoding of turbo-MIMO systems.* Edinburgh, UK: University of Edinburgh.

Barbero, L., & Thompson, J. (2006). A fixed-complexity MIMO detector based on the complex sphere decoder. In *Proceedings of the IEEE International Workshop on Signal Processing Advances for Wireless Communications* (pp. 1-5). New York, NY: IEEE.

Barrenechea, M. (2012). *Design and implementation of multi-user MIMO precoding algorithms.* Mondragon, Spain: University of Mondragon.

Barrenechea, M., Barbero, L., Jiménez, I., Arruti, E., & Mendicute, M. (2011). High-throughput implementation of tree-search algorithms for vector precoding. In *Proceedings of the IEEE International Conference on Acoustics, Speech and Signal Processing* (pp. 1689-1692). New York, NY: IEEE.

Barrenechea, M., Barbero, L., Mendicute, M., & Thompson, J. (2012). Design and implementation of a low-complexity multiuser vector precoder. *International Journal of Embedded and Real-Time Communication Systems, 3*(1), 1–18. doi:10.4018/jertcs.2012010102

Barrenechea, M., Burg, A., & Mendicute, M. (2012). Low complexity vector precoding for multiuser systems. In *Proceedings of the Asilomar Conference on Signals, Systems and Computers* (pp. 453-457). New York, NY: IEEE.

Barrenechea, M., Mendicute, M., & Arruti, E. (2013). Fully-pipelined Implementation of Tree-search Algorithms for Vector Precoding. *International Journal of Reconfigurable Computing.* doi:10.1155/2013/496013

Barrenechea, M., Mendicute, M., Del Ser, J., & Thompson, J. S. (2009). Wiener filter- based fixed-complexity vector precoding for the MIMO downlink channel. In *Proceedings of the IEEE Signal Processing Advances in Wireless Communications* (pp. 216–220). New York, NY: IEEE.

Bhagawat, P., Wang, P., Uppal, M., Choi, G., Xiong, Z., Yeary, M., & Harris, A. (2008). An FPGA implementation of dirty paper precoder. In *Proceedings of the IEEE International Conference on Communication.* (pp. 2761–2766). New York, NY: IEEE.

Burg, A., Seethaler, D., & Matz, G. (2007). VLSI implementation of a lattice-reduction algorithm for multi-antenna broadcast precoding. In *Proceedings of the IEEE International Symposium on Circuits and Systems* (pp. 673 – 676). New York, NY: IEEE.

Burg, A., Wenk, M., Zellweger, M., Wegmueller, M., Felber, N., & Fichtner, W. (2004). VLSI implementation of the sphere decoding algorithm. In *Proceedings of the European Solid-State Circuits Conference* (pp. 303-306). New York, NY: IEEE.

Costa, M. H. M. (1983). Writing on dirty paper. *IEEE Transactions on Information Theory, 29*(3), 439–441. doi:10.1109/TIT.1983.1056659

Damen, M. O., El Gamal, H., & Caire, G. (2003). On maximum-likelihood detection and the search for the closest lattice point. *IEEE Transactions on Information Theory, 49*(10), 2389–2402. doi:10.1109/TIT.2003.817444

Dietrich, F. A. (2008). *Robust Signal Processing for Wireless Communications*. Berlin, Germany: Springer.

Fincke, U., & Pohst, M. (1985). Improved methods for calculating vectors of short length in a lattice, including a complexity analysis. *Mathematics of Computation*, *44*(170), 463–471. doi:10.1090/S0025-5718-1985-0777278-8

Fischer, R. F. H., Windpassinger, C., Lampe, A., & Huber, J. B. (2002). Space-time transmission using Tomlinson-Harashima precoding. In *Proceedings of the 4th ITG Conference on Source and Channel Coding* (pp. 139 – 147). Frankfurt, Germany: VDE Verlag.

Gesbert, D., Kountouris, M., Heath, R., & Chae, C. (2007). From single user to multi user communications: Shifting the MIMO paradigm. *IEEE Signal Processing Magazine*, *24*(5), 36–46. doi:10.1109/MSP.2007.904815

Habendorf, R., & Fettweis, G. (2007). Vector precoding with bounded complexity. In *Proceedings of the IEEE Workshop on Signal Processing Advances in Wireless Communications* (pp. 186-190). New York, NY: IEEE.

Harashima, H., & Miyakawa, H. (1972). Matched-transmission technique for channels with inter-symbol interference. *IEEE Transactions on Communications*, *CM-20*(4), 774–780. doi:10.1109/TCOM.1972.1091221

Hess, C., Wenk, M., Burg, A., Luethi, P., & Studer, C. Felber, N., & Fichtner, W. (2007). Reduced-complexity MIMO detector with close-to ML error-rate performance. In *Proceedings of the ACM Great Lakes Symposium on VLSI* (pp. 200-203). New York, NY: ACM.

Hochwald, B. M., Peel, C. B., & Swindlehurst, A. L. (2005). A vector-perturbation technique for near-capacity multiantenna multiuser communication: perturbation. *IEEE Transactions on Communications*, *53*(3), 537–544. doi:10.1109/TCOMM.2004.841997

Joham, M., Brehmer, J., & Utschick, W. (2004). MMSE approaches to multiuser spatio-temporal Tomlinson-Harashima precoding. In *Proceedings of the International ITG Conference on Source and Channel Coding* (pp. 387- 394). New York, NY: IEEE.

Joham, M., Utschick, W., & Nossek, J. A. (2005). Linear transmit processing in MIMO communications systems. *IEEE Transactions on Signal Processing*, *53*(8), 2700–2712. doi:10.1109/TSP.2005.850331

Kusume, K., Joham, M., Utschick, W., & Bauch, G. (2005). Efficient Tomlinson-Harashima precoding for spatial multiplexing on flat MIMO channel. In *Proceedings of the IEEE International Conference on Communications* (pp. 2021-2025). New York, NY: IEEE.

Kusume, K., Joham, M., Utschick, W., & Bauch, G. (2007). Cholesky factorization with symmetric permutation applied to detecting and precoding spatially multiplexed data streams. *IEEE Transactions on Signal Processing*, *55*(6 II), 3089- 3103.

Lenstra, A. K., Lenstra, H. W., & Lovász, L. (1982). Factoring polynomials with rational coefficients. *Mathematische Annalen*, *216*(4), 513–534.

Mohaisen, M., & Chang, K. (2011). Fixed-complexity sphere encoder for multiuser MIMO systems. *Journal of Communications and Networks*, *13*(1), 36–39. doi:10.1109/JCN.2011.6157253

Mondal, S., Ali, W., & Salama, K. (2008). A novel approach for K-Best MIMO detection and its VLSI implementation. In *Proceedings of the IEEE International Symposium on Circuits and Systems* (pp. 936 - 939). New York, NY: IEEE.

Peel, C. B., Hochwald, B. M., & Swindlehurst, A. L. (2005). A vector-perturbation technique for near-capacity multiantenna multiuser communication: channel inversion and regularization. *IEEE Transactions on Communications*, *53*(1), 195–202. doi:10.1109/TCOMM.2004.840638

Schmidt, D., Joham, M., & Utschick, W. (2008). Minimum mean square error vector precoding. *European Transactions on Telecommunications*, *19*(3), 219–231. doi:10.1002/ett.1192

Schnorr, C. P., & Euchner, M. (1994). Lattice basis reduction: Improved practical algorithms and solving subset sum problems. *Mathematical Programming*, *66*(2), 181–199. doi:10.1007/BF01581144

Tomlinson, M. (1971). New automatic equaliser employing modulo arithmetic. *Electronics Letters*, *7*(5-6), 138–139. doi:10.1049/el:19710089

Verdú, S. (1998). *Multiuser Detection*. Cambridge, UK: Cambridge University Press.

Viterbo, E., & Boutros, J. (1999). A universal lattice code decoder for fading channels. *IEEE Transactions on Information Theory*, *45*(5), 1639–1642. doi:10.1109/18.771234

Wenk, M. (2010). MIMO-OFDM Testbed: Challenges, Implementations, and Measurement Results. Zürich, Switzerland: Eidgennössische Technische Hochschule (ETH).

Wenk, M., Zellweger, M., Burg, A., Felber, N., & Fichtner, W. (2006). K-Best MIMO detection VLSI architectures achieving up to 424 Mbps. In *Proceedings of the IEEE International Symposium on Circuits and Systems* (pp. 1150- 1154). New York, NY: IEEE.

Windpassinger, C., Fischer, R. F. H., & Huber, J. B. (2004). Lattice-reduction-aided broadcast precoding. *IEEE Transactions on Communications*, *52*(12), 2057–2060. doi:10.1109/TCOMM.2004.838732

Wübben, D., Böhnke, R., Rinas, J., Kühn, V., & Kammeyer, K. (2001). Efficient algorithm for decoding layered space-time codes. *IEEE Electronic Letters*, *37*(22), 1348–1350. doi:10.1049/el:20010899

Zhang, J., & Kim, K. J. (2005). Near-capacity MIMO multiuser precoding with QRD-M algorithm. In *Proceedings of the Asilomar Conference on Signals, Systems, and Computers* (pp. 1498–1502). New York, NY: IEEE.

ADDITIONAL READING

Airy, M., Bhadra, S., Heath, R. W., & Shakkottai, S. (2006). Transmit precoding for the multiple antenna broadcast channel. In *Proceedings of the IEEE Vehicular Technology Conference* (pp. 1396-1400). New York, NY: IEEE.

Barbero, L., & Thompson, J. (2006). FPGA design considerations in the implementation of a fixed-throughput sphere decoder for MIMO systems. In *Proceedings of the IEEE International Conference on Field Programmable Logic and Applications* (pp. 1-6). New York, NY: IEEE

Barbero, L., & Thompson, J. (2006). Rapid prototyping of a fixed-throughput sphere decoder for MIMO systems. In *Proceedings of the IEEE International Conference on Communications* (pp. 3082-3087). New York, NY: IEEE.

Barbero, L., & Thompson, J. (2008). Extending a fixed-complexity sphere decoder to obtain likelihood information for turbo-MIMO systems. *IEEE Transactions on Vehicular Technology*, *57*(5), 2804–2814. doi:10.1109/TVT.2007.914064

Burg, A., Borgmann, M., Wenk, M., Zellweger, M., Fichtner, W., & Bölcskey, H. (2005). VLSI implementation of MIMO detection using the sphere decoding algorithm. *IEEE Journal of Solid-State Circuits*, *40*(7), 1566–1577. doi:10.1109/JSSC.2005.847505

Chen, S., Zhang, T., & Xin, Y. (2007). Relaxed K-Best MIMO signal detector design and VLSI implementation. *IEEE Transactions on Very Large Scale Integration (VLSI). Systems*, *15*(3), 328–337.

Foschini, G., & Gans, M. (1998). On limits of wireless communications in a fading environment when using multiple antennas. *Wireless Personal Communications*, 6(3), 311–335. doi:10.1023/A:1008889222784

Guo, Z., & Nilsson, P. (2006). Algorithm and implementation of the K-Best sphere decoding for MIMO detection. *IEEE Journal on Selected Areas in Communications*, 24(3), 491–503. doi:10.1109/JSAC.2005.862402

Habendorf, R., Riedel, I., & Fettweis, G. (2006). Reduced complexity vector precoding for the multiuser downlink. In *Proceedings of the IEEE Global Telecommunications Conference* (pp. 1-5). New York, NY: IEEE.

Joham, M., & Utschick, W. (2005). Ordered spatial Tomlinson Harashima precoding. In T. Kaiser, A. Bourdoux, H. Boche, J. R. Fonollosa, J. Bach Andersen, & W. Utschick (Eds.), *Smart Antennas State-of-the-Art* (pp. 401–422). New York, NY: Hindawi Publishing Corporation.

Li, Q., & Wang, Z. (2006). Improved K-Best Sphere Decoding Algorithms for MIMO Systems. In *Proceedings of the IEEE International Symposium on Circuits and Systems* (pp. 1159-1162). New York, NY: IEEE.

Lin, K., Lin, H., Chang, R., & Wu, C. (2006). Hardware architecture of improved Tomlinson-Harashima precoding for downlink MC-CDMA. In *Proceedings of the IEEE Asia Pacific Conference on Circuits and Systems* (pp. 1200-1203). New York, NY: IEEE.

Mahdavi, M., Shabany, M., & Vahdat, V. (2010). A modified complex K-best scheme for high-speed hard-output MIMO detectors. In *Proceedings of the IEEE International Midwest Symposium on Circuits and Systems* (pp. 845-848). New York, NY: IEEE.

Mennenga, B., & Fettweis, G. (2009). Search Sequence Determination for Tree Search Based Detection Algorithms. In *Proceedings of the IEEE Sarnoff Symposium* (pp. 1-6). New York, NY: IEEE.

Park, K., Cha, J., & Kant, J. (2008). A computationally efficient stack-based iterative precoding for multiuser MIMO broadcast channel. In *Proceedings of the IEEE Vehicular Technology Conference* (pp. 1-5). New York, NY: IEEE.

Shabany, M., & Gulak, P. (2008). Scalable VLSI architecture for K-best lattice decoders. In *Proceedings of the IEEE International Symposium on Circuits and Systems* (pp. 940-943). New York, NY: IEEE.

Shabany, M., Su, K., & Gulak, P. (2008). A pipelined scalable high-throughput implementation of a near-ML K-Best complex lattice decoder. In *Proceedings of the IEEE International Conference on Acoustics, Speech and Signal Processing* (pp. 3173-3176). New York, NY: IEEE.

Shen, C., & Eltawil, A. (2010). A radius adaptive K-best decoder with early termination: algorithm and VLSI architecture. *IEEE Transactions on Circuits and Systems*, 57(9), 2476–2486. doi:10.1109/TCSI.2010.2043017

Tsai, P., Chen, W., & Huang, M. (2010). A 4x4 64-QAM Reduced-Complexity K-Best MIMO Detector up to 1.5 Gbps. In *Proceedings of the IEEE International Symposium on Circuits and Systems* (pp. 3953-3956). New York, NY: IEEE.

Wenk, M., Bruderer, L., Burg, A., & Studer, C. (2010). Area- and throughput optimized VLSI architecture of sphere decoding. In *Proceedings of the VLSI and System-on-Chip* (pp. 189-194). New York, NY: IEEE.

Wong, K., Tsui, C., Cheng, R., & Mow, M. (2002). A VLSI architecture of a K-best lattice decoding algorithm for MIMO channels. In *Proceedings of the IEEE International Symposium on Circuits and Systems* (pp. 273-276). New York, NY: IEEE.

Yu, W., & Cioffi, J. M. (2001). Trellis precoding for the broadcast channel. In *Proceedings of the IEEE Global Communications Conference* (pp. 1344-1348). New York, NY: IEEE.

KEY TERMS AND DEFINITIONS

Broadcast Channel: Downlink channel. Multiuser MIMO scenario where only the transmission from base station to user terminals in considered.

Fixed Sphere Encoder: Vector precoder at which the tree branches to be evaluated are set beforehand, allowing a fixed complexity and rate of the precoder.

Matrix Preprocessing: User ordering. Process used to select the order in which the user streams will be evaluated with a specific tree-search algorithm.

Multiuser MIMO: Multiuser multiple-input multiple-output system. Wireless multiuser system where both the transmitters and receivers are equipped with multiple antennas.

Precoder: Element of the transmission chain that modifies the symbols to be transmitted in order to improve their reception at the end user terminals.

Sequential Best-Node Expansion: Depth-first implementation of an originally breadth-first algorithm that also includes a sphere constraint for additional pruning, reducing the average number of evaluated candidates.

Sphere Encoder: Tree-search algorithm that limits the candidate points to a hypersphere around a reference signal.

Tree-Search: Search algorithm used to obtain the closest point in a lattice, named after its graphical representation.

Vector Precoder: Non-linear precoding scheme that adds a perturbation vector to the transmission symbols.

Chapter 7
Information Filter–Assisted Indoor Bluetooth Positioning

Liang Chen
Finnish Geodetic Institute, Finland

Jingbin Liu
Finnish Geodetic Institute, Finland

Heidi Kuusniemi
Finnish Geodetic Institute, Finland

Jian Tang
Wuhan University, China

Yuwei Chen
Finnish Geodetic Institute, Finland

Laura Ruotsalainen
Finnish Geodetic Institute, Finland

Ling Pei
Shanghai Jiao Tong University, China

Ruizhi Chen
Finnish Geodetic Institute, Finland & Texas A&M University – Corpus Christi, USA

ABSTRACT

This chapter studies wireless positioning using a network of Bluetooth signals. Fingerprints of Received Signal Strength Indicators (RSSI) are used for localization. Due to the relatively long interval between the available consecutive Bluetooth signal strength measurements, the authors applied an information filter method with speed detection, which combines the estimation information from the RSSI measurements with the prior information from the motion model. Speed detection is assisted to correct the outliers of position estimation. The field tests show the effectiveness of the information filter-assisted positioning method, which improves the horizontal positioning accuracy of indoor navigation by about 17% compared to the static fingerprinting positioning method, achieving a 4.2 m positioning accuracy on the average, and about 16% improvement compared to the point Kalman filter. In RSSI fingerprinting localization, building a fingerprint database is usually time-consuming and labour-intensive. In the final section, a self-designed autonomous SLAM robot platform is introduced to be able to carry out the Bluetooth RSS data collecting.

INTRODUCTION

Location and navigation technologies have been researched in a pervasive way in recent years. Diverse location based applications have been

DOI: 10.4018/978-1-4666-6034-2.ch007

fast deployed in our daily life. GNSSs (Global Navigation Satellites Systems) have been widely accepted for positioning and navigation outdoors. The built-in GPS (Global Positioning System) on a handset is capable of providing location information in open signal environments. However, for indoor positioning, GPS is unable to provide the desired level of accuracy or is even unavailable (Kaplan, 1996; Liu et al, 2012a). As an alternative, multiple sensors and signals of opportunity (SoOP) have been used for indoor positioning and navigation (Bahl & Padmanabhan, 2000; Pei et al., 2010a). Different with GNSS, SoOP are not originally intended for the purpose of navigation. With suitable methods, currently, research that utilize SoOP for navigation to achieve the positioning results includes WiFi, Bluetooth (Gomes & Sarmento, 2009; Chen et al., 2013), RFID (radio-frequency identification) (Ni et al., 2004), UWB (ultra wideband) (Pahlavan et al., 2006), long range signals of such as, GSM (global system for mobile communications) (Syrjärinne, 2001), and DTV (digital television) (Chen et al., 2012). In addition, multiple sensors can also be used to assist navigation indoors. Examples of such sensors include accelerometers, gyroscopes, compasses, cameras, proximity sensors (Pei et al., 2012).

In this work, we will focus on one of the SoOP, the Bluetooth, for the research of indoor positioning. Bluetooth is a technology for short-range wireless data and voice communication with low power consumption. It has been utilized in the communication and proximity market for a long time. As widely supported by mobile devices, Bluetooth is a potential technology to become an alternative for indoor positioning. However, indoor positioning using Bluetooth signals has not been widely studied so far. Bandara et al. (2004) developed a multi-antenna Bluetooth Access Point (AP) for location estimation based on received signal strength indicators (RSSI). The test obtained 2 meters of error in a 4.5 m×5.5 m area with four antennas. Sheng & Pollard (2006) modified the Bluetooth standard to estimate the distance between a reference transmitter and a mobile receiver, using RSSI measurements and a line-of-sight radio propagation model within a single cell. A high-density Bluetooth infrastructure is necessary to achieve an accurate position in the above two approaches. In order to minimize the Bluetooth infrastructure, Kelly et al. (2008) used only one class 1 Bluetooth AP for a home localization system, which combined the measurements of the link quality, RSSI, and cellular signal quality to obtain room-level accuracy. Pei et al. (2010b) present a Bluetooth locating solution in a reduced Bluetooth infrastructure area by using RSSI probability distributions. Other topics related to Bluetooth positioning can be found in (Hay & Harle, 2009; Anastasi et al., 2003; Bargh & Groote, 2008; Jevring et al., 2008; Naya et al., 2005). New specifications and products have been developed for a relatively longer range of transmission. Compared with the class 2 device (e.g. the Bluetooth module in a smart phone), which has only the range of about 20-30 meters (Bluetooth, 2010), a class 1 Bluetooth device (e.g. the Bluegiga AP 3201) has an effective range up to 200 meters and the newly developed Bluegiga AP 3241 can even achieve an effective range of 800 m in an open area without obstructions (Bluegiga, 2010).

In this study, for the indoor positioning test, a Bluetooth network including 13 long range APs (Bluegiga 3201 and 3241) have been deployed in the area of interest. For a low cost receiver, RSSIs are the feasible observables that could be used for positioning. Generally, there are two basic approaches used for the estimation of locations with RSSI measurements. The trilateration-based approach first translates RSSI measurements into the distances between a mobile user and multiple access points (APs) based on a radio propagation model and then calculates the user's location using the obtained distances and AP coordinates (Liu et al., 2012b). The major challenges in this approach include the large errors associated with estimated distances and difficulties in system deployment, e.g., the trouble associated with obtaining the AP coordinates indoors. In contrast, the fingerprint approach determines a user's position by matching RSSI measurements with a fingerprint database in a deterministic or stochastic way. The k-nearest neighbors (KNN) method employs a deterministic approach to estimate a location (Liu, 2012), which is the centroid of the k closest neighbors, whose RSSI measurements in the database have the shortest Euclidean distance to the online RSSI measurements. The stochastic methods impose a probabilistic model on the online RSSI measurements and calculate the posterior probability distribution (Liu & Chen, 2011). Different probabilistic models have been used in previous studies, ranging from a simple Gaussian model to more complex kernel functions (Liu et al., 2012b).

In this work, utilizing the RSSI from Bluetooth APs, we consider a fingerprinting method for position estimation. In order to achieve accurate positioning results, we take into the following two issues: first, as a major challenge in the fingerprint approach, the variance of RSSI observables is very large, which is caused by the significantly non-stationary nature of received signals. In signal-point positioning approach, where positions were considered as a series of isolated points (Liu et al., 2012b), the positioning results are easily vulnerable to RSSI variance. As a result, the positioning accuracy and reliability are degraded significantly. Considering the restricted dynamics of indoors users and their locations highly correlated over time, the sequential estimation for the position can be therefore introduced to mitigate the impact of RSSI variance. Second, the indoor environment usually restricts the pedestrian activity within a defined area and the user dynamics and activities are limited, which will be beneficial to detect the erroneous estimation when extremely large dynamics is found. Therefore, by considering the factors mentioned above, in this work, we present a method of information filtering with speed detection. The position is sequentially estimated by combining the information from the measurements and the prior motion model. Speed detection is further assisted to correct for the outliers of position estimation.

For RSSI fingerprint positioning, to manually build a fingerprint database is usually labour intensive and time consuming, especially in a large mapping area with high resolution of calibration points, which is required being stored in the database. Moreover, since signals sensitive to environment change, the dataset should be maintained timely by recalibration. Obviously, this maintenance is a high-cost labour work (Chen et al., 2010). In this study, we also introduce a self-designed autonomous SLAM (simultaneous localization and mapping) robot platform to be able to carry out the Bluetooth RSS data collecting for indoor positioning.

The book chapter is organized as follows: the system model and the problem of indoor positioning are formulated in the Section of system description. Section 3 considers the information filtering with speed detection as assistance. In Section 4, the experimental platform is described in detail and Section 5 shows the numerical results and a performance comparison is presented and discussed. In Section 6, a self-designed autonomous robot platform for collecting the RSSI is presented in brief. In Section 7, conclusions are drawn as well as future research discussed.

SYSTEM DESCRIPTION

In this work, we focus on the fingerprint positioning method in a Bluetooth network. Fingerprinting is a feasible technique for positioning using RSSI measurements. It works in two phases: the training phase and the online positioning phase. In the training phase, the radio map is constructed based on the calibration point within the area of interest. The radio map implicitly characterizes the RSSI position relationship through training measurements at the calibration points with known coordinates. In the online positioning phase, the mobile device measures RSSI observations and the positioning system uses the radio map to provide position estimation.

The fingerprinting method has been widely discussed for indoor positioning. To name a few, (Honkavirta et al., 2009) gives a thorough summary and analysis for different steps and factors that affect fingerprints, and in (Bar-Shalom et al., 2001), different positioning algorithms are compared using wireless local network (WLAN) fingerprints.

The basic fingerprinting method only compares the current RSSI measurements with the radio map to estimate the position. The accuracy and reliability of single-point positioning solutions are degraded by the non-stationary nature of RSSI due to the multipath and non-line-of-sight propagation of WLAN signals. The studies presented in (Liu 2012; Liu & Chen, 2011; Liu et al., 2012b) show that positioning accuracy can be improved by the incorporation of current RSSI measurements in conjunction with knowledge of a pedestrian motion model and historical measurements in time series. In this work, we consider sequential estimation of the pedestrian movements indoors.

Motion dynamics describe the correlation of the spatial coordinates of user positions over time. In previous studies, two approaches have been proposed to use motion dynamics information for improving positioning accuracy. One approach represents a user's motion with a set of predefined motion models, which describes the time evolution of the user's positions (Liu, 2012). The other approach uses a map to restrict the potential direction of motion and the space of the user. Based on the both approaches, a form of Bayesian filters has been used to perform position estimation. The applicability of these solutions is restricted by the fidelity of motion models. When considering general users, for example, in an office or in a shopping center, the common models are insufficient for the representation of indoor user motion, which may involve abrupt turns or stops. The motion dynamics of a pedestrian user are especially complex: user motion is governed by decision models, purpose of the movement, choice of destination, and interactions with other people or objects in the environment. An incorrect model results in inaccurate estimates (Liu et al., 2012b). Map data can only provide static

information, e.g., potential movement directions and intersections and are incapable of presenting real-time motion status. For example, a pedestrian may turn around suddenly in a linear corridor. Further, in the specific cases given in previous studies, the utilization of map data was based on a unique layout of the indoor environment. As a result, these solutions must process the map data of different indoor environments on a per-case basis, and they cannot be applied universally until a unified method for obtaining indoor map information exists. To make our solution widely applicable, this book chapter does not include the utilization of map data in the proposed solution, although map data also can be used to improve the positioning accuracy further.

State Model

A mobile device carried by a pedestrian is assumed to move on a two-dimensional Cartesian plane. The state at time instant t_k is defined as the vector

$$\mathbf{x_k} = \begin{bmatrix} x_k & y_k & \dot{x}_k & \dot{y}_k \end{bmatrix}^T,$$

where $\begin{bmatrix} x_k & y_k \end{bmatrix}^T$ corresponds to the East and North coordinates of the mobile position; $\begin{bmatrix} \dot{x}_k & \dot{y}_k \end{bmatrix}^T$ are the corresponding velocities. The mobile state can be modelled as (Bar-Shalom et al, 2001):

$$\mathbf{x_{k+1}} = \mathbf{F}_k + \mathbf{x}_k \mathbf{w}_k \tag{1}$$

where the state transition matrix:

$$\mathbf{F} = \begin{bmatrix} \mathbf{I}_2 & \Delta t \mathbf{I}_2 \\ 0 & \mathbf{I}_2 \end{bmatrix},$$

with $\mathbf{I}_2$ being the 2×2 identity matrix and Δt the sampling period. The random process $\mathbf{w}_k$ is a white zero mean Gaussian noise, with covariance matrix:

$$\mathbf{Q} = \begin{vmatrix} \dfrac{\Delta t^4}{4}\Omega & \dfrac{\Delta t^3}{2}\Omega \\ \dfrac{\Delta t^3}{2}\Omega & \Delta t^2\Omega \end{vmatrix} \tag{2}$$

where:

$$\Omega = \begin{vmatrix} \sigma_x^2 & 0 \\ 0 & \sigma_y^2 \end{vmatrix}.$$

Measurements

1. **Measurements to Build a Radio Map:** In the training phase, the RSS values of the radio signals transmitted by Bluetooth APs are collected in the calibration points for a certain period of time and stored into the radio map. Denote the ith fingerprint as $\mathbf{R}_i$ with the form:

$$\mathbf{R}_i = \left(\mathbf{c}_i, \left\{\mathbf{a}_{i,j}\right\}\right), j \in \{1, \cdots, N\}$$

where $\mathbf{c}_i$ is the coordinate of th ith calibration point and $\mathbf{a}_{i,j}$ holds the l RSSI samples from the access point AP_j, i.e. $\mathbf{a}_{i,j} = \{a_{i,j}^1, a_{i,j}^2, \cdots, a_{i,j}^l\}$, N is the total number of Bluetooth APs. The set of all fingerprints is denoted as $\mathbf{R} = \{\mathbf{R}_1, \cdots, \mathbf{R}_M\}$, where M is the total calibration points.

2. **Measurements for Position Estimation:** In the positioning phase, denote $\mathbf{z}_k$ as RSS values measured from the different Bluetooth APs at time t_k. Then, the measurement sequence is $\mathbf{z}_{1:k} = \{\mathbf{z}_1, \cdots, \mathbf{z}_k\}$.

Problem Formulation

The problem of tracking the pedestrian indoors is to infer the mobile state $\mathbf{x}_k$ from the measurement sequence $\mathbf{z}_{1:k}$ and the constructed radio map

$\mathbf{R}$. Within the Bayesian estimation framework, solving this problem corresponds to computing the posterior $p(\mathbf{x}_k \mid \mathbf{z}_{1:k}, \mathbf{R})$. By applying the Bayes' Rule, the posterior can be calculated as:

$$p(\mathbf{x}_k \mid \mathbf{z}_{1:k}, \mathbf{R}) = \frac{p(\mathbf{z}_k \mid \mathbf{x}_k, \mathbf{R})p(\mathbf{x}_k \mid \mathbf{z}_{1:k-1}, \mathbf{R})}{p(\mathbf{z}_k \mid \mathbf{z}_{1:k-1}, \mathbf{R})}$$

Due to the complex electromagnetic environment indoors, it is not easy to give an explicit measurement function $\mathbf{z}_k = h_k(\mathbf{x}_k)$ within the whole positioning area. Thus, the likelihood $p(\mathbf{z}_k \mid \mathbf{x}_k, \mathbf{R})$ could not be exactly calculated.

An alternative approximation is to compute $p(\mathbf{z}_k \mid \mathbf{R})$, which is based on the assumption that the whole area of interest is divided into M small cells and the RSSI distribution on the ith calibration point represents the distribution of all the points within the corresponding cell. However, $p(\mathbf{z}_k \mid \mathbf{R})$ is the discrete probability distribution on the M coordinates of calibration points, based on which the mean and covariance of the position can be estimated, while $p(\mathbf{x}_k \mid \mathbf{z}_{1:k-1}, \mathbf{R})$ predicts of the position and velocity. Therefore, the posterior $p(\mathbf{x}_k \mid \mathbf{z}_{1:k}, \mathbf{R})$ relates to fusing two state estimations with different dimensions, which is not straightforward to update.

INFORMATION FILTERING WITH VELOCITY DETECTION

Decompose the mobile state $\mathbf{x}_k$ into position $\mathbf{u}_k = \begin{bmatrix} x_k, y_k \end{bmatrix}^T$ and velocity $\mathbf{v}_k = \begin{bmatrix} \dot{x}_k, \dot{y}_k \end{bmatrix}^T$. The state model (1) can be written as:

$$\begin{cases} \mathbf{v}_k = \mathbf{v}_{k-1} + \boldsymbol{\mu}_{k-1} \\ \mathbf{u}_k = \mathbf{u}_{k-1} + \mathbf{H}\mathbf{v}_{k-1} + \frac{1}{2}_{k-1} \end{cases} \tag{3}$$

where

$$\mathbf{H} = \Delta t \mathbf{I}_2, \ \boldsymbol{\mu}_{k-1} \sim \mathbf{N}(0, \mathbf{Q}_{\boldsymbol{\mu}}), \ \frac{1}{2}_{k-1} \sim \mathbf{N}(0, \mathbf{Q}_{\frac{1}{2}}).$$

According to (2):

$$\mathbf{Q}_{\frac{1}{2}} = \frac{\Delta t^4}{4}\Omega, \ \mathbf{Q}_{\boldsymbol{\mu}} = \Delta t^2 \Omega.$$

In addition, the process noise $\boldsymbol{\mu}_k, \nu_k$ are correlated:

$$\mathbf{C} = \mathbf{E}\left[\boldsymbol{\mu}_k \frac{1}{2}_k^T\right] = \frac{\Delta t^3}{2}\Omega.$$

The posterior of mobile state $p(\mathbf{x}_k \mid \mathbf{z}_{1:k}, \mathbf{R})$ can be factorized as:

$$p(\mathbf{x}_k \mid \mathbf{z}_{1:k}, \mathbf{R}) = p(\mathbf{u}_k, \mathbf{v}_k \mid \mathbf{z}_{1:k}, \mathbf{R}) = p(\mathbf{v}_k \mid \mathbf{u}_k, \mathbf{z}_{1:k}, \mathbf{R})p(\mathbf{u}_k \mid \mathbf{z}_{1:k}, \mathbf{R}) \quad (4)$$

According to the Bayes rule, $p(\mathbf{u}_k \mid \mathbf{z}_{1:k}, \mathbf{R})$ can be computed as

$$p(\mathbf{u}_k \mid \mathbf{z}_{1:k}, \mathbf{R}) = \eta \cdot p(\mathbf{z}_k \mid \mathbf{u}_k, \mathbf{R})p(\mathbf{u}_k \mid \mathbf{z}_{1:k-1}, \mathbf{R}) \approx \eta \cdot p(\mathbf{z}_k \mid \mathbf{R})p(\mathbf{u}_k \mid \mathbf{z}_{1:k-1}, \mathbf{R}) \quad (5)$$

where $\eta = 1 / p(\mathbf{z}_k \mid \mathbf{z}_{1:k-1}, \mathbf{R})$ is the normalization factor.

Position Prediction by Motion Model

Suppose at time t_{k-1}, the estimated mean and covariance of $\mathbf{u}_{k-1}$ are $\{\hat{\mathbf{u}}_{k-1|k-1}, \mathbf{P}^{\mathbf{u}}_{k-1|k-1}\}$, and the estimated mean and covariance of $\mathbf{v}_{k-1}$ are $\{\hat{\mathbf{v}}_{k-1|k-1}, \mathbf{P}^{\mathbf{v}}_{k-1|k-1}\}$. Then, according to the linear model (3) the predicted mean $\hat{\mathbf{u}}_{k|k-1}$ and covariance $\mathbf{P}^{\mathbf{u}}_{k|k-1}$ of the position are:

$$\hat{\mathbf{u}}_{k|k-1} = \hat{\mathbf{u}}_{k-1|k-1} + \mathbf{H}_k \hat{\mathbf{v}}_{k-1|k-1}$$
$$\mathbf{P}^{\mathbf{u}}_{k|k-1} = \mathbf{E}\left[\mathbf{u}_{k|k-1}\mathbf{u}^T_{k|k-1}\right] = \mathbf{H}(\mathbf{P}^{\mathbf{v}}_{k-1|k-1} + \mathbf{Q}_\epsilon)\mathbf{H}^T + \mathbf{Q}_\nu + \mathbf{H}^T\mathbf{C} + \mathbf{H}\mathbf{C}^T \quad (6)$$

$\mathbf{P}^{\mathbf{u}}_{k|k-1}$ is calculated by considering the correlated process noise ϵ_t and ν_t

Position Estimation from Measurements

To compute $p(\mathbf{z}_k \mid \mathbf{R})$, we use the statistical information from the radio map. From the ith fingerprints $\mathbf{R}_i$, the mean $\bar{a}_{i,j}$ and the variance $\sigma^2_{i,j}$ can be obtained from the measurements $\mathbf{a}_{i,j}$. In the positioning phase, based on the RSS measurement $z_{k,j}$ and the Gaussian approximation to the histogram of $\mathbf{a}_{i,j}$, the weight $w_{i,j}$ can be computed as:

$$w_{i,j} = \mathbf{N}(z_{k,j}; \bar{a}_{i,j}, \sigma^2_{i,j}), j \in \{1, \cdots, N\} \quad (7)$$

When the measurement $z_{k,j}$ does not exit, which means the device did not hear the AP_j at time t_k, set $w_{i,j} = w_0$, where w_0 is a very low value. Assume the measurements $\mathbf{z}_k$ from different AP_j are independent, then:

$$p(\mathbf{z}_k \mid \mathbf{R}_i) = \prod_{j=1}^{N} w_{i,j} \quad (8)$$

and the normalized weight w_i equal to:

$$w_i = \frac{\prod_{j=1}^{N} w_{i,j}}{\Sigma_{i=1}^{M} \prod_{j=1}^{N} w_{i,j}} \qquad (9)$$

Thus, from (7-9), $p(\mathbf{z}_k \mid \mathbf{R})$ is the discrete probability distribution with the weight w_i on the i th coordinate of the calibration point $\mathbf{c}_i$. Accordingly, the first two moments are:

$$\hat{\mathbf{u}}_k^r = \Sigma_{i=1}^{M} w_i \mathbf{c}_i$$
$$\mathbf{P}_k^r = \Sigma_{i=1}^{M} w_i \left[(\hat{\mathbf{u}}_k^r - \mathbf{c}_i)(\hat{\mathbf{u}}_k^r - \mathbf{c}_i)^T \right] \qquad (10)$$

Position Update by Information Filtering

When fusing the measurement estimation (10) and the model prediction (6), the estimated mean $\hat{\mathbf{u}}_{k|k}$ and covariance $\mathbf{P}_{k|k}$ of the position $\mathbf{u}_k$ can be updated as

$$(\mathbf{P}_{k|k}^{\mathbf{u}})^{-1} = (\mathbf{P}_{k|k-1}^{\mathbf{u}})^{-1} + (\mathbf{P}_k^r)^{-1} \hat{\mathbf{u}}_{k|k} =$$
$$(\mathbf{P}_{k|k}^{\mathbf{u}}) \left[(\mathbf{P}_{k|k-1}^{\mathbf{u}})^{-1} \hat{\mathbf{u}}_{k|k-1} + (\mathbf{P}_k^r)^{-1} \hat{\mathbf{u}}_k^r \right]^{-1}$$
$$\hat{\mathbf{u}}_{k|k} = (\mathbf{P}_{k|k}^{\mathbf{u}}) \left[(\mathbf{P}_{k|k-1}^{\mathbf{u}})^{-1} \hat{\mathbf{u}}_{k|k-1} + (\mathbf{P}_k^r)^{-1} \hat{\mathbf{u}}_k^r \right]^{-1}$$
$$\qquad (11)$$

Velocity Update by Kalman Filter with Correlated Noise

Based on state model (3) and the position estimation (11), update the estimation of velocity $\mathbf{v}_k$ by Kalman filter with correlated noise:

$$\hat{\mathbf{v}}_{k|k} =$$
$$\hat{\mathbf{v}}_{k|k-1} + \mathbf{P}_{k|k-1}^{\mathbf{uv}} (\mathbf{P}_{k|k-1}^{\mathbf{u}})^{-1} (\hat{\mathbf{u}}_k - \hat{\mathbf{u}}_{k-1} - \mathbf{H}\hat{\mathbf{v}}_{k|k-1})$$
$$\mathbf{P}_{k|k}^{\mathbf{u}} = \mathbf{P}_{k|k-1}^{\mathbf{u}} - \mathbf{P}_{k|k-1}^{\mathbf{uv}} (\mathbf{P}_{k|k-1}^{\mathbf{v}})^{-1} (\mathbf{P}_{k|k-1}^{\mathbf{uv}})^T$$
$$\qquad (12)$$

where:

$$\hat{\mathbf{v}}_{k|k-1} = \hat{\mathbf{v}}_{k-1|k-1}$$
$$\mathbf{P}_{k|k-1}^{\mathbf{v}} = \mathbf{P}_{k-1|k-1}^{\mathbf{v}} + \mathbf{Q}_\epsilon$$
$$\mathbf{P}_{k|k-1}^{\mathbf{uv}} = \mathrm{E}\left[\mathbf{u}_{k|k-1} \mathbf{v}_{k|k-1}^T \right] = \qquad (13)$$
$$(\mathbf{P}_{k-1|k-1}^{\mathbf{v}} + \mathbf{Q}_\epsilon)\mathbf{H}^T + \mathbf{C}$$

$\mathbf{P}_{k|k-1}^{\mathbf{u}}$ is given in (6). $\mathbf{P}_{k|k-1}^{\mathbf{uv}}$ is also calculated by considering the process noise ϵ_t and ν_t are correlated.

Velocity Detection and Position Correction

Exploiting the fact that the indoor pedestrian has a limited walking speed, we set the constraint that if:

$$\left\| \hat{\mathbf{v}}_{k|k} \right\| > \mathrm{V}_{\max} ,$$

Then:

$$\alpha = \arctan\left(\hat{y}_k / \hat{x}_k \right)$$
$$\hat{\mathbf{v}}_{k|k} = [\mathrm{Vmax} \cos \alpha, \mathrm{Vmax} \sin \alpha]^T$$
$$\mathbf{P}_{k|k}^{\mathbf{v}} = \mathbf{P}_{k|k}^{\mathbf{v}} \cdot (\mathrm{V}_{\max} / \left\| \hat{\mathbf{v}}_{k|k} \right\|)^2 \qquad (14)$$
$$\hat{\mathbf{u}}_{k|k} = \hat{\mathbf{u}}_{k-1|k-1} + \hat{\mathbf{v}}_{k|k} \cdot \Delta t$$
$$\mathbf{P}_{k|k}^{\mathbf{u}} = \mathbf{P}_{k|k}^{\mathbf{u}} \cdot (\mathrm{Vmax} / \left\| \hat{\mathbf{v}}_{k|k} \right\|)^2$$

The Algorithm 1 describes the whole scheme.

Algorithm 1 Information fusion with speed detection

1. Initial state at time t_0:

$$\hat{\mathbf{u}}_{0|0} = \hat{\mathbf{u}}_0, \mathbf{P}_{0|0}^{\mathbf{u}} = \mathbf{P}_0^{\mathbf{u}}, \hat{\mathbf{v}}_{0|0} = \hat{\mathbf{v}}_0, \mathbf{P}_{0|0}^{\mathbf{v}} = \mathbf{P}_0^{\mathbf{v}}$$

2. Input

Fingerprints: $\mathbf{R}$, RSS measurement at time t_k:

$\mathbf{z}_k$

Position and velocity estimation at time t_{k-1}:

$\hat{\mathbf{u}}_{k-1|k-1}, \ \mathbf{P}_{k-1|k-1}^{\mathbf{u}}, \ \hat{\mathbf{v}}_{k-1|k-1}, \ \mathbf{P}_{k-1|k-1}^{\mathbf{v}}$

3. Output

for $k = 1$ to T

a. position prediction by motion model: (6)
b. position estimation by RSS measurement: (7), (10)
c. position update by information fusion: (11)
d. velocity update by Kalman filter: (12), (13)
e. velocity detection and position correction: (14), (10)

end for

The method of information fusion with speed detection (IFSD) simultaneously considers the information from the RSS measurements and the prior positioning from the motion model. The indoor pedestrian speed is further set as a constraint to detect the outliers. Compared with the point Kalman filter, which only puts the mean:

$\hat{\mathbf{u}}_k^r$

as the input to the Kalman filter, the information fusion step in the IFSD considers the covariance estimation of the RSS measurements:

$((\mathbf{P}_{k|k-1}^r))$.

EXPERIMENTS

This section introduces results of experiments conducted in a Bluetooth network with the motion restricted fusion filter.

Bluetooth RSS Data Collecting System

In this study, a Bluetooth RSS data collecting system is developed for indoor positioning. The system consists of a Bluetooth evaluation kit and a data collecting program developed during the research work (Figure 1). The basic function of the system is to scan the Bluetooth Access Points (APs) nearby, collect the RSS from the detected APs, and then send the measurements to the laptop via a serial port. The sampling interval can be adjusted within $1.2 - 11.5$ seconds with a resolution of 1.28 seconds according to the priority chosen.

The evaluation kit is equipped with a RS-232 and an USB interface, an on-board Pulse Code Modulation (PCM) codec, a 16-pin I/O interface, and a Serial Peripheral Interface (SPI) for upgrad-

Figure 1. Bluetooth RSS data collecting system

ing the firmware and parameters. It is powered by a laptop through a USB connection. The core component of the evaluation kit is the Bluegiga WT41 module (Figure 2), a class 1 Bluetooth® 2.1 plus an Enhanced Data Rate (EDR) module, which contains all the necessary elements from Bluetooth radio to antenna and fulfils a fully implemented protocol stack within the embedded firmware. The module is optimized for long range applications and the effective scanning range is approximately 800 m.

The data collecting program is developed based on a Visual C++ IDE. The flow chart of the program is presented in Figure 3. The program uses a multi-thread application, with one thread for communicating with the Bluetooth evaluation kit in real-time to collect the RSS from the Bluetooth AP and the other thread for showing the results in a user interface.

SPAN High Accuracy GPS-IMU System

To evaluate the positioning accuracy of different algorithms, a reference trajectory, used as the ground truth, is obtained via NovAtel's high-accuracy SPAN (Synchronized Position Attitude Navigation) system. SPAN technology is a tightly couple solution of a precision Global Navigation Satellite System (GNSS) receiver with a robust

Inertial Measurement Unit (IMU) from NovAtel. It can provide reliable, continuously available measurements including position, velocity, and attitude even through short periods of time when no GNSS satellites are available. During the tests, the GPS receiver of the SPAN system is a NovAtel DL-4plus, containing the NovAtel OEM-G2 engine. A dual-frequency NovAtel GPS-702 antenna is applied. The inertial measurement unit is a tactical-grade, ring laser gyro based IMU. The SPAN system can operate either in Real-Time Kinematic (RTK) mode or Virtual Reference Station (VRS) mode for real-time and post-processing applications to get trajectories with centimetre level of accuracy.

Field Test

Indoor tests were carried out in a corridor of an office-building at the Finnish Geodetic Institute. During the test, the indoor corridor was equipped with 13 Bluetooth access points. Figure 4 shows

Figure 3. The flow chart for Bluetooth RSS data collecting system

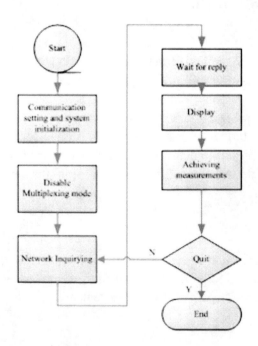

Figure 2. Evaluation kit

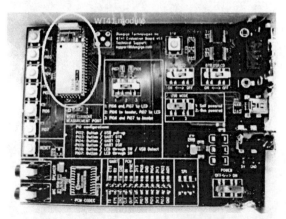

the corridor from the inside. Figure 5 shows the testing platform with the floormap, the position of the Bluetooth APs, and the test route in the corridor.

Two tests were carried out in the scenario. In both tests, a tester walked along the corridors back and forth with the test cart. Test 1 lasts about 6 minutes, while with a relatively faster speed, test 2 only lasts for 3 minutes. The Bluetooth priority

Figure 4. Interior of the glass, concrete, and steel office building, which is the pedestrian navigation testing environment

Figure 5. Floormap, the position of the Bluetooth APs and the testing equipment

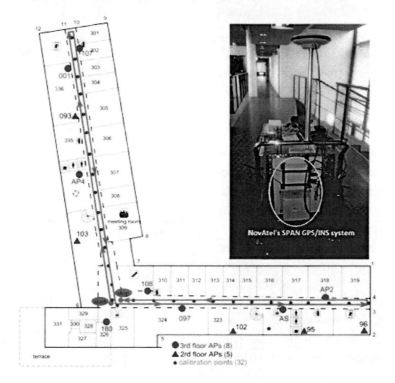

scale is set to 6, which corresponds to a sampling interval $\Delta t \approx 9$ s. The maximum horizontal speed is set to 2 m/s, which is typical for pedestrian indoor motion.

Results

We compare the proposed information filter with speed detection (IFSD) with the Bayes static estimation (BSE) method and the point Kalman filter (PKF). The BSE method only uses the RSSI measurements at current epoch to estimate the posterior mean and covariance of the position. The estimation are derived from (7)-(10). The PKF uses a Kalman filter to further smooth the position estimated from the BSE. In the PKF, a stationary motion model is used to formulate the movement and the posterior mean of the position estimated from the BSE, i.e. $\hat{\mathbf{u}}_k^r$ in (10), is used as the observation in the measurement model. In our tests, the covariance of the process noise in the PKF is set as $\mathbf{Q}_{PKF} = (\mathbf{V}_{max} \cdot \Delta t)^2 \cdot \mathbf{I}_2 = 18^2 \cdot \mathbf{I}_2$ and the measurement covariance $\mathbf{R}_{PKF} = 9 * \mathbf{I}_2$. In the IFSD, we set $\sigma_x = \sigma_y = 1 / \Delta t$, which means that the changes in the velocity over a sampling interval are in the order of 1 m/s in each direction. The initial position of the IFSD and the PKF is obtained from the first output of the BSE. The initial covariance for the position estimation is $9 * \mathbf{I}_2$ and the initial velocity for the IFSD is 0 m/s with covariance $\mathbf{I}_2$.

Figure 6 and 8 represent the estimated trajectories of the 3 different algorithms in a North-East coordinate frame including also the SPAN reference track as the ground truth. Figure 7 and 9 show the position error vs. time epoch of the three algorithms. Position errors are compared in Table I.

From the results, the mean error of the BSE is about 5 m. The PKF smoothes the positions obtained by the BSE. Based on the stationary motion model, the estimation errors are reduced

in several epochs, e.g. k = 8, 10, 13 in Figure 7 and k = 4, 5, 7, 9 in Figure 9. However, the improvement of the PKF is very slight, only 0.1 m. The reason for this may lie in the fact that the $\mathbf{Q}_{PKF}$ is relatively large due to the long sampling interval of the Bluetooth. Thus, the prior information from the motion model has little impact on the position estimation at each epoch. In comparison, the positioning error of the IFSD is 4.2 m on the average, about 1 m less than BSE and 0.8 m less than the PKF. Significant improvements

Figure 6. Position estimate in Test 1

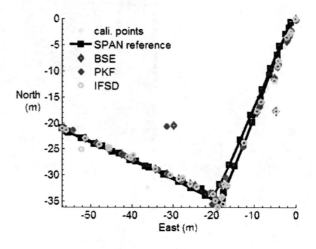

Figure 7. Position error vs. time epoch in Test 1

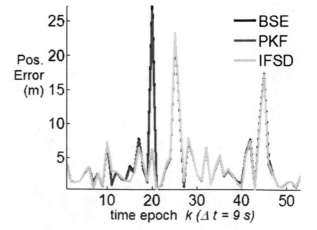

Figure 8. Position estimate in Test 2

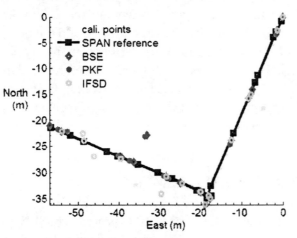

the complexity of the three algorithms, the PKF and IFSD both uses the Kalman filter to further smooth the position estimated from the BSE. Thus, BSE has the lowest complexity within the three, while PKF and IFSD have the same order of the time complexity.

Lidar-Based Robot Platform for a Bluetooth RSSI Fingerprint Collecting System

This section introduces briefly how to generate the fingerprint database of Bluetooth with a LiDAR based robot platform.

When RSSI fingerprinting is used for positioning, the accuracy of the calibration points in the fingerprint database affects the position results significantly. It has been recognized that

Figure 9. Position error vs. time epoch in Test 2

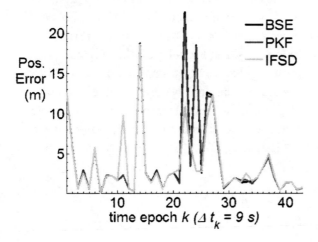

Figure 10. LiDAR SLAM robot platform for a fingerprint collection system

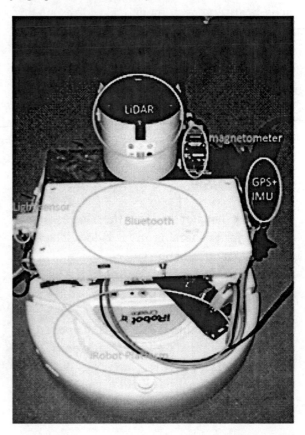

can be observed at k = 13 in test 1 and k = 4 to 10 in test 2, where the large errors are detected by the indoor pedestrian speed constraint and filtered out.

Thus, according to the test results, it is clear that the proposed IFSD can effectively correct the large outliers in position estimation and achieves the best position accuracy among the three algorithms. This is achieved by combining the estimation information (mean and covariance) from the measurements with the prior information from the motion model, together with the speed detection and position correction. Considering one of the most important challenges is to build and maintain the database for RSSI fingerprinting. The traditional method is to collect the RSSI signal with a portable device on every calibration point, while the coordinate of each calibration point is measured manually pre-hand or after. In this work, we introduce a self-designed LiDAR (LIght Detection And Ranging) based Simultaneous Localization and Mapping (SLAM) (Leonard & Durrant-Whyte, 1991) robot platform that is used for fingerprint database collection, as shown in Figure 10. The advantage of this system is to simultaneously construct the geometry map, provide high accurate position and collect fingerprints of SoOP for indoor positioning, e.g. Bluetooth, WiFi, etc.

The self-designed platform is integrated with a laser scanning device (SICK LMS151 or IBEO LUX), a mobile phone, a Bluetooth RSS data collecting system, a magnetometer, a light sensor and a spectrometer and an iRobot platform. The iRobot includes a gyroscope, an odometer, an on-board control computer and corresponding accessories for data collecting. The laser SLAM platform is expected to achieve sub-meter position accuracy and simultaneously construct a grid map indoors. With the SLAM platform, a fingerprint database with high density of calibration points is also expected to be easily built.

CONCLUSION

This book chapter studied the fingerprinting methods for positioning in an indoor Bluetooth network. An information filter with speed detection method was proposed for positioning. The method simultaneously considered the Bayesian static estimation results and the pedestrian motion model. By exploiting the fact that the pedestrian indoors has a limited walking speed, speed detection was further assist to correct the outliers of the position estimation. Field tests showed that

the proposed method was effective and achieved an improved accuracy when compared with the Bayesian static estimation method and the point Kalman filter. In addition, a Bluetooth fingerprint collecting system with a Lidar-based robot platform is also introduced and the robot platform is expected to provide means to build up a fingerprint database in a cost-efficient way.

ACKNOWLEDGMENT

This work was a part of the project INOSENSE (INdoor Outdoor Seamless Navigation for SEnsing Human Behavior) funded by the Academy of Finland. Liang Chen would like to acknowledge the Postdoctoral Research Fellowship granted by the Academy of Finland (No. 254232). This work was partly supported by the project "Research on Indoor Position Model based on Signals of Opportunity and PDR of smartphone" (41304004) funded by National Nature Science Foundation of China. This work is also a part of the project Centre of Excellence in Laser Scanning Research (CoE-LaSR) (272195) funded by the Academy of Finland.

REFERENCES

Anastasi, G., Bandelloni, R., Conti, M., Delmastro, F., Gregori, E., & Mainetto, G. (2003). Experimenting an Indoor Bluetooth-Based Positioning Service. In *Proceedings of the 23rd International Conference on Distributed Computing Systems Workshops* (pp. 480-483). Providence, RI: IEEE.

Bahl, P., & Padmanabhan, V. N. (2000). Radar: An In-Building RF Based User Location and Tracking System. In *Proceedings of Infocom—Nineteenth Annual Joint Conference of the IEEE Computer and Communications Societies* (pp. 775–784). Tel-Aviv, Israel: IEEE.

Bandara, U., Hasegawa, M., Inoue, M., Morikawa, H., & Aoyama, T. (2004). Design and Implementation of a Bluetooth Signal Strength Based Location Sensing System. In *Proceedings of IEEE Radio and Wireless Conference* (pp. 319 – 322). Atlanta, GA: IEEE.

Bar-Shalom, Y., Li, R. X., & Kirubarajan, T. (2001). *Estimation with Applications to Tracking and Navigation, Theory Algorithms and Software.* John Wiley & Sons. doi:10.1002/0471221279

Bargh, M., & Groote, R. (2008). Indoor Localization Based on Response Rate of Bluetooth Inquiries. in *Proceedings of the first ACM international workshop on Mobile entity localization and tracking in GPS-less environments* (pp. 49-54). San Francisco: ACM.

Bluegiga. (2010). *Specification of the Bluegiga System, WT41-A/WT41-N Preliminary Data Sheet, v. 7.* Retrieved 7 October 2010 from http://www.bluegiga.com

Bluetooth. (2010). *Specification of the Bluetooth System, Core Specification v2.0 + EDR, Bluetooth SIG.* Retrieved 18 August 2010 from http://www.bluetooth.org/

Chen, L., Pei, L., Kuusniemi, H., Chen, Y., Kröger, T., & Chen, R. (2013). Bayesian Fusion for Indoor Positioning Using Bluetooth Fingerprints. *Wireless Personal Communications, 70*(4), 1735–1745. doi:10.1007/s11277-012-0777-1

Chen, L., Yang, L., & Chen, R. (2012). Time Delay Tracking for Positioning in DTV Networks. In *Proceedings of Ubiquitous Positioning, Indoor Navigation, and Location Based Service (UPIN-LBS)* (pp. 1–4). Helsinki: IEEE. doi:10.1109/UPINLBS.2012.6409784

Chen, Y., Chen, R., Pei, L., Kröger, T., Chen, W., Kuusniemi, H., & Liu, J. (2010). Knowledge-based Error Detection and Correction Method of a Multi-sensor Multi-network Positioning Platform for Pedestrian Indoor Navigation. In *Proceedings of IEEE/ION PLANS 2010* (pp. 873-879). Indian Wells, CA: IEEE.

Gomes, G., & Sarmento, H. (2009). Indoor Location System Using ZigBee Technology. In *Proceedings of Third International Conference on Sensor Technologies and Applications* (pp. 152–157). Athens, Greece: IEEE.

Hay, S., & Harle, R. (2009). Bluetooth Tracking without Discoverability. *Lecture Notes in Computer Science, 5561*, 120–137. doi:10.1007/978-3-642-01721-6_8

Honkavirta, V., Perälä, T., Ali-Löytty, S., & Piché, R. (2009). Location Fingerprinting Methods in Wireless Local Area Network. In *Proceedings of the 6th workshop on Positioning, Navigation and Communication (WPNC'09)* (pp. 243-251). Hannover, Germany: IEEE.

Jevring, M., Groote, R., & Hesselman, C. (2008). Dynamic Optimization of Bluetooth Networks for Indoor Localization. In *Proceedings of the 5th international conference on Soft computing as transdisciplinary science and technology* (pp. 663-668). Paris, France: ACM.

Kaplan, E. D. (Ed.). (1996). *Understanding GPS: Principles and Applications.* Norwood, MA: Artech House.

Kelly, D., McLoone, S., & Dishongh, T. (2008). A Bluetooth-Based Minimum Infrastructure Home Localization System. in *Proceedings of 5th IEEE International Symposium on Wireless Communication Systems* (pp. 638-642). Reykjavik, Iceland: IEEE.

Kjærgaard, M. B. (2007). A taxonomy for radio location fingerprinting. In *Location-and context-awareness* (pp. 139–156). Berlin: Springer. doi:10.1007/978-3-540-75160-1_9

Leonard, J. J., & Durrant-whyte, H. F. (1991). Simultaneous Map Building and Localization for an Autonomous Mobile Robot. In *Proceedings of IEEE/RSJ International Workshop on Intelligent Robots and Systems, Intelligence for Mechanical Systems* (pp. 1442–1447). IEEE.

Liu, J. (2012). Hybrid Positioning with Smart-phones. In *Ubiquitous Positioning and Mobile Location-Based Services in Smart Phones* (pp. 159–194). IGI Global. doi:10.4018/978-1-4666-1827-5.ch007

Liu, J., & Chen, R. (2011). *Smartphone Positioning Based on a Hidden Markov Model*. Tech Talk Blog. Retrieved March 2011 from http://gpsworld.com/tech-talk-blog/smartphone-positioning-based-a-hidden-markov-model-11385-0

Liu, J., Chen, R., Chen, Y., Pei, L., & Chen, L. (2012a). iParking: An Intelligent Indoor Location-Based Smartphone Parking Service. *Sensors (Basel, Switzerland)*, *12*(11), 14612–14629. doi:10.3390/s121114612 PMID:23202179

Liu, J., Chen, R., Pei, L., Guinness, R., & Kuus-niemi, H. (2012b). A Hybrid Smartphone Indoor Positioning Solution for Mobile LBS. *Sensors (Basel, Switzerland)*, *12*(12), 17208–17233. doi:10.3390/s121217208 PMID:23235455

Naya, F., Noma, H., Ohmura, R., & Kogure, K. (2005). Bluetooth-Based Indoor Proximity Sensing for Nursing Context Awareness. In *Proceedings of the 9th IEEE International Symposium on Wearable Computers* (pp. 212-213). Osaka, Japan: IEEE.

Ni, L. M., Liu, Y., Lau, Y. C., & Patil, A. P. (2004). Landmarc: Indoor Location Sensing Using Active RFID. *Wireless Networks*, *10*, 701–710. doi:10.1023/B:WINE.0000044029.06344.dd

Pahlavan, K., Akgul, O. F., Heidari, M., Hatami, A., Elwell, M. J., & Tingley, D. R. (2006). Indoor Geolocation in the Absence of Direct Path. *IEEE Transactions on Wireless Communications*, *13*(6), 50–58. doi:10.1109/MWC.2006.275198

Pei, L., Chen, R., Liu, J., Tenhunen, T., Kuus-niemi, H., & Chen, Y. (2010a). Inquiry-Based Bluetooth Indoor Positioning via RSSI Probability Distributions. In *Proceedings of the Second International Conference on Advances in Satellite and Space Communications (SPACOMM 2010)* (pp. 151–156). Athens, Greece: IEEE.

Pei, L., Chen, R., Liu, J., Tenhunen, T., Kuusniemi, H., & Chen, Y. (2010b). Using Inquiry-based Bluetooth RSSI Probability Distributions for Indoor Positioning. *Journal of Global Positioning Systems*, *9*(2), 122–130.

Pei, L., Liu, J., Guinness, R., Chen, Y., Kuusniemi, H., & Chen, R. (2012). Using LS-SVM Based Motion Recognition for Smartphone Indoor Wireless Positioning. *Sensors (Basel, Switzerland)*, *12*(5), 6155–6175. doi:10.3390/s120506155 PMID:22778635

Sheng, Z., & Pollard, J. K. (2006). Position Measurement Using Bluetooth. *IEEE Transactions on Consumer Electronics*, *52*(2), 555–558. doi:10.1109/TCE.2006.1649679

Syrjärinne, J. (2001). *Studies on Modern Techniques for Personal Positioning*. (Ph.D. Thesis). Tampere University of Technology, Tampere, Finland.

KEY TERMS AND DEFINITIONS

Signals of Opportunity: Refers to those radio frequency signals that are not intended for positioning purposes, but are freely available all the time.

Bluetooth: A commonly used technology used for short-range wireless communication with low power consumption.

Fingerprinting: A positioning process of finding a location from which the fingerprints stored in database has the best match to the snapshot of the observed measurements.

Received Signal Strength Indicator (RSSI): An indication of the power level received by an antenna.

Bayesian Framework: Statistical modelling and recursive state estimation based on the Bayes rule to obtain the posterior distribution of the model parameters.

Kalman Filtering: A common fusion algorithm that uses a dynamics model, known control inputs to the system, and measurements over time to form an estimate of the system's state.

Wireless Localization: Obtaining a position solution utilizing wireless network measurements.

Chapter 8
Analysis of Real–Time Hybrid–Cooperative GNSS–Terrestrial Positioning Algorithms

Francesco Sottile
Istituto Superiore Mario Boella, Italy

Zhoubing Xiong
Istituto Superiore Mario Boella, Italy

Claudio Pastrone
Istituto Superiore Mario Boella, Italy

ABSTRACT

This chapter analyzes some hybrid and cooperative GNSS-terrestrial positioning algorithms that combine both pseudorange measurements from satellites and terrestrial range measurements based on radio frequency communication to improve both positioning accuracy and availability. A Simulation Tool (ST) is also presented as a viable tool able to test and evaluate the performance of these hybrid positioning algorithms in different scenarios. In particular, the ST simulates devices belonging to a Peer-to-Peer (P2P) wireless network where peers, equipped with a wireless interface and a GNSS receiver, cooperate among them by exchanging positioning aiding data in order to enhance the overall performance. Different hybrid and cooperative algorithms, based on Bayesian and least squares approaches proposed in the literature, have been implemented in the ST and simulated in different simulation scenarios including the vehicular urban one. Moreover, all these algorithms are compared in terms of computational complexity to better understand their feasibility to achieve a real-time implementation. Finally, the sensitivity of the hybrid and cooperative algorithms when pseudorange measurements are affected by large noise and in presence of malicious peers in the P2P network is also assessed by means of the ST.

INTRODUCTION

Nowadays, various applications in wireless communication networks are based on mobile positioning. In fact, data collected or communicated by a wireless node is often useless if it is not associated to the node's location. For example, a generic wireless sensor network (WSN) monitoring application (Garcia-Hernandez, Ibarguengoytia-Gonzalez & Perez-Diaz, 2007), where nodes collect data such as temperature, humidity, etc., requires the nodes' location to recognize where the data come

DOI: 10.4018/978-1-4666-6034-2.ch008

from. Moreover, location-based technology allows the deployment of many other applications, including asset tracking, intruder detection (Yan, He & Stankovic, 2003), healthcare monitoring (Budinger, 2003), emergency 911 services (Caffery & Stuber, 1995) and so forth (Liu, Darabi, Banerjee & Liu, 2007).

Typically, positioning techniques infer the position of unknown nodes by applying a two-step estimation process (Caffery & Stuber, 1998). In the first step, ranging estimation is performed by means of various techniques such as received signal strength (RSS) (Hashemi, 1993; Laitinen, Juurakko, Lahti, Korhonen & Lahteenmaki, 2007), time of arrival (ToA) or time difference of arrival (TDoA) (Caffery & Stuber, 1998). Then, the position of the unknown nodes are inferred in a two or three dimensional plane by applying a localization algorithm, which takes as inputs the range measurements and the positions of anchor peers, whose coordinates are known a priori.

Recently, much attention is focused on the signal-of-opportunity (SoO) approach (Yang & Nguyen, 2009) which consists in the exploitation of terrestrial communication systems (*e.g.,* WSN, ultra wide-band (UWB), Wi-Fi, cellular networks, etc.) with a purpose other than navigation in order to guarantee and improve the global navigation satellite systems (GNSS)-based services and enhance the robustness of the overall GNSS end-user performance. In general, GNSS can provide accurate position estimation when the receiver's antenna is able to acquire at least four satellites (Del Re, 2011). This condition is generally verified in open sky outdoor environments. On the contrary, in GNSS-challenged environments such as urban canyons, under dense foliage and indoors, the *line of sight (LoS)* between the satellites and the receiver's antenna is often obstructed and the GNSS-based localization performance degrades or fails completely. Therefore, in such environments, different aiding/augmentation approaches have been proposed and adopted. Satellite-based augmentation systems (SBAS), such as wide

area augmentation system (WAAS), European geostationary navigation overlay service (EG-NOS) and multi-functional satellite augmentation system (MSAS), use a number of known global positioning system (GPS) receiving stations to improve the positioning accuracy (Diggelen, 2009). Ground-based augmentation systems, such as Assisted-GPS (A-GPS) and differential-GPS (D-GPS), adopt ground based networks to locate the users (Monteiro, Moore & Hill, 2005) with enhanced performance. In addition, (Retscher & Fu, 2007) investigated active RFID in combination with GNSS and dead reckoning solutions by means of odometers or inertial sensors for pedestrian navigation. However, these systems rely on the deployment of an infrastructure that provides the augmentation service. Moreover, in most cases, they require a unidirectional information flow from infrastructures to receivers. The main focus of this chapter is to study and analyse GNSS assisting schemes that rely on the *cooperative paradigm* and do not require a fixed infrastructure. In cooperative localization systems, each node voluntarily shares its own positioning data and information with its neighbours, with the aim to enhance the location accuracy and availability.

The rest of this chapter is organized as follows: first section 2 overviews the state of art regarding hybrid and cooperative positioning algorithms; section 3 presents the main structure of the ST including simulation results and complexity analysis; after that section 4 presents future research directions while conclusions are drawn in section 5.

BACKGROUND

Recently, many hybrid and cooperative GNSS/terrestrial localization approaches, which combine both pseudorange measurements from GNSS satellites and wireless-based ranging measurements from neighbouring peers, have been proposed in the literature to improve both positioning availability and accuracy. (Caceres, Penna, Wymeersch

& Garello, 2010; Caceres, Sottile, Garello & Spirito, 2010; Sottile, Wymeersch, Caceres & Spirito, 2011) proposed different positioning algorithms that fuse GNSS pseudorange and terrestrial distance measurements from ultra-wide band (UWB) sensors. (De Angelis, Baruffa, & Cacopard, 2013) proposed a hybrid GNSS/cellular positioning system that uses cellular network data to increase the location accuracy when the number of visible satellites is not adequate. (Alam, Kealy, & Dempster, 2013) proposed a cooperative inertial navigation method that shares the inertial based positioning data with neighbouring vehicles traveling in the opposite direction.

This chapter extends the work presented in (Sottile, Caceres & Spirito, 2011) and (Sottile, Caceres and Spirito, 2012), which proposed a simulation tool (ST) able to test hybrid "GNSS-terrestrial" and cooperative positioning algorithms in realistic and GNSS-challenged scenarios. The ST is a system level simulator. In fact, it simulates both the range and the position velocity and time (PVT) layers except the physical layer. In particular, range and pseudorange measurements are generated according to realistic models, described in more details in (Sottile, Caceres & Spirito, 2011). Concerning the PVT layer, the ST is able to test cooperative positioning algorithms, where peers help each other by exchanging positioning aiding packets including, for instance, position and covariance matrix estimates. At each time simulation step, apart from the execution of the selected cooperative positioning algorithm, the ST simulates also communication among peers. This approach makes the ST different from other proposed tools available in the literature. For instance, the simulator presented in (Ma, 2009) focuses on a multi-stage simulation framework, where in a first step the communication is simulated through the ns-2 engine, and in a second step the corresponding results are loaded into Matlab, which in turn runs a localization algorithm based on extended Kalman filter. Other positioning tools exist, for instance, GPS Toolbox (L3Nav)

and Wi-Fi Positioning Access Point Simulator (Spirent). The former includes a set of software modules developed in Matlab environment but the simulator is able to test and analyse GPS-only applications. In fact, it does not take into account terrestrial range measurements. The latter is a tool to design and test Wi-Fi-based positioning solutions with the possibility to integrate GPS signal as well. However, it does not take into account the cooperation among users.

This chapter presents also an approach to make hybrid and cooperative positioning algorithms more robust to inaccurate information flowing in the P2P network. In particular, the proposed method selects the most reliable neighbouring peers on the basis of the exchanged estimated position covariance matrix. According to simulation results, this approach improves both estimation convergence and positioning accuracy while lowering network traffic and latency. Some results concerning vehicular navigation in urban canyon scenarios are also presented. Moreover, this chapter compares different state of the art hybrid and cooperative algorithms in terms of performance and computational complexity. Finally, some results show the sensitivity of the hybrid and cooperative positioning algorithms in presence of high level pseudorange noise and against malicious peers.

MAIN FOCUS OF THE CHAPTER

This section includes the description of the simulation tool, the implementation of real-time cooperative positioning algorithms and the analysis of complexity and sensitivity. Real-time Cooperative Positioning

In order to implement a real-time cooperative positioning algorithm, each peer executes three different tasks as showed in Figure 1. As it can be observed, the position estimation timeline is divided into time slots (TS). Each TS corresponds to the time interval between two consecutive final

Figure 1. Example of position estimation timeline with I = 3 position iterations

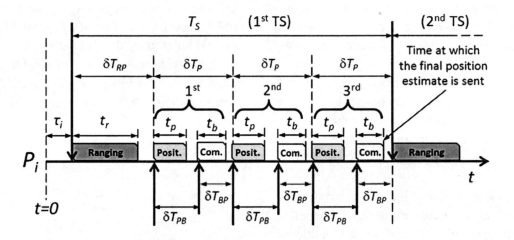

position estimations or equivalently to the time interval between two range measurement tasks. The TS duration, denoted with T_s, is the final position estimation period, and its inverse, denoted with f_s $= 1/T_s$, is the final position estimation frequency. The following tasks are executed within each TS at each peer:

1. **Ranging:** A peer performs both terrestrial range measurements from the neighbouring peers, and pseudorange measurements from the visible satellites. Note that this task involves also communication between terrestrial peers. In fact, depending on the selected ranging technique (*e.g.*, a ToA-based method (Lee & Scholtz, 2002)), peers may transmit their identifier and time stamp information. The duration of this task, denoted with t_r, strongly depends on the number of neighbours, while, in general, t_r is independent from the number of satellites as pseudorange measurements are performed in parallel by different code-divided channels, each one synchronized to a single satellite.

2. **Positioning:** A peer updates its position estimate according to both the last received aiding data (*e.g.*, estimated position and covariance matrix accuracy) and the last range

and pseudorange measurements (Sottile, Wymeersch, Caceres & Spirito, 2011; Caceres, Sottile, Garello & Spirito, 2010). The duration of this task, denoted with t_p, depends on the intrinsic complexity of the positioning algorithm and the number of input measurements from both neighbouring peers and visible satellites.

3. **Communication:** A peer broadcasts to its one-hop neighbours the updated aiding data. In accordance to the position estimation timeline, an aiding data is considered transmitted and correctly received – if collision on the radio channel does not take place – at the end of this task, see Figure 1. The duration of this task, denoted with t_b, strongly depends on the MAC layer and on the amount of aiding data to be transmitted (Sottile, Vesco, Scopigno & Spirito, 2012).

Moreover, a peer can be configured to perform a certain number of position iterations within a single TS to increase the position accuracy. Figure 1 shows an example where three position iterations are executed, $I = 3$, in single TS after the ranging task. Note that a positioning iteration is composed of a positioning task and a communication task. It is worth remarking that the configuration of the position estimation timeline is crucial for the real-time cooperative positioning. The time line must take into account the mobility degree of peers and the positioning accuracy requirement ε_p. The former influences the position estimation period T_S: the higher the mobility, the shorter T_S must be. Moreover, the requirement on ε_p strongly influences the choice of the maximum number of position iterations to be performed. In fact, a larger number of iterations improve the position accuracy.

A possible design of the position estimation timeline is here reported considering a generic time slot duration T_S and number of position iterations I. Let denote with δT_{RP} the time available to perform the ranging task (see Figure 1); it corresponds to the time interval between the beginning of the ranging task and the following beginning of the positioning task. Let denote with δT_P the time interval available to perform one position iteration (*i.e.*, the interval between the beginnings of two consecutive positioning tasks as shown in Figure 1). The choice of δT_{RP} and δT_P is not unique as both of them depend on different factors such as average number of neighbours, mobility of peers, ranging technique adopted, number of iterations required and computation complexity. Heuristically, a possible choice for δT_{RP} and δT_P could be:

$$\delta T_{RP} = \delta T_P = \frac{T_S}{I+1}, \tag{1}$$

where the following equality holds:

$$\delta T_{RP} + I \cdot \delta T_P = T_S. \tag{2}$$

Let us denote with δT_{PB} the time interval available to perform the positioning task (corresponding to the interval between the beginning of the positioning task and the subsequent beginning of the communication task). Moreover, let denote with δT_{BP} the time interval available to perform the communication task (*i.e.*, the interval of time

between the beginning of the communication task and the following beginning of the positioning task). Again, the choice of δT_{PB} and δT_{BP} is not unique since they depend on the computation complexity, average number of neighbours, mobility of peers and MAC layer latency. Again, a heuristic choice for these two parameters could be:

$$\delta T_{PB} = \frac{2}{3} \cdot \delta T_P = \frac{2}{3} \cdot \frac{T_S}{I+1}, \tag{3}$$

$$\delta T_{BP} = \frac{1}{3} \cdot \delta T_P = \frac{1}{3} \cdot \frac{T_S}{I+1}, \tag{4}$$

where the following equality holds:

$$\delta T_{PB} + \delta T_{BP} = \delta T_P. \tag{5}$$

As mentioned above, the positioning timeline design is not unique, but in order to have a proper real-time implementation, it is important that consecutive tasks do not overlap between each other, hence: $t_r < \delta T_{RP}$, $t_p < \delta T_{PB}$ and $t_b < \delta T_{BP}$.

To sum up, t_r depends on the complexity of the ranging algorithms and the number of neighbouring peers, t_p depends on the complexity of the positioning algorithms, finally, t_b depends on the communication technology adopted as well as the number of neighbours as these might compete to accessing the radio channel.

Simulation Tool Structure

The functional architecture of the ST is depicted in Figure 2. As it can be observed, every software module is labelled with a number, which indicates the execution order once the simulation is started. Modules in grey colour manages functionalities related to the generation of the terrestrial P2P network while striped modules manage functionalities related to the satellite section, *e.g.* visibility and pseudorange measurements. The main functionalities of each block are summarized as follows:

Figure 2. Architecture of the simulation tool

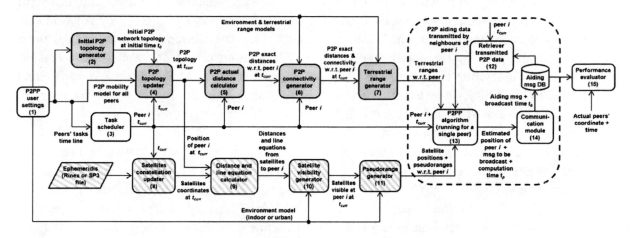

1. **P2P User Settings:** This module represents the graphical user interface (GUI) of the ST. It allows the user to set all the simulation parameters, such as topology, mobility model, range model and so forth.

2. **Initial P2P Topology Generator:** This module generates the initial topology of the P2P network at the initial time t_0 according to the parameters set by the user. Different choices are available, such as grid topology, uniform random topology, or specified by the user from file.

3. **Task Scheduler:** This module manages the three different tasks (ranging, positioning and communication) assigned to each peer according to the peer's timeline. The outputs of this module are: peer identifier, denoted with i, the assigned task to be executed, and the time t_{curr} at which the task must be executed.

4. **P2P Topology Updater:** This module updates the whole P2P topology at time t_{curr} according to the mobility model.

5. **P2P Actual Distance Calculator:** Given as input the exact P2P topology at time t_{curr}, this module computes the set of Euclidean distances from the peer i with respect to all peers present in the network.

6. **P2P Connectivity Generator:** Given as input the set of actual P2P distances, this module generates the connectivity with respect to peer i at time t_{curr} according to both the connectivity model and the selected environment, indoor or urban.

7. **Terrestrial Range Generator:** Given as input the set of actual P2P distances related to the connected peers, this module generates range measurements with respect to peer i at time t_{curr} according to the selected propagation models. In addition, it provides as output the time t_r required to perform the whole ranging task. This time depends on the number of neighbours and the selected range estimation technique.

8. **Satellites Constellation Updater:** Given as input the ephemeris of the considered GNSS constellation (stored either in a RINEX or a SP3 file), this module generates the satellites coordinates (in earth centred earth fixed (ECEF) coordinate system) at time t_{curr}, by interpolating the predicted or measured constellation orbits.

9. **Distance and Line Equation Calculator:** Given as input both satellites' coordinates and the position of peer i at time t_{curr}, this module calculates both the set of exact

distances from the peer i with respect to all satellites. Moreover, it calculates the equations of the connecting lines which will be subsequently used by the next module to determine the satellite visibility.

10. **Satellites Visibility Generator:** Given as input the previous set of line equations and the environment model description (indoor or urban), this module determines the satellites that are in LoS with peer i at time t_{curr}.

11. **Pseudorange Generator:** Given as input the set of satellites visible by peer i at time t_{curr}, this module generates the pseudorange measurements related to peer i according to the selected pseudorange model.

12. **Retriever Transmitted P2P Data:** Given as input the following parameters: identifier of the peer i to be estimated, time t_{curr} and all previous P2P transmitted aiding data, this module retrieves the last received aiding data transmitted by the neighbours of peer i before time t_{curr}.

13. **P2P Algorithm:** Given as input the range and pseudorange measurements, coordinates of the visible satellites, the last P2P aiding data received from the neighbouring peers before time t_{curr}, this module executes the positioning algorithm running for the single peer i and provides the following outputs: peer's position estimate, aiding data to be transmitted and computation time t_p. This time depends both on the complexity of the algorithm which in turns depends on the number of connected peers, number of visible satellites and on the computation processing capability available at peer i.

14. **Communication Module:** Given as input the P2P aiding data to be transmitted, this module simply generates the broadcast delay t_b according to the selected MAC model, number of neighbouring peers and the amount of data to be transmitted.

15. **Performance Evaluator:** Given as input both the set of estimated peer's positions and the actual ones, this module evaluates

the performance of the selected algorithm in terms of positioning accuracy and availability. A comparison with the theoretical lower bound is provided as well by the ST based on the *Cramér-Rao lower bound (CRLB)* derived in (Penna, Caceres & Wymeersch, 2010).

In order to reduce simulation time, data generated by the ST, such as ranging measurements, satellites' positions and all inputs to the P2P algorithm module are saved in a proper data structure and made available for several tests. Moreover, the ST can test different P2P algorithms using the same set of generated data. Moreover, following this approach, the ST performs a fair comparison among different algorithms while saving simulation time. In fact, the ST does need to run again modules from 2 to 11 but it has to execute only modules inside the dashed block showed in Figure 2.

Scenario Configurations

The ST includes different environment models and different mobility models.

Environment Models

As already mentioned above, the ST is able to test cooperative P2P algorithms in two different realistic scenarios, namely indoor and urban, which are presented on next.

1. **Indoor Environment Model:** As depicted in Figure 3a, the indoor environment is simply modelled with a rectangular area whose sides are parallel to the North-South and East-West directions on the Earth surface. The building is composed of four pillars located at the corners while the four sides are assumed to be large windows that do not obstruct the GPS signal. On the contrary, the ceiling is assumed to completely block the GPS signal. Moreover, a deep indoor area

inside the building has been defined, where there is not satellite visibility at all. Through the Matlab GUI, the user can set the following indoor environment parameters: ECEF coordinates of the building's floor centre, base's sizes, and building's height.

2. **Urban Environment Model:** The urban environment is modelled with a series of blocks with squared basis aligned parallel to the North-South and East-West directions on the Earth surface (see example depicted in Figure 3b). The number of blocks is the same for both sides and blocks are separated by roads where peers can move. Through the GUI, the user can set the following parameters: ECEF coordinates of the urban ground centre, number of blocks per side, block size, height and street width. Blocks are assumed to completely obstruct both satellite and terrestrial links.

Mobility Models

Mobility models are implemented in the topology updater module (see module 4 in Figure 2). For the indoor scenario only static and pedestrian mobility are allowed, while for the urban scenario only static and vehicular mobility can be set. The exact knowledge of the topology along the time is important not only for the evaluation of the position performance but also for the generation of the following data: range and pseudorange measurements, P2P connectivity and satellite visibility. Moreover, apart from the ranging task, it is necessary to generate the P2P connectivity also during the communication task. In fact, it may happen that if T_S is large and the mobility of peers is high, peers' connectivity occurring during the range task is different from the one occurring during the subsequent broadcast tasks. As a consequence, within a TS, the topology is updated both at the beginning of the range tasks and at the beginning of communication tasks.

In the following it is reported a description of mobility models implemented on the ST.

1. **Pedestrian Mobility:** Two pedestrian mobility models, namely Brownian Motion and Random Waypoint, have been implemented in the ST. In one hand, the Brownian motion model updates peers' positions randomly according to the following additive Gaussian relation:

$$x\left(t+1\right) = x\left(t\right) + \Delta x, \tag{6}$$

$$\Delta x \sim \mathcal{N}\left(\mu_m, \Sigma_m\right), \tag{7}$$

where μ_m is usually a null vector and the Σ_m is a diagonal covariance matrix expressed in the easting northing up (ENU) frame:

$$\Sigma_m = \operatorname{diag}\left(\sigma_E^2, \sigma_N^2, \sigma_U^2\right). \tag{8}$$

In the other hand, the random waypoint model (Johnson & Maltz, 1996) is a random-based mobility model mainly used in mobility management schemes for mobile communication systems. According to this model, each peer, initially placed at a random position within the simulation area, stays at its current location for a random period uniformly selected within the interval $[0, T_{\text{pause}}]$ seconds; then, the peer randomly chooses a new location to move and a random velocity within $[v_{\min}, v_{\max}]$ at which to move there. Once arrived, each peer continues this behaviour alternatively staying and moving to a new location for the whole duration of the simulation.

2. **Vehicular Mobility:** In the urban scenario, peers' positions are updated according to a simple mobility model. In particular, every road of the urban scenario is composed of two main lanes along which peers can move according to the right-hand drive (see Figure 3b). Peers move along one of the main lane

with constant velocity; once the 'vehicle' arrives at the end of the road, it goes back on the other lane using a lower velocity.

Optimization of Hybrid and Cooperative Positioning Algorithms

An optimization approach is here presented that aims to make the hybrid and cooperative positioning algorithms more robust to inaccurate information flowing in the network and less complex in terms of computation capacity and network load. The proposed approach selects the most reliable cooperating neighbour peers on the basis of the estimated position covariance matrices related to their estimated positions. Several solutions have been proposed in the literature, *e.g.*, the censoring approach presented in (Das & Wymeersch, 2010) based on the Cramér-Rao bound, which minimizes the amount of information shared between peers while maintaining acceptable performance. On the contrary, the approach proposed here selects reliable cooperative neighbouring peers based on the estimated position accuracy in order to achieve faster convergence and improved final accuracy while reducing network traffic and latency.

After the neighbour selection step, the positioning task will be executed only if a certain number of robust measurements are available at the unknown peer. In particular, the whole *robust process* is summarized as follows.

1. An unknown peer i uses the aiding data from a generic neighbouring peer p if and only if p is classified as *"robust peer"*, *i.e.* the corresponding estimated position accuracy, denoted with $\hat{\sigma}_p^{(k)}$ and available at time k, is less than a prefixed threshold σ_{thr}:

$$\hat{\sigma}_p^{(k)} \leq \sigma_{\text{thr}}, \qquad (9)$$

where

$$\hat{\sigma}_p^{(k)} = \sqrt{\text{trace}\left(\hat{\Sigma}_p^{(k)}\right)}, \qquad (10)$$

and $\hat{\Sigma}_p^{(k)}$ is the estimated covariance matrix of the neighbouring peer's position. Equivalently, in order to lower network traffic, a peer can decide to broadcast the aiding data only if condition (9) holds.

2. $r_i^{(k)}$ denotes the *"number of robust measurements"* available at peer i at time k defined as:

$$r_i^{(k)} = a_i^{(k)} + p_i^{(k)} + s_i^{(k)}, \qquad (11)$$

where $a_i^{(k)}$ is the number of anchor peers connected to peer i, $p_i^{(k)}$ is the number of *"robust peers"* connected to peer i and selected according to (9), and $s_i^{(k)}$ is the number of satellites visible by peer i.

3. The position of the unknown peer i is estimated if and only if the *number of robust information* satisfies the following observability condition:

$$\begin{cases} r_i^{(k)} \geq N_{\text{thr}} & \text{if } s_i^{(k)} = 0, \\ r_i^{(k)} \geq N_{\text{thr}} + 1 & \text{if } s_i^{(k)} > 0, \end{cases} \qquad (12)$$

where N_{thr} is a prefixed threshold related to the number of input measurements used by the positioning algorithm. As it can be observed from equation (12), when an unknown peer uses measurements from satellites (*i.e.*, $s_i^{(k)} > 0$), the number of robust measurements is increased by one because there is an additional unknown to be estimated, *i.e.*, the bias of the GNSS receiver.

To sum up, the above robust process uses two parameters (N_{thr}, σ_{thr}) whose values need to be tuned on the basis of the network connectivity and quality of range and pseudorange measurements.

This approach represents a viable solution for cooperative algorithms, in fact, as proved through simulations (shown in the following sections), it assures a faster estimation convergence and improved accuracy while lowering computational complexity and network traffic.

Simulation Results in Different Scenarios

This section presents positioning performance related to four state-of-the-art hybrid and cooperative algorithms that have been implemented in the ST, namely *hybrid-cooperative particle filter (HC-PF)* (Sottile, Wymeersch, Caceres & Spirito, 2011*), hybrid sum-product algorithm over a wireless network (H-SPAWN)* (Caceres, Penna, Wymeersch & Garello, 2010), *hybrid-cooperative unscented Kalman filter (HC-UKF)* (Caceres, Sottile, Garello & Spirito, 2010) and *hybrid-cooperative weighted least squares (HC-WLS)*. In particular, the HC-WLS algorithm has been implemented according to the iterative descent algorithm proposed in (Wymeersch, Lien & Win, 2009) and extended to the hybrid GNSS and terrestrial scenario.

Indoor Small-Scale Network

This section reports the performance of the above mentioned algorithms simulated in the small 3D network scenario showed in Figure 3a. In particular, the indoor scenario used is of size 50×50 m and height 8 m, where a total of seven unknown static peers are deployed, six in light indoor and one in deep indoor visibility. Note that no anchor peers are used in this scenario. The centre of the building is located at the same position of Istituto Superiore Mario Boella (ISMB) premises in Turin, 45.06° latitude, 7.66° longitude. Satellites positions are updated at each TS according to real GPS orbits.

Pseudorange measurements $\rho_i^{(k)}$ are generated by using the following additive Gaussian noise model:

$$\rho_i^{(k)} = d_{ij}^{(k)} + b_i^{(k)} + n_{ij}^{(k)}, \qquad (13)$$

where $d_{ij}^{(k)}$ is the Euclidean distance between a generic peer i and a generic satellite j, $b_i^{(k)}$ is the bias of peer i (expressed in meters) with respect to the GPS satellite clock and $n_{ij}^{(k)}$ is an additive Gaussian noise, $n_{ij}^{(k)} \sim \mathcal{N}(0, \sigma_{ij}^2)$.

Realistic biases are generated uniformly in the interval ±0.5 milliseconds of clock misalignment (*i.e.*, duration of one GPS C/A code), correspond-

Figure 3. Example of indoor scenario (a) and urban scenario (b) dotted lines represent connectivity between peers

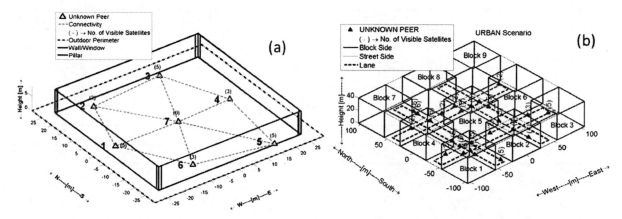

ing to about ± 150 km if expressed in distance units. The standard (std) deviation value σ_{ij} depends on the carrier-to-noise ratio (C/N_0), or equivalently, on the elevation angle at which the peer i sees the satellite j, see reference (Sottile, Caceres & Spirito, 2011) for more details. The larger elevation angle, the larger σ_{ij}, since the signal travels across a longer path on the ionosphere and troposphere layers, whose atmospheric effects are harder to mitigate with standard models. Both satellite visibility and std deviation of the pseudorange measurements for each peer are reported in Table 1 at the initial TS t_0, where the minimum std deviation value observed is $\sigma_{ij}(min) = 6.25$ m, while the maximum one is $\sigma_{ij}(max) = 18.3$ m.

In total, $K = 20$ TSs are simulated, with $T_s = 1$ s, hence, the simulated time window is [0, 20] s. Terrestrial range measurements are generated using an additive Gaussian model noise with zero mean and constant std deviation $\sigma_r = 0.2$ m, a typical value for terrestrial ranging based on the UWB technology.

New terrestrial range and pseudorange measurements are generated at each t_{curr}, and all algorithms iterated 3 times per TS ($I = 3$). Initial peers' position distributions are set uniform inside a sphere of 100 m radius centred at the origin of the environment. In order to perform a compari-

son among the implemented algorithms, 50 *Monte Carlo (MC)* simulations of the above scenario are performed. Then, the localization error, *i.e.*, the Euclidean distance between the estimated position and the true one, $\varepsilon_m^{(k)} = \left\| \hat{\boldsymbol{p}}_m^{(k)} - \boldsymbol{p}_m^{(k)} \right\|$, is computed for every peer m in the network at the end of each TS, indexed with k.

First of all, some simulations are performed in order to test the impact of the robust parameters on the performance of cooperative algorithms. In particular, Table 2 lists the performance of the HC-PF as a function of the robust parameters N_{thr} and σ_{thr} reported in the first and second columns, respectively. Sequentially, the other columns report, the root mean square (RMS) of the horizontal position error, the RMS of the vertical position error (both calculated at TS equal to 20) and the position estimation availability, expressed in percentage.

As it can be observed, the tested threshold N_{thr} goes from 1 (relaxed observability condition) to 4 (strict observability condition). Moreover, for each tested N_{thr}, three values for σ_{thr} are tested: ∞, 20 and 14. Note that $\sigma_{thr} = \infty$ means that all aiding data are broadcast regardless the value of $\hat{\sigma}_p^{(k)}$.

In general, as it can be observed from the 3-rd and 4-th columns, the positioning performance

Table 1. Satellite visibility and corresponding pseudorange std deviation values (expressed in meters) for the hybrid scenario showed in Figure 3a. Symbol '–' indicates non satellite visibility

		Satellite PRN										
		2	4	5	7	8	10	13	15	24	25	28
Peer Id	P1	11.5	18.3	6.5	–	–	–	–	17.3	12.9	–	–
	P2	–	–	–	–	–	–	–	–	12.9	9.2	–
	P3	–	–	6.5	6.7	–	–	–	17.3	12.9	9.2	–
	P4	–	–	–	–	–	–	11.1	–	–	9.2	12.0
	P5	11.5	18.3	–	–	6.3	6.3	–	–	–	–	12.0
	P6	11.5	18.3	–	–	–	–	–	–	–	–	12.0
	P7	–	–	–	–	–	–	–	–	–	–	–

Table 2. Performance analysis of the HC-PF algorithm as a function of the robust parameters (N_{thr}, σ_{thr})

N_{thr}	σ_{thr} [m]	RMSE(Hor) [m]	RMSE(Ver) [m]	Posit. Availability [%]
1	∞	26.50	18.80	100
	20	25.67	19.60	99.7
	14	28.27	18.49	99.7
2	∞	23.51	22.44	99.3
	20	13.38	12.95	97.2
	14	8.52	8.34	94.3
3	∞	20.32	15.86	99.3
	20	**8.24**	**7.95**	94.6
	14	8.63	9.04	95.5
4	∞	9.83	9.11	97.9
	20	9.45	8.86	92.0
	14	**7.99**	9.35	82.7

improves as N_{thr} increases and σ_{thr} decreases. However, such improvement comes at the cost of less position availability.

Note that, in the indoor scenario showed in Figure 3a, where every peer has a number of visible satellites larger or equal than 2, the setting $N_{thr} = 1$ and $\sigma_{thr} = \infty$ corresponds to the case in which the robust estimation approach does not take place. In fact, all peers perform position calculation regardless of aiding data quality. In this particular case, it can be observed that even if more TSs are simulated, the performance does not improve along the estimation process because initial peers' positions are too inaccurate (*i.e.*, $\hat{\sigma}_p^{(0)}$ values are large) and some peers get stacked in wrong positions. Thus, in order to obtain good convergence properties, it is better to wait some TSs until more accurate aiding date are available from the neighbours, *i.e.*, $\hat{\sigma}_p^{(k)} \leq \sigma_{thr}$, rather than using from the beginning bad aiding data. To sum up, the lower σ_{thr} the more accurate position performance, but the lower initial position availability.

Table 2 highlights in bold the best horizontal accuracy equal to 7.99 m *root mean squared error (RMSE)* achieved by setting (N_{thr}, σ_{thr}) = (4, 14)

and the best vertical accuracy equal to 7.95 m RMSE, achieved by setting (N_{thr}, σ_{thr}) = (3, 20). Finally, between the two mentioned pairs, the table highlights in bold the pair (3, 20) as this is the one that provides the best global position accuracy ($\sqrt{H_{err}^2 + V_{err}^2}$ = 7.95 m RMSE) among all the simulated cases. Thus, according to these results, the following simulations use the settings: (N_{thr}, σ_{thr}) = (3, 20) as suboptimal robust parameters.

Figure 4 shows an example of graphical simulation results provided as output by the ST. In particular, Figure 4a shows the RMS of the horizontal position error for each peer as a function of the time obtained with the HC-PF algorithm, while Figure 4b shows the performance of the non-cooperative version of the PF algorithm. Note that the non-cooperative algorithm is able to estimate only 3 peers out of 8. Thus, by comparing Figure 4a with Figure 4b, it can be concluded that the HC-PF algorithm improves both position accuracy and position availability. Finally, Figure 4c shows the performance of all algorithms in terms of CDF of the global horizontal position error and Figure 4d the corresponding RMSE as a function of the time. It can be observed, HC-PF provides the best accuracy among all.

Urban Medium-Scale Vehicular Network

This section presents simulation results of hybrid and cooperative positioning algorithms simulated in the urban scenario depicted in Figure 3b. The centre of the environment is the city centre of Singapore, 1.2381° latitude, 76.1557° longitude. The vehicular network is composed of 25 unknown

vehicles and the urban environment consists of 9 blocks and 4 streets. Similar to the indoor case, no anchor peers are deployed. The exact vertical heights of unknown vehicles are uniformly chosen within [0, 2] m.

The communication range of UWB transceivers is set to 80 m, which is a typical value of UWB commercial modules (P400 2011). The selected mobility of each peer is vehicular, that is 10 m/s

Figure 4. Simulation results of indoor small-scale network. (a) Horizontal position error per peer as a function of the TS for the HC-PF, (b) horizontal position error per peer as a function of the TS for the non-cooperative particle filter, (c) CDF of the global horizontal positioning error, (d) RMS of the global horizontal position error as a function of the TS.

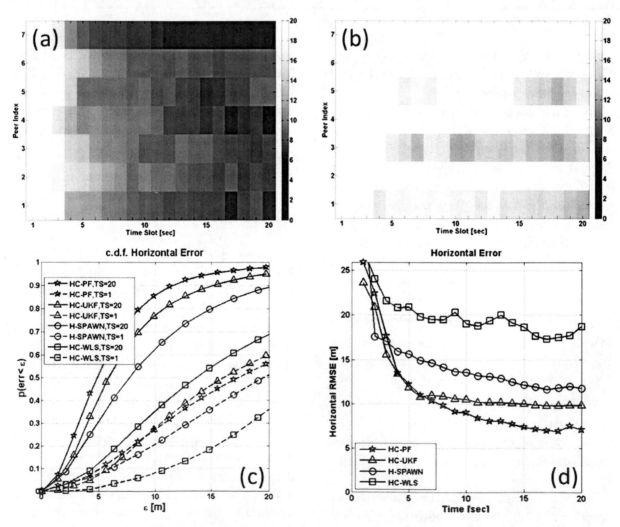

on the lane and 2 m/s during the turning. Based on the simulated scenario, the average number of visible satellites is 4.9.

The pseudorange measurements are generated based on the real GPS orbits and the exact vehicles' positions in the environment while the standard deviation and bias are generated as before. Concerning the satellite visibility, it is assumed that the blocks completely obstructed the GPS signal. Hence, a specific GPS satellite is visible only if the line that connects the satellite to the vehicle is not intersected any building.

Figure 5 shows performance of different algorithms in scenario Figure 3b. Figure 5a plots the CDF of the global horizontal errors while Figure 5b plots the horizontal errors at each TS. As it can be seen, all the algorithms show good positioning availability due to the peer cooperation. However, neither HC-UKF nor H-SPAWN proves good convergence, mainly because they are not properly tuned for this urban mobility scenario. It can be observed that after the tenth time slot, errors remain stable around 25 m for H-SPAWN while it goes below 10 m for HC-UKF. The HC-PF

instead, is able to track all vehicles and achieves a final accuracy of about 5 m. In the HC-LS algorithm, large oscillations are observed, which may be driven by few peers with large errors, because they may have lost some key satellites or peers.

Moreover, the HC-PF is chosen to assess the beneficial effect of using a fixed infrastructure made of anchor peers. In the same scenario of Figure 3b, some fixed anchors are deployed, in which two configurations have been tested, one with 6 anchors and another one with 12 while the other settings remain the same as before. The 2D view of this urban scenario is shown in Figure 6a, where the anchors in the 6-anchor configuration are represented by diamonds while anchors in the 12-anchor configuration are represented by squares. All the anchors are evenly distributed on the roof of the blocks. The vehicular peers perform range measurements with both anchors and mobile neighbours while moving.

As showed in Figure 6b, the hybrid use of GPS and UWB range measurements can significantly improve both positioning availability and accuracy. It is worth observing that the case without fixed

Figure 5. Simulation results of Urban medium-scale vehicular network. (a) CDF of the global horizontal positioning error, (b) RMS of the global horizontal position error as a function of the TS

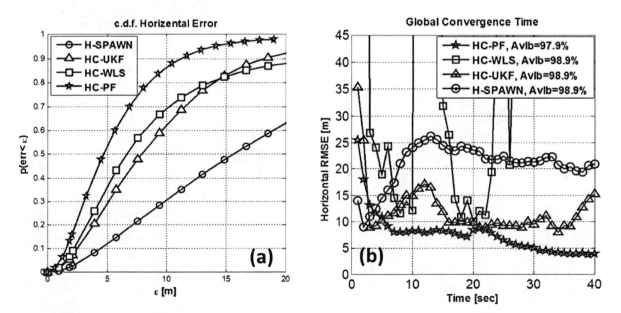

infrastructure (*i.e.*, no anchor), the hybrid and cooperative positioning algorithm outperforms the GPS-only solution, in fact, the positioning errors are 4 m and 10 m, respectively. Moreover, the position availability is increased by 19%.

Of course, by using anchors the positioning errors decreases dramatically. In particular, by using 6 anchors, the final positioning error is around 2 m while using 12 anchors the error is around 0.8 m.

Since installing and maintaining a fixed UWB infrastructure in urban environment is expensive, the hybrid and cooperative localization without anchors is a viable solution for vehicular navigation in urban environments.

Complexity Analysis

As far as the computational complexity is concerned, all the considered hybrid and cooperative algorithms belong to the Non-deterministic Polynomial-time (NP) class. Their different orders of complexity depend mainly on the number of available measurements and intrinsic state representation (*i.e.*, number of samples in particle based methods) at each position estimate. Such orders of complexity can be expressed in terms of the number of neighbour peers P (including both anchors and unknown peers), the number of visible satellites S, the number of particles M, the space dimension D and the state vector dimension V.

The simplest (and less accurate) of the simulated hybrid and cooperative algorithms is the HC-WLS, whose main complexity lies in a pseudo-inverse computation, which involves the inversion of a square matrix: $C_{HC\text{-}WLS} = O((P+S)^{2.376})$. Note that the Coppersmith-Winograd method for computing a matrix inversion has a reduced complexity with respect to the classical Gauss-Jordan elimination which is $O(n^3)$. More details about the complexity budget are presented in Table 3, where FLO stands for floating-point operations.

The second algorithm considered is the HC-UKF, whose complexity is driven by two operations: the Cholesky decomposition needed for the sigma-point generation, and the sample weighted mean and covariance matrix estimators, with complexity $O((V+S+PD)^3)$. The detailed com-

Figure 6. (a) Simulation scenario of two fixed-anchor configurations. (b) RMS of the global horizontal position error as a function of the TS

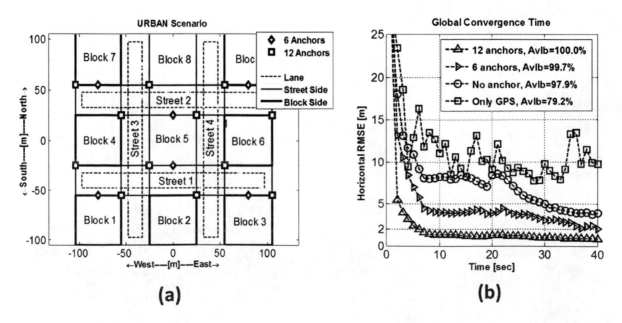

(a)

(b)

plexity analysis is in Table 4, where L denotes the length of the augmented state vector in unscented transformations.

The complexity of the HC-PF algorithm is dominated by the Sequential Importance Resampling (SIR) approach employed, which not only evaluates the likelihood of each measurement per particle, but estimates the mean and covariance

matrix of the *a posteriori* estimate and regenerates the samples at each step. The order of its complexity is proportional to the number of particles and the cube of the state vector dimension: $C_{HC-PF} = O((P + S)V^3M)$. More details of this result are in Table 5.

Similar to the HC-PF, the main contribution to H-SPAWN's complexity is given by the evalu-

Table 3. Complexity of HC-WLS

Computation	Operation	No. of FLO	Complexity
$\Delta z = z_u - h\left(\hat{x}_u, X_{su}, \hat{X}_{pu}\right)$	$S + P$ evaluations of $h(\cdot)$	$(D^2 + 1)(S + P) + S$	$\mathcal{O}\left(D^2(S + P)\right)$
$H = \begin{vmatrix} \frac{\partial h_s}{\partial x} & \frac{\partial h_s}{\partial y} & \frac{\partial h_s}{\partial z} & \frac{\partial h_s}{\partial b} \\ \frac{\partial h_n}{\partial x} & \frac{\partial h_n}{\partial y} & \frac{\partial h_n}{\partial z} & \frac{\partial h_n}{\partial b} \end{vmatrix}$	$S + P$ partial derivatives of $h(\cdot)$	$2D(S + P)$	$\mathcal{O}(D(S + P))$
$W = \begin{vmatrix} \frac{1}{\hat{\sigma}_{su}^2} & \cdots & 0 \\ \vdots & \ddots & \vdots \\ 0 & \cdots & \frac{1}{\hat{\sigma}_{pu}^2 + u\,(\hat{P}_P)} \end{vmatrix}$	$S + P$ divisions & $P(D + 1)$ sums	$S + P(D + 2)$	$\mathcal{O}(S + PD)$
WH	A matrix multiplication	$V(S + P)^2$	$\mathcal{O}\left(V(S + P)^2\right)$
$J = \left(H^T WH\right)^{-1}$	A matrix multiplication & inverse	$V^2(S + P) + (S + P)^3$	$\mathcal{O}\left((S + P)^3\right)$
$G = JH^T W$	Two matrix multiplications	$V(S + P) + V(S + P)^2$	$\mathcal{O}\left(V(S + P)^2\right)$
$\hat{x}_u = \hat{x}_u + \delta G\Delta z$	A matrix sum & multiplication	$2(S + P) + V(S + P)^2$	$\mathcal{O}\left(V(S + P)^2\right)$
$\hat{\Sigma}_u = \frac{J\Delta z^T (W - WHG)\Delta z}{(S + P - \text{rank}(H))}$	Several matrix multiplications	$2V^2(S + P) + 2(S + P)^2$	$\mathcal{O}\left((S + P)^2\right)$

Table 4. Complexity of HC-UKF

Computation	Operation	No. of FLO	Complexity
Generate $\chi_{u,i}^{(k-1)} = \hat{x}_u^{(k-1)} \pm \left(\sqrt{(L + \lambda)\hat{\Sigma}_u^{(k-1)}}\right)$	σ pts. $L = 2V$	$\frac{8}{3}V^3 + 4V^2$	$\mathcal{O}(V^3)$
Predict $\chi_{u,i}^{(k\|k-1)} = f(\chi_{u,i}^{(k-1)})$	$2L + 1$ scalar ops.	$12V + 3$	$\mathcal{O}(V)$
Estimate $\hat{x}_u^{(k\|k-1)} = \sum_{i=0}^{2L} w_i^\mu \chi_{u,i}^{(k\|k-1)}$	$2L + 1$ vector ops.	$16V^2 + 4V$	$\mathcal{O}(V^2)$
$\hat{\Sigma}_u^{(k\|k-1)} = \text{cov}\left(\chi_{u,i}^{(k\|k-1)}, w_i^\Sigma\right)$	$2L + 1$ matrix ops.	$48V^3 + 20V^2 + 2V$	$\mathcal{O}(V^3)$
Generate $\chi_{u,i}^{(k\|k-1)} = \hat{x}_u^{(k\|k-1)} \pm \left(\sqrt{(L + \lambda)\hat{\Sigma}_u^{(k\|k-1)}}\right)$	σ pts. $L = V + S + P + PD$	$\frac{1}{3}(V + S + P + PD)^3 + 2(V + S + P + PD)^2$	$\mathcal{O}\left((V + S + PD)^3\right)$
Propagate $\zeta_{u,i}^{(k)} = h(\chi_{u,i}^{(k\|k-1)})$	$2L + 1$ scalar ops.	$(2(V + S + P + PD) + 1) \cdot (2D + 2)$	$\mathcal{O}(V + S + PD)$
Estimate $\hat{z}_u^{(k)} = \sum_{i=0}^{2L} w_i^\mu \zeta_{u,i}^{(k)}$	$2L + 1$ vector ops.	$4(V + S + P + PD) + 2$	$\mathcal{O}(V + S + PD)$
$S_u^{(k)} = \text{cov}\left(\zeta_{u,i}^{(k)}, w_i^\Sigma\right)$	$2L + 1$ matrix ops.	$(2(V + S + P + PD) + 1) \cdot ((S + P) + 3(S + P)^2)$	$\mathcal{O}\left((S + P)^3\right)$
$P_{x\|z}^{(k)} = \text{cov}\left(\chi_{u,i}^{(k\|k-1)}, \zeta_{u,i}^{(k)}, w_i^\Sigma\right)$	$2L + 1$ matrix ops.	$(2(V + S + P + PD) + 1) \cdot (S + P + (3S + 3P + 1) \cdot (V + S + P + PD))$	$\mathcal{O}\left((S + P)^3\right)$
Calc $K^{(k)} = P_{x\|z}^{(k)} S^{(k)^{-1}}$	A matrix mult. & inv.	$(V + S + P + PD) \cdot (S + P)^2 + (S + P)^3$	$\mathcal{O}\left((S + P)^3\right)$
Correct $\hat{x}_u^{(k)} = \hat{x}_u^{(k\|k-1)} + K^{(k)}(z_u^{(k)} - \hat{z}_u^{(k)})$	A matrix mult. & sum	$(2(V + S + P + PD) + 1) \cdot ((S + P) + 3(S + P)^2)$	$\mathcal{O}\left((S + P)^3\right)$
$\hat{\Sigma}_u^{(k)} = \hat{\Sigma}_u^{(k\|k-1)} - K^{(k)} S^{(k)} K^{(k)^T}$	$2L + 1$ vector ops.	$(V + S + P + PD) \cdot (S + P + 1)$	$\mathcal{O}\left((S + P)^2\right)$

ation of the incoming particle-based distributions needed for the message multiplication, keeping the same linearity with the number of particles and state dimension: $C_{\text{H-SPAWN}} = O((P + S)V^3M)$, thus $P + 1$ times more complex than the HC-PF. The detailed complexity budget is listed in Table 6.

More details about the above algorithms and related performance are presented in (Sottile, Wymeersch, Caceres & Spirito, 2011; Caceres, Penna, Wymeersch & Garello, 2010; Caceres, Sottile, Garello & Spirito, 2010).

Sensitivity Analysis

This section presents simulation results obtained through the ST about the sensitivity of cooperative positioning algorithms in two different conditions: first, in presence of high level pseudorange noise, and then, in presence of a malicious peer which broadcasts misleading positioning aiding data to the network.

Sensitivity of P2P Positioning as a Function of Pseudorange Measurements Noise

This section reports the behaviour of hybrid and cooperative positioning algorithms (in particular HC-PF) when a peer of the P2P network is subject to high level of pseudorange noise. Simulations are performed on the small scale scenario depicted in Figure 3a. During simulations, it is assumed that P1 believes to perform pseudorange measurements according to the Gaussian model and std deviation values listed in Table 1, but actually, P1's measurements are generated with larger std deviation values. In particular, different simulations are run by using std deviation values increased by step of

Table 5. Complexity of HC-PF

Computation	Operation	No. of FLO	Complexity
$\mathbf{x}_u^{(k)'} = A\mathbf{x}_u^{(k-1)'} + B_f\mathbf{f}_u^{(k-1)'}$	M evaluations of $\mathbf{x}_u^{(k)'}$	$3V^2M$	$\mathcal{O}(V^2M)$
$p\left(\mathbf{z}_u^{(k)} \middle\| \mathbf{x}_u^{(k)'}\right)$	M evaluations of $p(\cdot)$	$(D+1)(P+S)V^3M$	$\mathcal{O}((P+S)V^3M)$
$w_u^{(k)'} = \tilde{w}_u^{(k)'}/\sum_{i=1}^M \tilde{w}_u^{(k)'}$	M divisions	M	$\mathcal{O}(M)$
$\hat{\mathbf{x}}_u^{(k)} = \sum_{i=1}^M w^{(k)'}\mathbf{x}^{(k)'}$	M products and sums	$2VM$	$\mathcal{O}(VM)$
$\hat{\Sigma}_u^{(k)} = \frac{\sum_{i=1}^M w_u^{(k)'}\left(\mathbf{x}_u^{(k)'} - \hat{\mathbf{x}}_u^{(k)}\right)^T\left(\mathbf{x}_u^{(k)'} - \hat{\mathbf{x}}_u^{(k)}\right)}{1-\sum_{i=1}^M w_u^{(k)'^2}}$	Covariance estimate	$M + V^2M$	$\mathcal{O}(V^2M)$
$b_u^{(k)'} = \frac{1}{\|S_u\|}\sum_{s \in S_u}\left(\tilde{\rho}_{u,s}^{(k)} - \mathrm{d}(\mathbf{p}_u^{(k)'}, \mathbf{p}_s^{(k)})\right)$	M evaluations of $b_u^{(k)'}$	$(D+1)SM$	$\mathcal{O}(DSM)$
Resampling	M drawings from $\mathcal{N}(\cdot)$	V^2M	$\mathcal{O}(V^2M)$

Table 6. Complexity of H-SPAWN

Computation	Operation	No. of FLO	Complexity
Predict $\hat{\tilde{x}}_u^{(k\|k-1)}, \tilde{\Sigma}_u^{(k\|k-1)}, \mathcal{M}_{f_u \to u}\left(\hat{x}_u^{(k)}\right)$	Some matrix multiplications	$2V^2 + 2V^3$	$\mathcal{O}(V^3)$
The following operations are repeated $P+1$ times and iterated I times			
Draw $\chi_{u,i}^{(k\|k-1)} \sim \mathcal{N}(\hat{\tilde{x}}_u^{(k\|k-1)}, \tilde{\Sigma}_u^{(k\|k-1)})$	M samples from $\mathcal{N}(\cdot)$	V^2M	$\mathcal{O}(V^2M)$
Evaluate $q(\chi_{u,i}^{(k)})$ and $p_j(\chi_{u,i}^{(k)})$	$S + P + 2$ evaluations of $p(\cdot)$	$(S+P+2)V^3M$	$\mathcal{O}(V^3M)$
Compute $w_i \propto \frac{\prod_{j=1}^{S+P+1}p_j(\chi_{u,i}^{(k)})}{q(\chi_{u,i}^{(k)})}$	M products	$(S+P+4)M$	$\mathcal{O}((S+P)M)$
Estimate $\hat{\tilde{x}}_u^{(k)} = \sum_{i=1}^M w_i\chi_{u,i}^{(k)}$	M products and sums	$2VM$	$\mathcal{O}(VM)$
$\hat{\Sigma}_u^{(k)} = \frac{\sum_{i=1}^M w_i\left(\chi_{u,i}^{(k)} - \hat{\tilde{x}}_u^{(k)}\right)^T\left(\chi_{u,i}^{(k)} - \hat{\tilde{x}}_u^{(k)}\right)}{1-\sum_{i=1}^M w_i^2}$	Covariance estimate	$M + V^2M$	$\mathcal{O}(V^2M)$

5 up to 20 m with respect to the corresponding values listed in Table 1. Position performance is shown in Figure 7, where it can be observed how the global positioning error (*i.e.*, the RMSE for any node in the network) increases as the additional noise on the pseudorange measurements of peer P1 increases.

Moreover, Figure 8 shows the Contour plot for RMS of the horizontal positioning error in three different cases of additional std deviation values on peer P1. In particular, Figure 8a shows the baseline with no increment on the std deviation value; in Figure 8b the increment is of 5 m; while in Figure 8c the increment is of 10 m. Making a comparison between Figure 8b and Figure 8a, it can be noticed that the higher level of std deviation in peer P1 (+5 m) affects significantly its own performance, and slightly the convergence of its neighbouring peers P2 and P6, while P7, even being a neighbour of P1, is not affected thanks to its higher connectivity. Since peers P3, P4 and P5 are peers not direct connected to peer P1, they

are marginally affected. However, in the case of 10-meter additional std deviation in peer P1 (see Figure 8c), it can be observed that all peers in the network are affected, in particular peer P6. To sum up, errors in measurement modelling do impact proportionally on the propagated error.

In addition to the above results, the sensitivity of the hybrid and cooperative algorithms is assessed when large level of multipath affects pseudorange measurements. Pseudorange measurements are modelled as follows:

$$\hat{\rho}_i^{(k)} = d_{ij}^{(k)} + b_i^{(k)} + n_{ij}^{(k)} + m_{ij}^{(k)}. \tag{14}$$

This model differs from the previous Gaussian one reported in (13) only for the element m_{ij} which represents the multipath contribution. In particular, m_{ij} is modelled as a uniformly random variable drawn from an interval between zero and a maximum value, $m_{ij} \sim U[0, m_{max}]$. In order to asses such sensitivity, several simulations are

Figure 7. Global horizontal error position as a function of the TS considering different additional pseudorange std deviation values affecting P1

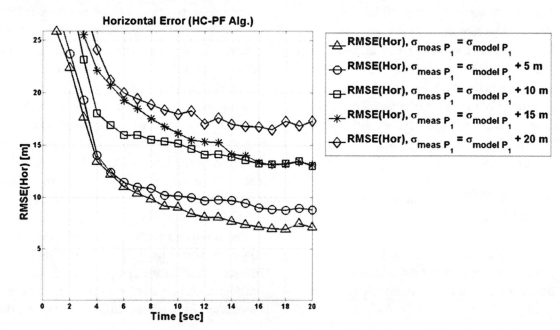

Figure 8. Contour plots of the horizontal error per peer as a function of the TS: (a) none additional std deviation in P1, (b) additional 5m in std deviation of P1, (c) additional 10m in std deviation of P1

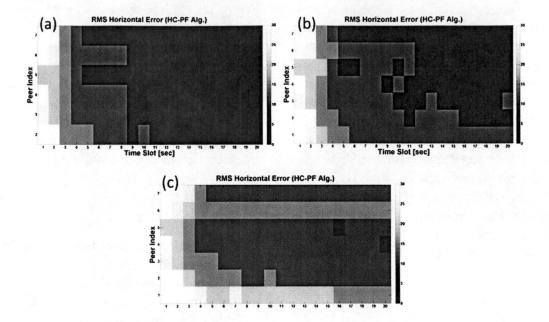

performed increasing the interval, *i.e.*, m_{max}, in steps of 20 up to 100 m.

Figure 9 shows the performance of HC-PF in presence of the multipath contribution . As it can be seen, there are no significant variations on the global performance when the maximum multipath contribution is lower or equal to 40 m. However, a noticeable degradation is observed when the maximum multipath bias is 60 m, which is more accentuated to higher biases. For instance, when the interval is of 100 m, there is an increment on the global positioning error of about 7 m, which is comparable with the one obtained with just the additive Gaussian model with an additional std deviation of 15 m.

Sensitivity of P2P Positioning Against a Malicious Peer

This section reports simulation results obtained by testing the sensitivity of hybrid and cooperative positioning algorithms to malicious peers (*i.e.*,

peers which periodically broadcasts misleading positioning aiding data to its 1-hop neighbours).

Recall that the aiding data is composed of two elements: the estimated position and its corresponding estimated covariance matrix. It is assumed that the malicious peer broadcasts an estimated position whose position error is kept constant and controlled by the simulator. In particular, the estimated position of the malicious peer is generated on the same horizontal plane of the actual position and at a distance from it equal to the prefixed position error. Moreover, the fake position lies on a line whose angle with respect to the positive direction of the East-axis is uniformly generated in the interval [0, 2π]. The malicious peer sets its estimated covariance matrix containing small values such that its neighbours believe that the broadcast aiding data is reliable.

The simulated scenario is the one depicted in Figure 3a, where peer P1 is the malicious peer. Different simulations are performed where the positioning error related to the malicious peer P1 is increased in steps of 5 m from 10 up to 20 m.

Figure 9. Global horizontal error position as a function of the TS considering different additional multipath contribution levels affecting P1

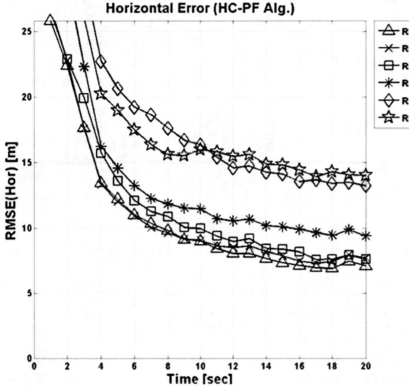

Horizontal Error (HC-PF Alg.)

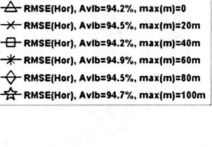

Figure 10 shows the global performance including peers from 2 to 7 as a function of the time in different cases: (1) in presence of cooperative peer P1 (non-malicious behaviour), (2) without peer P1 and finally (3) in presence of the malicious behaviour of peer P1 with different positioning errors. Note that the positioning error related to the malicious peer P1 is removed from the global performance. As expected, the best situation observed is when peer P1 cooperated with non-malicious behaviour (solid line with circle marks). Global performance degraded when peer P1 does not cooperate at all (line with triangle marks), and it gets worse when peer P1 cooperated maliciously broadcasting aiding data with a position errors from 10 to 20 m. It is worth observing that vertical errors are not affected significantly since the positioning error of malicious peer P1 is generated

only on the horizontal plane. In addition, Figure 11 shows the corresponding contour plot for the RMS of the horizontal positioning error in three different cases: cooperative peer P1 (Figure 11a), malicious peer P1 with 10 m of positioning error (Figure 11b) and, finally, positioning error of 15 m (Figure 11c). It can be observed from Figure 11b and c, in most affected peers, there are P2 and P6 that are direct neighbours of peer P1. Note that performance of malicious peer P1 is not reported in the error contour plots.

Finally, the sensitivity of the hybrid and cooperative algorithms in a large scale P2P network scenario composed of 100 peers (shown in Figure 12) has been tested, where P12 is selected as the malicious peer.

Figure 13 displays a detailed analysis of the position error at a peer level. In particular, Figure

Figure 10. Global horizontal and vertical position error as a function of the TS considering different levels of positioning error injected by the malicious peer P1 in the small scale scenario

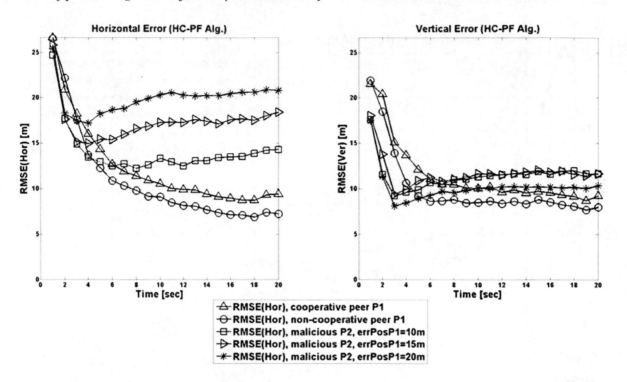

Figure 11. Contour plots of the horizontal error per peer as a function of the TS: (a) P1 is cooperative and non-malicious, (b) P1 is malicious with position error equal to 10 m, (c) P1 is malicious with position error equal to 15 m

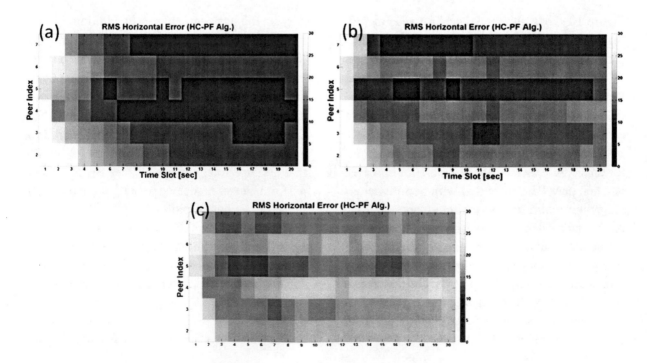

13a shows the performance of the one-hop neighbours of P12 (*i.e.* peers: 1, 2, 11, 13, 14, 21, 22 and 23) when P12 is cooperative and non-malicious, while Figure 13b shows the same performance when P12 is malicious with position error equal to 15 m. In addition, Figure 13c and Figure 13d show the same analysis but considering the 2-hop neighbours of peer P12 (*i.e.* peers: 3, 4, 5, 15, 24, 31, 32, 33 and 34), where the impact of the malicious peer P12 is less evident. To sum up, the propagation of malicious position information fades out getting far from the malicious source in terms of number of hops.

FUTURE RESEARCH DIRECTIONS

Currently, research activities focused on hybrid and cooperative positioning approaches are still on the theoretical phase. In fact, they are lack of real implementation and testing. The hybrid and cooperative approaches could be adopted to enable various applications in the field of the Internet of Things (IoT), which often require high position accuracy and low energy consumption. Therefore, the future research topics should include the adoption of more terrestrial communication technologies, the study of less complex positioning algorithms and the realizations of cooperative P2P networks.

At present, cellular communication, UWB devices, inertial sensors are adopted separately to augment the GNSS-based positioning system, but they can be combined together to further improve the localization performance. Furthermore, other future positioning techniques might be considered to support Intelligent Transportation Systems (ITS) applications, in combination with the adoption of the IEEE 802.11p protocol (IEEE P802.11p, 2009), which is an approved amendment to the

Figure 12. Large scale P2P network (the number between brackets represents the number of visible satellites for each peer)

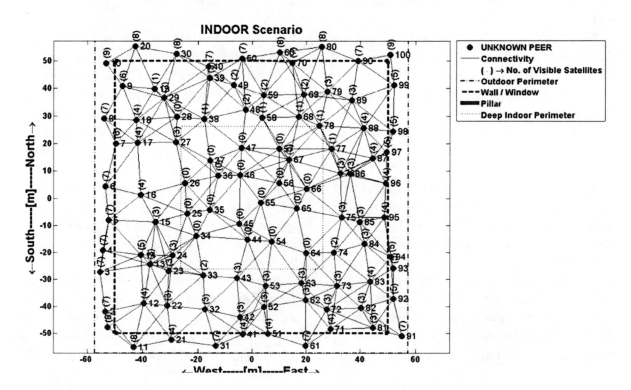

Figure 13. Horizontal error per peer as function of TS: (a) 1-hop neighbours of P12 when P12 is cooperative and non-malicious, (b) 1-hop neighbours of P12 when P12 is malicious, (c) 2-hop neighbours of P12 when P12 is cooperative and non-malicious, (d) 2-hop neighbours of P12 when P12 is malicious.

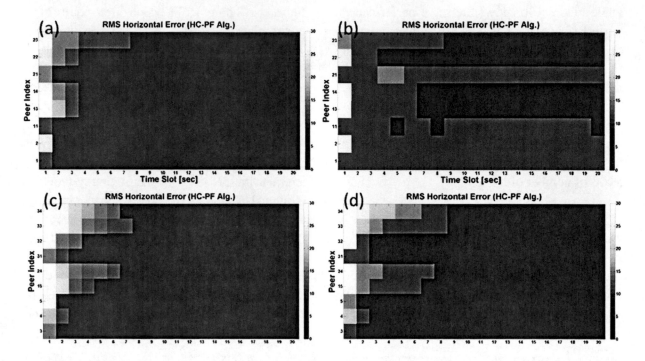

IEEE 802.11 standard to add wireless access in vehicular environments (WAVE).

The hybrid and cooperative positioning algorithms, like HC-PF and H-SPAWN, which show large computation complexity, may not be suitable to be implemented in cheap devices. Therefore, simpler algorithms, such as improved least square and belief propagation based on Kalman filter, should be studied to reduce the complexity while keeping the same accuracy.

Moreover, some issues regarding the practical realizations should be considered, for instance, the setup of the cooperative P2P networks. More in details, the future research activities should include the study of a suitable communication protocol for exchanging aiding packets, the design of the aiding data format that takes into account energy consumption and scalability aspects. Important are also the study of incentive schemes to encourage cooperation of users as well as security aspects.

CONCLUSION

This chapter analysed hybrid and cooperative GNSS-terrestrial positioning algorithms for real-time implementation. First it presented and described a ST able to test hybrid and cooperative positioning algorithms at system level, where peers help each other by exchanging positioning aiding data. Furthermore, it presented an approach to enhance the robustness of cooperative algorithms to inaccurate information flowing inside the network. Simulation results proved that the proposed approach enhanced both estimation convergence and position accuracy while saving some computational complexity and network traffic.

Simulation results in the vehicular urban scenario demonstrated that, thanks to cooperation, the hybrid positioning algorithms could accurately track the movement of vehicles in presence of urban canyons even without fixed terrestrial infrastructure.

Moreover, it presented a detailed complexity analysis of different state of the art hybrid and cooperative positioning algorithms, which is useful to evaluate the feasibility for a real-time implementation.

Finally, the sensitivity of the implemented hybrid and cooperative positioning schemes has been tested by means of the ST in two different scenarios: first in presence of high levels of pseudorange noise, and then in presence of a malicious peer in the P2P network. Concerning the sensitivity of hybrid and cooperative P2P positioning algorithms to malicious peer behaviour, simulation results showed that the corresponding harmful effects are less evident in the peers that are farther than the malicious peer in terms of number of hops.

REFERENCES

L3Nav. (n.d.). *GPS Toolbox*. Retrieved from http://www.l3nav.com/gps_toolbox.htm

P400. (n.d.). *Data Sheet*. Retrieved 2011 from http://www.timedomain.com/p400.php

Alam, N., Kealy, A., & Dempster, A. G. (2013). Cooperative Inertial Navigation for GNSS-Challenged Vehicular Environments. *IEEE Transactions on Intelligent Transportation Systems, 14*(3), 1370–1379. doi:10.1109/TITS.2013.2261063

Budinger, T. (2003). Biomonitoring with wireless communications. *Annual Review of Biomedical Engineering, 5*(1), 383–412. doi:10.1146/annurev.bioeng.5.040202.121653

Caceres, M. A., Penna, F., Wymeersch, H., & Garello, R. (2010). *Hybrid GNSS terrestrial cooperative positioning via distributed belief propagation*. Paper presented at IEEE Global Communication Conference. Miami, FL.

Caceres, M. A., Sottile, F., Garello, R., & Spirito, M. A. (2010). *Hybrid GNSS ToA localization and tracking via cooperative unscented Kalman filter*. Paper presented at IEEE PIMRC. Istanbul, Turkey.

Caffery, J., & Stuber, G. (1995). Radio location in urban CDMA microcells. In *Proceedings of Sixth IEEE International Symposium on Personal, Indoor and Mobile Radio Communications, Wireless: Merging onto the Information Superhighway* (pp. 858–862). Toronto, Canada: IEEE.

Caffery, J. J., & Stuber, G. L. (1998). Overview of radiolocation in CDMA cellular systems. *IEEE Communications Magazine, 36*(4), 38–45. doi:10.1109/35.667411

Das, K., & Wymeersch, H. (2010). Censored cooperative positioning for dense wireless networks. In *Proceedings of IEEE 21st International Symposium on Personal, Indoor and Mobile Radio Communications Workshops* (pp. 262-266). Instanbul: IEEE.

De Angelis, G., Baruffa, G., & Cacopardi, S. (2013). GNSS/Cellular Hybrid Positioning System for Mobile Users in Urban Scenarios. *IEEE Transactions on Intelligent Transportation Systems, 14*(1), 313–321. doi:10.1109/TITS.2012.2215855

Del Re, E. (2011). Integrated satellite-terrestrial NAV/COM/GMES system for emergency scenarios. In *Proceedings of the 2nd International Conference on Wireless Communication, Vehicular Technology, Information Theory and Aerospace & Electronic Systems Technology* (pp. 1-5). Chennai: IEEE.

Diggelen, F. (2009). *A-GPS: Assisted GPS, GNSS, and SBAS*. Norwood, MA: Artech House.

Garcia-Hernandez, C. F., Ibarguengoytia-Gonzalez, P. H., & Perez-Diaz, J. A. (2007). Wireless Sensor Networks and Applications: A Survey. *International Journal of Computer Science and Network Security, 7*(3), 264–273.

Hashemi, H. (1993). The Indoor Radio Propagation Channel. *Proceedings of the IEEE, 81*(7), 943–968. doi:10.1109/5.231342

Johnson, D. B., & Maltz, D. A. (1996). Dynamic source routing in ad-hoc wireless networks. In T. Imielinski, & H. F. Kluwer (Eds.), *Mobile Computing* (pp. 153–181). New York, NY: Springer US.

Laitinen, H., Juurakko, S., Lahti, T., Korhonen, R., & Lahteenmaki, J. (2007). Experimental evaluation of location methods based on signal-strength measurements. *IEEE Transactions on Vehicular Technology, 56*(1), 287–296. doi:10.1109/TVT.2006.883785

Lee, J. Y., & Scholtz, R. A. (2002). Ranging in a dense multipath environment using an UWB radio link. *IEEE Journal on Selected Areas in Communications, 20*(9), 1677–1683. doi:10.1109/JSAC.2002.805060

Liu, H., Darabi, H., Banerjee, P., & Liu, J. (2007). Survey of wireless indoor positioning techniques and systems. *IEEE Transactions on Systems, Man and Cybernetics. Part C, Applications and Reviews, 37*(6), 1067–1080. doi:10.1109/TSMCC.2007.905750

Ma, Y. (2009). *Deliverable D2.2 version 1.0 of the WHERE project: Cooperative positioning (intermediate report)*. Retrieved 2009 from http://www.kns.dlr.de/where/documents/Deliverable22.pdf

Monteiro, T., Moore, T., & Hill, C. (2005). What is the Accuracy of DGPS? *Journal of Navigation, 58*(2), 207–225. doi:10.1017/S037346330500322X

Penna, F., Caceres, M. A., & Wymeersch, H. (2010). Cramér-Rao bound for hybrid GNSS-terrestrial cooperative positioning. *IEEE Communications Letters, 14*(11), 1005–1007. doi:10.1109/LCOMM.2010.091310.101060

Retscher, G., & Fu, Q. (2007). Integration of RFID, GNSS and DR for Ubiquitous Positioning in Pedestrian Navigation. *Journal of Global Positioning Systems, 6*(1), 56–64. doi:10.5081/jgps.6.1.56

Sottile, F., Caceres, M. A., & Spirito, M. A. (2011). *A Simulation tool for hybrid-cooperative positioning*. Paper presented at IEEE International Conference on Localization and GNSS. Tampere, Finland.

Sottile, F., Vesco, A., Scopigno, R., & Spirito, M. A. (2012). *MAC layer impact on the performance of real-time cooperative positioning*. Paper presented at IEEE Wireless Communications and Networking Conference. Paris, France.

Sottile, F., Wymeersch, H., Caceres, M. A., & Spirito, M. A. (2011). *Hybrid GNSS-ToA cooperative positioning based on particle filter*. Paper presented at IEEE Global Communication Conference. Houston, TX.

Spirent. (n.d.). *Wi-Fi Positioning Access Point Simulator*. Retrieved from http://www.spirent.com/~/media/Datasheets/Positioning/GSS5700.ashx

Wymeersch, H., Lien, J., & Win, M. Z. (2009). Cooperative localization in wireless networks. *Proceedings of the IEEE, 97*(2), 427–450. doi:10.1109/JPROC.2008.2008853

Yan, T., He, T., & Stankovic, J. (2003). Differentiated surveillance for sensor networks. In *Proceedings 1st International Conference on Embedded Networked Sensor Systems* (pp. 51–62). Los Angeles, CA: ACM.

Yang, C., & Nguyen, T. (2009). Self-calibrating position location using signals of opportunity. In *Proceedings of the 22nd International Technical Meeting of The Satellite Division of the Institute of Navigation* (pp. 1055-1063). Savannah, GA: The Institute of Navigation.

ADDITIONAL READING

Arnaud, D., Simon, G., & Christophe, A. (2000). On Sequential Monte Carlo Sampling Methods for Bayesian Filtering. *Statistics and Computing, 10*(3), 197–208. doi:10.1023/A:1008935410038

Arulampalam, M. S., Maskell, S., Gordon, N., & Clapp, T. (2002). A Tutorial on Particle Filters for Online Nonlinear/Non-Gaussian Bayesian Tracking. *IEEE Transactions on Signal Processing, 50*(2), 174–188. doi:10.1109/78.978374

Bertsekas, D. P. (1994). Incremental least squares methods and the extended Kalman filter. *In the Proceedings of the 33rd IEEE Conference on Decision and Control* (pp.1211-1214). Lake Buena Vista, FL: IEEE.

Chen, J., Wang, Y., Maa, C., & Chen, J. (2003). Mobile position location using factor graphs. *IEEE Communications Letters, 7*(9), 431–433. doi:10.1109/LCOMM.2003.817337

Chen, J., Wang, Y., Maa, C., & Chen, J. (2006). Network-side mobile position location using factor graphs. *IEEE Transactions on Wireless Communications, 5*(10), 2696–2704. doi:10.1109/TWC.2006.03401

Chun, Y., Thao, N., Venable, D., White, M., & Siegel, R. (2009). Cooperative position location with signals of opportunity. *In the Proceedings of the IEEE 2009 National Aerospace & Electronics Conference* (pp.18-25). Dayton, OH: IEEE.

Conti, A., Guerra, M., Dardari, D., Decarli, N., & Win, M. Z. (2012). Network Experimentation for Cooperative Localization. *IEEE Journal on Selected Areas in Communications, 30*(2), 467–475. doi:10.1109/JSAC.2012.120227

Dammann, A., Sand, S., & Raulefs, R. (2012). Signals of opportunity in mobile radio positioning. *In Proceedings of the 20th European Signal Processing Conference* (pp.549-553). Bucharest: EURASIP.

Falsi, C., Dardari, D., Mucchi, L., & Win, M. (2006). Time of arrival estimation for UWB localizers in realistic environments. *EURASIP Journal on Applied Signal Processing, 2006*, 1–13. doi:10.1155/ASP/2006/32082

Gleason, S., & Gebre-Egziabher, D. (2009). *GNSS Applications and Methods.* Norwood, MA: Artech House.

Hasik, J. M., & Michael, R. R. (2001). GPS at war: a ten-year retrospective. *In Proceedings of the 14th International Technical Meeting of the Satellite Division of the Institute of Navigation* (pp. 2406–2417). Salt Lake City, UT: The Institute of Navigation.

Heinrichs, G., Mulassano, P., & Dovis, F. (2004). A hybrid positioning algorithm for cellular radio networks by using a common rake receiver. *In the Proceedings of 15th IEEE International Symposium on Personal, Indoor and Mobile Radio* (pp. 2347–2351). Barcelona: IEEE.

Ihler, A. T., Fisher, J. W., Moses, R. L., & Willsky, A. S. (2004). Nonparametric belief propagation for self-calibration in sensor networks. *In proceedings of Third International Symposium on Information Processing in Sensor Networks* (pp.225-233). Berkeley: ACM.

Kalman, R. E. (1960). A new approach to linear filtering and prediction problems. *Journal of Basic Engineering, 82*, 35–45. doi:10.1115/1.3662552

Kaplan, E., & Hegarty, C. (2006). *Understanding GPS: Principles and Applications.* Norwood, MA: Artech House.

Kschischang, F. R., Frey, B. J., & Loeliger, H. (2001). Factor graphs and the sum-product algorithm. *IEEE Transactions on Information Theory, 47*(2), 498–519. doi:10.1109/18.910572

Kuusniemi, K., Lachapelle, G., & Takala, J. (2004). Position and Velocity Reliability Testing in Degraded GPS Signal Environments. *GPS Solutions*, *8*(4), 226–237. doi:10.1007/s10291-004-0113-7

MacGougan, G., O'Keefe, K., & Klukas, R. (2010). Tightly-coupled GPS/UWB Integration. *Journal of Navigation*, *63*(1), 01–22. doi:10.1017/S0373463309990257

Mensing, C., & Plass, S. (2007). Positioning Based on Factor Graphs. *EURASIP Journal on Advances in Signal Processing*, *2007*, 1–12. doi:10.1155/2007/41348

Mensing, C., Sand, S., & Dammann, A. (2009). GNSS Positioning in Critical Scenarios: Hybrid Data Fusion with Communications Signals. *In Proceedings of 2009 IEEE International Conference on Communications Workshops* (pp. 14-18). Dresden: IEEE.

Mensing, C., Sand, S., & Dammann, A. (2010). Hybrid Data Fusion and Tracking for Positioning with GNSS and 3GPP-LTE. *International Journal of Navigation and Observation*, *2010*, 1–12. doi:10.1155/2010/812945

Patwari, N., Ash, J. N., Kyperountas, S., Hero, A. O., Moses, R. L., & Correal, N. S. (2005). Locating the nodes: cooperative localization in wireless sensor networks. *IEEE Signal Processing Magazine*, *22*(4), 54–69. doi:10.1109/MSP.2005.1458287

Ristic, B., & Arulampalam, S. (2004). *Beyond the Kalman Filter: Particle Filters for Tracking Applications*. Norwood, MA: Artech House.

Shen, Y., Wymeersch, H., & Win, M. Z. (2010). Fundamental Limits of Wideband Localization—Part II: Cooperative Networks. *IEEE Transactions on Information Theory*, *56*(10), 4981–5000. doi:10.1109/TIT.2010.2059720

Soloviev, A. (2008). Tight coupling of GPS, laser scanner, and inertial measurements for navigation in urban environments. *In the Proceedings of 2008 IEEE/ION Position, Location and Navigation Symposium* (pp.511-525). Monterey, CA: IEEE.

Sottile, F., Caceres, M. A., & Spirito, M. A. (2012). A Simulation Tool for Real-Time Hybrid-Cooperative Positioning Algorithms. *International Journal of Embedded and Real-Time Communication Systems*, *3*(3), 67–87. doi:10.4018/jertcs.2012070105

KEY TERMS AND DEFINITIONS

Computer Science: Scientific and practical approach to computation and its applications.

GNSS: Global navigation satellite system, currently include GPS (America), GLONASS (Russia), Galileo (Europe), Beidou (China), IRNSS (India).

GPS: Global positioning system, a space-based satellite navigation system developed by the Department of Defense of America.

Radio Frequency: A rate of oscillation in the range of about 3 kHz to 300 GHz, which corresponds to the frequency of radio waves that carry radio signals.

UWB: Ultra-wide band, a radio technology communication technology using a large portion of the radio spectrum. It provides good precision time-of-flight-based ranging measurements.

Wireless Localization: Localize unknown users or agents by using wireless communication technology.

WSN: Wireless sensor network, is a collection of nodes organized into cooperative network. A sensor node accommodates different types of sensors that are capable of measuring or monitoring parameters of interest, such as temperature, sound, pressure, etc.

Section 3
System Specification and Modeling

Chapter 9
Co–Modeling of Embedded Networks Using SystemC and SDL:
From Theory to Practice

Valentin Olenev
St. Petersburg State University of Aerospace Instrumentation, Russia

Irina Lavrovskaya
St. Petersburg State University of Aerospace Instrumentation, Russia

Pavel Morozkin
St. Petersburg State University of Aerospace Instrumentation, Russia

Alexey Rabin
St. Petersburg State University of Aerospace Instrumentation, Russia

Sergey Balandin
Open Innovations Association FRUCT, Finland

Michel Gillet
Nokia Research Center, Finland

ABSTRACT

This chapter gives an overview of a modeling application in the general embedded systems design flow and presents two general approaches for the embedded networks simulation: network modeling and protocol stack modeling. The authors select two widely used modeling languages, which are SDL and SystemC. The analysis shows that both languages have a number of advantages that could be combined by the joint use of SystemC and SDL. Thus, the authors propose an approach for the SystemC and SDL co-modeling. This approach can be used in practice to perform protocol stack simulation as well as simulation of network operation. Therefore, the authors give examples of co-modeling practical applications.

INTRODUCTION

The embedded systems design is a very complicated process. It consists of multiple steps of development which should result in a physical implementation. Some of these steps are closely related to modeling, which simplifies development, reduces its cost and helps to avoid serious bugs and errors and fix them before the implementation in hardware. Modeling can be done

DOI: 10.4018/978-1-4666-6034-2.ch009

by a variety of languages. Among them there are such widely used languages as SDL and SystemC. The main purpose of this chapter is to find out a methodology for embedded systems modeling by SDL and SystemC joint use. To achieve the defined above result we firstly should determine the place of a modeling task in the embedded system design flow and identify main objectives which modeling can solve. Then we find out the possible directions of model's representation to obtain different simulation results. So finally, we focus on each direction, show their distinctive features and give examples for their application on the basis of the proposed SDL/SystemC co-modeling approach.

EMBEDDED SYSTEMS DESIGN FLOW

An *embedded system* is a specific combination of computer hardware and software which is specifically designed to perform a particular function (or a range of functions) of a larger system. It usually has strict real-time computational, size and energy restrictions (Heath, 2003), (Barr, 2006), (Kamal, 2008).

There are many implementation steps needed to build an embedded system, and each step is a set of complex actions. Performance modeling helps to understand and establish the major characteristics of the future product. The result of the functional modeling is a specification of the product's functional behavior. During the design and synthesis step the developers implement the specified mechanisms in details and check them. Validation and verification step ensure that the final implementation behaves in a strong accordance to the specification. All these activities operate on models and not on the real physical object. The reasons for using a model are that, firstly the real product is not available on a de-

velopment stage, and secondly it is much cheaper to test the specification of a model that on a real device prototype. (Jantsch, 2004).

The embedded systems design encounters a number of difficulties caused by increasing complexity of projects, increasing requirements to products reliability, power consumption and demand to speed-up the project design phase. So the modern approach to the system design implies the parallel execution of some design tasks. It is illustrated by Figure 1 and includes the following stages:

Figure 1. General design flow

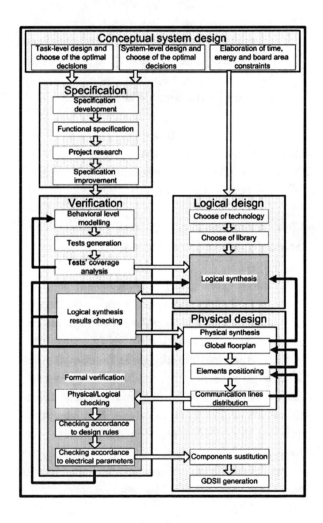

- **Conceptual System Design:** The primary goal of this stage is the main system description, analysis of the system mechanisms and development of the first specification draft;
- **Specification:** This stage is targeted to get the final version of the system specification and the system model in a high-level language (usually in C/C++, SDL);
- **Logical (Architectural) Design:** Includes translation of the executable project specification to the register level (in Verilog/ VHDL) and further at the gate level;
- **Project Verification:** Verification of the design decisions on conformity to the specification and other requirements;
- **Physical Design:** This stage begins from the selection of technological and library basis and it is completed when everything is ready for the final product production.

The main purpose of Figure 1 is to show the modeling role in the general design flow. Also this picture proves that modeling could be used on every design stage (Olenev 3, 2009).

Modeling for the Product Specification Development

The main objective of the specification development is to define and specify the functions of the system and to develop the executable system model. This model is used to verify correctness of the system operation from the functional point of view, estimate required hardware resources and define the system architecture. It is used also to choose the right ways for the specification evolution and check various specified mechanisms on correctness and compatibility.

At this stage the following problems have to be solved:

- Creation of the system functional model, i.e. description of the target system in terms of algorithms and functions, which should be implemented;
- Simulation of the system in its operational environment;
- Development of the system specification depending on the simulation results (Olenev 3, 2009), (Bykhteev, 2008).

Another interesting application of the modeling is development of the specification reference model. This is a model that clearly describes all the mechanisms and interactions of the system with full correspondence to the specification. Such a model could be used as an appendix for the textual specification for the clarification of some complex mechanisms.

Modeling for the Specification Validation

Functional verification becomes more and more important in the design process because of the growth of systems' complexity. The system specification has to be verified before the device production stage to minimize the costs. The verification process needs some specialized tools or some testing and verification environment. This environment could be implemented by a specification model developer, or could be taken as a separate tool, that gives an ability to test the model of the system.

It is very important to define right algorithms for model validation. These algorithms are needed to find bugs in the model and though – in the specification, e.g. dead-locks, buffer overflow, event processing errors and so on.

Modeling for Product Testing

The important stage of the embedded systems design is testing of the real devices by using of the system reference model. At this stage the real device communicates with a reference model of the system via special software called a Tester. Tester consists of:

- Some inputs and outputs that are defined in one of the simulation languages,
- Test generators which can generate different kinds of traffic, make some settings etc.,
- System reference model.

By using such kind of software, the designer can:

- Validate the standard specification, e.g. during the standard development and evolution;
- Validate the model for correspondence to the specification;
- Test prototypes or boards in production;
- Certify products, verify products for conformance to the standard.

There is no standard or a guide for building such testing environments. So proper implementation of the Tester is a challenging task that requires use of flexible programming languages or a combination of these languages (Olenev 3, 2009), (Gillet, 2008), (Suvorova, 2007).

Two Approaches for the Embedded Networks Simulation

Modeling of the embedded networks is usually related to the modeling of the protocol mechanisms that this network uses for the communication. The model of the communication protocol clearly describes the behavior of the node in the network and also behavior of the whole network with multiple nodes. So during the embedded network development the communication protocol model is one of the major things that should be very well tested.

The first basic way for testing the protocol stability and work characteristics is by implementing the network model. This model consists of models of devices that use this protocol. For example, if you need to simulate the network, and, particularly, packets exchange, then the interaction of components and processes inside the device (e.g. between levels of the stack) is not so important for simulation. The real interest is mainly focused on the mechanism of inter-devices communication, for example, all steps of packets transmission, routing, etc. The general idea of network modeling is illustrated by Figure 2.

Network simulation gives information on the communication between devices: sending and receiving, transfer of packets in the physical media, delivery speed and routing mechanisms. As a result many research questions, e.g. algorithms for checking message delivery such as flow control, any type of symbol-coding, packet analysis are outside of the scope of consideration of this method. This helps to remove additional complications from the model and make it less resource-consuming.

Figure 2. The network modeling approach

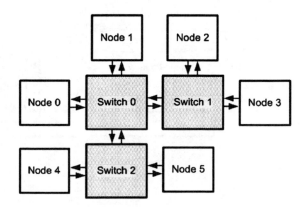

There is another way to study things that are not covered in first approach – protocol stack modeling (see Figure 3). The purpose of creation of such a model is to check the protocol characteristics. This approach is good to study mechanisms of packets formation, intra-device operations, inter-layers services, internal timers, layers management and so on. It gives an ability to check the protocol specification and algorithms on errors and inconsistencies.

This way of modeling is better for the protocol testing, because we model the internal protocol mechanisms in details. For the protocol stack modeling it is enough to model the point-to-point communication of two nodes for most of the purposes. It gives an ability to check the mechanisms of the protocol that need the other end of the transmission, such as quality of service, flow control, reliable data delivery and so on (Olenev 1, 2009).

OVERVIEW OF THE SYSTEMC MODELING LANGUAGE

The *SystemC* (Open SystemC Initiative, 2005) modeling becomes one of the most efficient and widely used methods for studying, analysis and constructing multi-component systems, such as embedded networks of a large number of nodes, systems-on-chip, networks-on-chip, etc (Gipper, 2007).

SystemC is a set of C++ classes and macros that provide an event-driven simulation engine. It qualifies as a language for modeling, design and verification of systems, and it is specifically designed for modeling parallel systems. This library allows describing multi-component systems and modeling their operation. The next advantage of SystemC is the modeling time. It is the best solution, when the model of the system requires a notion of delays, clocks or time features which are not present in C++. Communication mecha-

Figure 3. The protocol stack modeling approach

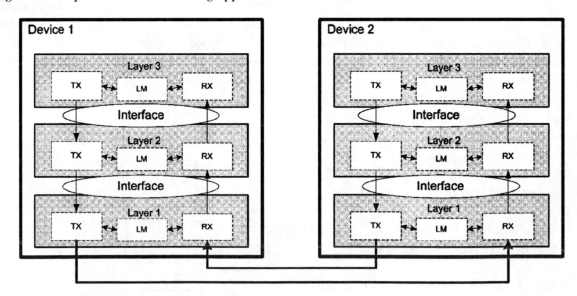

nisms used in hardware, such as signals and ports, are very different from those used in software programming. Data types of C++ are too remote from the actual hardware implementation. So the SystemC kernel defines specific data types and such kernel units as processes, threads, signals, events, FIFOs and ports (Olenev, 2008).

Moreover, SystemC has the following advantages:

- It is feasible to integrate all kinds of C++ based models;
- SystemC gives high flexibility in parallel use of multiple models;
- SystemC based on C++ language is the most widely known by both software and hardware experts;
- SystemC supports modern methodologies for verification by means of SystemC verification library;
- SystemC provides an ability to build IP-blocks, it allows reusing of the design and verification components;
- Modeling of the on-chip communications is naturally supported in SystemC;
- It is open-source.

The ability to cleanly support such a wide range of system design tasks and features within a single language is unique to SystemC (Black, 2004), (Nemydrov, 2004), (Swan, 2003).

OVERVIEW OF SDL MODELING LANGUAGE

The SDL (Specification and Description Language) (ITU, 2011) is a language for unambiguous specification and description of the telecommunication systems behavior. The "*specification*" term means the description of its required behavior; "*description*" of a system means the description of its actual behavior; i.e., its implementation (ITU, 2011).

The SDL model covers the following five main aspects: structure, communication, behavior, data, and inheritance. The behavior of components is captured by partitioning the system into a series of hierarchies. Communication between the components takes place through gates connected by channels. It should be mentioned that each process instance in an SDL system represents an independent asynchronously executing communicating extended finite state machine (Mitschele-Thiel, 2001).

SDL language is intended for description of structure and operation of the distributed real-time systems. The functionality of a SDL system is defined through the dependence of the output signals sequence from the input signals sequence. Thus, first of all, SDL is intended for a description of the systems containing components interpretable in a form of finite state machines. Therefore, the interaction between these components as state machines is very important.

The SDL state machine is represented by SDL *processes*. It is the main functional component of the SDL language that describes the behavior of the SDL system and operates as a state machine.

Sets of processes are grouped into *blocks*, which provide strong structural facilities. These blocks, in turn, are connected by channels with each other and with the environment. SDL processes communicate with each other by means of SDL signals. The architectural view of SDL system (IBM, 2009) is presented in Figure 4.

SDL structuring features are its key advantages as they allow simplifying the description of large and compound systems. They allow splitting the system into separate units, which could be studied independently. Derived units could be split into subunits and so on. In the end, it leads to creation of a multilevel hierarchical structure. The description of system operation could be located at bottom layers.

SDL can be used at various stages of system creation: from its initial design to the development and maintenance (IBM, 2009), (Karabegov, 1993).

Figure 4. General structure of SDL system

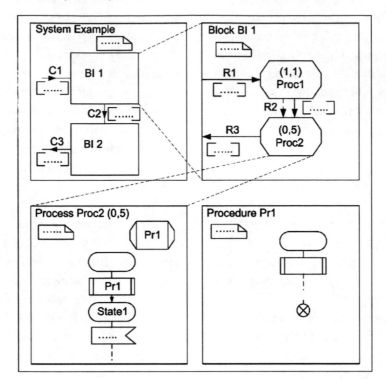

Analysis of SDL and SystemC Applicability FOR the Protocols Simulation

The selection of the right modeling language is the key factor of success and must be carefully thought before the system modeling starts. This choice depends on a large number of factors, which is primarily driven by the vision of what kind of problems need be solved by the developed model.

Dynamic Object Creation

One of the important factors to be considered is a possibility to dynamically determine parameters of the system during modeling of devices. Assume there is a need to create a model of a switch to simulate a network of devices. SDL language unfortunately does not allow giving number of switch ports as a parameter. Designer has to define different switches with various numbers of ports.

While in SystemC the programmer can define the switch class and set the number of ports as a parameter. Thus, in this case SystemC is better fulfilling the system of devices description.

Simulation Time

The next criterion to take into account is different approaches for the simulation of time use. SDL has two data types that can be used to work with time: *Time* and *Duration*. The first one is used to define the "point of time" and the second for "time interval". Generally both types are used in conjunction with timers. There is only one operator "now" that returns value of type *Time*. This value could be used by setting the timer if a delay with type *Duration* is added. Note that SDL specification does not define the system time unit, so it depends on the implementation of specific SDL environment. Thus both of these data types are types with floating point (IBM, 2009).

In SystemC modeling the absolute integer time is used. The modeling time is encoded by an unsigned integer of at least 64 bits. In contrast to the continuous astronomical time the modeling time is discrete.

Modeling time starts at 0 and increments at the course of simulation. There are two parameters related to the modeling time, they could be configured by the developer. These parameters are time resolution and the default time unit. Time resolution defines the time interval that corresponds to one unit of modeling time. All events could be planned only for the edges of these intervals. Thus the potential accuracy of the modeling depends on the size of the interval. However this can cause the limitation of the maximal simulation time in a particular modeling environment by the reason of the required actual time increase for simulation. The default value for time resolution in SystemC is one picosecond. The default time unit corresponds to one nanosecond.

The modeling time in SystemC models provides broader and more flexible opportunities for its application. In addition to the definition of timers, similarly to SDL, the SystemC gives a possibility of modeling per cycles and setting of clock signal for all blocks of modeling system. It is possible to set the clock signal for each block separately and the sensitivity of the block can be put to its positive or negative edge. Duration of the clock cycle can also be specified for each device differently. Moreover, it is possible to start system modeling for the time specified in time parameter by setting of *sc_start (time, ...)* function parameters. This significantly simplifies testing of the whole model and its elements.

Testing of the Models

The last point of our analysis is opportunities for testing of the models. This point is the common point to both SDL and SystemC languages. The implementation of the model should be competently done by use of decomposition principles, defining data types as well as considering the implementation details before the actual development starts. These concepts provide an opportunity to test and investigate certain parts of the model independently.

OVERVIEW OF RELATED STUDIES

Unfortunately there are not many publications in the field of SystemC and SDL co-modeling, however a few good papers can be found. For example, there is a good publication by M. Haroud and L. Blazevic which gives superficial comparison of SDL and SystemC languages and proposes a method for SDL translation into SystemC code (Haroud, 2006). This method is used to translate the SDL implementation of protocol specification into the SystemC model. It allows using rigorous protocol modeling and verification provided by SDL and FPGA netlist synthesis enabled by SystemC. This paper gives a comparison and defines general use of SystemC and SDL for the protocols modeling.

Another interesting paper was published by T. Josawa at.el. from Nokia Research Center (Jozawa, 2006). This work is more closely correlated to this study and the proposed methods for the co-modeling have a lot in common with our "SystemC and SDL parallel use" co-modeling approach. These methods are illustrated by Figure 5.

THE SDL→SYSTEMC CO-MODELING APPROACH

Different Ways for SDL and SystemC Co-Modeling

As we discussed before, SDL and SystemC are very popular tools for modeling embedded networks and both have clear strong sides, but are there any chance to use these two languages together and combine the benefits? The next question, if

Figure 5. Two approaches for SDL and SystemC co-modeling proposed in (Jozawa, 2006)

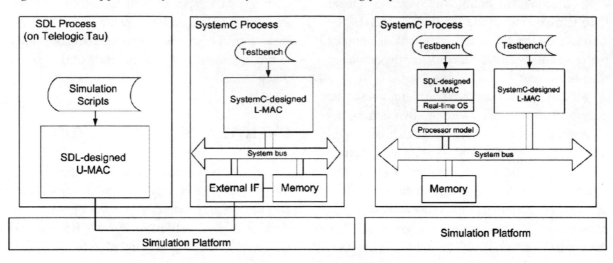

joint use even is possible will it give an ability to use both modules at the same time by setting rules of direct interaction and internal operations of the simulation?

During the research we considered three possible co-modeling solutions for SDL and SystemC. The first approach was to insert SystemC modules into the SDL model by including the corresponding C header files into SDL model. The second approach was to insert SDL module into SystemC by "teaching" the SDL model how to process the requests and commands of the SystemC model. And the third approach assumed running both SDL and SystemC independently in parallel in the operating environments and make a special tool that will interface SDL and SystemC.

The applicability of each approach depends on how well its features match to the developer requirements. For example, SystemC→SDL approach inherits maximum of SystemC advantages, i.e. clocking and threads. In the SDL↔SystemC approach both languages get complete set of tools to exchange data, control and synchronization information, including clock signals. However analysis of all factors showed that inserting of SDL into SystemC gives the best solution for co-modeling of SystemC and SDL. So let us consider SDL→SystemC approach in details (Balandin, 2011).

Tool Choice

The approach of SDL/SystemC co-modeling considered in this paper assumes that we have a SystemC project, which corresponds to the whole model. This model contains the SDL and SystemC parts. Consequently, this approach uses C/C++ representation of the SDL system (Olenev 2, 2009). Before starting a description of the discussed approach, we need to introduce general principles of co-modeling with some requirements and important notions for modeling. Consider some abstract SDL tool which should meet the following requirements:

1.   Provide a possibility to generate C/C++ code from the implemented model, which will be the equivalent of the SDL model.
2.   The generated C/C++ code operation should be controlled by some kind of a manager engine (SDL kernel).
3.   The SDL kernel should provide a number of functions for initialization and simulation of the SDL model. The function which enforces simulation should operate in such a way that ne SDL transition is executed during each call of this function. One SDL transition corresponds to moving from one system state to another.

It should be pointed out that such kind of the SDL tool already exists: all these features are provided by the IBM Rational SDL Suite (IBM, 2009).

General Principles of Co-Modeling

This section discusses main principles when SDL modules are inserted into SystemC. The integration of SDL model into SystemC model can be divided into the following general stages:

- Preparation of the SDL system to be the part of the whole model.
- Generation of C/C++ code on basis of the created SDL system. The SDL code is not used hereafter.
- Insertion of this C/C++ code to the SDL kernel.
- Preparation of the SystemC part of the model.
- Integration of the SDL kernel with the generated C code into the whole model.

The overall procedure for this approach is illustrated by Figure 6.

As it was mentioned in the previous section IBM Rational SDL Suite allows generating C code based on SDL model and provides environment for SDL model analysis, simulation and validation to the SDL kernel. Now let us step-by-step discuss the process of the model integration.

First of all SDL part of the model should be prepared. Then using the IBM Rational SDL Suite we generate C code out of SDL model and add it to the predefined *SDL kernel*. After that C code is a part of the joint SDL model.

SDL kernel includes a number of functions that facilitates simulation of a SDL system. The most important are functions *xMainInit()* and *xMainLoop()*. Function *xMainInit()* performs initialization of the SDL kernel and it is called before a start of the simulation. Function *xMainLoop()* is a scheduler and a launcher for transitions between SDL processes. It is defined as an infinite loop where one SDL transition is performed per iteration. SDL transition is one step of a SDL process from current to the next state. If any output signal

Figure 6. Insert of SDL into SystemC (SDL SystemC)

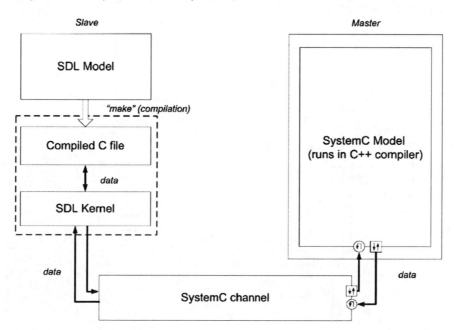

is sent during this step then the receiving process will be informed and the next transition of the receiver is scheduled.

Then SystemC model has to be implemented. It could contain a number of SystemC threads. A couple of these threads is presented in a SystemC part of the model and others in the channel part. One thread should be kept to work with SDL (*sdl_thread*). Sdl_thread calls *xMainLoop()* function when a SDL part gets incoming data events. The corresponding initialization of a SystemC part requires call of *xMainInit()* function.

The clock signal is initialized in SystemC (*sc_clock*) and SDL uses it only for synchronization.

Figure 7 shows a simple example of co-modeling SDL and SystemC. This is an example when two nodes communicate with each other, but one node is implemented in SDL, another in SystemC.

Notion of Delta-Cycles

Another key aspect for discussing modeling principles of different languages is *delta-cycles*, as this notion is very important for elaboration of co-modeling approaches.

The first solution for organization of delta-cycles is provided in SystemC and requires declaration of a notion for *delta-delay* (Δ), which is used to define cause-effect relation between different events. In terms of modeling time the delta-delay triggers in zero time. If two events occur at the same time, but the first event causes the second one then it is considered as there is a "delta-delay" between these events.

Figure 8 shows the cause-effect relation between events in SystemC. The scheduled *event#1* is performed at T1 modeling time. While the event is in processing the function *notify(ZERO)* for *event#2* is called. In terms of modeling time *event#1* and *event#2* should be processed at *T1* moment. But from the cause-effect relation point of view *event#1* causes *event#2*. So *event#2* should be performed after a delta-delay (at *T1+Δ* modeling time).

An execution of the delta-cycle is divided into the several steps. At the first step a scheduler defines a list of processes, which are to be executed at this moment of modeling time.

The second step is evaluation of the system values. At this step all scheduled processes are

Figure 7. SDL/SystemC co-modeling example

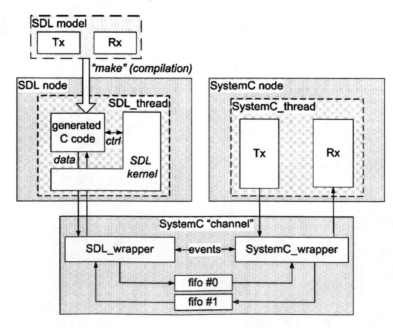

Figure 8. Cause-effect relation of events in SystemC

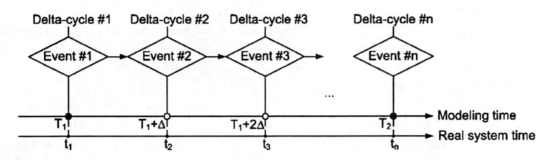

executed sequentially. New values are computed and saved but not assigned. All processes use the same values of the objects independently of the processing order.

The third step is an update of the values that were computed at the previous step.

Then the list of new notify events scheduled for the current moment of modeling time is formed. They will be executed after a delta-delay.

If the list of notify events is not empty after the check for delta-cycles then the sequence of steps is performed again after one delta-delay by moving to the next delta-cycle.

The typical flow chart for handling delta-cycles is shown on Figure 9.

Let us take modeling time *T1* at which all scheduled events are performed. Then all events that were scheduled for *T1* with zero delay will be executed at *T1+Δ*. At the moment *T1+2Δ* will be executed all events scheduled at *T1+Δ* with zero delay, and so on. Finally the modeling time is set to *T2* only when all events scheduled for the moment of time *T1* will be completely processed (Suvorova, 2003).

SDL specification does not define notions of the delta-cycles and delta-delays. However, we have to introduce these structures to describe the principles of SDL systems modeling using SDL Tool for elaboration of co-modeling approaches.

Let us take certain moment of modeling time, when there is a number of scheduled events, where each SDL event represents transition of the process from one state to another. Each delta-

cycle contains one execution of SDL transition and consequently during SDL delta-cycle a set of tasks is performed, one of which is scheduling of the new events. There are two ways for events scheduling: signals and timers.

Using of signals means that the event should be performed at the current moment of modeling

Figure 9. Algorithm of handling delta-cycles

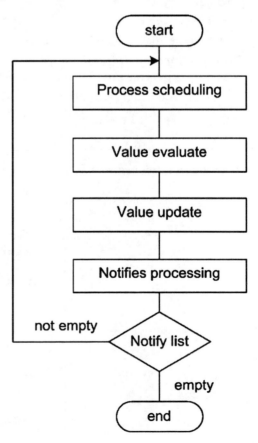

time. Such event is processed during the next delta-cycles after a delta-delay and this delta-delay triggers in zero time.

The timer expiration is scheduled at another moment of modeling time. So it causes a new event, which is processed when all current time events will be performed.

Thus, to sum up the information above, the SDL/SystemC co-modeling requires using a common events scheduler.

Way to Address the Co-Modeling Approach

Interaction between SDL and SystemC parts (see system on Figure 7) is organized in the following way (Stepanov 1, 2010):

- Each node generates data and sends it (as *data.request*) to the remote one via a channel;
- Local *data.request* and *data.response* become *data.indication* and *data.confirm* for the remote node correspondingly;
- When node gets the *data.indication*, it generates and sends *data.response*;

- When node gets the *data.confirm*, it generates and sends new data (*data.request*).
- Read and write operations to SystemC channel are processed at different delta-cycles (Black, 2004).

Figure 10 shows an example of interaction between SDL and SystemC parts.

A description of interaction between SDL and SystemC parts according to the example above is presented in Table 1.

TESTER FOR THE PROTOCOL STACK MODELING

As it was mentioned before, modeling is a useful tool for the specification development, protocol validation and device testing. The last point is an area where co-modeling of SystemC and SDL could play the key role. SDL is used to describe the specification. The SDL model of the specification can be used for verification and validation of the protocol. But as was shown by Figure 2, proper device testing requires not only protocol model, but also testing environment and dynamic modeling software, that cannot be done in SDL,

Figure 10. Interaction between SDL and SystemC parts

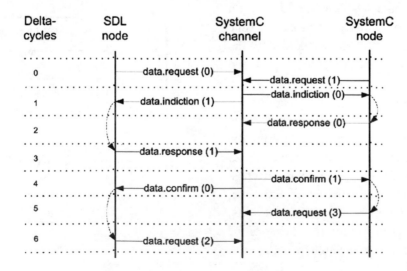

Table 1. Description of SDL and SystemC interaction for the example

Delta-Cycle	SDL Part Work	SystemC Part Work
0	SDL thread calls *xMainLoop()* function. During this call SDL Tx generates *data.request* and writes it to the SystemC channel. After this function *xMainLoop()* is completed.	SystemC Tx generates *data.request* and writes it to the channel.
1	SDL thread calls *xMainLoop()* function. During this call SDL Rx reads *data.indication* from the channel and sends a SDL signal to an intermediate block between Tx and Rx. So a new event is scheduled at the next delta-cycle.	SystemC Rx reads *data.indication* from the channel and sends incoming data to the Tx part. So a new event is scheduled at the next delta-cycle.
2	According to the scheduled event, SDL thread calls *xMainLoop()*. During this call SDL intermediate block sends signal to Tx. So a new event is scheduled at the next delta-cycle.	According to the scheduled event SystemC Tx receives data from Rx, generates new *data.response* and then writes it to the channel.
3	According to the scheduled event, SDL thread calls *xMainLoop()*. During this call SDL Tx receives signal from intermediate block, generates a new *data.response* and writes it to the channel.	
4	SDL thread calls *xMainLoop()* function. During this call SDL Rx reads *data.confirm* from the channel and sends SDL signal to intermediate block between Tx and Rx. So a new event is scheduled at the next delta-cycle.	SystemC Rx reads *data.confirm* from the channel and sends incoming data to the Tx part. So a new event is scheduled at the next delta-cycle.
5	According to the scheduled event, SDL thread calls *xMainLoop()*. During this call SDL intermediate block sends signal to Tx. So a new event is scheduled at the next delta-cycle.	According to the scheduled event SystemC Tx receives data from Rx and generates new *data.request* and writes it to the channel.
6	According to the scheduled event, SDL thread calls *xMainLoop()*. During this call SDL Tx receives signal from intermediate block and generates new *data.request* and writes it to the channel.	

but easily implementable in C/C++. It would be great help for developers to allow co-execution of the code prepared in two different languages.

Moreover, SDL and SystemC models can be originally done for different purposes. For example, SDL model for verification and SystemC model for the studies of further hardware development. In this case it might be useful to allow these models work together to compare results and check on correctness of implementation.

The Protocol Model Tester is a good example of a practical applicability of SDL/SystemC co-modeling. It is used for simulation and investigation of data transfer protocols (Stepanov 2, 2010). A general scheme of the protocol SDL models Tester is depicted in the Figure 11. The Tester contains three parts: the Test Engine, the Modeling

Core and the Medium. Descriptions of all these modules and interconnections between them are discussed in details in the following subsections.

Modeling Core

The Modeling Core consists of the testing SDL model, the SDL/SystemC Wrapper and the Communication Wrapper. Note that implementation of the Communication Wrapper depends on the SDL model. The features of the SDL model which should be taken into account are a number of protocol layers in the node and interface descriptions of each protocol layer.

The SDL model of a five-layer protocol stack is depicted in Figure 12. This protocol stack contains five layers: Physical Layer (L1), Data Link

Figure 11. General scheme of the protocol SDL models Tester

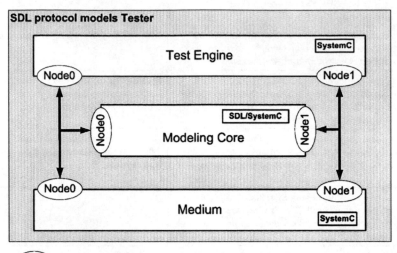

Node0 - interface for interconnection between modules

Layer (L2), Network Layer (L3), Transport Layer (L4) and Application Layer (L5) (Tanenbaum, 2011). Each Layer of the stack can communicate with adjacent layers through its interfaces. The interface defines which primitive operations and services the lower layer provides to the upper one (Tanenbaum, 2011).

Models of the Layers from the first to the fourth are implemented in SDL. The Test Engine can represent each of five layers described above in accordance with the following rule: when the Test Engine can communicate with a peer object via a protocol Layer X+1 by consuming a Layer X service through the interface X it means that the Test Engine is a representation of the Layer X+1 for the current stack.

The discussed protocols stack includes five interfaces provided by the following layers: the Medium, the Physical Layer, the Data Link Layer, the Network Layer and the Transport Layer. The Communication Wrapper can contain a number of special modules called Intermediate Blocks (IB). Each Intermediate Block addresses to a particular interface, so this IB can be used in the following cases:

- Interconnection between the Test Engine and a layer providing the addressed interface;
- Interconnection between a layer using the addressed interface and the Medium;
- Interconnection between a user and a provider of the addressed interface.

The interconnection between the Test Engine and a layer providing the addressed interface should be established when the Test Engine is a consumer of the defined interface. An example of this case is shown in the Figure 13. There is the Test Engine which represents the Transport Layer because it uses the N interface to access to a Network Layer service. It is implemented by introduction of the IB_N block. This block works as a service provider for the Test Engine and as a service user for the Network Layer. Note, that all Intermediate Blocks are implemented in SystemC.

The interconnection between a layer using the addressed interface and the Medium should be established when the layer is the lowest layer of the stack and its protocol is used for data exchange between the nodes. The Medium implements a connection to the peer node and provides an

Figure 12. The SDL model of five-layer protocol stack and Communication and SDL/SystemC Wrappers

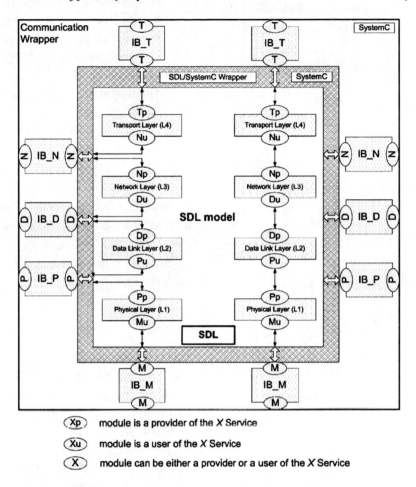

interface for defined layer. In the Figure 13 the Data Link Layer is the lowest layer of the stack and communicates with the Medium by the P interface through the IB_P block. In this case the data exchange between the nodes is performed by Data Link Layer data units. Consequently an object of the peer node connected to the Medium should also support the Data Link Layer protocol.

The interconnection between the user and the provider of the addressed interface leads to two sub cases:

- Both user and provider are represented by layers of the stack;

- The user is represented by the Test Engine and a provider is represented by the Medium.

Two adjacent layers of the SDL stack can communicate with each other in two ways: via SDL channels or via an Intermediate Block corresponding to an interface of the lower layer. The last way can be chosen for getting features of an Intermediate Block defined below. This addresses to the first sub case which is depicted in the Figure 13. There are the Network Layer and the Data Link Layer communicating through the IB_D block.

The second sub case is used when there is only one node in the SDL model and the Test Engine corresponds to the second node of the

Figure 13. Example of the interconnection between the Test Engine, layers of the SDL model and the Medium through the Communication Wrapper

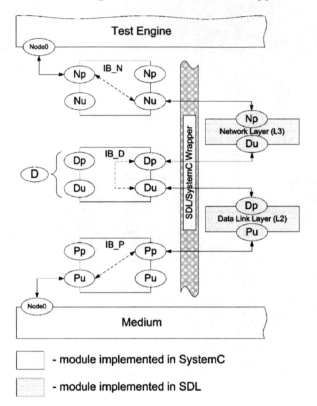

   ☐ - module implemented in SystemC

   ▨ - module implemented in SDL

- If purposes of the test does not require getting of Intermediate Block features for the considered connection, two SDL blocks can communicate through SDL channel only;
- Otherwise, each block should be connected to the SDL system environment (ITU, 2011). So in this case both layers will interconnect with an appropriate Intermediate Block.

Note that independently of test purposes the highest and lowest layers of both nodes have to connect to the SDL environment for communication with the Test Engine and the Medium correspondently.

Medium

The Medium module contains a set of channels and the Control block as it is shown in the Figure 14. Each channel is responsible for connection between two nodes by a particular data transfer protocol. So each channel provides a service for data transmission and an interface to access to this service. The Control block sets the following features of the used channel:

- Delay introduction – to block a transmitted data unit for a predefined time interval;
- Error injection – to corrupt a transmitted data unit according to a predefined condition and value;
- Logs – to write information about particular actions into a specific out stream.

An example represented in the Figure 14 corresponds to the five-layer protocol stack which was discussed above (see description of Figure 12). Under requirements of this stack testing the Medium has to implement four channels to perform data exchange by protocols of the following layers: the Physical Layer, the Data Link Layer, the Network Layer and the Transport Layer. So each

tested system. So, the Test Engine should send data to the first (SDL) node directly, i.e. through the Medium. In this case the Test Engine is a user of the communication interface and the Medium is a provider.

Each Intermediate Block can provide the following features: managing data flows transmitted through the Communication Wrapper, parsing transmitted data, making logs and error injection.

A set of key issues are necessary to take into account for implementation of the SDL model. Firstly, according to the stack structure each layer shall be represented by one SDL block. Then all required layers should be joined to one SDL system. Each two adjacent layers of one node can be connected in two ways:

Figure 14. The Medium scheme

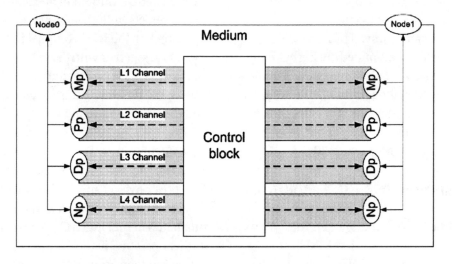

of these layers can use the Medium for data transmission and access the Medium service through an interface correspondent the underlying layer.

Test Engine

The Test Engine module is responsible for control of the SDL model simulation. In terms of this requirement this module performs a number of specific tasks.

The first one is configuration of the SDL model Communication Wrapper before the start of test sequence. In accordance with the Intermediate Block features the Test Engine defines the transmission mode of each IB, enables/disables the data parsing, the making logs and the error injection.

The second task is configuration of the Medium. During this phase channel parameters such as the channel delay and the error injection are defined.

The main task is data exchange with the tested SDL model. If the tested model does not contain a special traffic generator module the Test Engine represents a certain layer of the tested stack. In other words the Test Engine operates in accordance with a protocol of the chosen layer and uses services of the layer below. The implementation of these entire tasks is a core of each particular test.

Main Tester Abilities

Thereby in terms of simulation and modeling the discussed Tester gives the following abilities:

- Representation of tested network layers by means of finite state machine;
- Access not to the whole SDL model only but also to certain layers of the stack through appropriate Intermediate Blocks;
- Getting all necessary test results by SystemC implementation of the test environment.

The similar concept of the protocol SDL models testing has been already implemented in practice.

NETWORK SIMULATION ON THE BASIS OF SDL/ SYSTEMC CO-MODELING

The protocol Model Tester primarily provides a strong means for exploration of functional characteristics of the protocols. Due to its structure it does not provide facilities for performance analysis and simulation of network operation. The task of network modeling can be also solved on the basis of the SDL/SystemC co-modeling.

This problem can be solved by means of a special library, which can be applied during networks simulation in SystemC. This library should implement the original SDL model and SDL simulation kernel as well as provide special services for the user (term *'user'* stands for the SystemC developer, who uses the library) (Morozkin, 2013).

Different Approaches for Solution

Our goal is to develop an approach that allows creating several instances of the SDL model. Moreover, since we use the IBM Rational SDL Suite, our approach is tool specific. Especially for the SDL/SystemC co-modeling we use the CAdvanced code generator and a source code of the SDL model, which has been generated by it.

Since SystemC is based on C++ and since CAdvanced generates the plain C code, there appears a task of combining C and C++ parts of the model. The use of available C++ code generators can probably solve some problems. But in case of using our codebase, embedding a new tool will require making global changes in our projects.

First of all a proper solution requires understanding of how the original SDL model is implemented in the source code and, moreover, what are the source code equivalents for different elements of the SDL language. Another question is how we can reuse the code to have an opportunity to instantiate more than one SDL model. There are not many publications which describe architecture of the generated source code of the SDL model and principles of its communication with SDL simulation kernel. The paper, which observes some details of generated code and principles of its functionality, is (Haroud, 2005). Another publication is a thesis (Dietterle, 2009) proposing an integrated design flow for embedded systems. Author also uses the CAdvanced code generator and describes several mechanisms that

are used in SDL Simulator and how the generated code interacts with the simulation kernel.

Basing on the IBM Rational SDL Suite documentation together with the presented above publications we have conducted a research in order to understand the architecture of generated model to find the ways for solving our specific task.

The first approach is *integration of C Code into C++ Environment*. To solve the problem we need to localize the SDL model instances in a memory. The most obvious approach is integration of the generated C code of the SDL model and the SDL simulation kernel into a C++ environment for its further operation in the user's code. If integration is possible, then the target library can be developed with the use of different OOP patterns. However, practical application of this approach has shown that this way entails a number of technical problems. Since the generated code of the SDL model is represented by a C code which strictly conforms the ANSI C standard, the most complex problem is integration of the C code for operation in the C++ project. Consequently, the significant part of the SDL simulation kernel and generated code of the SDL model should be changed. Therefore, it can be concluded that the implementation of this approach takes a considerable time.

Another way of solving the problem is *postprocessing of the generated C code* of the SDL model in order to have an opportunity of creation of a different number of SDL model instances with use of dynamic memory allocation. The main feature of the CAdvanced code generator is that the implementation of an SDL model represents the hierarchical structure which is called symbol table and organized as a tree (Dietterle, 2009). Symbol table contains objects which represent SDL entities (system, blocks, processes, signals, etc). These objects are global variables. So, the static memory allocation is used.

To have an opportunity of creation of different number of SDL model instances we need to change a memory allocation mechanism from static memory allocation to dynamic memory allocation. This can be solved by the code post-processing. Ideally, we need to change the implementation of the CAdvanced code generator. However, this is almost impossible as we use an existing industrial tool. On the other hand, it is a well known approach to develop an auxiliary toolchain for existing products.

Basing on the analysis above the code post-processing approach seems more flexible and advantageous solution for implementation of the network simulation model on the basis of SDL/SystemC co-modeling.

The Library Development Flow

The solution is aimed to develop an environment that allows creating the target library. This library provides an ability to use a different number of SDL model instances in the SystemC user's project and contains both the SDL model and the SDL simulation kernel. The library development flow and library usage in a project is shown in Figure 15.

These are the steps of the proposed library development flow (Morozkin, 2013):

1. Analysis of requirements and implementation of an SDL model.
2. Obtaining a PR-model using the GR-to-PR converter.
3. Obtaining C code of the SDL model with use of CAdvanced. The code consists of three parts: a symbol table, which corresponds to the SDL model architecture, a set of initialization functions and a set of PAD (Process Activity Description) functions (IBM, 2009) which implement the behavior of SDL processes.
4. Code post-processing of the obtained C code. Generation of initialization functions and patching of some parts of PAD functions.
5. Building a target library according to the proposed approach. Creation of the symbol table selector. Development of a user's code interface, which is a set of C++ classes.

Figure 15. Library development and usage

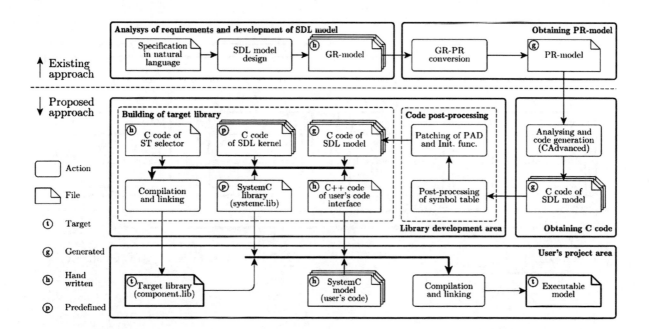

Then all the generated source code is compiled and linked, so the user gets a target library *'component.lib'*. The implementation of the SDL kernel stays unchanged during the library development flow, but the new functionality for operating with a different number of SDL model instances is added. User's project operates with the target library and the SystemC library simultaneously. User interface is intended for using services provided by the library.

Application Structure

Let us consider a simple example. Figure 16 shows the architectural diagram of the SystemC model of SpaceWire MCK-01 switch (ELVEES, 2005).

The model contains a Switch module and four ports connected to four independent SDL model instances. The Switch and Port modules are implemented in SystemC. According to the proposed approach it is possible to design a switch model, where each port includes the implementation of the full protocol stack in SDL. In this case the network layer is implemented in SystemC while the bottom ones – in SDL. The structure of the

application implemented in accordance with the proposed approach is shown in Figure 17. It consists of the following parts:

1. *The SystemC library*, which includes SystemC kernel.
2. *The SystemC model* implemented by a user.
3. *The target library*, which provides an ability to create a number of different SDL model instances. The library is divided into three parts: the user's model interface, which describes the services for communication between the users SystemC model and the SDL kernel; the SDL kernel, which performs scheduling of the generated SDL model and the SDL model itself. For communication with C++ classes a basic *xInEnv/ xOutEnv* (IBM, 2009) mechanism is used. Implementation of the SDL model has four parts:
    a. A set of PAD functions. These functions implement behavior of SDL model processes.
    b. A selector of a symbol table (ST).

Figure 16. SystemC model structure

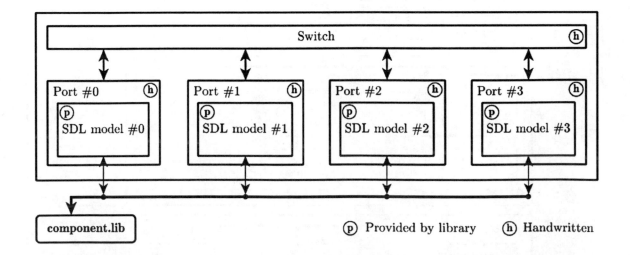

Co-Modeling of Embedded Networks Using SystemC and SDL

Figure 17. Application structure

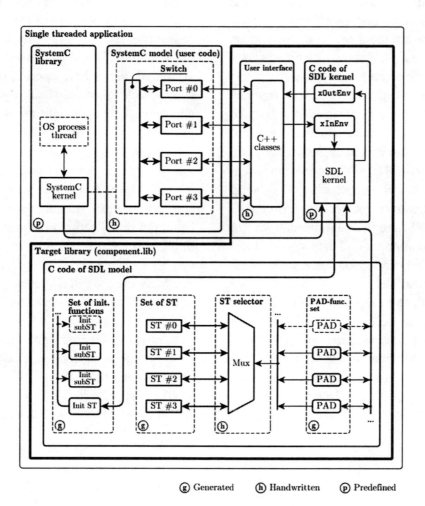

(g) Generated   (h) Handwritten   (p) Predefined

c. A set of SDL model instances. Each of them has its own symbol table, but all instances have a set of common PAD functions.

d. A set of initialization functions. These functions are used for instantiation of each new symbol table with the use of the dynamic memory allocation.

User's SystemC model is a single threaded application, which is controlled by the SystemC kernel. The Switch module with its ports communicates with the SDL kernel via user's interface. The SDL kernel is responsible for scheduling of SDL model processes. The kernel calls different PAD functions and each PAD function chooses an SDL model instance by means of ST selector. ST selector uses a symbol table identifier, which is generated by a user's C++ object which represents the SDL model (for example *Port #0*). The ST selector clearly indicates the required SDL model instance.

Main Principles of Model Memory Organization and Management

As an example we take the simple SDL model to clearly explain the proposed approach. The

SDL model, which is taken as the basis for the example, consists of one block communicating with the environment by means of signals *'sig.req'* and *'sig.rsp'*. This block contains only one process, which operates with the same signals.

Firstly we should generate the C code from the SDL model. It is done by means of CAdvanced code generator. This C code contains a set of interacting data structures and each of them could be a huge hierarchical tree. These structures represent the SDL model symbol table. Thereafter, it is possible to convert the generated C code to the XML. The XML representation contains 226 nodes (a node comprises a C data structure and its fields) with 251 connections between them for such a simple example.

According to the proposed approach the C code should be divided into a number of post-processing steps. Initialization functions were generated and PAD functions were patched. Initialization functions are called each time a new SDL model instance is initialized. In case when original SDL model is designed with use of pack-

ages the CAdvanced generates an implementation of each package and places them into separated source file. Each package has its own initialization function. Therefore, we need to post- process all source files to have opportunity of build the symbol table in memory using dynamic memory allocation. After the initialization each SDL model instance is separately stored in the heap. Since initialization is performed with the use of the same function, which does not return any value, it is not possible to get access to all symbol tables (as each new instance is initialised by a consecutive calling of initialization functions). To solve this problem some special nodes of the SDL model symbol table are added to special arrays. These arrays are used to determine the necessary nodes of the symbol table of SDL model instance while sending signals from environment (ITU, 2011) to SDL model or sending signals from SDL model to an environment. Communication mechanisms are shown in Figure 18.

The heap stores two symbol tables of the SDL model. A set of arrays, which are global variables,

Figure 18. Communication mechanisms

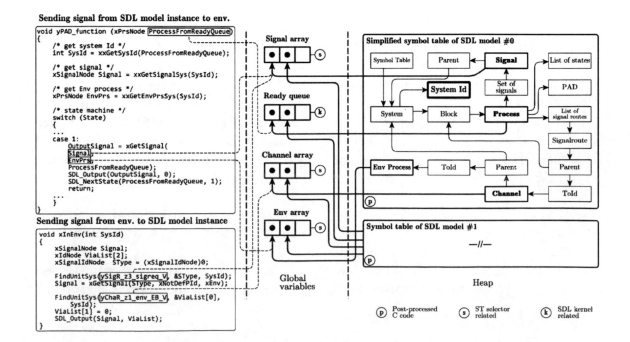

is stored in a data segment and contains pointers to signals, channels and environment processes, since every SDL model instance has its own environment. The SDL kernel extracts the first process from the ready queue (IBM, 2009) and calls the correspondent PAD function.

The PAD function must obtain information about SDL model instance before it can send a signal to any process. It is done by using a multiplexer, which is able to choose an instance depending on the information from arrays. The signal array, the channel array and the environment processes array are used for a required SDL model instance choice. Multiplexer does it by using traverse of hierarchical part of a symbol table. The system identifier is stored on a system level of the hierarchy of SDL entities (ITU, 2011). Such an identifier is associated with each new SDL model during the initialization stage.

So during the execution of a PAD function and sending a signal to environment by *xOutEnv* function or to other process by *SDL_Output* function, it uses the multiplexer to determine a symbol table of SDL model instance. When signal is sent from environment (from user's SystemC model) to SDL model instance by *xInEnv* function, it uses the multiplexer which performs a search for

a channel and a signal in arrays for identification of the SDL model instance.

An Example of the Approach Application

This example will give more details of the proposed approach and will show how the SystemC developer can use it in his project. Let us assume that we need to create a network model in SystemC and also we need to use it for an exploration of non-functional properties of a protocol while the SDL model of a protocol has already been implemented. To simplify the SDL model we use the same SDL model as we used in the previous section. The example is shown in Figure 19.

The SystemC model includes the *Source* node, three switches and the *Destination* node. The *Source* node is responsible for data generation while the *Destination* node is responsible for its reception. Each switch contains the SystemC module, which includes a number of SystemC threads, each of which corresponds to an independent instance of the SDL model. All these instances are created by the user in C++. The library controls all of them.

Figure 19. Example of the approach application

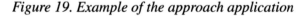
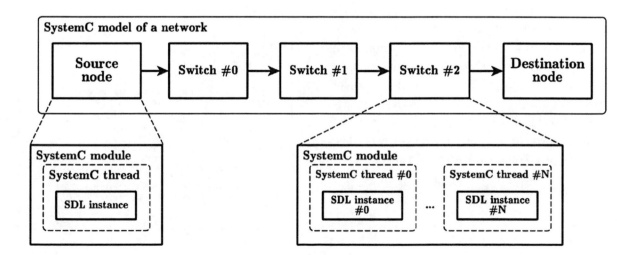

A fragment of source code of the SystemC module of a *Switch* is shown in Figure 20. This fragment shows the part of source code of the *switch_module* class and *sdl_model* class. Constructor of the *switch_module* class is responsible for initialization of a module and creation of a corresponding new instance of the SDL model. The *sdl_model* class contains functions which form the user's interface and which are used in the *switch_module* class. The *sdl_model_thread* firstly waits for a request event. After the event has been generated the thread handles it and the signal is sent to SDL model instance using function call. The function is provided by the library and it is one of a set of functions of the user's interface. Therefore, the instance can be used in such a manner as if it is a SystemC component. Thus, SystemC developer can work with any instance of the SDL model of a protocol not knowing anything about its implementation.

The proposed approach gives an opportunity to focus on implementation of the SystemC model of a network comprising hundreds of nodes rather than on implementation of the SDL model of a protocol.

Benefits of the Approach and Open Questions

The proposed approach provides an opportunity for network modeling on the basis of the SDL/SystemC co-modeling. The proposed approach is expected to reduce the project work effort and to help in achieving better quality of the simulation results. However, there are still a number of open questions and tasks for future work: definition of rules for code post-processing, memory management, proving of a model implementation and behavior correctness. The future work would be mostly focused on the creation of a special tool.

CONCLUSION

In current paper we proposed a methodology for embedded systems modeling by SDL and SystemC joint use. The paper specifically discusses the main problems of the modeling and sets the main requirements for the structure of functional design and verification. In practice modeling is used at different steps of the embedded system design flow and, therefore this makes the modeling one of the key development facilities.

There are two basic ways for testing the embedded system protocol stability and work characteristics which are the network model implementation and protocol stack modeling. Each of these ways

Figure 20. Listing of a part of SystemC module source code

```
/**** part of user's interface ****/
class sdl_model {
public:
    // initialization of a new instance
    sdl_model ();

    // function for sending sig.req to the instance
    void send_sig_reg ();
    ...
};

/**** part of user's code ****/
class switch_module : ... {
public:
    ...
    void sdl_model_thread ();
private:
    sc_event sig_req_event;
};

// switch module ctor
switch_module::switch_module(sc_module_name name) :
sc_module(name){
    // thread creation and event setting
    SC_THREAD (sdl_model_thread);
    sensitive << sig_req_event;

    // creation of a new instance of the SDL model
    sdl_model_instance = new sdl_model;
    ...
}

// switch module thread
void switch_module::sdl_model_thread (){
    while(1){
        wait(sig_req_event); // waiting for input event

        // sending sig.req signal to the instance
        sdl_model_instance->send_sig_reg();
    }
}
```

is applied for their specific purposes and gives an ability to obtain different simulation results.

Then the paper provided a general description of SDL and SystemC languages and provides an analysis of SDL and SystemC applicability for the protocols simulation. SystemC is a set of C++ classes that is used for modeling parallel systems. The mechanism of internal events allows to efficiently models distributed in time operation of the system under study. The main conclusion of the provided summary is that SystemC qualifies as a proper language for modeling, design and verification of the systems. The purpose of SDL is to make an unambiguous specification and description of the system behavior. SDL language is intended for the description of the structure and operation of the distributed real-time systems, which has components that are interpretable in a form of finite state machines. The analysis proved that choice of the modeling language depends on the modeling purposes.

This paper proposed an approach for SDL and SystemC co-modeling to take the best features from both languages and get the better modeling results. The approach for SDL/SystemC co-modeling concerned in this paper assumes to have a SystemC project, which corresponds to the system model. This model contains the SDL and SystemC parts. SDL part is represented by C/C++ code which is generated from the SDL model during the "make" stage.

Moreover, this paper described two ways of practical application the SDL/SystemC co-modeling. First one is the protocol stack model tester which can be used for simulation and investigation of data transfer protocol stacks providing a flexible test environment and logging facilities. Another way is network simulation which gives opportunities for performance analysis and simulation of network operation. This way is based on post-processing of the generated C code of the SDL model in order to have an ability of creation of a different number of SDL model instances with use of dynamic memory allocation.

We consider that our future work would be mostly focused on the creation of a special tool for network modeling. This tool is planned to be applied during implementation and validation of spacecraft and on-board communication protocols specifications.

ACKNOWLEDGMENT

We would like to sincerely thank our colleagues Alexander Stepanov and Konstantin Nedovodeev for taking part at different stages of this research.

REFERENCES

Balandin, S., Gillet, M., Lavrovskaya, I., Olenev, V., Rabin, A., & Stepanov, A. (2011). Co-Modeling of Embedded Networks Using SystemC and SDL. *International Journal of Embedded and Real-Time Communication Systems*, 2(3), 24–49.

Barr, M., & Massa, A. (2006). *Programming Embedded Systems: With C and GNU Development Tools* (2nd ed.). O'Reilly Media, Inc.

Black, D., & Donovan, J. (2004). [*From the Ground Up*. New-York: Springer Science+Buisness Media, Inc.]. *System*, C.

Bykhteev, A. (2008). Methods and facilities for systems-on-chip design. *ChipInfo microchip manual*. Retrieved from http://www.chipinfo.ru/literature/chipnews/200304/1.html

Dietterle, D. (2009). *Efficient Protocol Design Flow for Embedded Systems*. Brandenburg University of Technology. Retrieved from http://systems.ihp-microelectronics.com/uploads/downloads/diss_dietterle.pdf

ELVEES R&D Center. (2005). *SpaceWire switch MCK-01*. Retrieved from http://multicore.ru/index.php?id=850

Gillet, M. (2008). Hardware/software co-simulation for conformance testing of embedded networks. In *Proceedings of 6ᵗʰ Seminar of Finnish-Russian University Cooperation in Tele-communications (FRUCT) Program*. Retrieved October 31, 2008, from http://fruct.org/index. php?option=com_content&view=article&id=6 8&Itemid=73

Gipper, J. (2007). SystemC the SoC system-level modeling language. *Embedded computing Design*. Retrieved from www.embedded-computing.com/ pdfs/OSP2.May07.pdf

Haroud, M., & Biere, A. (2005). SDL Versus C Equivalence Checking. In *Proceedings of 12ᵗʰ International SDL Forum* (LNCS), (vol. 3530, pp. 323-339). Springer.

Haroud, M., & Blazevic, L. (2006). HW accelerated Ultra Wide Band MAC protocol using SDL and SystemC. In *Proceedings of Fourth IEEE International Conference on Pervasive Computing and Communications Workshops (PERCOMW'06)*. IEEE. Retrieved from http://fmv.jku. at/papers/HaroudBlazevicBiere-RAWCON04.pdf

Heath, S. (2003). *Embedded Systems Design* (2nd ed.). Newnes.

IBM. (2009). *SDL Suite and TTCN Suite Help*. IBM Rational SDL and TTCN Suite.

International Telecommunication Union. (2011). *Recommendation Z.100. Specification and Description Language (SDL)*. Geneva: Author.

Jantsch, A. (2004). *Modeling Embedded Systems and SoCs*. Stockholm: Morgan Kaufmann Publishers.

Jozawa, T., Huang, L., Sakai, E., Takeuchi, S., & Kasslin, M. (2006). Heterogeneous Co-simulation with SDL and SystemC for Protocol Modeling. In *Proceedings of IEEE Radio and Wireless Symposium 2006*, (pp. 603–606). Retrieved from http:// research.nokia.com/node/5789

Kamal, R. (2008). *Embedded systems: architecture, programming and design* (2nd ed.). New Delhi: Tata McGraw-Hill Publishing Company Limited.

Karabegov, A., & Ter-Mikaelyan, T. (1993). *Introduction to the SDL language*. Moscow: Radio and Communication.

Mitschele-Thiel, A. (2001). *System Engineering with SDL*. Chichester, UK: John Wiley & Sons, Ltd. doi:10.1002/0470841966

Morozkin, P., Lavrovskaya, I., Olenev, V., & Nedovodeev, K. (2013). Integration of SDL Models into a SystemC Project for Network Simulation. *Lecture Notes in Computer Science*, *7916*, 275–290. doi:10.1007/978-3-642-38911-5_16

Nemydrov, V., & Martin, G. (2004). *Systems-on-chip: Design and evaluation problems*. Moscow: Technosphera.

Olenev, V. (2009). Different approaches for the stacks of protocols SystemC modeling analysis. In *Proceedings of the Saint-Petersburg University of Aerospace Instrumentation scientific conference* (pp. 112-113). Saint-Petersburg, Russia: Saint-Petersburg University of Aerospace Instrumentation (SUAI).

Olenev, V., Onishenko, L., & Eganyan, A. (2008). Connections in SystemC Models of Large Systems. In *Proceedings of the Saint-Petersburg University of Aerospace Instrumentation scientific student's conference* (pp. 98-99). Saint-Petersburg, Russia: Saint-Petersburg University of Aerospace Instrumentation (SUAI).

Olenev, V., Rabin, A., Stepanov, A., & Lavrovskaya, I. (2009). SystemC and SDL Co-Modeling Methods. In *Proceedings of 6ᵗʰ Seminar of Finnish-Russian University Cooperation in Telecommunications (FRUCT) Program* (pp. 136-140). Saint-Petersburg, Russia: Saint-Petersburg University of Aerospace Instrumentation (SUAI).

Olenev, V., Sheynin, Y., Suvorova, E., Balandin, S., & Gillet, M. (2009). SystemC Modeling of the Embedded Networks. In *Proceedings of 6th Seminar of Finnish-Russian University Cooperation in Telecommunications (FRUCT) Program* (pp. 85-95). Saint-Petersburg, Russia: Saint-Petersburg University of Aerospace Instrumentation (SUAI).

Open SystemC Initiative (OSCI). (2005). *IEEE 1666™-2005 Standard for SystemC*. Retrieved from http://www.systemc.org

Stepanov, A., Lavrovskaya, I., & Olenev, V. (2010). SDL and SystemC Co-Modelling: The Protocol SDL Models Tester. In *Proceedings of the 8th Conference of Open Innovation Framework Program FRUCT*. Saint-Petersburg, Russia: Saint-Petersburg University of Aerospace Instrumentation (SUAI).

Stepanov, A., Lavrovskaya, I., Olenev, V., & Rabin, A. (2010). SystemC and SDL Co-Modelling Implementation. In *Proceedings of the 7th Conference of Finnish-Russian University Cooperation in Telecommunications (FRUCT) Program*. Saint-Petersburg, Russia: Saint-Petersburg University of Aerospace Instrumentation (SUAI).

Suvorova, E. (2007). A Methodology and the Tool for Testing SpaceWire Routing Switches. In *Proceedings of the first International SpaceWire Conference*. Retrieved September 19, 2007, from http://spacewire.computing.dundee.ac.uk/proceedings/Papers/Test and Verification 2/suvorova2.pdf

Suvorova, E., & Sheynin, Y. (2003). *Digital systems design on VHDL language*. Saint-Petersburg, Russia: BHV-Petersburg.

Swan, S. (2003). *A Tutorial Introduction to the SystemC TLM Standard*. Retrieved July 7, 2008, from http://www-ti.informatik.uni-tuebingen.de/~systemc/Documents/Presentation-13-OSCI_2_swan.pdf

Tanenbaum, A. S., & David, J. W. (2011). *Computer Networks* (5th ed.). Prentice Hall.

KEY TERMS AND DEFINITIONS

Co-Modeling: Modeling performed on the basis of joint use of two or more different modeling languages.

Delta-Cycle: A part of one modeling time unit which corresponds to execution of events which are not in a cause-effect relation with each other. Moreover, delta-cycle is referred to an evaluation of new values.

Embedded Network: A specific combination of computer hardware and software which is specifically designed to perform a particular function (or a range of functions) of a larger system.

Modeling: A representation of an object by a model in order to obtain information about the object. This information is usually obtained through experiments with the object's model.

Modeling Time: Is a notion which is used for imitation of system clocks. It starts at 0 and increments at the course of simulation.

SDL: Specification and Description Language intended for unambiguous specification and description of telecommunication systems.

SystemC: A system design and modeling language based on C++. This language evolved to meet a system designer's requirements for designing and integrating today's complex electronic systems very quickly while assuring that the final system will meet performance expectations.

Chapter 10
Model–Based Testing of Highly Configurable Embedded Systems

Detlef Streitferdt
Technische Universität Ilmenau, Germany

Holger Kaul
ABB Corporate Research, Germany

Florian Kantz
ABB Corporate Research, Germany

Thomas Bauer
Fraunhofer IESE, Germany

Philipp Nenninger
ABB Automation Products, Germany

Tanvir Hussain
The Mathworks GmbH, Germany

Thomas Ruschival
Datacom Telematica, Brazil

Robert Eschbach
ITK Engineering AG, Germany

ABSTRACT

This chapter reports the results of a cycle computer case study and a previously conducted industrial case study from the automation domain. The key result is a model-based testing process for highly configurable embedded systems. The initial version of the testing process was built upon parameterizeable systems. The cycle computer case study adds the configuration using the product line concept and a feature model to store the parameterizable data. Thus, parameters and their constraints can be managed in a very structured way. Escalating demand for flexibility has made modern embedded software systems highly adjustable. This configurability is often realized through parameters and a highly configurable system possesses a handful of those. Small changes in parameter values can often account for significant changes in the system's behavior, whereas in some other cases, changed parameters may not result in any perceivable reaction. The case studies address the challenge of applying model-based testing to configurable embedded software systems in order to reduce development effort. As a result of the case studies, a model-based testing process was developed. This process integrates existing model-based testing methods and tools such as combinatorial design and constraint processing as well as the product line engineering approach. The testing process was applied as part of the case studies and analyzed in terms of its actual saving potentials, which turned out to reduce the testing effort by more than a third.

DOI: 10.4018/978-1-4666-6034-2.ch010

INTRODUCTION

In the automation domain, large and complex systems like chemical or power plants are common practice. The products of these plants are part of our daily lives, and our living standard depends directly on their reliable supply. This dependency accounts for the *high quality* requirements for these plants, which adds to the burden of voluminous costs for engineering and operation. Of course, such *high quality* is required for almost all of the components of a plant in order to ensure the proper functioning up to the point that even certain failures should not lead to unbearable consequences. On the upper level, *control systems* based on workstation platforms (e.g. Microsoft Windows®) are used to control the overall function of the plant, for example the generation of energy in a power plant. Between the control system layer and the lowest sensor and actuator level, several layers of embedded systems of varying complexity are used to collect and pass on sensor data (like temperature or pressure values), monitor the proper function of plant sub modules and actuate upon requests from the upper level *control system* (e.g., close a valve).

The main challenge in the application of model-based testing for embedded systems is their simple behavior visible from the outside, which internally gets dramatically complex due to configurable features and parameters. Each system has many parameters and within this system, a *configuration* is a set of parameters with concrete values selected for each parameter. Such configurations are intended for various purposes, for example for dealing with different modes of operation, different types of user interactions, error and exception handling etc. Different kinds of system behavior are directly related to configurations and as a result, the verification of the system is cumbersome and difficult as the number of available configurations rises.

This article presents the results of the industrial automation domain case study of the ITEA2-project D-MINT (http://www.d-mint.org), driven by ABB. The resulting testing process was and the elaborated model based testing approach applied to a cycle computer at the Technische Universität Ilmenau. Here, the concept of product lines has been added to the testing process resulting in a reduction of relevant parameters for the parameter model since those parameter can be moved into the feature model.

The case studies aimed at answering questions regarding the most promising model-based testing methods and tools as a way of addressing the goal of reduced testing efforts. In addition, the questions of how and when to apply model-based testing were answered and ultimately led to a new and holistic view on model-based testing for embedded systems, based on (Bauer, Eschbach, Groessl, Hussain, Streitferdt, & Kantz, 2009). Finally, the case studies deliver an analysis and precise numbers of the actual savings as a result of applying the developed model-based testing process.

In the section "Example Case Studies", the softstarter and the cycle computer are introduced as examples of embedded devices and a basis for the case studies. In the section "Model-Based Testing Process for Embedded Devices", the integrated testing process is discussed as a key concept of the case study. In the section "Evaluation of the Approach", the results of applying the process in the case study are presented. In section "Related Work", an overview of the relevant testing technologies and methods is given. Finally, this article concludes with a brief summary and topics for further research.

EXAMPLE CASE STUDIES

The following two case studies have been elaborated subsequently. In the first industrial case study, the softstarter, a single system testing process was developed.

Based on this testing process the second case study a cycle computer has been developed in a University environment and resulted in an enhanced testing process with product line support.

This case study covers multiple domains (computer science, electrical engineering, mechanical engineering) for the best possible learning curve of the future software engineers. These learning demands are a direct result of the initial softstarter case study. Secondly, the cycle computer is a multi-platform project. Many operating systems, different languages and different hardware platforms are in use.

Softstarter Case Study

Electric motors are common actuators in process automation. For this article, the starting and the stopping of an electric motor is taken as example. The device used in this article is a softstarter as shown in Figure 1, which is used to smoothly ramp up/down a motor. This functionality is needed for large motors where the peak current consumption from the power grid may cause a breakdown in voltage or for conveyor belts where sudden steep acceleration ramps may damage the transported goods.

Besides ramping the motor up and down, a softstarter monitors the motor to detect, e.g., a locked or overheated motor or disturbances in the power supply network that might damage the motor.

The softstarter is able to control a wide range of electric motors in different scenarios (e.g., stone crushers, conveyor belts, fans or water pumps). Besides the mechanical installation of the softstarter, its behavior can be configured by a set of user changeable parameters for different scenarios. During commissioning, the softstarter parameters like the ramp up time need to be set according to the desired usage scenario – the softstarter is configured.

This kind of configuration is done by setting the values of 150 parameters; a selection of these parameters is shown in Table 1. This complexity poses a challenge for testing the softstarter. To ensure complete coverage, the $1.0 \cdot 10^{110}$ possible combinations of parameter values (see section "Parameter Modeling") would have to be tested. Without a doubt, this is not testable within an acceptable time frame. The only way to handle this complexity is the structured selection of a subset of parameter configurations.

Cycle Computer Case Study

Within the second case study, the model based testing approach was applied to a cycle computer development project to further refine the approach.

Figure 1. Softstarter for different motor sizes

Table 1. Softstarter Parameters, adaptable by the user based on the Softstarter User Manual, page 81

Param. Number	Description	Display Text	Setting Range	Default Value
1	Setting current	Setting Ie	9,0...1207A	Individual
2	Start ramp	Start Ramp	1...30s, 1...120s	10s
3	Stop ramp	Stop Ramp	0...30s, 0...120s	0s
4	Initial voltage	Init Volt	30...70%	30%
5	End voltage	End volt	30...70%	30%
6	Step down voltage	Step down	30...100%	100%
7	Current limit	Current Lim	2,0...7,0xle	4,0xle
8	Kick start	Kick Start	Yes, No	No
9	Kick start level	Kick Level	50...100%	50%
10	Kick start time	Kick Time	0,1...1,5s	0,2s
11	Start ramp range	Start Range	1-30s, 1-120s	1-30s
12	Stop ramp range	Stop Range	0-30s, 0-120s	0-30s
13	Overload protection type	Overload	No, Normal, Dual	Normal
14	Overload protection class	OL Class	10A, 10, 20, 30	10
15	Overload class, dual type, start class	OL Class S	10A, 10, 20, 30	10
16	Overload class, dual type, run class	OL Class R	10A, 10, 20, 30	10
17	Overload protection, type of operation	OL Op	Stop-M, Stop-A, Ind	Stop-M
...	...	...	...	...

The cycle computer is a multi-domain and multi-platform system.

The cycle computer is based upon two main components, the permanently attached sensor unit (right part in Figure 2) and the detachable processing unit on the handlebar (middle part in Figure 2). The five-way button (left part in Figure 2) is optional. The sensor unit is built using an embedded system to pre-process all the sensors attached to the bicycle and is equipped with additional hardware to control the dynamo as energy provider for the complete system. Many different sensors can be attached to the sensor unit such as, temperature of tires, break-pads or the rider of the bike, or the pressure of tires or the suspension. The processing unit receives all the sensor values, processes, analyzes and displays the required values. The information flow between all three components is realized by a cycle computer data protocol. Using such a protocol decouples the components very efficiently. Thus, arbitrary components can be integrated in the systems, such as the PC simulator, which is a full implementation of the cycle computer on a PC platform.

The processing unit of the cycle computer system has a four-layer architecture as shown in Figure 3. The core of this system is designed according to the well known model-view-controller pattern (Gamma, Helm, Johnson, & Vlissides, 1994). The SensorValues hold all the sensor values separated in direct raw values (such as the speed) and processed values (such as the average speed). The DataModel_Facade exposes the complete data model API to the outside for accessing and manipulating the actual data.

The controller handles all the events of this event-driven system. The active receiving and sending classes in the controller manage the serial communication to this system via the cycle computer data protocol. The data packets of this

Figure 2. Cycle computer global architecture

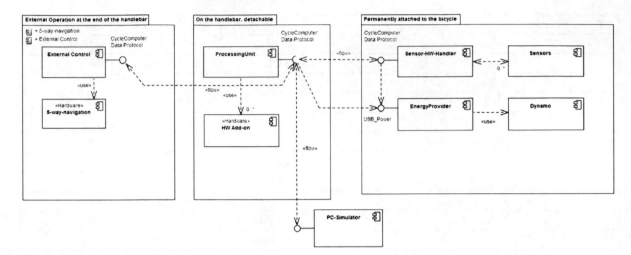

protocol deliver the sensor data from the sensor unit, indicate any operation of the five-way navigation button or issue debugging commands to actually test the system from the outside.

Finally, the presentation layer offers a user configurable view on subsets of the data in the data model.

The configurability of this system is the result of the cycle computer being used with different platforms and different user selectable features. The architecture is implemented for an embedded board with the FreeRTOS (Barry, 2010) operating system, as an *App* for Android Smartphones and as a simulation version for Microsoft Windows®. For the different platforms and for the different features of the system we use the product line (Pohl, Böckle, & van der Linden, 2005) approach to handle the variabilities on the model level. The implemented software systems binds these variabilities using a feature model (Kang, Lee, & Donohoe, 2002).

TEST ENVIRONMENT

The model-based testing approach described in this article relies on a test environment needed to control and execute the test cases for the softstarter and cycle computer respecively. In Figure 4, the test environment is visualized. The *Developer PC* is used to implement the test process, generate the test cases, and finally analyze the test results. The *Test Execution Hardware* is important for the test environment since the analysis of the test results for the softstarter depend on timing constraints, which cannot be met on an office PC platform with a general-purpose operating system.

The cycle computer is testable via its data protocol, using specifically defined debugging daa packets. The data protocol is based on RS232 for the majority of the tests. Highly time dependent and critical measurements need to be done with a hardware setup (we use a *National Instruments CompactRIO* System) comparable to the softstarter testing hardware.

Figure 3. Software architecture of the cycle computer processing unit

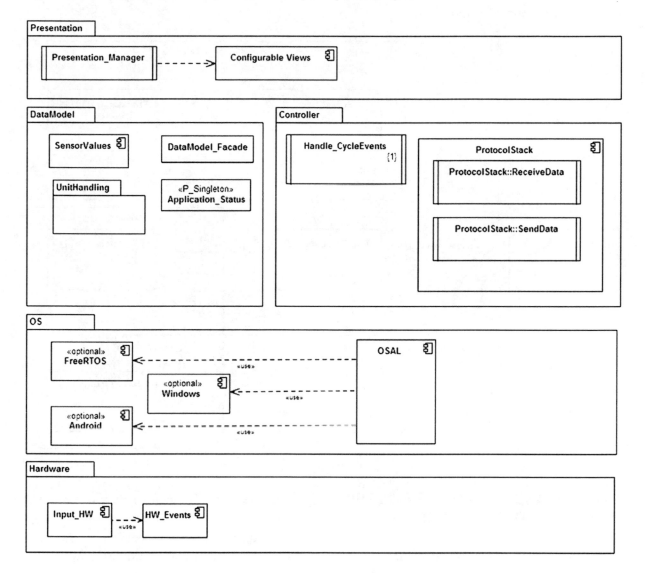

The generated test cases are transferred to the *Test Execution Hardware*, which in turn operates the *Softstarter/cycle computer components*, e.g., it emulates the press of a button on the *External Keypad*, as prescribed by the test cases. Besides measuring voltage and current, the *Test Execution Hardware* tracks the menu content of the *External Keypad* via a tapped serial connection or the cycle computer data protocol contents. With this information, it is possible to monitor any press of a button.

On the *DeveloperPC*, test cases are assembled out of simple test steps, which are methods of a test interface specifically developed for testing the softstarter as well as the cycle computer components. The test interface offers methods for operating the softstarter as well as the test environment, like Start_Motor(), Read_Temp() or Set_int_Parameter().

Figure 4. Test Environment for the automated generation and execution of test cases for the softstarter and the cycle computer components

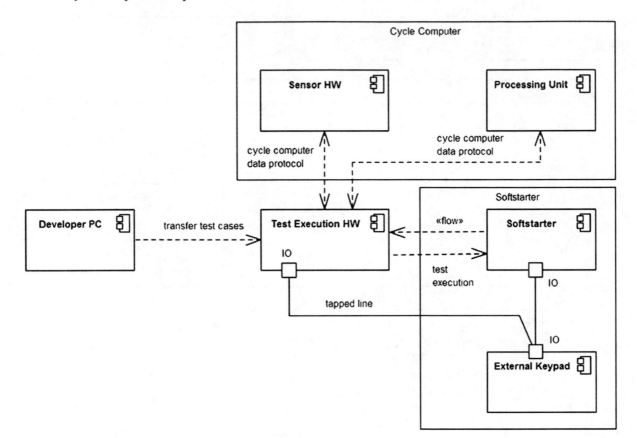

MODEL-BASED TESTING PROCESS FOR EMBEDDED DEVICES

The development of embedded devices in the automation domain is aligned along the V-Model (Reinhold, 2003). As shown in Figure 5, the model-based testing activities discussed here focus on the upper levels "System Requirements Definition" and "System Validation".

In the softstarter case study, the inclusion of behavioral changes connected to the parameters in the device led to an explosion of states in the test model. Thus, the test model was not manageable any more. To keep the test model at a manageable level of complexity and address parameter-related behavioral changes, existing approaches were composed into a development process that is new

to the automation domain. This test process is shown in Figure 6. It is a black-box test process addressing the user-visible behavior of the system. Within the domain analysis phase, many sources of information are analyzed, such as requirements documents, use-case models, or the user manual. This information is structured so as to be ready for referencing in the following phases.

As for many other domains (van der Linden, Schmid, & Rommes, 2007) the product line approach is a good choice to handle constraints amongst many features of a system where feature may also refer to concrete parameter values. Either the features are parameters themselves (e. g. the user may set the overall recording time of a video cam on the cycle computer) or a feature manages a set of parameters and thus keeps track of the

Figure 5. Integration of this model-based testing approach into the V-Model

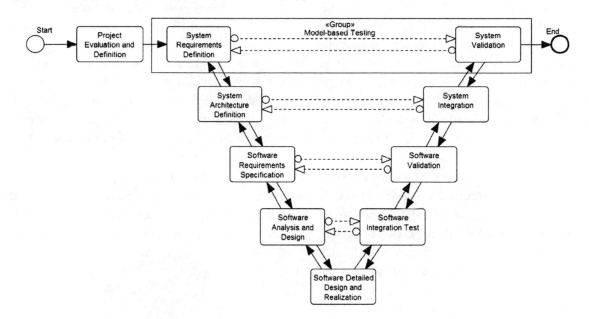

Figure 6. Detailed model-based testing process

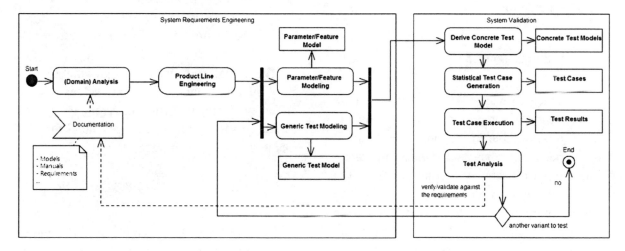

constraints relevant for this set of parameters. In the video cam example above the recording time feature will be part of the feature video for a complete high tech television system. Here, managing the recording time means setting the needed persistent storage size accordingly which is also based on the selected video codecs.

For this article the *Product Line Engineering* step in Figure 6 is an abstraction of the full version as described in (Schmid & Santana de Almeida, 2013). Here, we supplement this process with the parameter/feature model. Testing has to be done for each application derived from the product

line. After the *Test Analysis* step the corresponding decision select the next variant and starts the testing process again.

In the early development stage of system requirements definition, a *parameter/feature model* and a *generic test model* are added to the development process. The *generic test model* formally describes the valid system usage scenarios independent of the system configurations.

Based on the type of development, the *parameter/feature model* and the *generic test model* are roughly sketched and further refined in the subsequent development phases of the V-Model for developments starting from scratch. Or, the model can be fully reused for some development such as an update of existing products or existing product line variants.

In the last stage of the development, the system validation, concrete test models are derived from the generic test model and the parameter model. The concrete test models serve as input for the statistical test case generation, which results in executable test cases. While the test cases are executed, several measurements are made and stored. They are used in the final step, the test analysis.

Although the documents are concise and detailed, the task of developing a parameter model and a test model constitutes advanced engineering work, where experience and expert knowledge are an advantage. The following sections correspond to the development steps in Figure 6, starting with parameter modeling.

PARAMETER MODELING

Many parameters (in the case of the softstarter case 150, for the cycle computer 28), partly shown in Table 1, are available inside the device to adapt it to a specific application, such as operating a fan, a stone crusher, or a conveyor belt. The behavior visible from the outside is simple. An electric motor can be started and stopped again (after a while), which would be one important test case. To achieve complete coverage, this test case would have to be executed with all possible permutations of the parameters. In real life, this

is not possible due to the huge number of parameter permutations. The following sections briefly explain the method of clustering, the application of constraints, and pairwise testing. Finally, the approaches are integrated.

Table 2 describes the parameter types typically used in devices in the automation domain together with the number of permutations for a single parameter of the given type.

Based on the number of possible values for each parameter, it is easy to calculate $n_{complete}$ as the mathematically complete permutations of all parameters in the softstarter using Equation 1.

$$n_{complete} = \prod_{i=1}^{N_{params}} n_i \qquad (1)$$

N_{param} : Number of parameters

n_i : Number of possible values of parameter i

The softstarter uses 150 parameters: 65 Boolean parameters, 55 integer values and 30 enumerated values. Assuming about 100 values for integers using allowed ranges and 7 for enumerations, $n_{complete}$ is reduced to $n_{reduced}$.

$$n_{reduced} = 2^{65} \cdot 100^{55} \cdot 7^{30} = 8.3 \cdot 10^{154}$$

It takes between five seconds and five minutes (our assumption is an average of 1 minute of testing time, $t_{testcase}$) to execute a single test case for the softstarter. The combination of this execution

Table 2. Parameter types

Type	Permutations	Comment
Boolean	2	Standard type
Integer	2^8, 2^{16} or 2^{32}	Standard type
String	$\approx 50^{20}$	Assuming a character set with 50 chars and strings with an average of 20 chars
Enumeration	≈ 7	Assuming an average of 7 values

time with the set of parameter permutations results in the testing time t_{test} of:

$$t_{test} = n_{reduced} \cdot t_{testcase} \approx 1.6 \cdot 10^{149} a$$

a : Years

The testing time t_{test} exceeds the age of the universe of about $14 \cdot 10^9$ years dramatically. Thus, further reduction of the parameter permutations to be tested is required, which is accomplished using methods described in the following sections, which are based on (Kantz, Ruschival, Nenninger, & Streitferdt, 2009). These methods as well as the corresponding tools have been chosen or developed based on the state of the art and on experiences of the testing domain experts in the D-MINT project, our own assessments of the usability and acceptance of the methods in our industrial domain, and our own expert knowledge in applying and integrating scientific methods towards industrial usage.

First, the parameters will be organized into sub-sets referred to as *clustering*. A cluster is a set of parameters according to the following rules:

1. All parameters within a cluster must be independent of all parameters outside the cluster.
2. Parameters inside a cluster may be dependent on each other.

The calculation of the overall permutations is the sum of the permutations of all individual clusters. In order to group the parameters into the clusters, expert knowledge is necessary to ensure the requirement of independence (see first rule above). This knowledge is highly dependent on the device functionality and on the specific device design.

Applying the clustering approach to the soft-starter, we obtained 15 functionally independent groups with about 10 parameters each. For the

running example and a simplified calculation, the clusters *Protections* (containing parameters like locked rotor protection, phase imbalance protection, or high current protection), *Warnings* (containing parameters like high current warning, overload warning, or thyristor overload warning), and *Faults* (containing parameters like phase loss fault, fieldbus fault, or frequency fault) were selected. All three clusters together contain 50 parameters. By clustering, the permutations for these parameters can be reduced from $n_{simplified}$ as calculated above with Eq.1 and the parameter types, to $n_{cluster}$.

$$n_{simplified} = 4.92 \cdot 10^{37}$$

$$n_{cluster} = 4.67 \cdot 10^{17}$$

Taking this reduction rate as an average for the whole parameter set, it is possible to reduce the number of permutations to:

$$\frac{n_{cluster}}{n_{simplified}} = \frac{4.67 \cdot 10^{17}}{4.92 \cdot 10^{37}} \approx 9.49 \cdot 10^{-19}\%$$

of the entire configuration space. Applying this huge reduction to the initially calculated $1.0 \cdot 10^{110}$ parameter permutations $n_{reduced}$, $1 \cdot 10^{90}$ permutations still remain.

A second approach, *parameter constraints*, has been used to further reduce the parameter permutations. The dependencies of parameters inside a group can be used for further reduction of the number of permutations needed to test the device. After computing the possible permutations (brute force) per parameter group, each parameter set needs to fulfill the constraints; otherwise, it is discarded. The following three constraints were analyzed.

The first constraint, *Mutual Exclusion*, is present if the selected value of a single parameter switches between parameter sets. For the parameter "Start Mode", two values "Volt" or "Torque" are possible. Either all Start/Stop cycles will be controlled by the voltage passed on to the motor or all Start/Stop cycles will be controlled by the torque the electric motor delivers. Each mode is further parameterized by a set of sub-parameters (e.g., ramp-up time or initial start voltage level if the motor is voltage-controlled) belonging only to this mode and therefore the parameters of other modes are mutually excluded (e.g., torque limit if the motor is torque-controlled). As soon as a mode is selected, other parameters are not taken into account for test parameter permutations. In most cases, expert knowledge is needed to identify mutual exclusions and all affected parameter sets.

The second constraint is a specialized form of mutual exclusion, referred to as *Function Switching Parameters*. In this case, a parameter switches a mode or functionality *on* or *off* whereas further parameters are used for a detailed configuration of the mode or functionality. The parameter "Locked rotor protection" can be switched *on* or *off*. This type of protection is triggered by the current flow through the motor and can be parameterized by the maximum duration of this current flow and the desired action (e.g., complete stop or automatic restart). Without this protection, all the permutations for the sub-parameters are irrelevant.

The third constraint is the *Selection of Ranges*. It directly influences the possible permutations of a parameter. While calculating the permutations, a parameter that rules over the range of another parameter forms the trigger for the reduction of the parameter space depending on the range selected. Of course, it is useful to create values slightly outside the defined range while testing the limits on the selected range.

The application of parameter constraints in each cluster led to a reduction of the permutations to $1.45 \cdot 10^{17}$. In the case study, scripting and spreadsheets were used to realize this reduction.

Finally, the *Pairwise Testing* approach assumes (based on empirical data of 329 error reports) that a majority of faults occur when changing values in a pair of parameters independently of further parameters. Thus, the testing effort can again be reduced significantly (Kuhn et al., 2004).

Due to the number of parameters in the clusters, the following example is presented only with three parameters, A: *Kick-Start* (before ramping up, kick-start the motor if, e.g., stones are blocking a stone crusher), B: *External By-Pass* (by-pass the softstarter once the motor is running at full speed), and C: *Fieldbus Control* (remote control the softstarter). All parameters can have two values, *yes* or *no*. Considering such a system, eight parameter sets are possible, see Table 3 leftmost

Table 3. n-Way testing

A	B	C		A	B		A	C		B	C		A	B	C		Parameters
n	n	n		n	n		n	n		n	n		n	n	n		A : Kick-Start
n	n	y		n	y		n	y		n	y		n	n	y		B : External By-Pass
n	y	n		y	n		y	n		y	n		n	y	n		C : Fieldbus Control
n	y	y		y	y		y	y		y	y		n	y	y		
y	n	n											y	n	n		
y	n	y											y	n	y		
y	y	n											y	y	n		
y	y	y											y	y	y		

part. Each of the eight parameter sets would have to be tested. By building pairs as in the three middle parts of Table 3, the possible permutations of the pairs A-B, A-C and B-C need to be present in the final testing parameter set. The gray parameter sets, rightmost part in Table 3, are enough to cover all individual permutations of the three middle pairs. Thus, five instead of eight parameter sets will be tested.

Complex errors depending on the combination of three or more parameter values cannot be found systematically by using the approach of pairwise testing. In order to enhance the coverage and detect these faults, the application of 3-wise testing with the softstarter parameter set results in a reduction of the parameter permutations to $3.7 \cdot 10^9$.

In the sections above, possible techniques for reducing the parameter permutations are presented. Although the resulting number of permutations still remains large, the *combination of the different approaches* resulted in a reasonable reduction.

The initial step for the optimization will be clustering. In the second step, a decision between two possibilities needs to be made. Either parameter constraints or the pairwise testing approach can be applied. Figure 7 gives an overview of the number of combinations that could be obtained with the different approaches.

Starting with the simplified subset of parameters $n_{simplified}$, the clustering approach results in a reduced configuration space containing n_{clust} permutations. By applying constraints to the clustered parameters, the number of permutations can be further reduced to $n_{clust+constr}$. Finally, using the pairwise testing approach on the clustered parameters, the number of permutations can be reduced to $n_{clust+pair}$. The reduction for the paths Clustering-Constraints or Clustering-Pairwise in Figure 7 is between $10^{-27}\%$ and $10^{-19}\%$. Despite this huge reduction, the remaining configuration space is still large. Thus, future efforts will be spent on the combination of clustering, constraints, and the pairwise testing approach (the dotted line in Figure 7).

Figure 7. Combination with resulting permutations of parameter permutation reduction approaches

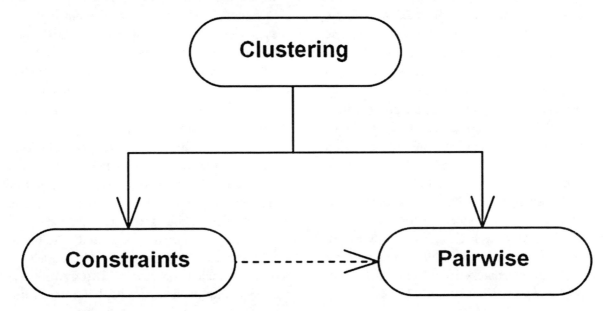

Figure 8. Cycle computer feature model

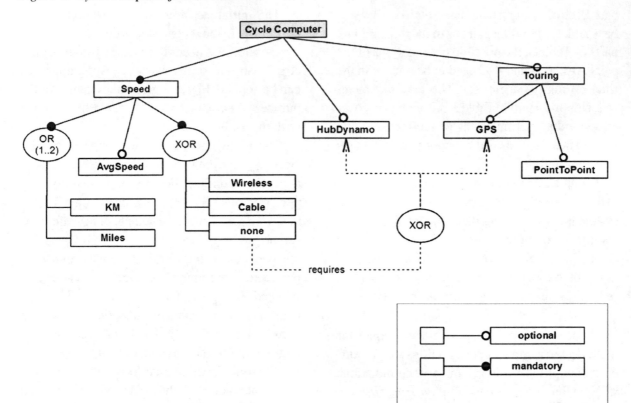

FEATURE MODEL

Several of the features of embedded systems are bound to the product line features of this system. Here, a feature is something a user is willing to pay for and thus the feature decides whether a functionality / a behavior will be part of the finally generated application based on the product line.

The cycle computers feature model is shown partly in Figure 8. As guideline for the inclusion of parameters in the feature model: include as many parameters as possible. In Figure 8 a cycle computer might have a video cam (cycle cam) which is modeled as optional feature. As soon as a user selected the Cycle Cam feature an SD-Card is mandatory to store the videos. The parameter feature Size is mandatory as well. The Rec. Length parameter feature is left and may be set by users

who find the recording length important. The calculate constraint between the parameter feature Rec. Length and Size holds the formula to derive the needed size of the SD-Card with a given and required recording length.

As already mentioned we use feature models to configure the resulting cycle computer applications to the user needs. Using this approach the parameters present in the feature model are modeled in their semantic context. Thus, the user can easily understand the parameter and select a desired value. Now, the parameter is not subject of any permutation of values as in the last section. The parameter is bound to a user selected value.

In the cycle computer case study the initial configuration had 28 parameters. Using the feature model approach we could move 18 (64%) of these parameter into the feature model with the corresponding constraints. Thus, the performance of the last sections Parameter Modeling is increased since the number of remaining parameters can be reduced.

GENERIC TEST MODELING

Model-based statistical testing (MBST) is a *black-box* testing technique that enables the generation of representative or failure-sensitive test cases from the tester's or user's perspective (Prowell et al., 1999; Prowell, 2005). The central element of the approach is a state-based test model, which describes the relevant system inputs and usages, and the expected system responses. Test models can be annotated with probabilistic weights to express frequency of use, costs, or criticality of inputs, outputs, and usages. Models that incorporate frequency of use are also called *usage models*. The underlying modeling notation is discrete time Markov chains. MBST allows the estimation of the system's reliability considering the given usage profile. The approach has been extended for risk-based testing (Zimmermann et al., 2009; Bauer et al., 2008; Zimmermann et al., 2009) and applied to safety-critical embedded systems (Kloos et al., 2009; Bauer, Böhr, Landmann, Beletski, Eschbach, & Poore, 2007). MBST was used to construct generic test models and to automatically generate test cases from the configuration-specific test models.

Figure 9 shows the steps of the model-based statistical testing approach with feature model support. As stated in (Farrag, Fengler, Streitferdt, & Fengler, 2010) a test model can be derived using the feature model as input. A test model for domain test cases, the core of the product line, and application specific test cases are needed for the overall test cases. In (Cai & Zeng, 2013) variation points are used to select the application specific set of test cases for derived applications based on a product line. The complexity of current systems did not leave any space for a reduction of test cases based on the fact that all variants of a product line are built with the same core. All derived applications need to be tested. Thus, our approach makes use of the reduction potential in the starting configuration phase. After this only the remaining parameters of an application need to be

Figure 9. Process steps of the model-based statistical testing approach

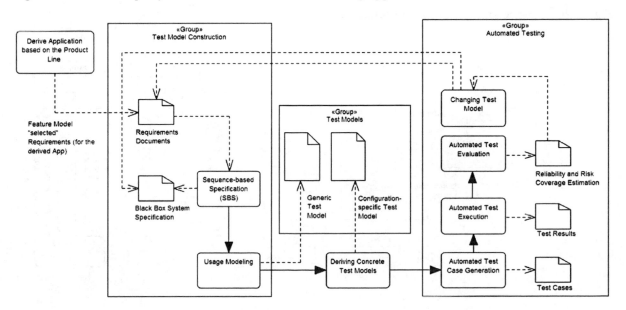

managed. We make use of the trace links between the features and the requirements of our system. By choosing the feature set the corresponding subset of requirements is also identified. These "feature selected" requirements are the input for the first stage in Figure 9.

In the first stage after the product line application and requirements generation on the left side, *Test Model Construction*, a test model is built from the system requirements, which represents relevant system inputs, usages, and the expected system responses. The second stage comprises the *Automated Testing*, which is divided into automated generation of test cases from the test model, automated execution of test cases in the test environment, and automated evaluation of test results. The derivation of configuration-specific test models from the generic model is an additional step, which is not part of the original MBST approach.

The technique for the systematic construction of the model is called sequence-based specification (SBS, Prowell, & Poore, 2003). SBS is a systematic approach for formalizing textual requirements and transforming them into a consistent and complete black-box specification represented as a finite

state machine. The black-box specification is a mapping, which associates a stimulus history with its associated response. It defines the required external behavior of a system. During the SBS, the original system requirements are inspected with respect to system stimulation, usages, and responses. Every modeling decision in the SBS has to be justified and linked to the original system requirements. This assures the construction of a consistent, complete, and traceable black-box specification.

In the first step of the SBS the system boundary is identified, i.e., interfaces and associated stimuli and responses. Consequently, a sequence enumeration is performed, leading to a well-defined blackbox specification mapping each stimulus sequence to a sequence of responses.

The process of sequence enumeration aims at systematically writing down each possible stimulus sequence, starting with sequences of length one. A sequence of stimuli represents one history of the usage of the system under test. For every sequence, it is determined whether the sequence is valid or invalid related to the system requirements and whether corresponding expected responses are given. Empty responses are also possible. Only

Figure 10. Generic test model for the softstarter

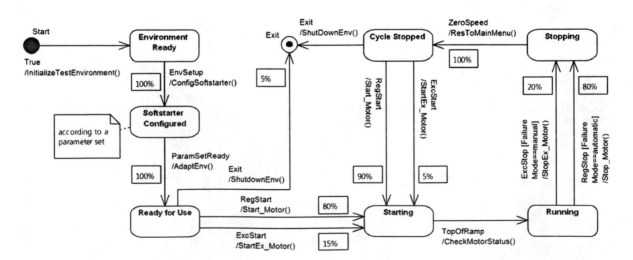

valid enumerated sequences will be extended and analyzed again in the next step. Every sequence is compared to previously analyzed sequences in terms of equivalence. Two sequences are equivalent if their responses to future stimuli are identical. Equivalent sequences are reduced to the shorter equivalent sequence and not extended.

A valid sequence that cannot be reduced to a shorter equivalent sequence is called a *canonical sequence*. A canonical sequence corresponds to the shortest sequence of stimuli, which leads to a state in the test model. The set of canonical sequences corresponds to the set of usage states identified for the test object. Usage states are named according to their meaning in the test model, see Figure 10. Alternatively, state variables can be introduced to label the states of the model. The enumeration stops if all sequences are invalid or reduced to equivalent sequences. All construction decisions for validity, equivalence, responses, and requirement coverage are traced back to the original system requirements.

The enumeration step provides a finite state machine that can be used further for usage modeling. The enumeration procedure helps in revealing inconsistencies, missing specifications, unclear requirements, and vague formulations in the original requirements document.

Like most of the models used in any model-based testing approach, the models constructed using the abovementioned technique also contain certain abstractions. The abstractions help in restricting the size of the model and thus help in understanding, analyzing as well as in test case generation.

For each selected product configuration of the test object, a particular configuration specific test model is required which is later used for automated test case generation. In our approach, a generic test model is first built using the above mentioned steps, as depicted in Figure 9. The transitions in the model use a guard expression for the "FailureMode" parameter. The values of the parameter "FailureMode" are "manual" and

Figure 11. Creation of a configuration-specific test model from a generic model

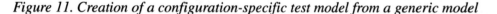

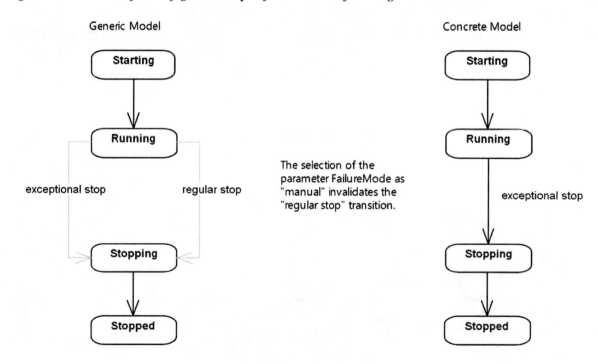

"automated". In the configuration specific test models, the variability of parameters and multiple transitions are resolved. The resulting models do not contain variables and guards, which facilitates automated test case generation.

The generic model describes the configuration-independent functional aspects of the device, i.e., the mapping from stimuli sequences to responses. Specific models related to each configuration are then derived from that model using guards that exclude certain transitions and unreachable states from the generic model. A simple example of this concept is illustrated in Figure 11. Here, the configuration is meant to define the behavior of the software with regard to failure handling. When configured for manual handling of a failure, the regular stop transition labeled "RegStop" (which is the automatic handling) is not valid any more. The selection of a value for the "FailureMode" parameter removes the "RegStop" transition, as shown in the right part of Figure 11.

The resulting state machine represents the structure of the usage model and describes all possible usages identified from the system requirements. By adding probabilities to the transitions, the state machine becomes a discrete-time Markov chain. The probabilities reflect the frequency or criticality of the system usage. Data for probability distributions is sometimes available from domain experts or monitoring data from similar systems, partially or completely, for usage environments. Without such data, uniform probabilities are assumed. This ultimately implies that all outcomes are equally likely. In our case, we used risk analysis data to define the usage profile of the test object, as shown in Figure 10.

AUTOMATED GENERATION OF TEST CASES

Test models have particular states for the initialization (START) and finalization (EXIT) of test cases. A test case is an arbitrary path through the model from START to EXIT traversing a sequence of states and transitions. The state START describes the system state at the beginning of a test case. The state EXIT marks the end of a test case and can be reached from all states where a test case can end. Different strategies for automated test case generation exist, e.g.:

- *Model coverage tests* make sure that the whole model is covered by test cases. This means that each transition and each state of the test model is tested.
- *Random tests* are randomly generated paths through the test model based on an operational profile, e.g., frequency or criticality.

Figure 12. Enterprise architect testmodel generation plug-in

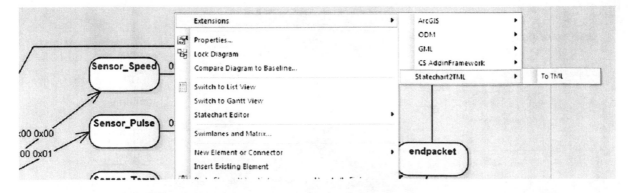

From every configuration-specific test model, a set of test cases was generated: one set of test cases to cover all model elements and one test set of test cases with a number of random tests. The tool-supported generation of test cases takes just seconds, but the execution of the test cases is dependent on the planned and available test effort.

The cycle computer project resulted in a plugin development for *Enterprise Architect* to generate Markov chain test models in the TML notation (test modeling language, Prowell 2000).

TEST CASE EXECUTION AND ANALYSIS OF RESULTS

The transitions of the test model are annotated with scripts for the test runner, which is part of the execution hardware. Hence, during the generation of a specific test case, the scripts of the transitions on the path will be aggregated one after another in order to build a concrete test case that is executable in the selected test environment as shown in Figure 2.

Figure 13. Test case execution sequence with the test environment elements

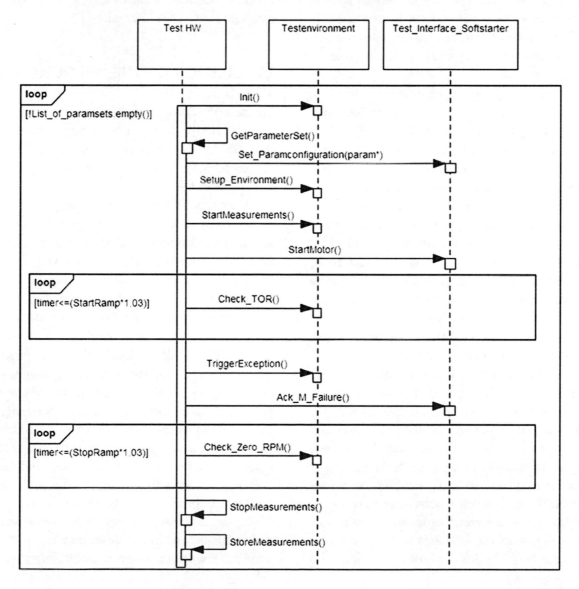

As an example, Figure 13 shows a test case generated from a concrete test model with manual acknowledgement for failures due to an exceptional stop of the motor. The sequence includes the testing hardware, the environment, and the test interface towards the softstarter itself. The test cases and the corresponding parameter set are transferred to the testing hardware. The test case will be executed for each parameter set of the reduced parameter permutations according to the section "Parameter Modeling".

First, the test execution hardware is initialized and the parameter set is prepared for use with this test case. The current parameter configuration is transferred to the softstarter via the external keypad, after which the test execution hardware will set up the environment according to the parameters (e.g., surrounding temperature). The measurements of voltage and current, for example, are started before the actual test sequence is started. After issuing the start command to the softstarter, the test execution hardware checks (with a time limit) whether the connected motor has reached the running state and whether the test sequence can continue. The concrete test model now only allows for an exceptional stop, for which an exception needs to be triggered. This forces the softstarter to stop the motor, which is checked again by the test execution hardware. Finally, the test execution hardware stops and stores the measurements for the upcoming analysis of the test case. This sequence is repeated for all parameter sets in the current list of parameter permutations.

The test analysis evaluates the observable reactions of the system at each test step. Each test case corresponds to a specific transition of the concrete test model and describes the expected system responses. Responses can be discrete (e.g., a relay-switch output for locked rotor) or continuous quantities (e.g., start/stop ramp voltage variation). For continuous outputs, reference shapes were used to ensure efficient test evaluation. In the next step, the test cases were evaluated. A test case fails if at least one of its test steps has detected a failure. A test run is passed if all test cases have not revealed any failure. Finally, the reliability of the test object is estimated based on the test results. During test execution, the number of executions with and without failure is counted for each transition. Based on the usage profile and the failure statistics, the reliability of the test object is estimated after each test run. Reliability in MBST is the probability of a no-failure operation (Miller, Morell, Noonan, Park, Nicol, Murril, & Voas, 1992). In the case of a criticality profile, safety compliance is estimated instead of reliability. The system is released as soon as the reliability reaches a defined value and stays within a certain confidence interval.

EVALUATION OF THE APPROACH

For the evaluation of our test approach in the industrial case study, a measurement program based on the GQM (goal-question-metric, Basili, Caldiera, & Rombach 1994) approach was prepared. The GQM approach allows determining case study- and company-specific goals and refining them into measurable metrics. Metrics are used to improve the software development and the testing process (and its resulting software products) while maintaining alignment with the organization's business and technical goals. In the first step, business and improvement goals are analyzed and metrics are defined according to the development and testing process. The result is a GQM plan that collects all goals, questions, and metrics together with details about who, when, and how data are to be collected. This information is then used for designing and defining the data collection procedures, i.e., instrumentation. Afterwards, measurement is performed and raw data are collected, validated, and analyzed according to the GQM plan. Initial feedback is provided to the interested parties. Then, the interested parties draw conclusions and consequences in a post-mortem analysis in accordance with their analysis and interpretations.

252

Finally, the analysis, interpretations, conclusions, and consequences are summarized and/or reported and collected as experience.

Two evaluation rounds were planned and conducted for measuring and assessing both approaches, the established test approach and the new model-based test approach. The main goal of the case study evaluation was the comparison of the model-based testing and non-model-based testing approaches in an industrial environment with respect to the categories development quality, test case quality, and test effort. For effort-related measurement, different metrics have been defined so that the different cost drivers for the two approaches can be taken into account. Measurement plans with responsibilities, schedules, and data collection methods for the categories development quality, test case quality, and test effort were prepared. Based on these plans the required data was collected in both evaluation rounds. Finally, the data was analyzed and interpreted together with the ABB research department and the soft-starter development team. The model-based test approach showed improvements in the categories shown in Table 4.

In the category "Development Quality", the hours to locate and fix a defect, including the effort for detecting failures, locating their causes (faults), and correcting them could be reduced by 20% with the model-based approach and the testing environment.

In the category "Test Case Quality", for the first metric "Test Case Reuse" the reuse of test cases was already present in the non-model-based testing approach. In addition to the existing test cases, formal models were used in the model-based testing approach as a basis for automated test case generation. With the existing test case reuse and the reuse of generated test cases, the overall test case reuse could be improved by 10%. For the metric "Model Reuse", modeling changed to be the central concept for the test phase, which resulted in an 80% improvement for the level of model reuse. In the non-model-based approach, models were used to refine test cases but not to derive test cases in an automated way.

In the category "Test Effort", for the metric "Test Case Definition" the effort needed for model construction and automated test case generation with the new approach remained constant due to the high effort for learning and technology transfer, compared to the effort for manual test case definition in the non-model-based approach. Thus, the improvement for test case definition was 0%. In further iterations, high effort savings for model construction are estimated by domain experts to be approximately 40% after the third iteration. For the metric "Manual Test Case Execution", the effort for test case execution and evaluation could be reduced by 50% for the model-based approach as a result of the highly automated execution and evaluation environment. For the metric "Manual Test Case Analysis", the effort for analyzing test

Table 4. Case study improvements

Quality Focus	Metric	Improvement
Development Quality	Hours to locate and fix a defect (old/new) [estimates of #defects and #hours to locate and fix all defects]	20%
Test Case Quality	Test Case Reuse [% Test Cases reused]	10%
	Model Reuse [% States Reused / % Edges Reused / % Annotations Reused]	80%
Test Effort	Test Case Definition [Person Hours]	0%
	Manual Test Case Execution [Person Hours]	50%
	Manual Test Case Analysis [Person Hours]	50%

case results and tracing them back to requirements and design elements could be reduced by 50% with the model-based approach and the test environment.

All the results from the measurements have the same importance for the ABB softstarter development team and the corporate research department. Thus, the calculated average improvement achieved in the softstarter case study was 35%. It is important to note that this improvement will be invested into the higher complexity of future devices. The development time can be kept stable, with more features and customer requirements being addressed within the same time frame.

The improvements for the cycle computer project are based on tracked student and PhD projects which were using the approach based on the cycle computer product line and the tool support described in the section Automated Generation of Test Cases. On average the parameter reduction due to their inclusion into the feature model and the tool support reduced the implementation effort by ½ day per test cycle (preparation and execution of all test cases) for their projects. Given the implementation durations between 6 to 14 weeks and between 4 and 18 test cycles the average effort reduction by 10% (10 weeks (50days) were reduced by 11 test cycles * ½ day = 5½ days). We are aware of the fact that a student environment is very special. Thus, we don't draw any lines between the results of the student / PhD projects so far and the above described industrial project. But it is nonetheless important to notice the further improvements by using feature models as supplement to the parameter modeling phase described in the corresponding section.

RELATED WORK

The automated generation of test cases is often achieved by using mathematical models. Model-based testing techniques have been developed that differ in terms of modeling notations and test case generation approaches (Utting, Pretschner & Legeard, 2012). Model-based testing refers to software and system testing where test cases are derived as a whole or in part from a model that describes selected, often structural, functional, or non-functional aspects of the test object. Examples of modeling notations are transition-based notations (e.g., finite state machines (Gill, 1962)), Pre/Post-models (e.g., OCL (Object Management Group, 2000), Z (Davies, & Woodcock, 1996)), history-based notations (e.g., sequence diagrams (Harel, & Thiagarajan, 2003)), operational notations (e.g., CSP (Hoare, 1985), Petri Nets (Peterson, 1981)), and statistical notations (e.g., Markov Chains (Prowell, 2005)).

An important task of the softstarter is to reduce the load and torque in the power train of the motor in the starting and stopping phase by controlling the voltage and the current of the electrical device. The control logic can be described well by a state-based model. The automated testing of non-functional quality properties such as reliability and safety is only supported by few model-based test approaches. A mature and systematic approach is model-based statistical testing (MBST, (Prowell, 2005), (Kloos, & Eschbach, 2009), (Prowell, Trammell, Linger, & Poore, 1999), (Zimmermann, Eschbach, Kloos, & Bauer, 2009)), which uses a state-based statistical test model to reflect importance, frequency of use, risk, and criticality. Test cases are automatically generated focusing on selected non-functional quality properties. Reliability testing is provided by considering the operational profile of the test object. The operational profile is a quantitative characterization of how a system will be used (Musa 1993). Testing according to the actual usage of the system allows predicting the future system's reliability in the field. Musa defined a stepwise approach for the determination of appropriate operational profiles from customer and user profiles. Approaches for incorporating existing risk analysis models are described in (Zimmermann, Kloos, Eschbach, & Bauer, 2009), (Bauer, Stallbaum, Metzger, & Eschbach 2008).

The systematic and efficient testing of configurable systems is a current research topic. Most of the work is related to product line testing. The main aspect in the work of (Cohen, Dwyer, & Shi, 2006) is the reusability of test cases for different configurations. In (McGregor, 2001), the focus is on specific variations points between products that are the basis for the composition of test cases. Abstract test scenarios are modeled as use cases representing the product requirements model. The scenarios are parameterized with characteristic values to instantiate the different product variants. A particular model-based testing approach is introduced in (Kamsties, Reuys, Pohl, & Reis, 2004) and (Olimpiew, & Gomaa, 2005). It describes the representation of use cases with UML extensions by considering the variability of software product lines. This method supports derivation of test cases from models. Another test approach (Reuys, Kamsties, Pohl, & Reis, 2005) provides the derivation of application-specific test scenarios describing the variation points of the product from use cases and activity charts on the domain level.

Other approaches use combinatorial techniques, especially in the field of experiment design (Cohen et al., 2006). The objective is to find an optimal set of configurations that satisfies maximal coverage criteria for each test case. The dependency between input parameters of system functions has been investigated in (Kuhn, Wallace, & Gallo, 2004). Their empirical study showed a dependency of two to six input parameters for different system functions.

As presented in the section "Example Systems", the behavior of the softstarter is tightly bound to its parameters. The behavioral part of the softstarter testing challenge can be covered by state-based statistical test models, whereas the parameters caused a state explosion. Thus, a second approach to handle the huge amount of parameter permutations was needed and found, by using combinatorial techniques.

For an extension of the application domain of this approach the feature models of the product line domain (Schmid & Santana de Almeida, 2013) have been choosen. A further reduction of the testing space could be achieved by moving some of the parameters into the feature model and using the feature model capabilities to validate the model before the derivation of a new application.

SUMMARY AND OUTLOOK

In this article, testing of highly configurable embedded systems has been discussed by means of two case studies with embedded systems. The basic behavior of such systems is rather simple but dependent on a large number of parameters. The permutation range resulting from the parameters is dramatically higher than the possible permutations of the few use cases of the system. Thus, a new testing process was developed supporting applications generated using the product line concept and two major models, a parameter model in combination with a test model, which is the basis for the automated generation of test cases. The testing process is a new combination and realization of state of the art methods. It was applied and evaluated in an industrial and a University case study. The resulting improvement of 35% shows the quality of the approach. While developing the new testing process, some issues arose and lead to the lessons learned. First, the testing process cannot replace expert knowledge nor will it replace existing test cases. Instead, generated test cases complement existing test cases to address coverage requirements efficiently. In addition, the testing process offers a structured way to handle the huge number of possible test cases. Second, the process requires initial efforts for its introduction. A further reduction of the development time seems feasible in case the testing process is used for more than three projects or project iterations. This requires a validation within a time frame beyond five years. Third, the parameter reduction approach of the testing process is also usable for existing and manually developed test cases, although the potential is reduced to address coverage requirements efficiently.

The cycle computer case study has shown a great potential by using feature models out of the product line domain to further reduce the number of parameters to be addressed by the model-based testing approach. The effort reduction of 10% for student projects emphasizes the usefulness of the integration of the product line approach into model based testing.

We face an increased presence of embedded product lines what requires further research and development work of this testing approach towards the software architecture modeling domain for product lines. The test model should be used on the product line level for analyses towards the expected testing effort, coverage for the derived applications and amongst these applications.

For analyzing defects that were not discovered during testing with the approach presented here, a method is discussed in (Ruschival, Nenninger, Kantz, & Streitferdt 2009). The contribution describes a systematic approach to modify test cases based on user stories, which led to identify root-causes of device failures. This approach needs to be further elaborated and integrated into a testing tool chain. Clearly, an important topic is continuous research on further automation potentials of the testing tool chain. Usability, efficiency, modularity, and ease of integration are the top issues for future tool chains. Finally, the integration of additional behavioral models currently used for simulation purposes is a promising approach for enhancing the quality of the test model (Zander, & Schieferdecker, 2010). The integration of such behavioral models will be the subject of future research. Additionally, the similarity of the concrete test models is an interesting fact that should be analyzed. The goal is to exploit the similarity of the model structure and use this information to reduce the set of test cases that have to be executed for a specific configuration.

REFERENCES

Barry, R. (2010). *Using the FreeRTOS™ Real Time Kernel*. Real Time Engineers Ltd. Retrieved from www.freertos.org

Barry, R. (2011). *The FreeRTOS™ Reference Manual*. Version 1.2.0. Real Time Engineers Ltd. Retrieved from www.freertos.org

Basili, V. R., Caldiera, G., & Rombach, H. D. (1994). The Goal Question Metric Approach. In *Encyclopedia of Software Engineering*. Wiley.

Bauer, T., Böhr, F., Landmann, D., Beletski, T., Eschbach, R., & Poore, J. H. (2007). From Requirements to Statistical Testing of Embedded Systems. In *Proceedings of Software Engineering for Automotive Systems - SEAS 2007, ICSE Workshops*. Minneapolis, MN: SEAS.

Bauer, T., Eschbach, R., Groessl, M., Hussain, T., Streitferdt, D., & Kantz, F. (2009). Combining Combinatorial and Model-Based Test Approaches for Highly Configurable Safety-Critical Systems. In *Proceedings of the 2nd Workshop on Model-based Testing in Practice at the 5th European Conference on Model-Driven Architecture Foundations and Applications*. Academic Press.

Bauer, T., Stallbaum, H., Metzger, A., & Eschbach, R. (2008). Risikobasierte Ableitung und Priorisierung von Testfällen für den modellbasierten Systemtest (in German). In *Proceedings of SE'08 – Software Engineering Konferenz*. München, Germany: SE.

Cohen, M. B., Dwyer, M. B., & Shi, J. (2006). Coverage and adequacy in software product line testing. In *Proceedings of the ISSTA 2006 workshop on Role of software architecture for testing and analysis*. ISSTA.

Davies, J., & Woodcock, J. (1996). *Using Z: Specification, Refinement and Proof*. Prentice Hall International Series in Computer Science.

Farrag, M., Fengler, W., Streitferdt, D., & Fengler, O. (2010). Test Case Generation for Product Lines based on Colored State Charts. In *Proceedings of 3rd Workshop on Model-based Testing in Practice* (MoTiP 2010) (pp. 31–40). MoTiP.

Gamma, E., Helm, R., Johnson, R., & Vlissides, J. M. (1994). *Design Patterns: Elements of Reusable Object-Oriented Softwaresystemen.* Addison-Wesley Professional.

Gill, A. (1962). *Introduction to the Theory of Finite-state Machines.* New York: McGraw-Hill.

Harel, D., & Thiagarajan, P. S. (2003). *Message Sequence Charts.* Retrieved March 17, 2010 from http://www.comp.nus.edu.sg/~thiagu/public_papers/surveymsc.pdf

Hoare, C. A. R. (1985). *Communicating Sequential Processes.* Prentice Hall.

Kamsties, E., Reuys, A., Pohl, K., & Reis, S. (2004). Testing Variabilities in Use Case Models. In *Software Product-Family Engineering* (pp. 6–18). Berlin: Springer.

Kang, K. C., Lee, J., & Donohoe, P. (2002). Feature-oriented product line engineering. *Software, IEEE, 19*(4), 58–65. doi:10.1109/MS.2002.1020288

Kantz, F., Ruschival, T., Nenninger, P., & Streitferdt, D. (2009). Testing with Large Parameter Sets for the Development of Embedded Systems in the Automation Domain. In *Proceedings of the 2nd International Workshop on Component-Based Design of Resource-Constrained Systems at the 33rd Annual IEEE International Computers, Software and Applications Conference.* IEEE.

Kloos, J., & Eschbach, R. (2009). Generating System Models for a Highly Configurable Train Control System Using A Domain-Specific Language: A Case Study. In *Proceedings of AMOST'09 - 5th Workshop on Advances in Model Based Testing.* Denver, CO: AMOST.

Kuhn, D., Wallace, D., & Gallo, A. M. J. (2004). Software fault interactions and implications for software testing. *IEEE Transactions on Software Engineering, 30*(6), 418–421. doi:10.1109/TSE.2004.24

McGregor, J. D. (2001). *Testing a software product line* (Tech. Rep. CMU/SEI-2001-TR-022). Pittsburgh, PA: Carnegie Mellon University.

Miller, K., Morell, L., Noonan, R., Park, S., Nicol, D., Murril, B., & Voas, J. (1992). Estimating the probability of failure when testing reveals no failures. *IEEE Transactions on Software Engineering, 18*(1), 33–43. doi:10.1109/32.120314

Musa, J. D. (1993). Operational Profiles in Software-Reliability Engineering. *IEEE Software, 10*(2), 14–32. doi:10.1109/52.199724

Object Management Group (OMG). (2000). *Object Constraint Language Specification.* OMG Unified Modeling Language Specification, Version 1.3. Retrieved March 17, 2010, from http://www.omg.org/spec/UML/1.3

Olimpiew, E. M., & Gomaa, H. (2005). Model-based testing for applications derived from software product lines. In *Proceedings of the 1st international workshop on Advances in model-based testing.* New York, NY: AMOST.

Peterson, J. L. (1981). *Petri Net Theory and the Modeling of Systems.* Prentice Hall.

Pohl, K., Böckle, G., & van der Linden, F. (2005). *Software Product Line Engineering: Foundations, Principles, and Techniques.* Springer-Verlag.

Prowell, S. (2005). Using markov chain usage models to test complex systems. In *Proceedings of the 38th Annual Hawaii International Conference on System Sciences.* Academic Press.

Prowell, S., Trammell, C., Linger, R., & Poore, J. (1999). *Cleanroom Software Engineering: Technology and Process.* Addison-Wesley-Longman.

Prowell, S. J. (2000). TML a description language for Markov chain usage models. *Information and Software Technology, 42*(12), 835–844. doi:10.1016/S0950-5849(00)00123-3

Prowell, S. J., & Poore, J. H. (2003). Foundations of Sequence-Based Software Specification. *IEEE Transactions on Software Engineering, 29*(5), 417–429. doi:10.1109/TSE.2003.1199071

Reinhold, M. (2003). *Praxistauglichkeit von Vorgehensmodellen: Specification of large IT-Systems – Integration of Requirements Engineering and UML based on V-Model'97.* Nordrhein-Westfalen: Shaker Verlag.

Reuys, A., Kamsties, E., Pohl, K., & Reis, S. (2005). *Model-Based System Testing of Software Product Families. Advanced Information Systems Engineering.* Berlin: Springer.

Ruschival, T., Nenninger, P., Kantz, F., & Streitferdt, D. (2009). Test Case Mutation in Hybrid State Space for Reduction of No-Fault-Found Test Results in the Industrial Automation Domain. In *Proceedings of the 2nd International Workshop on Industrial Experience in Embedded Systems Design (IEESD 2009) at the 33rd Annual IEEE International Computers, Software and Applications Conference.* IEEE.

Schmid, K., & Santana de Almeida, E. (2013). Product Line Engineering. *Software, IEEE, 30*(4), 24–30. doi:10.1109/MS.2013.83

Utting, M., Pretschner, A., & Legeard, B. (2012). A taxonomy of model-based testing approaches. *Software Testing. Verification & Reliability, 22*(5), 297–312. doi:10.1002/stvr.456

Van der Linden, F. J., Schmid, K., & Rommes, E. (2007). *Software Product Lines in Action: The Best Industrial Practice in Product Line Engineering.* Berlin: Springer.

Zander, J., & Schieferdecker, I. (2010). Model-based Testing of Embedded Systems Exemplified for the Automotive Domain. In Behavioral Modeling for Embedded Systems and Technologies: Applications for Design and Implementation (pp. 377-412). Idea Group Inc (IGI).

Zimmermann, F., Eschbach, R., Kloos, J., & Bauer, T. (2009). Risiko-basiertes statistisches Testen (in German). In TAV group Meeting of the GI (Gesellschaft für Informatik). Dortmund, Germany: GI.

Zimmermann, F., Kloos, J., Eschbach, R., & Bauer, T. (2009). Risk-based Statistical Testing: A refinement-based approach to the reliability analysis of safety-critical systems. In *Proceedings of EWDC'09, European Workshop on Dependable Systems.* Toulouse, France: EWDC.

ADDITIONAL READING

Ali, A., Nadeem, A., Iqbal, M. Z. Z., & Usman, M. (2007). Regression Testing Based on UML Design Models. In PRDC 2007. 13th Pacific Rim International Symposium on Dependable Computing,2007. (pp. 85-88).

Biswas, S., Mall, R., Satpathy, M., & Sukumaran, S. (2009). A model-based regression test selection approach for embedded applications. *SIGSOFT Software Engineering Notes, 34*(4), 1–9. doi:10.1145/1543405.1543413

Bradbury, J. S., Cordy, J. R., & Dingel, J. (2005). *An empirical framework for comparing effectiveness of testing and property-based formal analysis* (pp. 2–5).

Chen, L., Ravi, S., Raghunathan, A., & Dey, S. (2003). *A scalable software-based self-test methodology for programmable processors* (pp. 548–553). New York, NY, USA: ACM.

Chen, Y., & Probert, R. (2003). A Risk-based Regression Test Selection Strategy. In ISSRE 2003: Proceeding of the 14th IEEE International Symposium on Software Reliability Engineering (pp. 305-306). Denver, Colorado, USA.

Cibulski, H., & Yehudai, A. (2011). Regression Test Selection Techniques for Test-Driven Development. In Software Testing, Verification and Validation Workshops (ICSTW), 2011 IEEE Fourth International Conference on (pp. 115–124).

Delamare, R., Baudry, B., & Le Traon, Y. (2008). Regression test selection when evolving software with aspects. In Proceedings of the 2008 AOSD workshop on Linking aspect technology and evolution (pp. 71-75). New York, NY, USA: ACM.

Deng, D., Sheu, P. C.-Y., & Wang, T. (2004). Model-based testing and maintenance. In Multimedia Software Engineering, 2004. Proceedings. IEEE Sixth International Symposium on (pp. 278-285).

Farooq, Q. -u. -a., Iqbal, M., Malik, Z. I., & Riebisch, M. (2010). A Model-Based Regression Testing Approach for Evolving Software Systems with Flexible Tool Support. In Engineering of Computer Based Systems (ECBS), 2010 17th IEEE International Conference and Workshops on (pp. 41–49).

Farooq, Q. ul-ann. (2010). A Model Driven Approach to Test Evolving Business Process based Systems. In In proceedings of Doctoral Symposium at Models 2010.

Gonzalez-Sanchez, A., Piel, E., Abreu, R., Gross, H.-G., & van Gemund, A. J. C. (2011). Prioritizing tests for software fault diagnosis. *Software, Practice & Experience*, *41*(10), 1105–1129.

Harrold, M. J., Rosenblum, D., Rothermel, G., & Weyuker, E. (2001). Empirical studies of a prediction model for regression test selection. Software Engineering. *IEEE Transactions on*, *27*(3), 248–263.

Holler, J., Striz, A., Bretney, K., Kavett, K., & Bingham, B. (2007). Design, Construction, and Field Testing of an Autonomous Surface Craft for Engineering and Science Education (pp. 1–6).

Khan, T., & Heckel, R. (2011). On Model-Based Regression Testing of Web-Services Using Dependency Analysis of Visual Contracts. In D. Giannakopoulou & F. Orejas (Eds.), Fundamental Approaches to Software Engineering (Vol. 6603, pp. 341–355). Springer Berlin / Heidelberg.

Korel, B., Tahat, L. H., & Vaysburg, B. (2002). Model based regression test reduction using dependence analysis. In Software Maintenance, 2002. Proceedings. International Conference on (pp. 214–223).

Kuo, F.-C., Chen, T. Y., & Tam, W. K. (2011). Testing Embedded Software by Metamorphic Testing: a Wireless Metering System Case Study. 2011 IEEE 36th Conference on Local Computer Networks, 291–4.

Larsen, K. G., Mikucionis, M., Nielsen, B., & Skou, A. (2005). *Testing real-time embedded software using UPPAAL-TRON: an industrial case study* (pp. 299–306). New York, NY, USA: ACM.

Mansour, N., & Takkoush, H. (2007). UML based regression testing for OO software. In Proceedings of the 11th IASTED International Conference on Software Engineering and Applications (pp. 96-101). Anaheim, CA, USA: ACTA Press.

Mao, C., & Lu, Y. (2005). Regression Testing for Component-based Software Systems by Enhancing Change Information. In Proceedings of the 12th Asia-Pacific Software Engineering Conference (pp. 611–618). Washington, DC, USA: IEEE Computer Society.

Mehta, A., & Heineman, G. T. (2001). Evolving legacy systems features using regression test cases and components. In Proceedings of the 4th International Workshop on Principles of Software Evolution (pp. 190-193). New York, NY, USA: ACM.

Mei, L., Zhang, Z., Chan, W. K., & Tse, T. H. (2009). Test case prioritization for regression testing of service-oriented business applications. In Proceedings of the 18th international conference on World wide web (pp. 901-910). New York, NY, USA: ACM.

Modeling the Cost-Benefits Tradeoffs for Regression Testing Techniques. (2002). In Proceedings of the International Conference on Software Maintenance (ICSM'02) (pp. 204…). Washington, DC, USA: IEEE Computer Society.

Naslavsky, L., Ziv, H., & Richardson, D. J. (2009). A model-based regression test selection technique. In. ICSM 2009. IEEE International Conference on Software Maintenance, 2009 (pp. 515-518).

Nie, C., & Leung, H. (2011). A survey of combinatorial testing. ACM Comput. Surv., 43(2), 11:1–11:29.

Orso, A., Apiwattanapong, T., & Harrold, M. J. (2003). Leveraging Field Data for Impact Analysis and Regression Testing. In Proceedings of the 9th European software engineering conference held jointly with 11th ACM SIGSOFT international symposium on Foundations of software engineering (ESEC/FSE'03) (pp. 128–137). Helsinki, Finland.

Pilskalns, O., Uyan, G., & Andrews, A. (2006). Regression Testing UML Designs. In Proceedings of the 22nd IEEE International Conference on Software Maintenance (pp. 254-264). IEEE Computer Society.

Poll, S., Patterson-hine, A., Camisa, J., Garcia, D., Hall, D., & Lee, C. … Koutsoukos, X. (2007). Advanced diagnostics and prognostics testbed (pp. 178–185).

Pretschner, A., Loetzbeyer, H., & Philipps, J. (2001). Model Based Testing in Evolutionary Software Development. In Proceedings of the 12th International Workshop on Rapid System Prototyping (pp. 155). IEEE Computer Society.

Rosenblum, D., & Rothermel, G. (1997). A Comparative Study of Regression Test Selection Techniques. In In Proc. of the 2nd Int'l. Workshop on Empir. Studies of Softw. Maint (pp. 89–94).

Rothermel, G., & Harrold, M. J. (1994). Selecting Regression Tests for Object-Oriented Software. In Proceedings of the International Conference on Software Maintenance (pp. 14–25). Washington, DC, USA: IEEE Computer Society.

Shi, J., Zhang, T., & Wang, F. (May). A data preprocessing method for testability modeling based on first-order dependency integrated model (pp. 1–8). Presented at the 2012 IEEE Conference on Prognostics and System Health Management (PHM).

Spillner, A. (2000). From V-model to W-model - establishing the whole test process. In CONQUEST 2000. Proceedings 4th Conference on Quality Engineering in Software Technology and VDE-ITG Workshop on `Testing Non-Functional Software-Requirements' (pp. 222–31). Erlangen, Germany: Arbeitskreis Software Qualitat Franken.

Walid, S. Abd El-hamid, S. S. E., & Hadhoud, M. M. (2010). A General Regression Test Selection Technique. In E. World Academy of Science & Technology (Eds.)

Wendland, M.-F., Schieferdecker, I., & Vouffo-Feudjio, A. (2011). *Requirements-Driven Testing with Behavior Trees* (pp. 501–510).

White, L. J. (1996). Regression Testing of GUI Event Interactions. In Proceedings of the 1996 International Conference on Software Maintenance (pp. 350–358). Washington, DC, USA: IEEE Computer Society.

Willmor, D. (2006). *An Intensional Approach to the Specification of Test Cases for Database Applications* (pp. 102–111).

Wolff, M., Albrecht, M., & Gutsche, F. (2006). Test Driven Development. *JavaSpektrum, 06*, 36–39.

KEY TERMS AND DEFINITIONS

Automation Domain: Domain of designing, developing and integration components into system of system to fulfill automated tasks (e. g. produce goods).

Configuration Parameter: The parameters to configure a system towards the behavior and/ or look and feel.

Cycle Computer: Device to acquire and process sensor signals of a bicycle.

Feature Model: Model of a product line based on hierarchically organized features of the system in question. The feature model also includes constraints between features.

Model-Based Testing: Testing of system or software systems based on a test model. The test cases are generated in an automated manner based on the test model.

Product Line: Design and development of a system based on a framework with a core and variable components.

Softstarter: Hard- and software device to smoothly ramp up and down an electric motor.

Statistical Testing: Model-based statistical testing (MBST) is a *black-box* testing technique that enables the generation of representative or failure-sensitive test cases from the tester's or user's perspective.

Chapter 11
On Parallel Online Learning for Adaptive Embedded Systems

Tapio Pahikkala
University of Turku, Finland

Antti Airola
University of Turku, Finland

Thomas Canhao Xu
University of Turku, Finland

Pasi Liljeberg
University of Turku, Finland

Hannu Tenhunen
University of Turku, Finland

Tapio Salakoski
University of Turku, Finland

ABSTRACT

This chapter considers parallel implementation of the online multi-label regularized least-squares machine-learning algorithm for embedded hardware platforms. The authors focus on the following properties required in real-time adaptive systems: learning in online fashion, that is, the model improves with new data but does not require storing it; the method can fully utilize the computational abilities of modern embedded multi-core computer architectures; and the system efficiently learns to predict several labels simultaneously. They demonstrate on a hand-written digit recognition task that the online algorithm converges faster, with respect to the amount of training data processed, to an accurate solution than a stochastic gradient descent based baseline. Further, the authors show that our parallelization of the method scales well on a quad-core platform. Moreover, since Network-on-Chip (NoC) has been proposed as a promising candidate for future multi-core architectures, they implement a NoC system consisting of 16 cores. The proposed machine learning algorithm is evaluated in the NoC platform. Experimental results show that, by optimizing the cache behaviour of the program, cache/memory efficiency can improve significantly. Results from the chapter provide a guideline for designing future embedded multi-core machine learning devices.

INTRODUCTION

The design of adaptive systems is an emerging topic in the area of pervasive and embedded computing. Rather than exhibiting pre-programmed behaviour, it would in many applications be beneficial for systems to be able to adapt to their environment. Isoaho et al. (2010) analyze current key challenges in developing embedded systems. One of the outlined main challenges is

DOI: 10.4018/978-1-4666-6034-2.ch011

self-awareness, meaning that a system should be able to monitor its environment and own state, and based on this optimize its behaviour in order to meet service quality criteria. System security is outlined as another major challenge, as is making best use of parallelism that is increasingly present in modern embedded systems.

In order to meet these goals, a system should automatically learn to model its environment in order to choose correct actions, and over time improve its performance as more feedback is gained. Automatically constructed mathematical model of a system may be used to predict the future behaviour, given as input the current state and planned actions, an approach known as model predictive control. The behaviour patterns of software or human agents may be automatically observed in order to recognize possible security risks. And even more, imagine smart music players that adapt to the musical preferences of their owner, intelligent traffic systems that monitor and predict traffic conditions and re-direct cars accordingly, etc. Thus we motivate the need for a generic approach to learning predictive models in embedded environments, which are typically characterized by properties such as need for fast (and constant) response times, limited amount of memory resources and parallel computing architecture.

Machine Learning in Embedded Systems

Machine learning (ML) is a branch of computer science founded on the idea of designing computer algorithms capable of improving their prediction performance automatically over time through experience (Mitchell, 1997). Such approaches offer the possibility to gain new knowledge through automated discovery of patterns and relations in data. Further, these methods can provide the benefit of freeing humans from doing laborious and repetitive tasks, when a computer can be trained to perform them. This is especially important in problem areas where there are massive amounts of complex data available, such as in image recognition or natural language processing.

In the recent years ML methods have increasingly been applied in non-traditional computing platforms, bringing both new challenges and opportunities. The shift from the single processor paradigm to parallel computing systems such as multi-core processors, cloud computing environments, graphic processing units (GPUs) and the network on chip (NoC) has resulted in a need for parallelizable learning methods (Chu et al., 2007, Zinkevich et al., 2009, Low et al., 2010).

At the same time, the widespread use of embedded systems ranging from industrial process control systems to wearable sensors and smartphones have opened up new application areas for intelligent systems. Some such recent ML applications include embedded real-time vision systems for field programmable gate arrays (Farabet et al., 2009), personalized health applications for mobile phones (Oresko et al., 2010), sensor based videogame controls that learn to recognize user movements (Shotton et al., 2011), and classifying non-linear electrocardiogram signals (Sun and Cheng, 2012). For a thorough review of the design requirements of machine learning methods in embedded systems we refer to Swere (2008).

Majority of present-day machine learning research focuses on so-called batch learning methods. Such methods, given a data set for training, run a learning process on the data set and then output a predictor which remains fixed after the initial training has been finished. In contrast, it would be beneficial in real-time embedded systems for learning to be an ongoing process in which the predictors would be upgraded whenever new data become available. In machine learning literature, these kind of methods are often referred to as online learning algorithms (Bottou and Le Cun, 2004).

One of the principal areas of application of this type of adaptive learning systems are hand-held devices such as smart-phones that learn to adapt to their users preferences.

In this work we consider how to implement in parallel online machine learning methods in embedded computing environments. As a case study, we especially consider (multi-class) classification tasks, in which the system must assign a class label to a new object given the feature representation of the object. For example, in spam classification the features could be the words in an e-mail message, and possible classes consist of spam and not-spam, whereas in optical character recognition features could represent image scans and the set of available classes would encode different characters in the alphabet.

Our method is built upon the regularized least-squares (RLS) (Rifkin et al., 2003, Poggio and Smale, 2003), also known as the least-squares support vector machine (Suykens et al., 2002) and ridge regression (Hoerl and Kennard, 1970), a state-of-the art machine learning method suitable both for regression and classification. Compared to the ordinary least-squares method introduced by Gauss, RLS is known to often achieve better predictive performance, as the regularization allows one to avoid over-fitting to the training data. An important property of the algorithm is that it has a closed form solution, which can be fully expressed in terms of matrix operations. This allows developing efficient computational shortcuts for the method, since small changes in the training data matrix correspond to low-rank changes in the learned predictor.

An online version of the ordinary (non-regularized) least-squares method was presented more than half a century ago by Plackett (1950). The method has since then been widely used in real-time applications in areas such as machine learning, signal processing, communications and control systems. Online learning with RLS is also known in the machine learning literature (see e.g. Zhdanov and Kalnishkan (2010) and references

therein). In this work we extend online RLS for multi-output learning and present an implementation that takes advantage of parallel computing architectures and cache access optimization techniques. Namely, we present an online update algorithm that is considerably faster than an algorithm employing only the standard parallel matrix computation techniques due to its highly efficient cache accesses. Thus, the method allows adaptive learning efficiently in parallel environments, and is applicable to a wide range of problems both for regression and classification, and for single and multi-task learning.

Future Multi-Core Systems

In recent years, multi-core processors have been emerging with increasing number of cores. The reason is due to the constraints of chip clock frequency and power consumption - chip designers have to integrate more processor cores rather than to improve the performance of a single-core. Currently, even mobile phones and tablets are equipped with quad-core or octa-core processors. It is expected that digital devices with hundreds or even more cores on a chip may appear on the market in the future. However, as the number of components integrated on a chip die keeps increasing, traditional system design methods may encounter bottlenecks. In contrast with the increasing chip complexity, traditional system interconnects scale poorly in terms of performance and power consumption. Network-on-Chip (NoC) has become an emerging and promising solution in the field of Chip Multiprocessor (CMP) (Dally and Towles, 2001).

To address the scalability problems in conventional interconnects, NoC is proposed and endeavors to bring network communication methodologies to the on-chip communication. The fundamental design approach of a NoC-based chip is to create a communication infrastructure (routers and links) beforehand and then map the computational resources (processor cores) to it via resource dependent interfaces. Figure 1 shows a typical 2D-mesh (4x4) based NoC, in which Processing Elements (PE) are connected by Routers (R) and network links via Network Interfaces (NI), and data are transmitted in the form of network packets (or flits, in case of narrow links). A typical wormhole router is consisted of a Routing Computation Unit (RCU), a Virtual Channel Allocator (VCA), a Switch Allocator (SA), a Crossbar Switch (CS), several Virtual Channels (VC) and input buffers. Each basic architectural unit of a NoC is a tile or Node (N). By using this modular approach, more efficient communication can be achieved by leveraging computer network principles. Multiple transactions can occur simultaneously, and therefore the throughput of the system is increased. Moreover since nodes are connected by point-to-point links, system performance can be improved by pipelining communication among different nodes. NoC has been manufactured either for research or for commercial use. For example, Intel[1] has manufactured an experimental NoC (Single-chip Cloud Computer, or SCC) containing 48 cores using 4x6 mesh topology with 2 cores per node (Intel, 2010). The SCC processor is implemented using 45nm and consumes 125W power under nominal

conditions. Another 80 tile (8x10 mesh), 100M transistor, $275mm^2$ NoC was also demonstrated with 65nm processing technology (Vangal et al., 2007). Therefore NoC is chosen to be a platform for the study of the online machine learning algorithm.

Contributions of the Chapter

In this work we consider a parallel version of the online RLS method, and explore its suitability for implementing adaptive systems on the NoC platform. The proposed approach has the following key benefits:

- The system can automatically learn to perform a task given examples of the desired behavior, and can incorporate information from new training examples in an online fashion. This process may go on indefinitely, meaning lifelong learning.
- The system learns to predict several labels simultaneously which is beneficial, for example, in multi-class and multi-label classification as well as in more general forms of multi-task learning.
- Learning is parallelized and requires minimal storage and computational resources.

Figure 1. An example of 4x4 NoC using mesh topology

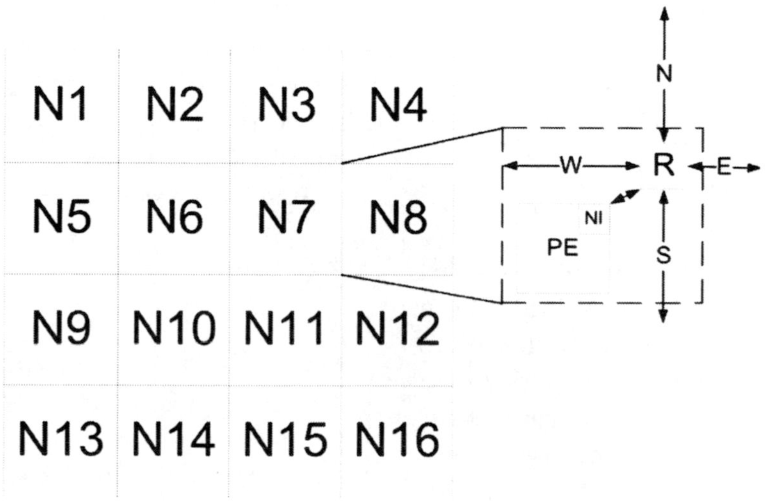

- The online update algorithm is designed and optimized in such a way that the memory consumption and cache efficiency are considerably better than those of a basic implementation constructed with standard parallel matrix computation techniques.
- The system is shown to work well on both on a quad-core platform and a NoC platform with 16 threads.

In our experiments, we evaluate the proposed system from the following perspectives:

- As a proof of concept demonstration, we show that the system can learn to solve a problem of hand-written digit recognition.
- The computational speed is measured with both a quad-core and a network-on-chip platform.
- We evaluate the memory and cache behavior of the proposed algorithm as well as their effect on speed and memory consumption and compare them with those of a basic implementation of the system constructed with standard parallel matrix computation techniques.

We expect that methods such as the online RLS have the capability to serve as the enabling technology for a wide range of applications in adaptive embedded systems.

ALGORITHM DESCRIPTIONS

Regularized Least-Squares

We start by introducing some notation. Let $\Re^m$ and $\Re^{m\times n}$ denote the sets of real valued column vectors and $m \times n$-matrices, respectively. To denote real valued matrices and vectors we use bold capital letters and bold lower case letters, respectively. Moreover, index sets are denoted with calligraphic capital letters. By denoting $\mathbf{M}_i$, $\mathbf{M}_{:,j}$, and $\mathbf{M}_{i,j}$, we refer to the ith row, jth column, and the value at position (i,j) of the matrix $\mathbf{M}$, respectively. Similarly, for index sets $\mathcal{R} \subseteq \{1,\ldots,n\}$ and $\mathcal{L} \subseteq \{1,\ldots,m\}$, we denote the sub-matrices of $\mathbf{M}$ having their rows indexed by $\mathcal{R}$, the columns by $\mathcal{L}$, and the rows by $\mathcal{R}$ and columns by $\mathcal{L}$ as $\mathbf{M}_{\mathcal{R}}$, $\mathbf{M}_{:,\mathcal{L}}$, and $\mathbf{M}_{\mathcal{R},\mathcal{L}}$, respectively. We use an analogous notation also for column vectors, that is, $\mathbf{v}_i$ refers to the ith entry of the vector $\mathbf{v}$.

Let $\mathbf{X} \in \Re^{m\times n}$ be a matrix containing the feature representation of the examples in the training set, where n is the number of features and m is the number of training examples. The entry of $\mathbf{X}$ at the position (i,j) contains the value of the jth feature in the ith training example. Moreover, let $\mathbf{Y} \in \Re^{m\times l}$ be a matrix containing the labels of the training examples. We assume each data point to have altogether l labels and the entry of $\mathbf{Y}$ at position (i,j) contains the value of the jth label of the ith training example. In multi-class or multi-label classification, the labels can be restricted to be either 1 or -1 depending whether the data points belongs to the class, for example, while they can be any real numbers in multi-label regression tasks (see e.g. Hsu et al. (2009) and references therein).

As an example, consider that we have a set of m images, each of which is represented by n features. In addition, each image is associated with an array of l binary labels of which the value of the jth label is 1 if the object indexed by j is depicted in the image and -1 otherwise. Our aim is to learn from the set of images to predict what is depicted in a any new image unseen in the set.

In this paper, we consider linear models of type:

$$f(\mathbf{x}) = \mathbf{W}^\mathrm{T}\mathbf{x}, \qquad (1)$$

where $\mathbf{W} \in \Re^{n\times l}$ is the matrix representation of the learned predictor and $\mathbf{x} \in \Re^n$ is a data point for which the prediction of l labels is to be made[2]. The computational complexity of making predictions with (1) and the space complexity of the predictor are both *O(nl)* provided that the feature vector representation x for the new data point is given.

Given training data **X,Y**, we find **W** by minimizing the RLS risk. This can be expressed as the following problem:

$$\arg \min_{\mathbf{W} \in \Re^{n \times l}} \left\{ \left\| \mathbf{XW} - \mathbf{Y} \right\|^2 + \lambda \left\| \mathbf{W} \right\|_F^2 \right\} \qquad (2)$$

where $\left\| \cdot \right\|_F$ denotes the Frobenius norm which is defined for a matrix $\mathbf{M} \in \Re^{n \times l}$ as:

$$\left\| \mathbf{M} \right\|_F = \sqrt{\sum_{i=1}^{n} \sum_{j=1}^{l} (\mathbf{M}_{i,j})^2}.$$

The first term in (2), called the empirical risk, measures how well the prediction function fits to the training data. The second term is called the regularizer and it controls the trade-off between the empirical error on the training set and the complexity of the prediction function.

Batch Learning for RLS

A straightforward approach to solve (2) is to set the derivative of the objective function with respect to **W** to zero. Then, by solving it with respect to **W**, we get:

$$\mathbf{W} = (\mathbf{X}^T \mathbf{X} + \lambda \mathbf{I})^{-1} \mathbf{X}^T \mathbf{Y}, \qquad (3)$$

where **I** is the identity matrix. We note (see e.g. Henderson and Searle (1981)) that an equivalent result can be obtained from

$$\mathbf{W} = \mathbf{X}^T (\mathbf{XX}^T + \lambda \mathbf{I})^{-1} \mathbf{Y}. \qquad (4)$$

If the number of features n is smaller than the number of training examples m, it is com-

putationally beneficial to use the form (3) while using (4) is faster in the opposite case. Namely, the computational complexity of (3) is:

$$O(n^3 + n^2 m + nml),$$

where the first term is the complexity of the matrix inversion, the second comes from multiplying $\mathbf{X}^T$ with **X**, and the third from multiplying the result of the inversion with the matrix **Y**. The complexity of (4) is:

$$O(m^3 + m^2 n + nml),$$

where the terms are analogous to those of (3). Putting these two together, the complexity of training a predictor is:

$$O(nm(\min\{n, m\} + l)).$$

It is also straightforward to see from (3) and (4) that the space complexity *O(m(n+l))* of RLS directly depends on the size of the matrices **X** and **Y**.

One of the benefits of RLS is that the number of labels per data point l can be increased up to the level of m or n until it starts to have an effect on the space and time complexities of RLS training. That is, we can solve several prediction tasks almost at the cost of solving only one. This is beneficial especially in multi-class and multi-label classification tasks, for example.

Online Learning for RLS

Next, we consider a computational short-cut for updating a learned RLS predictor when a new training example arrives. The short-cut is then used to construct an online version of the RLS algorithm. Similar considerations have already been presented in the machine learning literature (see e.g. Zhdanov and Kalnishkan (2010) and references therein) but here we formalize it for the first time for multiple outputs, that is, for the case in which the data points can have more than one label.

First, we present the following well-known result which is often referred to as the matrix inversion lemma or the Sherman-Morrison-Woodbury formula (see e.g. Horn and Johnson (1985, p. 18)). Let $\mathbf{M} \in \Re^{a \times a}$, $\mathbf{N} \in \Re^{b \times b}$, $\mathbf{P} \in \Re^{a \times b}$, and $\mathbf{Q} \in \Re^{b \times a}$ be matrices. If $\mathbf{M}$, $\mathbf{N}$, and $\mathbf{M} - \mathbf{PNQ}$ are invertible, then:

$$(\mathbf{M} - \mathbf{PNQ})^{-1} = \mathbf{M}^{-1} - \mathbf{M}^{-1}\mathbf{P}(\mathbf{N}^{-1} - \mathbf{QM}^{-1}\mathbf{P})^{-1}\mathbf{QM}^{-1}. \tag{5}$$

The main consequence of this result is that if we already know the inverse of the matrix $\mathbf{M}$ and if $b << a$, we can save a considerable amount of computational resources by using the right hand side of (5) instead of computing an inverse of an $a \times a$-matrix.

Assume that we have already trained an RLS predictor from the training set $\mathbf{X} \in \Re^{m \times n}$, $\mathbf{Y} \in \Re^{m \times l}$ with a regularization parameter value λ, and hence we have $\mathbf{W}$ stored in memory. In addition, let us assume that during the training process, we have computed the matrix:

$$\mathbf{C} = (\mathbf{X}^{\mathrm{T}}\mathbf{X} + \lambda\mathbf{I})^{-1}$$

and stored it in memory. According to (3), we have $\mathbf{W} = \mathbf{C}\mathbf{X}^{\mathrm{T}}\mathbf{Y}$. Moreover, let $\mathbf{x} \in \Re^n$, $\mathbf{y} \in \Re^l$ be a new data point unseen in the training set and let:

$$\hat{\mathbf{X}} \in \Re^{(m+1) \times n}, \hat{\mathbf{Y}} \in \Re^{(m+1) \times l}$$

be the new training set including the new training example. Now, since we already have the matrix $\mathbf{C}$ stored in memory, we can use the matrix inversion lemma to speed up the computation of the

matrix $\hat{\mathbf{C}}$ corresponding to the updated training set:

$$\begin{aligned}
\hat{\mathbf{C}} &= \left(\mathbf{X}^{\mathrm{T}}\mathbf{X} + \mathbf{xx}^{\mathrm{T}} + \lambda\mathbf{I}\right)^{-1} \\
&= \left(\mathbf{C}^{-1} + \mathbf{xx}^{\mathrm{T}}\right)^{-1} \\
&= \mathbf{C} - \mathbf{Cx}\left(\mathbf{x}^{\mathrm{T}}\mathbf{Cx} + 1\right)^{-1}\mathbf{x}^{\mathrm{T}}\mathbf{C}
\end{aligned} \tag{6}$$

where the calculation of last line requires only $O(n^2)$ time instead of the $O(n^3)$ time required in the fist line. The model corresponding to the updated training set can then be computed from

$$\hat{\mathbf{W}} \in \hat{\mathbf{C}}\left(\mathbf{X}^{\mathrm{T}}\mathbf{Y} + \mathbf{xy}^{\mathrm{T}}\right) \tag{7}$$

in $O(ln^2)$ time provided that $\mathbf{X}^{\mathrm{T}}\mathbf{Y}$ has already been computed during training with the original training set. If there are altogether m training examples, which are added into the training set one at a time, the overall computational complexity of this online variation would be $O(mln^2)$, which is slower than the batch RLS training if the number of labels per training examples l is large.

Next, we show how to improve the complexity even further. Let us first define some extra notation. Let:

$$\mathbf{v} = \mathbf{Cx} \tag{8}$$

$$c = \mathbf{x}^{\mathrm{T}}\mathbf{v}$$

$$d = (c + 1)^{-1}$$

$$\mathbf{p} = \mathbf{W}^{\mathrm{T}}\mathbf{x} \tag{9}$$

Continuing from (7), we get:

$$\hat{\mathbf{W}} \in \left(\mathbf{C} - \mathbf{Cx} \left(\mathbf{x}^T \mathbf{Cx} + 1 \right)^{-1} \mathbf{x}^T \mathbf{C} \right) \left(\mathbf{X}^T \mathbf{Y} + \mathbf{xy}^T \right)$$

$$= \left(\mathbf{C} - d\mathbf{vv}^T \right) \left(\mathbf{X}^T \mathbf{Y} + \mathbf{xy}^T \right)$$

$$= \mathbf{CX}^T \mathbf{Y} + \mathbf{Cxy}^T - d\mathbf{vv}^T \mathbf{X}^T \mathbf{Y} - d\mathbf{vv}^T \mathbf{xy}^T$$

$$= \mathbf{W} + \mathbf{vy}^T - d\mathbf{vp}^T - cd\mathbf{vy}^T$$

$$= \mathbf{W} + \mathbf{v} \left(\left(1 - cd \right) \mathbf{y}^T - d\mathbf{p}^T \right)$$

$$(10)$$

The computational complexity of calculating $\hat{\mathbf{W}}$ with (10) requires $O(n^2 + nl)$ time. Here, the first term is the complexity of calculating $\hat{\mathbf{C}}$ with (6) and $\mathbf{v}$ with (8). The second term is the complexity of calculating $\mathbf{p}$ with (9), multiplying $\mathbf{v}$ with $(1 - cd)\mathbf{y}^T - d\mathbf{p}^T$, and adding the result to $\mathbf{W}$ in (10). Multiplying the complexity of a single iteration with the number of training examples m, we get the overall training time complexity of online RLS $O(mn^2 + mnl)$. Thus, online training of RLS with a set of m data points is computationally as efficient as the training of batch RLS with (3) and provides exactly equivalent results. Batch learning with (4) is more efficient only if $m < n$ but this is rarely the case in the lifelong learning setting considered in this paper. The space complexity of online RLS is $O(n^2 + nl)$, where the first term is the cost of keeping the matrix $\mathbf{C}$ in memory and the second is that of the matrix $\mathbf{W}$.

Putting everything together, we present the online RLS in Algorithm 1. The algorithm first starts from an empty training set (lines 1-2). Next, it reads a feature representation of a new data point (line 4) and outputs a vector of predicted labels for the data point (line 5). After the prediction, the method is given a feedback in the form of the correct label vector of the data point (line 6). Finally, the features and labels of the new data point are used to update the predictor (lines 7-11). The steps 4-11 are reiterated whenever new data are observed.

Algorithm 1. Pseudo code for online RLS

1.	$\mathbf{C} \leftarrow \lambda^{-1} \mathbf{I}$
2.	$\mathbf{W} \leftarrow \mathbf{0} \in \Re^n$
3.	for $t = 1, 2, \ldots$ do
4.	Read data point $\mathbf{x}$
5.	Output prediction $\mathbf{p} \leftarrow \mathbf{W}^T \mathbf{x}$
6.	Read true labels $\mathbf{y}$
7.	$\mathbf{v} \leftarrow \mathbf{Cx}$
8.	$c \leftarrow \mathbf{x}^T \mathbf{v}$
9.	$d \leftarrow (c + 1)^{-1}$
10.	$\mathbf{C} \leftarrow \mathbf{C} - d\mathbf{vv}^T$
11.	$\mathbf{W} \leftarrow \mathbf{W} + \mathbf{v}((1 - cd)\mathbf{y}^T - d\mathbf{p}^T)$

Parallelized Online RLS

Because of the layout of matrices in memory and the nature of the basic matrix operations, it is often possible to gain considerable performance improvements with parallelization. Indeed, the parallelization of the batch RLS has been tested on graphics processing units by Do et al. (2008) who reported large gains in running speed. Here, we consider the parallelization of online RLS with multiple outputs.

The two most expensive parts in Algorithm 1 are the lines 7 and 10 which both require $O(n^2)$ time, and hence we concentrate primarily on those when designing a parallel version of online RLS. The parallelization of the lines 7 and 10 are presented in Algorithms 2 and 3, respectively. In both algorithms, the outer loop corresponds to distributing the work among p processors. The former algorithm is simply a parallelization of a matrix-vector product which is widely known in literature but we present if here for self-sufficiency. The parallelization of the outer product of two vectors considered in the latter algorithm is almost as straightforward. In both cases, there is no time

Algorithm 2. Parallel computation for $\mathbf{v} \leftarrow \mathbf{Cx}$

1. Split the index set $\{1, \ldots, n\}$ into p disjoint subsets
    $\mathcal{I}_1, \ldots, \mathcal{I}_p$
2. for $h = 1, \ldots, p$ do
3.   for $i \in \mathcal{I}_p$
4. $\mathbf{v}_i \leftarrow 0$
5.   for $j = 1, \ldots, n$ do
6. $\mathbf{v}_i \leftarrow \mathbf{v}_i + \mathbf{C}_{i,j}\mathbf{x}_j$
7. return $\mathbf{v}$

Algorithm 3. Parallel computation for $\mathbf{C} \leftarrow \mathbf{C} - d\mathbf{v}\mathbf{v}^{\mathrm{T}}$

1. Split the index set $\{1, \ldots, n\}$ into p disjoint subsets
    $\mathcal{I}_1, \ldots, \mathcal{I}_p$
2. for $h = 1, \ldots, p$ do
3.   for $i \in \mathcal{I}_p$
4.   for $j = 1, \ldots, n$ do
5. $\mathbf{C}_{i,j} \leftarrow \mathbf{C}_{i,j} - d\mathbf{v}_i\mathbf{v}_j$
6. return $\mathbf{C}$

wasted waiting for memory write locks, because every processor is updating different memory locations determined by the index sets. Moreover, if the processors have a sufficient amount of cache memory available, the progress can be accelerated even further, since the different processors require different portions of the matrix $\mathbf{C}$.

Finally, we note that if the number of label per data point l is large, the lines 5 and 11 in Algorithm 1 can be parallelized in similar way as the lines 7 and 10, respectively. This is, because the former contains a product of a matrix and a vector, and the latter contains an outer product of two vectors. Putting everything together, the computational complexity of a single iteration of parallel online RLS is $O(n^2 / p + nl / p)$, where p is the number of processors. The complexity of learning from a sequence of m data points is m times the complexity of a single iteration.

Memory Efficiency

While the amount of memory available for modern multi-core computers and embedded devices has increased drastically, it may nevertheless form an even tighter bottleneck than running time, especially when the cores employ complex cache architectures. The running speed of the algorithms can be increased considerably if the most frequently accessed memory locations can be stored in a high speed cache rather than in the slow performing main memory. Therefore, in this section we will give a closer look on the memory consumption and access optimization.

In the above considered machine learning algorithm, the memory consumption is dominated by the matrix $\mathbf{C}$, whose size grows quadratically with respect to the number of features n. We immediately observe that since $\mathbf{C}$ is symmetric, its memory consumption can be halved by only storing its upper triangular part, that is, its elements $\mathbf{C}_{i,j}$, where $i \leq j$. In addition, this reduces the number of computation steps required for updating the elements of $\mathbf{C}$. However, this complicates the algorithms dealing with the matrix to some extent. Next, we present a modification to the parallel online RLS which operates only on the upper triangular part of $\mathbf{C}$.

Let us first consider Algorithm 2, which performs the matrix vector product $\mathbf{Cx}$. At first glance, the only modification required would be to modify line 6 so that accessing $\mathbf{C}_{i,j}$ is switched to $\mathbf{C}_{j,i}$ whenever $j < i$. However, this may considerably increase the amount of cache misses, and hence we have to pay careful attention on the order the elements of $\mathbf{C}$ are stored in memory. Assuming that there is an N-word block of data for each cache entry, the N words in a cache entry will have consecutive memory addresses. Therefore, Algorithm 2 will work faster if the matrix $\mathbf{C}$ is stored in memory row-wise than if it is stored column-wise, because its inner loop goes through all elements of a single row. With row-wise storing, we mean that the elements in the same row and in adjacent columns have adjacent memory addresses. Now, if only the upper triangular part of $\mathbf{C}$ is stored, half of the access operations in line 6 of Algorithm 2, the ones whose row and column indices must be switched, will refer to memory locations that are not adjacent to those accessed in previous iteration of the inner loop, resulting to a cache miss with a high probability. In our experiment, this was indeed confirmed to increase the running time.

The problem of increased number of cache misses can be solved by storing $\mathbf{C}$ in a diagonal-wise fashion starting from the main diagonal and then continuing to the first, second, etc. super-diagonal. This indicates that two elements of $\mathbf{C}$ have consecutive memory addresses if they are adjacent elements in the main diagonal or in some of the super-diagonals of $\mathbf{C}$. The modification of Algorithm 2 that takes account of the diagonal-wise storage of $\mathbf{C}$ is presented in Algorithm 4. The inner loop is split in two parts. While the first inner loop goes through a super-diagonal of $\mathbf{C}$ containing $n - i$ elements, the corresponding second inner loop takes care of the sub-diagonal of $\mathbf{C}$ containing i elements. Note that the sub-diagonals are not stored in memory and the access operations always refer to the upper triangular area of $\mathbf{C}$. Now, in both inner loops, all accesses to $\mathbf{v}$, $\mathbf{x}$ and $\mathbf{C}$ are performed on such memory locations that are adjacent to those used in the previous iteration. The analogous modification of Algorithm 3 is illustrated in the modification of Algorithm 5.

In addition that the approach of storing only the upper triangular part of $\mathbf{C}$ leads to reduced memory consumption, the approach saves also computation time. This can be seen by comparing Algorithm 5 with Algorithm 3, where the former performs only the update operations concerning the upper triangular part of $\mathbf{C}$, while the latter updates the whole matrix. Note also that, when operating with multiple threads or processors, one must also pay attention on the partition of the indices done in line 1 of Algorithm 5. This

Algorithm 4. Parallel computation for $\mathbf{v} \leftarrow \mathbf{C}\mathbf{x}$ *with reduced memory consumption*

1. Split the index set $\{1, \ldots, n\}$ into p disjoint subsets $\mathcal{I}_1, \ldots, \mathcal{I}_p$
2. for $h = 1, \ldots, p$ do
3.   for $i \in \mathcal{I}_p$
4.   for $j = 1, \ldots, n - i$ do
5. $\mathbf{v}_j \leftarrow \mathbf{v}_j + \mathbf{C}_{j,i+j}\mathbf{x}_{i+j}$
6.   for $j = 1, \ldots, i$ do
7. $\mathbf{v}_{n-i+j} \leftarrow \mathbf{v}_{n-i+j} + \mathbf{C}_{n-i+j,j}\mathbf{x}_j$
8. return $\mathbf{v}$

Algorithm 5. Parallel computation for $\mathbf{C} \leftarrow \mathbf{C} - d\mathbf{v}\mathbf{v}^{\mathrm{T}}$ *with reduced memory consumption*

1. Split the index set $\{1, \ldots, n\}$ into p disjoint subsets $\mathcal{I}_1, \ldots, \mathcal{I}_p$
2. for $h = 1, \ldots, p$ do
3.   for $i \in \mathcal{I}_p$
4.   for $j = 1, \ldots, n - i$ do
5. $\mathbf{C}_{j,i+j} \leftarrow \mathbf{C}_{j,i+j} - d\mathbf{v}_j\mathbf{v}_{i+j}$
6. return $\mathbf{C}$

is because the inner loop in the algorithm has different number of iterations for each index, and hence the partition should be done so that the work load for the threads would be balanced. This can be achieved, for example, by ensuring that if the index i is under the responsibility of a certain thread, the same thread also handles the index $n - i$.

A possible pitfall in the above described approach is that it may increase the number of data access operations slightly. Firstly, this can be observed by again comparing the inner loops of Algorithm 2 and Algorithm 4. The former accesses a different element of $\mathbf{C}$ and $\mathbf{x}$ but operates with the same element of $\mathbf{v}$ in each iteration of the inner loop. In contrast, the latter also accesses a different element of $\mathbf{v}$ in each iteration of its both inner loops, which may lead to increased amount of cache misses. Similar differences can be observed between Algorithm 3 and Algorithm 5, while the number of update operations performed by the latter is smaller than that of the former.

In our experiments, we evaluate the effects of the memory and cache optimization by comparing the variation using the whole **C** with the one using the upper triangular part described in Algorithm 4 and Algorithm 5. In order to show that the approach of storing **C** in a diagonal-wise fashion really pays off, we also compare the two variations with a memory saving variation that does not take advantage of the diagonal-wise storing.

EXPERIMENTS

We implement the online RLS method in the C++ programming language, and parallelize it via the OpenMP parallel programming platform. Experiments are run both in a desktop environment with a multi-core Intel processor, as well as on a NoC simulation platform.

Recognition of Hand-Written Digits

As a proof of concept demonstration, we explore the behavior of the parallel online RLS on the MNIST handwritten digit database[3] benchmark. The database consists of 28 x 28 pixel black and white image scans of handwritten digits. The features of each image consist of pixel intensity values normalized to [0,1] range. The task is to be able to predict given the pixel intensity values of an image, which of the digits {0, ..., 9} it represents. We follow the original training-test split defined in the dataset, which ensures that the test characters have been written by different authors than the training ones. The training set consists of 60000 examples, and the test set of 10000 examples. The pixel intensity values are directly used as the features for the linear model. We note that developing a digit recognition system

of a state-of-the-art prediction performance would require the use of more sophisticated domain knowledge and feature engineering as well as a platform with much better computational resources but the consideration of such issues goes outside the scope of this paper and the current results already demonstrate the learning capability of the system. In particular, they demonstrate that a simple but efficient learning approach, suitable for deployment on resource constrained devices due to its computational efficiency, does already work reasonably well for the considered recognition task. The value of the regularization parameter is set to 100 in the experiments.

Baseline Method

As a point of comparison, we consider training RLS using stochastic gradient descent (SGD) (Bottou, 2010). In this approach, the gradient of the RLS loss function (2) is approximated by computing it for a single example at a time. After each example is processed, the current solution is updated by subtracting from it the approximated gradient, multiplied by a positive weight known as the learning rate. Compared to OnlineRLS, a single pass of SGD through the data will provide us only an approximation of the solution to the RLS optimization problem. However, we note that our final goal is not to have the most accurate possible solution to the optimization problem, but rather to find a model that generalizes well, and this goal is often reached also by SGD with reasonable amount of data and computation (Bottou, 2010). The SGD has an iteration cost of $O(nl)$, making it faster than OnlineRLS whenever $n >> l$. Techniques for parallelizing SGD have been considered for example by Langford et al. (2009).

Runtime and Classification Performance

First, we measure the effect of the parallelization on the runtime required for updating the

OnlineRLS predictor with a new example. We measure the average time spent on updating the learned model on a new training example, on the MNIST data set. The experiment is run on an Intel Core i7-950 processor machine. We compare three variants of the algorithm. P1 is the parallel implementation based on the method described in Algorithms 2 and 3, whereas P2 describes the memory and cache optimized variant described in Algorithms 4 and 5. To test the effects of cache usage optimization, we also make a comparison with P3, a naive version of the memory saving variant of OnlineRLS, whose the cache behaviour is not optimized.

The run-times are presented in Figure 2. On 4 cores the update operations are approximately 3 times faster than on a single core, demonstrating that substantial speedups can be gained in parallel hardware architectures for the method. The P1 and P2 implementations have virtually identical behavior, meaning that the memory efficient version can be implemented as efficiently

as the more straightforward parallel OnlineRLS implementation. The CPU time spent updating the predictor ranges from around 900 to below 300 microseconds per example, depending on the number of cores, which suggests that the method could be useful in systems, where only minimal computational times can be afforded. P3 scales much worse than P2 and P1, showing that in implementing the memory efficient variant it is important to ensure that the accessed entries in the matrices reside in consecutive memory locations.

In Figure 3 we plot the classification performance of the learned classifier as a function of number of training examples processed. The prediction performance is especially in the early phases of learning affected by the order in which the training examples are supplied. Therefore, we compute the average results over 10 repetitions of the experiment, where at the start of each experiment the order of the training examples is shuffled. The error rate measures the relative fraction of misclassified test examples. Since all classes are

Figure 2. Average computational cost of updating the predictor with new training example on Mnist

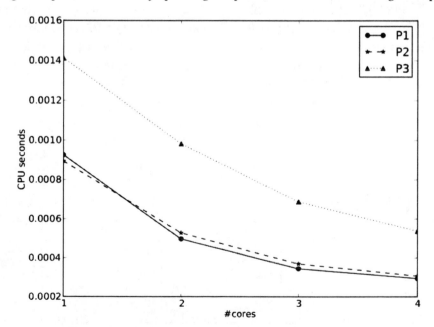

Figure 3. Error rate on the Mnist data as a function of training examples

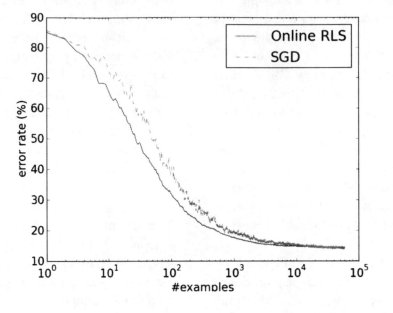

roughly equally represented in the test set, naive approaches such as predicting always the same class, or choosing the class randomly would lead to around 90% test error. The online RLS method also begins with predictive performance in this range, but as more examples are processed, we see significant improvements, with error rate reaching 14:7% once all the 60000 training examples have been processed. The curve demonstrates the ability of online RLS to improve its performance by adapting to examples provided over time. Compared to online RLS, the SGD method is clearly much less efficient in using the data for learning, reaching the same performance only after substantially more data has been processed.

Analysis of Memory and Cache Behaviors

To evaluate the memory and cache behavior of our program, we use a cache profiler (Nethercote and Seward, 2007) for the original program (P1), the program optimized for cache-memory hierarchy (P2) and the program (P3) that saves memory analogously with P2 but whose cache behavior is not optimized. The profiling results with 4 threads are summarized in Table 1. The problem size for our profiling is 5,000 inputs.

We note that, P2 and P3 consume less memory than P1 (65.3% of P1). As discussed above, this is primarily because P2 and P3 use only the up-

Table 1. Profiling results for P1, P2 and P3

Program	P1	P2	P3
Main memory usage Data references	3380KB	2208KB	2200KB
L1 misses	18,783M	26,359M	23,534M
L2 (512KB) misses	384M	290M	1,297M
L2 (1024KB) misses	384M	233M	236M
L2 (2048KB) misses	364M	105M	107M
	229M	0	0

per triangular part of the matrix C. For the same reason, P2 and P3 may also save computation time as the lower triangular part of C does not have to be updated.

As also expected according to the descriptions of Algorithms 4 and 5, the number of data accesses (including L1, L2 and main memory) in P2 is 40.3% higher than P1, while in P3 has about 89.3% data accesses compared with P2. It is noteworthy that, despite the fact that P2 has higher data accesses than P1 and P3, most of the accesses are hit. For example, assuming 32KB i+d L1 cache and 2MB L2 cache, the L1 miss rates are 2.0% and 1.1% for P1 and P2 respectively, the L2 miss rates are 59.6% and 0.01% for P1 and P2 respectively. For P3, the L1 miss rate is 5.5%, which is much higher than that of the other two programs, while the number of data references for all programs is almost the same. This introduces much higher accesses to the L2 cache (384.4M for P1, 290.6M for P2 and 1,296.9M for P3). For shared memory CMPs, assuming a two-level cache architecture, a large L2 cache is crucial. A L1 cache miss will require an access to the L2 cache, which is usually 5 times slower (e.g. 2 cycles for L1 and 10 cycles for L2). The latency between L2 cache and the off-chip memory is even higher, usually hundreds of cycles. We observe a 30% slow down of P3, compared with P1 and P2.

To further analyze the temporal locality of our algorithm, we estimate the cache miss rates with different L2 cache sizes. The size of L1 cache remains constant. Figure 4 show that, as the cache size increases, the cache miss rate decreases gradually for both programs (From 99.96% to 0.01% for P1, from 80.51% to 0.01% for P2 and from 18.2% to 0.01% for P3, respectively). The main reason is that, as the cache size increases, larger and larger portions of the matrix C, dominating the memory usage, will fit into the cache until eventually the whole matrix can fit in. Since P2 only uses the upper triangular part of C, the size of cache required to hold it wholly is about half of the size required by P1.

We note that, P1 requires 4MB of L2 cache to keep a low L2 miss rate (around 0.01%), while P2 just require 2MB. This is very important for an embedded system, since limited cache/memory is always a bottleneck for these systems. We also note that, while P3 requires even less L2 caches than other two, the size of L1 cache is a bottleneck for P3. For modern computer systems, the L1 private data cache is usually only 16KB to 32KB due to physical limitations, while the shared L2 cache is usually scalable from hundreds of kilobytes to dozens of megabytes.

The main insights from the profiling results can be summarized as follows. On one hand, due to its larger memory consumption, P1 also

Figure 4. L2 cache miss rate versus cache size for P1, P2 and P3

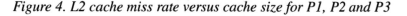

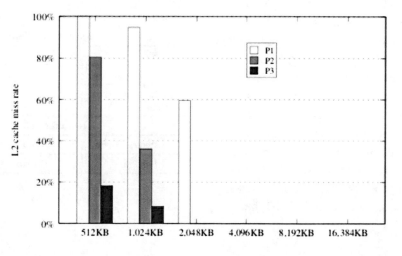

requires a larger L2 cache which makes it less attractive for embedded systems than the other two programs. On the other hand, the reduced memory consumption introduced in P3 leads to a considerably increased number of L1 misses which in turn makes the program less usable than P1. P2 fixes this problem with data access optimization, which makes it the most practical of the three programs.

NoC Simulation Environment

To evaluate our algorithm further, we use a cycle-accurate NoC simulator (see Figure 1). The simulation platform is able to produce detailed evaluation results. The platform models the routers and links accurately. The state-ofthe-art router in our platform includes a routing computation unit, a virtual channel allocator, a switch allocator, a crossbar switch and four input buffers. Routers in our NoC model have five ports (North, East, West, South and Local PE) and the corresponding virtual channels, buffers and crossbars. It is noteworthy that not all routers in a NoC require five ports, e.g. router of N1 in Figure 1 has only East, South and Local PE ports. Adaptive routing is used widely in off-chip networks, however deterministic routing is favorable for on-chip networks because the implementation is easier. In this paper, a dimensional ordered routing (DOR) (Sullivan and Bashkow, 1977) based X-Y deterministic routing algorithm is selected, in which a message packet (flit) is first routed to the X direction and last to the Y direction.

We use a 16-node network which models a single-chip CMP for our experiments. A full system simulation environment with 16 processors has been implemented. The simulations are run on the Solaris 9 operating system based on SPARC instruction set in-order issue structure. Each processor is attached to a wormhole router and has a private write-back L1 cache. The L2 cache shared by all processors is split into banks. The size of each cache bank node is 1MB; hence

the total size of shared L2 cache is 16MB. Each L2 cache bank is attached to a router as well. The simulated memory/cache architecture mimics SNUCA (Kim et al., 2002). A two-level distributed directory cache coherence protocol called MESI (Patel and Ghose, 2008) has been used in our memory hierarchy in which each L2 bank has its own directory. Four types of cache line status, namely Modified (M), Exclusive (E), Shared (S) and Invalid (I) are implemented. We use Simics (Magnusson et al., 2002) full system simulator as our simulation platform. The detailed configurations of processor, cache and memory configurations can be found in Table 2.

Table 2. System configuration parameters

Processor Configuration	
Instruction set architecture	SPARC
Number of processors	16
Issue width	1
Cache Configuration	
L1 cache	Private, split instruction and data cache, each cache is 16KB. 4-way associative, 64-bit line, 3-cycle access time
L2 cache	Shared, distributed in 4 layers, unified 2-16MB (16 banks, each 256KB-1MB). 64-bit line, 6-cycle access time
Cache coherence protocol	MESI
Cache hierarchy	SNUCA
Memory Configuration	
Size	4GB DRAM
Access latency	260 cycles
Requests per processor	16 outstanding
Network Configuration	
Router scheme	Wormhole
Flit size	128 bits

NoC Simulation Result Analysis

The normalized full system simulation results are shown in Figures 5 and 6. We use the machine learning algorithm with 16 threads. The simulation runs with 100 inputs. To analyze the temporal locality of our algorithm, we estimate the cache miss rate with different L2 cache sizes. For shared memory CMPs, a large last level cache is crucial, because a miss in the cache will require an access to the off-chip main memory. Figure 5 show that, as the cache size increases, the cache miss rate decreases gradually (From 2.79% to 1.48%). The main reason is that, in our algorithm, the data-set is pre-loaded into the cache-memory first. More data-sets and intermediate data can be stored with a larger cache. Average network latency represents the average number of cycles required for the transmission of all network messages. For each message, the number of cycle is calculated as, from the injection of a message header into the network at the source node, to the reception of a tail flit at the destination node. As illustrated in Figure 6, the 16MB configuration outperforms

others in terms of average network latency. The latency in 16MB configuration is 1.73%, 3.53% and 4.56% lower than the 8MB, 4MB and 2MB configurations, respectively. This is primarily due to the reduced cache miss rate in the 16MB configuration compared to the other configurations. We notice that, an off-chip access of the main memory will result a significantly higher network latency. However, since there are millions of cache/memory accesses, the impact is less significant as a whole.

CONCLUSION

We have introduced a machine learning system which is based on parallel online regularized least-squares learning algorithm. The system is specifically suitable for use in real-time adaptive systems due to the following properties it fulfills:

- The system is able to learn in online fashion, that is, it can update itself in real-time whenever new data is observed. This is an

Figure 5. Cache miss rate versus cache size

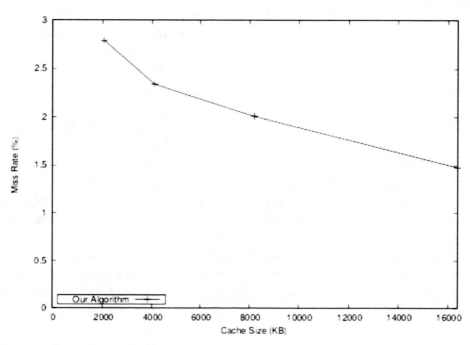

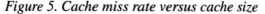

Figure 6. Normalized average network latency with different number of pillars

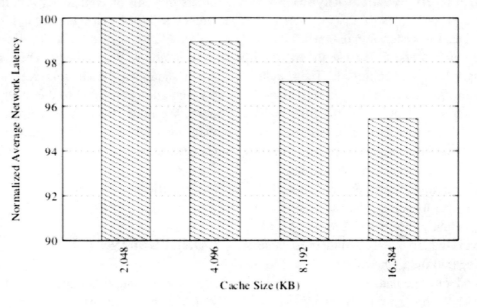

essential property in real-life applications of embedded machine learning systems, for example, in smart-phone applications that aim to adapt to their owners preferences.

- The learning system is parallelized and, due to memory usage and cache behavior optimization, works with limited processor time and memory. This opens the possibilities to deploy the system, for example, into small hand-held devices which operate in real-time.

- The system can carry out complex learning tasks involving simultaneous prediction of several labels per data point. Typical examples of this type of tasks are, for example, in multi-class and multi-label classification.

The run-time performance and cache behavior of the proposed system were evaluated using 1 to 4 threads, in a quad-core platform. It was shown that, as expected according to the theoretical considerations, the performance gain is roughly linear with respect to the number of cores. Moreover, we developed variants of the proposed algorithm, whose memory consumption and cache behavior are further optimized. In the experiments, it was shown that the optimization leads to a considerably decreased numbers of cache misses and requirements for L2 cache sizes. This, in turn, further increases the speed of the system and makes it even more attractive for real-time embedded systems. In an additional experiment, we used NoC platform to test the system in 16 threads. The NoC consists of a 4x4 mesh. The obtained results demonstrated that the system is able to learn with minimal computational requirements, and that the cache optimization and parallelization of the learning process considerably reduces the required processing time.

Altogether, the study sheds light on the possibilities of deploying modern machine learning methods into embedded systems based on future multi-core computing architectures. For example, the machine learning techniques that are able to operate in real time and in online fashion are promising tools for pursuing adaptivity of embedded systems. This is because they enable the real time updating of the system according to the data observed from the environment. While we used a digit recognition task as case study in this paper, the learning system can be applied on a wide range of other tasks such as energy efficiency or control of the embedded systems.

REFERENCES

Bottou, L. (2010). Large-Scale Machine Learning with Stochastic Gradient Descent. In *Proceedings of the 19th International Conference on Computational Statistics* (COMPSTAT'2010), (pp. 177–187). Paris, France: Springer.

Bottou, L., & Le Cun, Y. (2004). Large scale online learning. *Advances in Neural Information Processing Systems* 16.

Chu, C.-T., Kim, S. K., Lin, Y.-A., Yu, Y., Bradski, G., Ng, A. Y., & Olukotun, K. (2007). Map-reduce for machine learning on multicore. *Advances in Neural Information Processing Systems*, *19*, 281–288.

Dally, W. J., & Towles, B. (2001). Route packets, not wires: on-chip inteconnection networks. In *Proceedings of the 38th conference on Design automation*, (pp. 684–689). Academic Press.

Do, T.-N., Nguyen, V.-H., & Poulet, F. (2008). Speed up SVM algorithm for massive classification tasks. In *Proceedings of the 4th International Conference on Advanced Data Mining and Applications* (ADMA 2008) (LNCS), (vol. 5139, pp. 147–157). Springer.

Farabet, C., Poulet, C., & LeCun, Y. (2009). An FPGA-based stream processor for embedded real-time vision with convolutional networks. In *Proceedings of Fifth IEEE Workshop on Embedded Computer Vision* (ECV'09), (pp. 878–885). IEEE.

Henderson, H. V., & Searle, S. R. (1981). On deriving the inverse of a sum of matrices. *SIAM Review*, *23*(1), 53–60. doi:10.1137/1023004

Hoerl, A. E., & Kennard, R. W. (1970). Ridge regression: Biased estimation for nonorthogonal problems. *Technometrics*, *12*, 55–67. doi:10.1080/00401706.1970.10488634

Horn, R., & Johnson, C. R. (1985). *Matrix analysis*. Cambridge University Press. doi:10.1017/CBO9780511810817

Hsu, D., Kakade, S., Langford, J., & Zhang, T. (2009). Multi-label prediction via compressed sensing. *Advances in Neural Information Processing Systems*, *22*, 772–780.

Intel. (2010). *Single-chip cloud computer*. Retrieved from http://techresearch.intel.com/articles/Tera-Scale/1826.htm

Isoaho, J., Virtanen, S., & Plosila, J. (2010). Current challenges in embedded communication systems. *International Journal of Embedded and Real-Time Communication Systems*, *1*(1), 1–21. doi:10.4018/jertcs.2010103001

Kim, C., Burger, D., & Keckler, S. W. (2002). An adaptive, non-uniform cache structure for wire-delay dominated on-chip caches. In *Proceedings of ACM SIGPLAN*, (pp. 211–222). ACM.

Langford, J., Smola, A. J., & Zinkevich, M. (2009). Slow learners are fast. In Advances in Neural Information Processing Systems. MIT Press.

Low, Y., Gonzalez, J., Kyrola, A., Bickson, D., Guestrin, C., & Hellerstein, J. M. (2010). Graphlab: A new framework for parallel machine learning. In *Proceedings of the 26th Conference on Uncertainty in Artificial Intelligence* (UAI 2010). UAI.

Magnusson, P., Christensson, M., Eskilson, J., Forsgren, D., Hallberg, G., & Hogberg, J. et al. (2002). Simics: A full system simulation platform. *Computer*, *35*(2), 50–58. doi:10.1109/2.982916

Mitchell, T. M. (1997). *Machine Learning*. McGraw-Hill.

Nethercote, N., & Seward, J. (2007). Valgrind: a framework for heavyweight dynamic binary instrumentation. In *Proceedings of the 2007 ACM SIGPLAN conference on Programming language design and implementation*, PLDI '07, (pp. 89–100). New York, NY: ACM.

Oresko, J. J., Jin, Z., Cheng, J., Huang, S., Sun, Y., Duschl, H., & Cheng, A. C. (2010). A wearable smartphone-based platform for real-time cardiovascular disease detection via electrocardiogram processing. *IEEE Transactions on Information Technology in Biomedicine, 14*, 734–740. doi:10.1109/TITB.2010.2047865 PMID:20388600

Patel, A., & Ghose, K. (2008). Energy-efficient mesi cache coherence with pro-active snoop filtering for multicore microprocessors. In *Proceeding of the thirteenth international symposium on Low power electronics and design*, (pp. 247–252). Academic Press.

Plackett, R. L. (1950). Some theorems in least squares. *Biometrika, 37*(1/2), 149–157. doi:10.2307/2332158 PMID:15420260

Poggio, T., & Smale, S. (2003). The mathematics of learning: Dealing with data. [AMS]. *Notices of the American Mathematical Society, 50*(5), 537–544.

Rifkin, R., Yeo, G., & Poggio, T. (2003). Regularized least-squares classification. In Advances in Learning Theory: Methods, Model and Applications, (vol. 190, pp. 131–154). IOS Press.

Shotton, J., Fitzgibbon, A., Cook, M., Sharp, T., Finocchio, M., Moore, R., et al. (2011). Real-time human pose recognition in parts from single depth images. In *Proceedings of IEEE Computer Vision and Pattern Recognition (CVPR) 2011*. IEEE.

Sullivan, H., & Bashkow, T. R. (1977). A large scale, homogeneous, fully distributed parallel machine. In *Proceedings of the 4th annual symposium on Computer architecture*, (pp. 105–117). Academic Press.

Sun, Y., & Cheng, A. C. (2012). Machine learning on-a-chip: A high-performance low-power reusable neuron architecture for artificial neural networks in ECG classifications. *Computers in Biology and Medicine, 42*(7), 751–757. doi:10.1016/j.compbiomed.2012.04.007 PMID:22595230

Suykens, J., Van Gestel, T., De Brabanter, J., De Moor, B., & Vandewalle, J. (2002). *Least Squares Support Vector Machines*. World Scientific Pub. Co.

Swere, E. A. (2008). *Machine Learning in Embedded Systems*. (PhD thesis). Loughborough University.

Vangal, S., Howard, J., Ruhl, G., Dighe, S., Wilson, H., Tschanz, J., et al. (2007). An 80-tile 1.28tflops network-on-chip in 65nm cmos. In *Proceedings of IEEE International Solid-State Circuits Conference ISSCC 2007*, (pp. 98–589). IEEE.

Zhdanov, F., & Kalnishkan, Y. (2010). An identity for kernel ridge regression. In *Proceedings of the 21st international conference on Algorithmic learning theory* (LNCS) (vol. 6331, pp. 405–419). Berlin: Springer-Verlag.

Zinkevich, M., Smola, A., & Langford, J. (2009). Slow learners are fast. *Advances in Neural Information Processing Systems, 22*, 2331–2339.

On Parallel Online Learning for Adaptive Embedded Systems

KEY TERMS AND DEFINITIONS

Batch Learning: A machine learning approach, where all the training data used to learn the predictive model is processed at once in a batch fashion, after which the model remains fixed.

Machine Learning: Construction and study of algorithms that learn predictive models from data.

Multi-Label Learning: A form of machine learning, where multiple outcomes to be predicted are associated with each instance.

Network-on-Chip: an integration paradigm for large VLSI systems implemented on a single silicon chip.

Online Learning: A machine learning approach, where the model improves its performance over time via incremental processing of new data.

Parallel Computing: solving computational problems by subdividing them into smaller subtasks that are solved in parallel.

Regularized Least-Squares: an extension of the classical least-squares regression technique, that uses regularization to penalize too complex models.

ENDNOTES

1   Intel is a trademark or registered trademark of Intel or its subsidiaries. Other names and brands may be claimed as the property of others.

2   In the literature, the formula of the linear predictors often also contain a bias term. Here, we assume that if such a bias is used, it will be realized by using an extra constant valued feature in the data points.

3   Available at http://yann.lecun.com/exdb/mnist/

Chapter 12
Classification–Based Optimization of Dynamic Dataflow Programs

Hervé Yviquel
IRISA, France

Matthieu Wipliez
Synflow SAS, France

Emmanuel Casseau
IRISA, France

Jérôme Gorin
Telecom ParisTech, France

Mickaël Raulet
IETR/INSA, France

ABSTRACT

This chapter reviews dataflow programming as a whole and presents a classification-based methodology to bridge the gap between predictable and dynamic dataflow modeling in order to achieve expressiveness of the programming language as well as efficiency of the implementation. The authors conduct experiments across three MPEG video decoders including one based on the new High Efficiency Video Coding standard. Those dataflow-based video decoders are executed onto two different platforms: a desktop processor and an embedded platform composed of interconnected and tiny Very Long Instruction Word-style processors. The authors show that the fully automated transformations presented can result in a 80% gain in speed compared to runtime scheduling in the more favorable case.

INTRODUCTION

Dataflow programming paradigm was used for years to describe signal processing applications, since the representation of such application in a set of computational units interconnected by communication channel is quite straight forward. Consequently, several kinds of dataflow Models of Computation (MoC), defining the semantic of the programming language, were studied (Johnston, W. M., Paul Hanna, J. R., & Millar, R. J. (2004)). They can be split into two main classes: the static ones allowing a predictable behavior such that scheduling can be done at compile time, and others having a data-dependent behavior need a runtime scheduling (Lee, E. A., & Parks, T. M. (1995)).

DOI: 10.4018/978-1-4666-6034-2.ch012

Most of the studies on dataflow programming focus on the statically schedulable MoC due to efficient synthesis techniques on such models. Unfortunately, they do not take into consideration the flexibility and the expressiveness offered to the programmers by the dynamic dataflow MoC.

The challenge when optimizing the execution of an dynamic dataflow description is then to conserve their strong expressive power while reducing the overhead caused by run-time scheduling. This is the reason why we propose a process that reduces the number of actors that are required to be scheduled at run-time, by clustering network regions that have a locally static behavior. A locally static region is a set of connected actors in the description that has a firing order we can determine statically.

The chapter is organized as follows. First, the dataflow programming and the context of this study are introduced in Section DataFlow Programming. Then, we explore the work on actor classification in Section Actor Classification. In Section Execution of dynamic dataflow programs present the actor scheduling for dynamic dataflow programs and establish a clustering methodology to optimize it. Finally, we conclude and consider future works.

DATAFLOW PROGRAMMING

The concept of dataflow representation was introduced by Sutherland in 1966 as a visual way to describe a sequence of arithmetic statements (Sutherland, W. R. (1966)). Then, the first dataflow programming language was presented by Dennis in 1974 (Dennis, J. B. (1974)). In this language, a program is modeled as a directed graph where the edges represent the flow of data and the nodes describes control and computation.

Thus, a dataflow program is defined as a graph composed of a set of computational units interconnected by communication channels. The one presented in Figure 1 contains a network of five components interconnected using communication channels. In the dataflow approach, the communication corresponds to a stream of data composed of a list of tokens.

Properties

During the last twenty years, dataflow programming has been heavily used for the development of signal processing applications due to its consistency with the natural representation of the processing of digital signals. The emergence of parallel programming as a consequence of the frequency wall makes dataflow paradigm an alternative to the imperative paradigm for two reasons:

- The opportunity to use visual programming to describe the interconnection between its components. Such graphical approach is very natural and makes it more understandable by programmers that can focus on how things connect.
- Its ability to express concurrency (Johnston, W. M., Paul Hanna, J. R., & Millar, R. J.

Figure 1. A dataflow graph

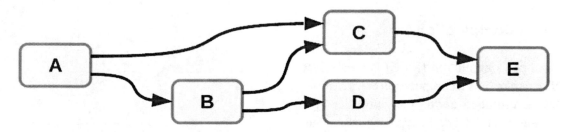

(2004)) without complex synchronization mechanism. The internal representation of the application is a network of processing blocks that only communicate through the communication channels. Consequently, the blocks are independent and do not produce any side-effect, which remove the potential concurrency issues that arise when asking the programmer to manage manually the synchronization between the parallel computations.

Modularity

The strong separation between the structural and behavioral modeling makes the application description very modular. For instance, a component can easily be replaced by another one while its interfaces (input and output ports) are strictly identical. Moreover, such descriptions are typically hierarchical, in that a component of the graph may represent another graph such as the one in the *Figure 2*.

Parallelism

A dataflow program states an abundance of parallelism thanks to the explicit exposition of the concurrency. In its structural view, the dataflow model presents three potential degrees of parallelism (task, data and pipeline) that can be applied to different granularities of description.

Additionally, these kinds of parallelism as well as the instruction-level parallelism, i.e. the potential overlap among instructions, can be potentially extracted from the internal algorithm of the components such as any procedural language.

Model of Computation

A Model of Computation (MoC) is an abstract specification of how a computation can progress. A MoC is useful to define the semantics of a programming model, i.e. the type of components

it can contain and the way they interact (Savage, J. E. (1998)). Classical examples of MoC are the Turing machine and Lambda calculus models. During the last twenty years, several dataflow MoC were studied due to the attractive use of dataflow programming for the development of signal processing applications.

Existing dataflow MoC can be split into two main classes: the static ones allowing a predictable behavior such that scheduling can be done at compile time, and others having a data-dependent behavior. Most of the studies on dataflow programming focus on the statically schedulable MoC due to efficient synthesis techniques on such models due to their analyzability. Unfortunately, they do not take into consideration the flexibility and the expressiveness offered to the programmers by the dynamic dataflow MoC.

Kahn Process Network

A KPN is represented as a graph $G = (V, E)$ such as V is a set of vertices modeling computational units that are called processes and E is a set of unidirectional edges representing unbounded

Figure 2. A subnetwork

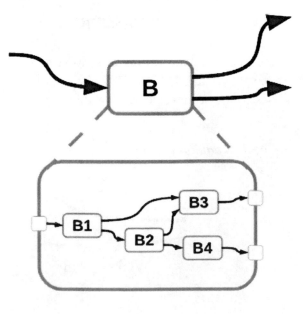

communication channels based on First In, First Out (FIFO) principles. The behavior of this model of computation can be described using the denotational semantic introduced by Kahn (Kahn, G. (1974)).

A FIFO channel $e \in E$ can be empty, denoted as $\perp$, or can carry a possibly infinite sequence of tokens:

$$X = \left[x_1, x_2, ... \right],$$

where each x_i is an atomic data called a token. A sequence X that precedes a sequence Y, e.g.:

$$X = \left[x_1, x_2 \right]$$

and:

$$Y = \left[x_1, x_2, x_3 \right],$$

is denoted $X \sqsubseteq Y$. The set of all possible sequences is denoted S, while S^p is the set of p-tuples of sequences on the p FIFO channels of a process. In other words,

$$\left[X_1, X_2, ..., X_p \right]$$

represents the sequence consumed/produced by a process. The length of a sequence is given by $|X|$.

A Kahn process with m inputs and n outputs is a continuous and monotonic function denoted as:

$$F : S^m \rightarrow S^n \tag{1}$$

A process is triggered when S^m appears on it inputs; it is activated iteratively as long as S^m exists. Conversely, the process is suspended when S^m does not exist on its input. In other terms, reading from a FIFO can be blocking for one process until S^m appears again.

The blocking reads insure that every program following this model of concurrency is deterministic. However, it also implies a thread-based implementation of KPN in a sequential environment to backup the current context of the blocked process before executing the next one. Using threads induces inefficiency due to the overhead of the context switching and discard the predictability and determinism of sequential computation (Lee, E. (2006)).

Dataflow Process Network

DPN (Lee, E. A., & Parks, T. M. (1995)), also known as Dynamic dataflow model (DDF), is closely related to KPN. The DPN model is Turing-complete that means it can model any algorithm even non-deterministic ones.

In this model, an application is represented as a graph $G = \left(V, E \right)$ within the vertices/processes are called actors. Additionally to the KPN model, it introduces the notion of firing. An actor firing, or action, is an indivisible quantum of computation which corresponds to a mapping function of input tokens to output tokens applied repeatedly and sequentially on one or more data streams. This mapping is composed of three ordered and indivisible steps: data reading, then computational procedure, and finally data writing. These functions are guarded by a set of firing rules R which specifies when an actor can be fired, i.e. the number and the values of tokens that have to be available on the input ports to fire the actor.

More formally, firings can be described using the denotational semantic extended by Dennis (Dennis, J. B. (1974)). Every actor $a \in V$ is associated with its own set of firing function F_a, and firing rules R_a such as:

$$F_a = \left[f_1, f_2, ..., f_M \right] \tag{2}$$

$$R_a = \left[\boldsymbol{R}_1, \boldsymbol{R}_2, ..., \boldsymbol{R}_N \right] \tag{3}$$

Within each function $f_i \in F_a$ is associated to a given firing rule $R_i \in R_a$.

A firing rule $\mathbf{R}_i$ defines a finite sequence of patterns, one for each input m of the actor such as:

$$\mathbf{R}_i = \left[P_{i,1}, P_{i,2}, ..., P_{i,m}, \right].$$

A pattern $P_{i,j}$ is an acceptable sequence of tokens in R_i on one input j from the input m of an actor. It is satisfied if and only if $P_{i,j} \sqsubseteq X_j$ where X_j is the sequence of tokens available on the jth FIFO channel. The pattern $P_{i,j} = \perp$ designates any empty list where any available sequence on input j is acceptable. The pattern $P_{i,j} = \left[* \right]$ is acceptable for any sequence containing at least one token. The length of a pattern $P_{i,j}$ is denoted $\left| P_{i,j} \right|$. We abuse of this notation by using $\left| R_i \right|$ to express the consumption rate of the firing rule R_i and $\left| f_i \right|$ the production rate of the firing function f_i.

An actor fires when at least one of its firing rules is satisfied. So that, DPN can describe nondeterministic algorithms when several firing rules are satisfied in the same time, which is not possible with the KPN model.

The strong encapsulation of the actors is described by Figure 3 that introduces the internal state of an actor. In fact, such an internal state is just a more convenient representation since it is strictly equivalent to a feedback loop, so it only depends on the ability of the language syntax to describe state variables.

Static Dataflow Model

The static dataflow model, or synchronous dataflow (SDF), can be seen as a simplification of the DPN model, in which an actor consumes and produces a constant number of tokens at each firing. It may have a single firing rule, which is valid for any sequence S^m of a certain size on its inputs (Lee, E., & Messerschmitt, D. (1987)). In the case where an actor has several firing rules, an actor is SDF if all its firing rules have the same consumption, which mean for $\mathbf{R}_a \in R$ and $\forall \mathbf{R}_a \in R$:

$$\mathbf{R}_a = \mathbf{R}_b. \tag{4}$$

All the firing functions of an SDF actor must also produce a fixed number of tokens at each firing, which means for $f_a \in F$ and $\forall f_b \in F$:

$$\left| f_a \left(s \right) \right| = \left| f_b \left(s \right) \right| \tag{5}$$

for any $s \in S^m$ and $s_b \in S^m$.

The cyclo-static dataflow MoC (CSDF) (Bilsen, G., Engels, M., Lauwereins, R., & Peperstraete, J. (1996)) extends SDF actors by allowing the number of tokens produced and consumed to

Figure 3. A self-contained actor with its own state, actions and firing rules

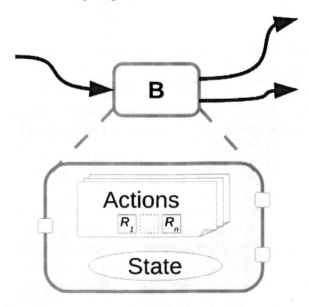

vary cyclically. This variation is modeled with a state in the actor, which returns to its initial value after a defined number of firing.

Quasi-Static Dataflow Model

Dataflow modeling is the question of striking the right balance between expressive power and analyzability: On the one hand, synchronous and cyclo-static dataflow limit the algorithms to be modeled as graphs with fixed production and consumption rates for their predictability and their strong properties that allow powerful optimizations to be applied. On the other hand, dynamic dataflow offers a large expressiveness, until Turing-completeness, able to describe complex algorithms with variable and data-dependent communication rate that makes their analyze and optimization ultimately harder.

The needs for a trade-off between expressiveness and predictability has brought the definition of so-called "quasi-static" dataflow models. Quasi-static dataflow differs from dynamic dataflow in that there are techniques that statically schedule as many operations as possible so that only data-dependent operations are scheduled at runtime (Buck, J. (1993), Buck, J., & Lee, E. (1993, Bhattacharya, B., & Bhattacharyya, S. S. (2001)).

Buck's Boolean Dataflow (BDF) model (Buck, J. (1993), Buck, J., & Lee, E. (1993) extends the SDF model with production/consumption rates that depends of a control port with a consumption rate statically fixed at one token by firing. Basically, the rate of a given port p of an actor can be controlled by its associated control port C_p, which means that the actor consumes a token from C_p and the value of this token varies the consumption/production rate of p. The fundamental dynamic actors of the BDF model are the Switch and Select that simply choose one of its two inputs or outputs according to the control token. The BDF model has been proven Turing-complete (Buck, J. (1993)) but it implies a very restrictive coding style that is not very useful for practical cases.

Parameterized dataflow (Bhattacharya, B., & Bhattacharyya, S. S. (2001)) is a higher-level approach to model quasi-static behavior by the extension of existing dataflow model using parameters modifiable at runtime. For example, Parameterized synchronous dataflow (PSDF) is a generalization of the initial SDF model that allows the expression of quasi-static behavior.

Dataflow MoCs are defined as subsets of the more general DPN model. The taxonomy shown on Figure 4 reflects the fact that MoCs are progressively restricted from DPN towards SDF with

Figure 4. Dataflow Models of Computation

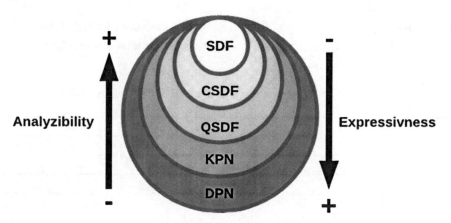

respect to expressiveness, but at the same time they become more amenable to analysis.

CASE STUDY: RECONFIGURABLE VIDEO CODING

During the last twenty years, several new programming languages based on dataflow models were proposed by the scientific community to solve a large panel of problems, from parallel programming to signal processing, but only a few have been emerging enough to be involved in the development of real world applications. One of them, the CAL Actor Language (CAL), is particularly interesting due to to the expressiveness offers by its ability to describe data-dependent behavior specified by the DPN MoC (Eker, J., & Janneck, J. (2003).).

The inclusion of a subset of CAL in the MPEG Reconfigurable Video Coding framework enables the development of several video decoders along other applications using dataflow programming. Such a collection of applications offers a great opportunity to study the scheduling of dynamic dataflow programs.

Overview

Reconfigurable Video Coding (RVC) (Mattavelli, M., Amer, I., & Raulet, M. (2010)) is an innovative framework introduced by MPEG to overcome the lack of interoperability between the several video codecs deployed in the market. The framework is dedicated to the development of video coding

tools in a modular and reusable fashion thanks to dataflow programming.

The RVC framework is supported by Orcc[1], an open-source programming toolset based on Model-Driven Engineering technologies, that contains a multi-target compiler as well as an integrated development environment. Orcc is able to translate a unique high-level dataflow program, written in RVC-CAL, into an equivalent description in both hardware and software languages.

Applications

For the experiments presented in this chapter, we use the dataflow descriptions of three MPEG video decoders developed by the RVC group: MPEG-4 Part 2 Simple Profile, MPEG-4 Part 10 Progressive High Profile (also known as AVC or H264) and the new MPEG-H Part 2 (better known as HEVC). Table 1 describes the properties of each description of these well-known decoders: Respectively, the profile of the decoder, the parallelization of the decoding for each component (Luma and Chromas), the number of actors and the number of FIFO channels.

The description of HEVC, even with a larger complexity, contains a lot less actors than the one of AVC for two reasons: The parallelization of the decoding for each component that induces a duplication of the residual and predication part; The AVC description was written in finer granularity with a lot of control communication.

The RVC-based video decoders are described with a fine granularity (at block level), contrary to the traditional coarse-grain dataflow (at frame

Table 1. Statistics about the RVC-CAL description of several MPEG video decoders

Decoder	Profile	YUV	#Actors	#FIFOs
MPEG-4 Part 2	SP	yes	41	143
MPEG-4 Part 10	PHP	yes	114	404
MPEG-H Part 2	Main	no	33	81

level). This fine-grain streaming approach induces a high potential in pipeline parallelism and the use of small communication channels, i.e. their size are defined between 512 and 8192.

Figure 5 presents the application graph of the MPEG-H Part 2 decoder which could be considered as a good reference of the existing RVC-based video decoders. The decoder is constituted of 4 distinct parts: an actor called Parser which performs the entropy decoding, a subnetwork known as Residual which decodes the prediction residual and another called Prediction which performs the motion compensation as well as the intra prediction and a last one, Filter, containing the deblocking filter and the sample adaptive offset filter. The subnetworks, Residual, Prediction and Filter can be duplicated for each component Y, U and V to increase the data parallelism.

ACTOR CLASSIFICATION

In the simplest case, structural information of an actor is enough to classify it, for instance the rules for an actor to be considered SDF only depend on the input and output patterns of actions. In more complicated cases, it is necessary to gather information from an actual execution of the actor.

Process

The literature introduces several algorithms (Zebelein, C., Falk, J., Haubelt, C., & Teich, J. (2008), Von Platen, C., Eker, J., Nilsson, A., & Arzen, K.-E. (2012), Wipliez, M. (2010), Wipliez, M., & Raulet, M. (2010)) to classify dynamic actors into restricted MoC that can be summed up as follow:

1. **Detection of Time-Dependent Actors:** DPN places no restrictions on the description of actors, and as such it is possible to describe a time-dependent actor in that its behavior depends on the time at which tokens are available. This happens when a given action reads tokens from input ports not read by a higher-priority action, and their firing rules are not mutually exclusive.

2. **Identification of Static Behavior:** Classification tries to classify each actor within models that are increasingly expressive and complex. The rationale behind this is that the more powerful a model, the more difficult it is to analyze. If an actor cannot be classified as a static actor, the method will try to classify it as cyclo-static, and then as quasi-static. An actor is classified as static iff it conforms to the SDF MoC, which means that all its actions have the same input and

Figure 5. RVC-based description of the MPEG-H Part 2 decoder

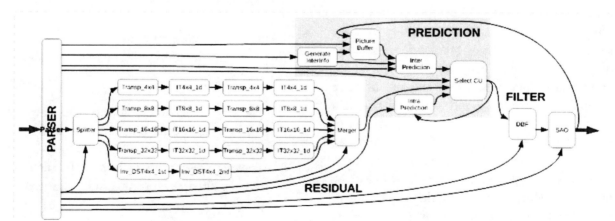

output patterns. A one-action actor is by definition static.

3. **Finding Cyclo-Static Behavior:** An actor has to meet two conditions to be a candidate for cyclo-static classification: it must have a state and there must be a fixed number of data-independent firings that depart from the initial state, modify the state, and return the actor to its original state. Once the actor was identified as a valid cyclo-static candidate, abstract interpretation can be used (see Section Abstract interpretation of Actors) to determine the sequence of actions characterizing its behavior, as well as its production and consumption rates.

4. **Determining Quasi-Static Behavior:** A quasi-static actor is informally described as an actor that may exhibit distinct static behaviors depending on data-dependent conditions. The algorithm is composed of two steps. First, the detection of the input FIFO channels used to control the behavior of the actor and their existing configuration. Then, the identification of static behavior for each configuration using abstract interpretation.

5. *If not* classified in a restricted MoC, the actor is defined as *dynamic*.

Abstract Interpretation of Actors

Classifying an actor within a MoC is based on checking that a certain number of MoC-dependent rules hold true for any execution of this actor. Some of these rules are verified solely from the structural information of the actor, for instance the rules for a static actor only depends on the input and output patterns of actions. In more complicated cases, we need to be able to obtain information from an actual execution. The actor must be executed so that the information obtained is valid for any execution of the actor, whatever its environment (the values of the tokens and the manner in which they are available). As a consequence it is not possible to simply execute the actor with a particular environment supplied by the programmer.

To circumvent this problem we use abstract interpretation (Cousot, P., & Cousot, R. (1977)). Abstract interpretation evaluates the computations performed by a program in an abstract universe of objects rather than on concrete objects. The abstract interpretation of an actor has the following properties:

- The set of values that can be assigned to a variable is

$$Values = Z \bigcup \{true, false\} \bigcup \{\bot\}$$

The value $\bot$ is used for variables whose value is unknown, e.g. for uninitialized variables.

- The environment is defined as an association of variables and their values:

$$Env : Ident \rightarrow Values$$

Env initially contains the state variables of the actor associated with their initial value if they have one, otherwise with $\bot$.

- When the interpreter enters an action, the environment is augmented with bindings between the name of the tokens in the input pattern and $\bot$. In other words, a token read has an unknown value by default.

The abstract interpreter interprets an actor by firing it repeatedly until either one of the conditions is met:

1. The interpreter is told to stop because analysis is complete as determined by the classification algorithm.

2. The interpreter cannot compute if an action may be fired because this information depends on a variable whose value is $\bot$.

To fire the actor, the interpreter starts by selecting one fireable action, which is an action that meets all its firing rules. As far as the quantity of tokens is concerned, the abstract interpretation models infinite FIFOs channel, which means an action always has enough tokens to fire. Other differences between concrete interpretation of an actor, and its abstract interpretation include the following. Any expression that references a variable v where $Env(v) = \perp$ has the value $\perp$. Conditional statements and loops that test an expression whose value is $\perp$ are not executed. However, guards evaluated as cause the abstract interpreter to stop as per condition 2.

Evaluation

We have tested the classification on the actors included in the RVC-based descriptions of the three tested video decoders. Table 2 shows the number of actors classified as static, cyclo-static, quasi-static, dynamic and time-dependent.

The results show that only a small percentage of the whole set of actors are time-dependent and quasi-static, since a vast majority is classified as static and dynamic. Most of the static actors correspond to algorithms that decode the residual picture. Only a few actors was detected with quasi-static behavior, this is a direct consequence of the classification algorithm that only handle simple case due to the difficulty to analyze it in all cases.

The granularity of the actors are very different: the static and cyclo-static actors are generally less than 50 lines of code, and most of dynamic actors are a lot larger, frequently between 500 and 1000 lines. This makes it unlikely to find actors that behave according to more restricted MoCs, and leads us to believe that the classification method will yield the best results on applications described with fine-grain, small actors.

EXECUTION OF DYNAMIC DATAFLOW PROGRAMS

The strong encapsulation of components in the dataflow execution model offers an explicit modeling of the concurrency within an application. A natural approach for handling concurrent execution on a sequential environment is the use of threads. However, thread-based implementations, on top of introducing non-determinism (Lee, E. (2006)), can lead to a large overhead when a large number of components are executed to the same processing unit (Carlsson, A., Eker J., Olsson, T., & Von Platen, C. (2010)).

Instead of relying on threads managed by the operating system kernel, the DPN model allows a continuous execution of the operations of a graph thanks to a user-level scheduler (Lee, E. A., & Parks, T. M. (1995)). This scheduler can sequentially test the firing rules from several actors, and fire an actor if a firing rule is valid. An efficient scheduling for dataflow programs consists in finding a, pre-defined or not, order of actor firings throughout the execution process

Table 2. Classification results for MPEG-4 Part 2 SP (a), MPEG-4 Part 10 PHP (b), MPEG-H Part 2 (c)

(a)		(b)		(c)		
#	%	#	%	#	%	
6	15	42	37	19	58	Static
8	20	5	4	1	3	Cyclo-static
3	1	5	4	0	0	Quasi-static
23	56	56	49	10	30	Dynamic
1	0.5	6	5	3	9	Time-dependent

capable of maximizing the use of all the processing units in one platform. Since actors in a DPN may have data-dependent behaviors, and the data are unknown in the system, determining an optimal schedule of a program is not possible at compile-time (equivalent to the halting problem (Parks, T. M. (1995))), i.e. the scheduling can be only done in the general case at run time.

The section describes several scheduling strategies designed to execute a DPN-based application on a single-core architecture, which handles the execution of only one actor at a given time. The strategies are then evaluated on two different platforms in order to be compared.

Round-Bobin

It is a simple scheduling strategy that continuously goes over the list of actors: The scheduler evaluates the firing rules of an actor, fires the actor if a rule is met and continues to evaluate the same actor until no firing rules are met then switches to the next actor. This scheduling policy guarantees to each actor an equal chance of being executed, and avoids deadlock and starvation. Contrary to classical round-robin scheduling, there is no notion of time slice: an actor is executed until it cannot fire anymore in order to minimize the number of actor switching and consequently the scheduling overhead. The reason of this actor switching is that in practice the FIFO channels will be finally full or empty because of their bounded sizes.

Figure 6 shows an application of this round-robin scheduling on the example of dataflow graph presented in *Figure 1*. The scheduler executes the actors in a circular order i.e. the five actors A, B, C, D and E are successively executed then the scheduler starts again from A and so on.

Data-Driven/Demand-Driven

This strategy is a more advanced runtime scheduling strategy. Indeed, the round-robin strategy schedules actors unconditionally i.e. the firing rules of an actor could be checked even if they are all invalid. In this case, the firing rules of the actor will be checked, but no computation will be performed, that is called a miss. As a result, the round-robin strategy becomes inefficient with

Figure 6. Example of round-robin scheduling with five actors

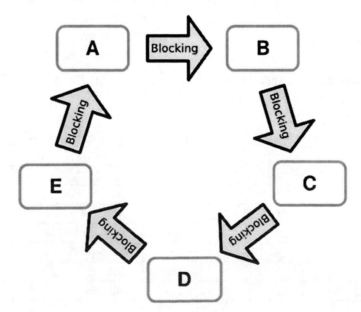

complex applications containing hundred of actors and a lot of control communications.

Data-driven / demand-driven scheduling (Yviquel, H., Casseau, E., Wipliez, M., & Raulet, M. (2011)) strategy is based on the well-known data driven and demand driven principles (Parks, T. M. (1995)). On the one hand, data-driven policy executes an actor when its input data have to be consumed to unblock the execution of the precedent actor. On the other hand, demand-driven executes an actor when its output is needed by one of its successor actor.

Two types of events can cause the blocking of an actor execution, each one is implying a different scheduling decision:

- When an actor is blocked because an input communication channel is empty, demand-driven policy is applied and ask the scheduler to execute the predecessor of this channel.
- When an actor is blocked because an output communication channel is full, data-driven policy is applied and ask the scheduler to execute the successor of this channel.

Contrary to the round-robin algorithm, a dynamic list of next schedulable actors is needed. The behavior of this schedulable list is illustrated with Figure 7. When an actor is blocked during its

Figure 7. Behavior of the dynamic list of next schedulable actor used by data-driven/demand-driven scheduling

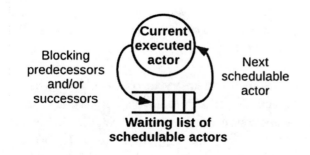

Waiting list of schedulable actors

execution, the empty or full FIFOs are identified and their associate predecessors or successors are added to the schedulable list. The actor to be executed next corresponds to the next entry in the schedulable list.

Actor Machine

Round-robin, as well as data-driven/demand-driven, strictly respects the execution model of DPN MoC defined by Lee and Parks (Lee, E. A., & Parks, T. M. (1995)). The execution of an actor is modeled by the repeated evaluation of the firing rules that are, in case of a success, followed by the firing of the associated firing function, also known as action. The firing rules can evaluate two kinds of condition, the amount of tokens available and the values of this tokens, that ultimately leads to execution of a large number of tests. A different approach, introduced by Janneck et Cedersjö, tries to reduce the number of test performed during the evaluation of firing rules using a new execution model, called actor machine (Cedersjo, G., & Janneck, J. W. (2012), Janneck, J. W. (2011)), that also considers the evaluation results of previous firing rules.

Actor machine deals with the memorization of the test results involved in the validation of previous firing rules to limit their reproduction. For instance, let two firing rules R_i and R_j tested successively such as:

$$R_i = \left[P_{i,1}, P_{i,2} \right]$$

and:

$$R_j = \left[P_{j,1}, P_{j,2} \right]$$

with:

$$P_{i,1} = P_{j,1} = \left[*, * \right];$$

if R_i is evaluated false such as:

$$R_i = \begin{bmatrix} true, false \end{bmatrix}$$

then $P_{j,1}$ could be already known valid during the evaluation of R_j and the evaluation of $P_{j,2}$ should be sufficient. To do so, the evaluations of previous patterns are preserved using an automaton mechanism. Several connected actor machines can also be composed in order to increase the potential possible reduction (Janneck, J. W. (2011)).

On the one side the scheduling of an actor machine could be more efficient compared to the traditional firing model thanks to the reduction of the number of tests performed; But on the other the translation to the actor machine execution model induces an explosion of the number of states in the scheduling algorithm due to the need of memorization. Moreover, a circular buffer implementation of the communication channel can permit an equivalent test reduction by mean of compiler optimization. Indeed common sub-expression elimination can search for identical patterns in firing rules evaluated successively, and replaces them with a single variable holding the result of their evaluation.

Quasi-Static Scheduling

In the previous sections, we have stated that one essential benefit of the DPN model lies in its strong expressive power, so as to simplify algorithm implementation for programmers. This expressive power includes: the ability to describe data-dependent computations through token production/consumption, where production/consumption may vary according to values of tokens; the ability to express non-determinism, which can be used to construct actors that respond to unpredictable sequences of tokens; and, the ability to produce time-dependent behaviors that rely on the time at which tokens are available on the input of an instance.

However, when dealing with the scalability of this model, we have stated that this strong expressive power incurs a cost on the efficiency of its implementation, as several operations may be scheduled at run-time on a single processing unit. The overhead caused by a scheduling strategy, along with its variable chance of success between test/validation of a firing rule for each operation, can create a succession of synchronization issues between the firing of actors in a description. This issue can ultimately lead to inefficient implementation of dataflow programs or to unsteady performance on their executions. The granularity of an application, i.e. the number of actors to schedule in the description, becomes an important factor that can prevent synchronization issue of instances.

The challenge when optimizing the execution of a dataflow description is then to conserve the strong expressive power of DPN while reducing the overhead caused by its required run-time scheduling. Quasi-static scheduling intends to make scheduling decisions as much as possible at compile-time by determining all static behavior and keeping only the necessary decision for run-time. The literature has introduced a large panel of methodologies to perform quasi-static scheduling of dynamic dataflow programs in different manner (Gorin, J., Wipliez, M., Prêteux, F., & Raulet, M. (2011), Gu, R., Janneck, J., Bhattacharyya, S., Raulet, M., Wipliez, M., & Plishker, W. (2009), Ersfolk, J., Roquier, G., Jokhio, F., Lilius, J., & Mattavelli, M. (2011), Ersfolk, J., Roquier, G., Lilius, J., & Mattavelli, M. (2012), Boutellier, J., Raulet, M., & Silven, O. (2013), Boutellier, J., Lucarz, C., Lafond, S., Gomez, V., & Mattavelli, M. (2009), Boutellier, J., Silven, O., & Raulet, M. (2011)).

Some of them try to prune all unreachable execution paths to remove all unnecessary tests using code instrumentation (Boutellier, J., Raulet, M., & Silven, O. (2013), Boutellier, J., Lucarz, C., Lafond, S., Gomez, V., & Mattavelli, M. (2009),

Boutellier, J., Silven, O., & Raulet, M. (2011)) or model checking (Ersfolk, J., Roquier, G., Jokhio, F., Lilius, J., & Mattavelli, M. (2011), Ersfolk, J., Roquier, G., Lilius, J., & Mattavelli, M. (2012)) to determined the possible executions. However, both of them are limited by their need of input data to perform their analysis. Such a requirement prevents the full support of all possible execution paths.

Another approach, based on the classification results, try to reduce the number of actors that are required to be scheduled at run-time, by clustering network regions that have a locally static behavior (Gorin, J., Wipliez, M., Prêteux, F., & Raulet, M. (2011), Gu, R., Janneck, J., Bhattacharyya, S., Raulet, M., Wipliez, M., & Plishker, W. (2009)). We mean by one locally static region a set of connected actors in the description that has a firing order we can determine statically, regardless the data stored in the FIFO channels of the description. The actor clustering approach is based on three existing algorithms that are applied sequentially as follow:

1. The actors with predictable behaviors present in the dataflow description are detect using actor classification as described in Section Actor Classification,

2. Predictable actors connected amongst themselves are clustered into a single node, the composite node to obtain a valid sequence of firing in it that can be determined at compile-time. As such, an essential condition to set a composite node is to determine whether such a sequence of firing is possible, the composition theorem described by Pino (Pino, J. L., & Lee, E. (1995)). The resulting cluster becomes a composite node in the graph of the dataflow description,

3. Actors grouped in a composite node are scheduled the Single-Appearance Scheduling (SAS) (Oh, H., Dutt, N., & Ha, S. (2005)), the optimum static scheduling strategy for code minimization where all repetitions of a same actor can be found side by side. The other remaining actors, along with the resulting composite nodes, are scheduled at run-time.

The methodology is illustrated on the Figure 8 on a dataflow example containing 5 actors. Each actor is firstly classified to determine, if possible, its production/consumption rates in order to detect the existing static region that can finally be scheduled.

Figure 8. Quasi-static scheduling using actor clustering

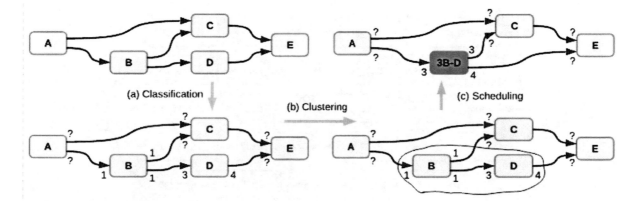

Evaluation

Quasi-static scheduling based on local regions clustering has been evaluated on the tested video decoders. Table 3 presents the number of clusters found in each description, as well as the number of actors and FIFO channels affected by the clustering.

As a more detailed example, we focus on the inverse transformation composing the description of the MPEG-H Part 2 decoder, see Figure 5. This Inverse Transformation is decomposed in five parallel paths. The first paths are dedicated to the prediction residual decoding of each existing transform units (TUs) sizes, respectively 4x4, 8x8, 16x16 and 32x32. The last one is used to compute the inverse discrete sine transform (DST) for 4x4 luma transform blocks that belong to an intra coded region.

All actors of the Inverse Transformation are classified static except the splitter and the merger. A clustering node is build for each paths with the schedule presented in Figure 9.

The experiments have been made from the following 720P sequences containing I/P/B frames:

- **MPEG-4 Part 2:** Old town cross - 25fps, 6Mbps
- **MPEG-4 Part 10:** A Place at the Table - 25fps, 6Mbps

Table 3. Clustering results.

	Clusters	Actors		FIFOs	
	#	#	%	#	%
MPEG-4 Part 2	3	6	15	3	2
MPEG-4 Part 10	8	30	26	22	5
MPEG-H Part 2	5	18	55	13	16

- **MPEG-H Part 2:** Four People - 60fps, 1Mbps

The Table 4 summarizes the results of the sequential execution of the tested application with different scheduling approaches, that are the round-robin and data-driven/demand-driven strategies with or without the clustering of locally static regions. The number of switches, firings and misses is detailed, additionally to the percentage of misses according to the total number of switches. A switch corresponds the execution of the next schedulable actor that occurs when the current actor is not fireable anymore due to the empty/full state of its communication channels; A firing corresponds to the execution of an action when all its firing rules are valid; And a miss is a switch that do not result to a firing.

Figure 9. Quasi-static scheduling of the Inverse Transformation of MPEG-H Part 2

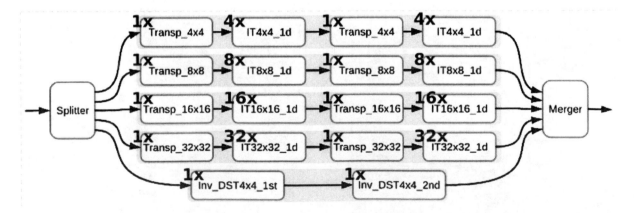

Table 4. Comparison of mono-processor scheduling strategies. The number of switches, firings and misses are expressed in .10^3

(A) MPEG-4 Part 2 SP				
Strategy	Switch	Firing	Miss	%
Robin	4306	2812677	801	19
Robin+Cluster	N/A	N/A	N/A	N/A
Driven	2420	2812654	252	10
Driven+Cluster	N/A	N/A	N/A	N/A

(B) MPEG-4 Part 10 PHP				
Strategy	Switch	Firing	Miss	%
Robin	522205	12526470	447065	86
Robin+Cluster	N/A	N/A	N/A	N/A
Driven	52096	12539650	25633	49
Driven+Cluster	50348	11824252	26369	52

(C) MPEG-H Part 2 Main				
Strategy	Switch	Firing	Miss	%
Robin	16389	232752	12855	78
Robin+Cluster	9929	226807	147	1
Driven	2808	233057	147	5
Driven+Cluster	2647	227228	91	3

From Desktop Processors

First of all, the scheduling strategies are evaluated on a desktop processor, an Intel Core i7 M640 clocked at 2.80GHz using the ANSI C backend of Orcc.

Table 5 presents the frame-rate of the tested decoder using the different scheduling strategies. The results show that even an impressive reduction of misses in the scheduling does not induce an improvement of the global performance of the application in term of frame-rate.

Past experimentations (Wipliez, M., & Raulet, M. (2010), Gorin, J., Wipliez, M., Prêteux, F., & Raulet, M. (2011), Gu, R., Janneck, J., Bhattacharyya, S., Raulet, M., Wipliez, M., & Plishker, W. (2009)) have presented interesting improvements of the performance of dynamic dataflow application using static scheduling of locally predictable regions. This difference can be explained by a more efficient implementation of the FIFO mechanism in the compiler. In the dataflow approach, the communication is one of the well-known bottleneck of the performance and one of the interest of actor clustering is the replacement of the channel by temporary variables. Consequently, the more efficient the implementation of those communication channel is, the less impressive the performance improvement induced by actor clustering is.

Such results can also be explained by the efficiency of desktop processors to handle the heavy control-oriented algorithm induce by dynamic dataflow models. In fact, contemporary processors use a large number of techniques to handle any unpredictability such as out-of-order execution and branch predicator.

Towards Embedded Platforms

The embedded multi-core platform is based on a VLIW-style architecture known as Transport-Trigger Architecture (TTA) (Corporaal, H. (1997)). TTA processors resemble VLIW processors in the sense that they fetch and execute multiple instructions each clock cycle. TTA processors do not support advanced hardware technology such as branch predicator or out-of-order execution. The architectural simplicity, joined with the extensive capacity of computation, makes it an interesting target for embedded platform. However, these properties make it very sensitive in term of performance to the control present in the application.

Run-time and quasi-static scheduling are both experimented on the inverse transformation, described in Figure 9, to compare their efficiency on a such platform. In one hand, the four actors composing the path are mapped to the same processor and scheduled dynamically with the round-robin policy. On the other hand, the four actors are clustered and scheduled using SAS in a composite node at compile-time then the resulting node is mapped to an equivalent processor. Table 6 presents the important acceleration rates obtained with the quasi-static scheduling for each TUs sizes. Such speed-ups show that the TTA processor benefits a lot more from to the compile-time predictability, offering more potential instruction-level parallelism, than the desktop processors.

CONCLUSION

The emergence of massively parallel architecture, along with the necessity of new parallel programming models, has revived the interest on dataflow programming due to its ability to express concurrency. As discussed throughout this chapter, although dynamic dataflow programming can be considered a flexible approach for the development of scalable application, there are still some open problems in concern of their execution. Since most of the literature stays focus on the study of predictable dataflow models due to their analyzability, the detection of static behavior in dynamic description aims to bridge the gap between both worlds.

As much of the experiments of this chapter demonstrate, the interest of quasi-static scheduling of dynamic dataflow programs is strongly dependent of the ability of the targeted platform to handle unpredictable behavior; a key feature therefore, is the ability of the analysis to determine the MoC of a given actor. Future research interests will include the improvement of the detection and scheduling of quasi-static MoCs, as well as, the study of such techniques in the context of a multi-core platform.

Table 5. Comparison of the frame-rate, in FPS, of the tested video decoders on 720p sequences using different scheduling approaches

Strategy	Robin		Driven	
Clustering	No	Yes	No	Yes
MPEG-4 Part 2	22.5	23	22.5	22.7
MPEG-4 Part 10	4.7	4.7	5.6	7.8
MPEG-H Part 2	14.7	14.6	15.3	15.1

Table 6. Scheduling results for xIT of HEVC

Region	Tokens	Round-Robin	Clustering	Acc
IT4x4	13696	385876	272982	1.41
IT8x8	38848	986408	664708	1.48
IT16x16	42752	2917961	1557331	1.87
IT32x32	19424	1747047	1567697	1.11

REFERENCES

Bhattacharya, B., & Bhattacharyya, S. S. (2001). Parameterized Dataflow Modeling for DSP Systems. *IEEE Transactions on Signal Processing, 49*, 2408–2421. doi:10.1109/78.950795

Bilsen, G., Engels, M., Lauwereins, R., & Peperstraete, J. (1996). Cyclo-static dataflow. *IEEE Transactions on Signal Processing, 44*(2), 397–408. doi:10.1109/78.485935

Boutellier, J., Lucarz, C., Lafond, S., Gomez, V., & Mattavelli, M. (2009). Quasi-static scheduling of CAL actor networks for Reconfigurable Video Coding. *Journal of Signal Processing Systems for Signal, Image, and Video Technology*, 1–12.

Boutellier, J., Raulet, M., & Silven, O. (2013). Automatic Hierarchical Discovery of Quasi-Static Schedules of RVC- CAL Dataflow Programs. *Journal of Signal Processing Systems for Signal, Image, and Video Technology, 71*(1), 35–40. doi:10.1007/s11265-012-0676-4

Boutellier, J., Silven, O., & Raulet, M. (2011). Scheduling of CAL actor networks based on dynamic code analysis. In *Proceedings of the IEEE International Conference on Acoustics, Speech, and Signal Processing (ICASSP)*. IEEE.

Buck, J. (1993). *Scheduling dynamic dataflow graphs with bounded memory using the token flow model*. (PhD thesis). University of California, Berkeley, CA.

Buck, J., & Lee, E. (1993). Scheduling dynamic dataflow graphs with bounded memory using the token flow model. In *Proceedings of IEEE International Conference on Acoustics, Speech, and Signal Processing*, (pp. 429-432). IEEE.

Carlsson, A., Eker, J., Olsson, T., & Von Platen, C. (2010). Scalable parallelism using dataflow programming. *Ericsson Review, 2*(1), 16–21.

Cedersjo, G., & Janneck, J. W. (2012). Toward Efficient Execution of Dataflow Actors. In *Proceedings of Signals, Systems and Computers (ASILOMAR)* (pp. 1465 – 1469). ASILOMAR.

Corporaal, H. (1997). *Microprocessor Architectures: from VLIW to TTA*. John Wiley & Sons.

Cousot, P., & Cousot, R. (1977). Abstract interpretation: a unified lattice model for static analysis of programs by construction or approximation of fixpoints. In *Proceedings of the 4th ACM Sigact-Sigplan Symposium on Principles of Programming Languages* (pp. 238–252). ACM.

Dennis, J. B. (1974). First Version of a Data Flow Procedure Language. In *Proceedings of Programming Symposium*, (pp. 362 – 376). Springer.

Eker, J., & Janneck, J. (2003). CAL Language Report. *Technical Report ERL Technical Memo UCB/ERL M03/48*. University of California at Berkeley.

Ersfolk, J., Roquier, G., Jokhio, F., Lilius, J., & Mattavelli, M. (2011). Scheduling of dynamic dataflow programs with model checking. In *Proceedings of Systems (SiPS)*, (pp. 37 – 42). SiPS.

Ersfolk, J., Roquier, G., Lilius, J., & Mattavelli, M. (2012). Scheduling of dynamic dataflow programs based on state space analysis. In *Proceedings of IEEE International Conference on Acoustics, Speech, and Signal Processing*. IEEE.

Gorin, J., Wipliez, M., Prêteux, F., & Raulet, M. (2011). LLVM-based and scalable MPEG-RVC decoder. *Journal of Real-Time Image Processing*. doi:10.1007/s11554-010-0169-2

Gu, R., Janneck, J., Bhattacharyya, S., Raulet, M., Wipliez, M., & Plishker, W. (2009). Exploring the concurrency of an MPEG RVC decoder based on dataflow program analysis. *IEEE Transactions on Circuits and Systems for Video Technology*.

Janneck, J. W. (2011). A machine model for dataflow actors and its applications. In Proceedings of Signals, Systems and Computers (ASILOMAR) (pp. 756 – 760). ASILOMAR.

Johnston, W. M., Paul Hanna, J. R., & Millar, R. J. (2004). Advances in dataflow programming languages. *ACM Computing Surveys*, *36*(1), 1–34. doi:10.1145/1013208.1013209

Kahn, G. (1974). The semantics of a simple language for parallel programming. [IFIP.]. *Proceedings of IFIP*, *74*, 471–475.

Lee, E. (2006). The problem with threads. *Computer*, *39*(5), 33–42. doi:10.1109/MC.2006.180

Lee, E., & Messerschmitt, D. (1987). Synchronous data flow. *Proceedings of the IEEE*, *75*(9), 1235–1245. doi:10.1109/PROC.1987.13876

Lee, E. A., & Parks, T. M. (1995). Dataflow Process Networks. *Proceedings of the IEEE*, *83*(5), 773–801. doi:10.1109/5.381846

Mattavelli, M., Amer, I., & Raulet, M. (2010). The Reconfigurable Video Coding Standard. *IEEE Signal Processing Magazine*, *27*(3), 159–167. doi:10.1109/MSP.2010.936032

Oh, H., Dutt, N., & Ha, S. (2005). Single appearance schedule with dynamic loop count for minimum data buffer from synchronous dataflow graphs. In *Proceedings of the 2005 international conference on Compilers, architectures and synthesis for embedded systems* (CASES '05), (pp. 157 – 165). ACM.

Parks, T. M. (1995). *Bounded Scheduling of Process Networks*. (Doctoral dissertation). Berkeley, CA.

Pino, J. L., & Lee, E. (1995). Hierarchical static scheduling of dataflow graphs onto multiple processors. In *Proceedings of International Conference on Acoustics, Speech, and Signal Processing* (ICASSP-95), (pp. 2643 – 2646). ICASSP.

Savage, J. E. (1998). *Models of computation: Exploring the power of computing*. Addison-Wesley Pub.

Sutherland, W. R. (1966). *The on-line graphical specification of computer procedures*. (PhD Thesis). MIT, Cambridge, MA.

Von Platen, C., Eker, J., Nilsson, A., & Arzen, K.-E. (2012). *Static Analysis and Transformation of Dataflow Multimedia Applications (Technical report)*. Lund University.

Wipliez, M. (2010). *Compilation Infrastructure for Dataflow Programs*. (Ph.D. Dissertation). National Institute of Applied Sciences (INSA).

Wipliez, M., & Raulet, M. (2010). Classification and Transformation of Dynamic Dataflow Programs. In *Design and Architectures for Signal and Image Processing (DASIP)*. Academic Press. doi:10.1109/DASIP.2010.5706280

Yviquel, H., Casseau, E., Wipliez, M., & Raulet, M. (2011). Efficient multicore scheduling of dataflow process networks. In *Proceedings of IEEE Workshop on Signal Processing Systems* (SiPS), (pp. 198 – 203). IEEE.

Zebelein, C., Falk, J., Haubelt, C., & Teich, J. (2008). Classification of General Data Flow Actors into Known Models of Computation. In *Proceedings of MEMOCODE*. Anaheim, CA: MEMOCODE.

KEY TERMS AND DEFINITIONS

Dataflow Models of Computation: A dataflow Model of Computation (MoC) is an abstract specification of how a computation can progress. A data MoC is useful to define precisely the semantics of a programming model, i.e. the type of components it can contain and the way they interact.

Dynamic Dataflow Model: Dynamic dataflow model, also known as Dataflow Process Network (DPN), is closely related to Kahn Process Network. Additionnaly, the DPN model introduce the concept of firings that makes it Turing-complete which means it can model any algorithm even non-deterministic ones.

Dynamic Scheduling: Since actors in a DPN may have data-dependent behaviors, and data are unknown in the system, determining an optimal schedule of a program is not possible at compile-time (equivalent to the halting problem), i.e. the scheduling can be only done in the general case at run time.

High Efficiency Video Coding: High Efficiency Video Coding (HEVC) is the last born video coding standard, developed conjointly by ISO / ITU. HEVC is improving the data compression rate, as well as the image quality, in order to handle modern video constraints such as the high image resolutions. Another key feature of this new video coding standard is its capability for parallel processing that offers scalable performance on the trendy parallel architectures.

MPEG Reconfigurable Video Coding: To overcome the lack of interoperability between all the video compression standards deployed in the market, MPEG has introduced an innovating framework, called Reconfigurable Video Coding, dedicated to the development of video coding tools in a modular and reusable fashion.

Quasi-Static Scheduling: Quasi-static scheduling involves techniques that statically schedule as many operations as possible so that only data-dependent operations are scheduled at runtime.

Synchronous Dataflow: The synchronous dataflow model (SDF) can be seen as a simplification of the DPN model, in which tokens consumption and production follow a predictable behavior, which means no data-dependent behavior.

Transport Trigger Architecture: Transport Trigger Architecture (TTA) processors resemble Very Long Instruction Word (VLIW) processors in the sense that they fetch and execute multiple instructions each clock cycle. A major difference, however, is that TTA processors have only one instruction: move, which simply transfers data from a processor internal place to another one.

ENDNOTES

[1] The Open RVC-CAL Compiler: A development framework for dataflow programs.

Chapter 13
Hierarchical Agent–Based Monitoring Systems for Dynamic Reconfiguration in NoC Platforms:
A Formal Approach

Sergey Ostroumov
Åbo Akademi University, Finland & TUCS - Turku Centre for Computer Science, Finland

Marina Waldén
Åbo Akademi University, Finland

Leonidas Tsiopoulos
Åbo Akademi University, Finland

Juha Plosila
University of Turku, Finland

ABSTRACT

A Network-On-Chip is a paradigm that tackles limitations of traditional bus-based interconnects. It allows complex applications that demand many resources to be deployed on many-core platforms effectively. To satisfy requirements on dependability, however, a NoC platform requires dynamic monitoring and reconfiguration mechanisms. In this chapter, the authors propose an agent-based management system that monitors the state of the platform and applies various reconfiguration techniques. These techniques aim at enabling uninterruptable execution of applications satisfying dependability requirements. The authors develop the proposed system within Event-B that provides a means for stepwise and correct-by-construction specification supported by mathematical proofs. Furthermore, the authors show the mechanism of decomposition of Event-B specifications such that a well-structured and hierarchical agent-based management system is derived.

INTRODUCTION

Modern applications are complex and demand a large number of resources for their intensive computations. Furthermore, the computations have to be carried out in an efficient manner. To fulfil these demands, *Network-on-Chip* (NoC) has been proposed as an efficient and scalable interconnect paradigm for many-core platforms (Benini & De Micheli, 2002). The cores of a plat-

DOI: 10.4018/978-1-4666-6034-2.ch013

form are interconnected with one structured net so that they can exchange data efficiently. However, the computations may be interrupted by various faults such as faults caused by high temperature of cores or hardware faults that occur in cores. This may lead to inappropriate results, especially for the applications from critical domains such as biomedical (Khatib, et al., 2006) or aerospace (Motamedi, Ionnides, Rümmeli, & Schagaev, 2009). To allow an application to complete its computations without interruption (i.e., produce the expected result), it is necessary to provide run-time means for monitoring the state of platform and reacting appropriately on changes.

These means are usually implemented in the form of *agents* (Rantala, Isoaho, & Tenhunen, 2007). The agents monitor the state of the platform and react to non-desired changes caused by faults by performing different procedures to tolerate these faults. Generally, the bigger the NoC platform the larger the number of agents it requires. In order for the system to manage a large number of agents, they are organised into a multi-level hierarchy. The hierarchy typically has a three-level structure for centralised schemes (Yin, et al., 2009) allowing agents to exchange the information about the state of the NoC platform efficiently and effectively.

The dependability of the system should also be maintained at a high level since inadequate behaviour of agents can cause dramatic effects. In other words, the proposed system has to be correct w.r.t. the requirements. One of the appropriate approaches for specifying reliable systems, as well as verifying their design is provided by *formal methods*. Formal development of NoC platforms and their agent-based monitoring systems allows us to prove their correctness mathematically by employing a stepwise and correct-by-construction approach. In this paper, we stepwise develop a specification of a hierarchical agent-based monitoring system for NoC platforms using the Event-B formalism (Abrial, 2010). Furthermore, we introduce various *dynamic reconfiguration* pro-

cedures into the specification on local (regional) and global (inter-regional) levels enabling effective tolerance of faults occurring in the platform. Event-B enables modelling a discrete transition system and proving the correctness of a model following the refinement-based approach. Moreover, it has a mature tool support via the Rodin platform (RODIN, 2011).

The rest of the paper is organised as follows. In the next section, we consider an agent-based management system for a NoC platform and briefly describe possible reconfiguration of the platform in case of faults. The section "The Event-B formalism" explains the essential parts of the Event-B framework used to specify the system. Furthermore, we show proof obligations that allow us to prove correctness of the specification. In addition, we describe possible decompositions of a model, which help us to cope with the model complexity. In the "Proposed approach" section, we describe the methodological steps for deriving correct agent-based systems. In the "Modelling an agent-based monitoring system for NoC platforms in Event-B" section, we develop the specification of the proposed system within the Event-B formalism. Finally, in the "Related work" section, we elaborate on related work and in the "Conclusion" section, we conclude this paper.

AGENT-BASED MONITORING SYSTEM FOR NOC PLATFORMS

A many-core platform requires special means to monitor the state of a system and take proper actions if inadequacies occur. These means are usually implemented as agents (Rantala, Isoaho, & Tenhunen, 2007). When the number of functions performed by agents and/or the number of agents grows, which is very likely in large-scale NoC, it is reasonable to form a multi-level hierarchy of the agents (Yin, et al., 2009). Typically, a three-level agent-based monitoring system for NoC platforms is considered efficient (Guang,

Yang, Plosila, Latif, & Tenhunen, 2010) and is shown in Figure 1, where a tile consists of a cell (a core) and a router. We assume NoC-based communication between tiles and do not consider its implementation details. Instead, we focus on the functionality of the agents, their communication in the hierarchy and the state of individual cells.

A *cell agent* is a local management unit attached to a particular cell (core). It monitors the state of the core by observing local parameters such as cell temperature as well as faults that may occur in the cell. The cell agent then promotes this information to higher-level agents.

A *cluster agent* monitors the state of a region (i.e., a cluster – a set of cores) where an application is mapped. It applies dynamic voltage and frequency scaling (DVFS) technique when necessary. The cluster agent also informs the platform agent about the situation in the region.

The *platform agent* manages the whole NoC platform. It performs initial mapping of applications to the platform. In addition, it may reallocate application tasks dynamically, i.e., at run-time, when necessary.

Upon a fault occurrence in a cluster, the corresponding cluster agent is the first agent that tries to take some actions to tolerate this fault. Its actions may not be sufficient in which case the platform agent takes control over the situation by reallocating tasks from faulty cells to free ones. It enables *run-time remapping* or *reallocation* allowing the computations to continue without interruption and produce the expected result.

The reallocation of a task from a faulty cell to a free one, on the one hand, allows completion of computations, on the other hand leads to reduced performance of the computations because of the longer communication distance between cells. Therefore, it is of high importance to specify the *local reconfiguration procedure* for faulty cells. This procedure aims at repairing the functionality of a faulty cell and enabling the platform agent to return the task to the reconfigured cell, which restores the performance of computations.

To guarantee correct behaviour of the agents under these circumstances, we use Event-B to stepwise develop the hierarchical specification of the agent-based monitoring and reconfiguration

Figure 1. An agent-based monitoring system for NoC

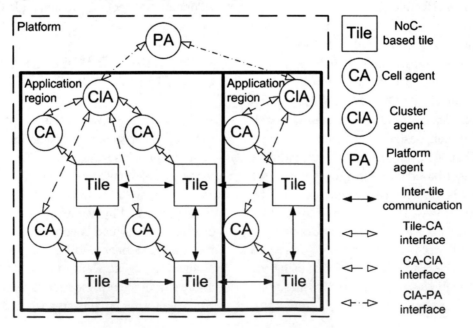

scheme. This formalism has already been used to model routing schemes for NoC platforms (Kamali, Petre, Sere, & Daneshtalab, 2011). Hence, Event-B constitutes a good basis for the correct-by-construction development of NoC-based systems.

THE EVENT-B FORMALISM

Event-B is a state-based formalism for developing systems in a correct-by-construction manner (Abrial, 2010). A system model (specification) within Event-B consists of two parts, namely a context and a machine. The detailed structure of contexts and machines and their relationships are shown in Figure 2.

The *context* defines the static part of the model – data types (*sets*), *constants*, and their properties given as a collection of *axioms*.

The *machine* describes the dynamic behaviour of the system in terms of its state (model variables) and state transitions, called *events*. The essential and guaranteed system properties are formulated as *invariants*. If the machine contains events that are executed several times in a row (convergent

events), one has to show that these events eventually terminate by providing a *variant*. The variant is a natural number whose value is decreasing each time a convergent event is executed.

The state variables of the machine are declared in the *variables* clause and initialized in the *initialisation* event. The variables are strongly typed by constraining predicates given in the *invariants* clause. The system invariant is then a conjunction of constraining predicates and the other predicates defining the system properties that should be preserved during system execution. The behaviour of the system is described by a collection of atomic events specified in the *events* clause. The syntax of an event is as follows:

e $\triangleq$ **ANY** x **WHERE** g **THEN** s **END**

where x is a list of local variables, the guard g is a conjunction of predicates over the state variables and the local variables and the action s is a collection of assignments to the state variables.

The guard is a predicate that defines the conditions under which the action can be executed, i.e., when the event is enabled. If several events are enabled simultaneously, then any of them can be chosen for execution non-deterministically. If none of the events is enabled, then the system deadlocks.

The action s of an event is a set of assignments to the state variables executed simultaneously (denoted by ||). The assignments can be either deterministic or non-deterministic. A deterministic assignment is denoted as x:= E(v), where x is a state variable and E(v) is an expression over the state variables v. A non-deterministic assignment is denoted as x:| Q(v, x'), where Q(v, x') is a predicate. As a result of a non-deterministic assignment, x gets such a value x' that Q(v, x') holds.

The semantics of Event-B events is defined using so-called *before-after predicates* (Métayer, Abrial, & Voisin, 2005). A before-after predicate describes a relationship between the system state before and after execution of an event. The formal semantics provides us with a foundation

Figure 2. The detailed structure of machine and context and their relationships

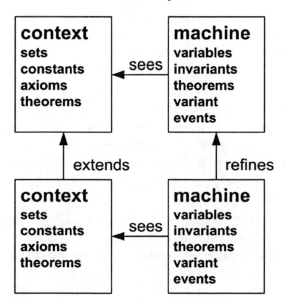

for establishing system correctness. To verify correctness (consistency) of a specification, one has to discharge two *proof obligations* for each event – *event feasibility* (FIS) and *invariant preservation* (INV):

$$Inv \wedge g_e \Rightarrow \exists v'. BA_e, (FIS)$$

$$Inv \wedge g_e \Rightarrow [BA_e]Inv (INV)$$

where Inv is a model invariant, g_e is the guard and BA_e is the before-after predicate of the event e. The primed variable v' stands for a new value assigned to the state variable v according to BA_e. The expression $[BA_e]Inv$ represents the substitution in the invariant Inv according to the before-after predicate BA_e of the event e.

Refinement in Event-B

The main development methodology of Event-B is *refinement* – the process of transforming an abstract specification into implementation. While refining an abstract specification, we introduce the required implementation details by adding new and/or replacing old data structures and events, thus bringing us closer to the eventual implementation. Proof of correctness of each refinement step is needed to establish that a more detailed machine refines its more abstract counterpart (*refines* relation in Figure 2), while its new context extends the corresponding abstract context (*extends* relation in Figure 2).

The connection between the newly introduced variables and the abstract variables is formally defined in the invariant of the refined model. For a refinement step to be valid, every possible execution of the refined machine must correspond to some execution of the abstract machine. The consistency of Event-B models as well as correctness of refinement steps should be formally demonstrated by discharging proof obligations. For the refined events, these proof obligations are *guard refinement* (GRD) and *simulation* (SIM) (Robinson, 2010):

Guard Refinement:

$$\forall S, C, S_r, C_r, V, V_r, x, x_r.$$

$$A \wedge A_r \wedge I \wedge I_r \Rightarrow (g_r \Rightarrow g), (GRD)$$

Simulation:

$$\forall S, C, S_r, C_r, V, V_r, x, x_r.$$

$$A \wedge A_r \wedge I \wedge I_r \wedge BA_{er} \Rightarrow BA_e, (SIM)$$

In a refinement, one can introduce new events that are always enabled preventing old (i.e., derived) events from execution. To avoid this, newly introduced events should disable their guards while executing their actions. Conversely, if these events modify a value of a variable in an iterative manner, one has to show that these events eventually terminate. This can be established by discharging the proof obligation (VAR):

$$\forall S, C, S_r, C_r, V, V_r, x, x_r.$$

$$A \wedge I \wedge I_r \Rightarrow Var \in \mathbb{N} \wedge [BA_{er}]Var < Var (VAR)$$

For the proof obligations mentioned above S depicts sets defined in contexts, C represents constants, V stands for a set of variables, A is a collection of axioms and I depicts a set of invariants. All letters with the subscript "r" express refined versions of the corresponding elements.

The Rodin platform (Event-B and the Rodin Platform, 2008), a tool supporting Event-B, generates the required proof obligations and attempts to prove (discharge) them automatically. Sometimes it requires user assistance provided via the interactive prover. However, in general, the tool achieves high level of automation (usually over 80%) in proving.

While refining a monolithic model, it can become rather complex to proceed with the development. To address this issue, model decomposition is proposed.

Decomposition of Event-B Models

A model can be decomposed into sub-models using the *shared-variable* (Abrial, 2009) or the *shared-event* (Butler, 2009) approach. Either of these approaches allows us to introduce a shared structure – variable(s) or event(s) – that represents an interface between machines and, therefore, sub-models (Figure 3). The decomposition of a model is rather automatic and is provided by a specific plug-in within the Rodin platform.

In this paper, we consider the shared-variable decomposition style. Consequently, each shared variable corresponds to data the agents exchange in the hierarchy.

PROPOSED APPROACH

Since an NoC platform can integrate hundreds of cores, its management should be performed effectively. To achieve this goal, we adopt the concept of agents where agents are organized into a three-level hierarchy as described in Section "Agent-based monitoring system for NoC platforms." We extend the monitoring functionality of each agent with a reconfiguration technique allowing effective management of faults. From the development point of view, a three-level hierarchy can be represented as a layered system (Waldén, 1998) as illustrated in Figure 4.

For developing complex systems such as dynamic and hierarchical agent-based management systems in a correct-by-construction manner, we propose a design approach using stepwise refinement and decomposition of a formal Event-B model. Event-B allows us to derive structured specifications in a correct-by-construction manner that can be verified with the help of the tool support. We proceed as follows with the development:

1. Specify the basic functionality of an abstract platform agent including resource request, abstract application mapping as well as reallocation procedures upon application faults.

2. Refine the abstract platform agent by introducing specific reallocation procedures.

3. Specify cluster agents by taking into account cluster parameters and faults.

4. Introduce cell agents to monitor state of cores and promote this data to higher-level agents.

5. Decompose the refined specification into platform, cluster and cell agent layers.

6. Refine the layers individually until the desired level of implementation details is achieved.

The abstract (basic) functionality of the platform agent includes requesting resources by an application in order to run its computations, mapping this application onto the NoC platform in a non-deterministic manner and performing remapping or reallocation procedures in case of faults. Here, we specify the connection between

Figure 3. Model decomposition

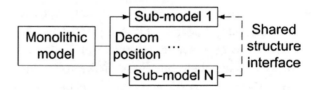

Figure 4. System development in a layered fashion

Contexts	Machines	
Applications and underlying platform	Basic functionality of the platform agent	Platform agent layer
Application mapping	Reallocation at the platform level	
Regional parameters	Cluster agents with DVFS	Cluster agents layer
Local parameters	Cell agents with local reconfiguration	Cell agents layer

applications and the platform agent. Then, we refine the abstract functionality of the platform agent by introducing a more specific mapping function and concrete reconfiguration (reallocation) mechanisms.

Since applications are mapped to regions, we introduce the functionality of the cluster agents at the next refinement step. It includes dynamic voltage and frequency scaling (DVFS) technique. We also specify the communication between the platform and cluster agents.

Finally, due to the fact that an NoC platform consists of individual cores, local monitors are required, namely cell agents. They monitor temperature and faults of the cells and promote the status to the higher-level agents. In addition, they can initiate local reconfiguration in order to restore the cell operability.

The platform agent and cell agents are fixed whilst cluster agents are dynamic. This is due to the fact that the cluster agents depend on the size of applications, i.e., on the number of cores the applications need to be mapped to. Hence, the platform agent creates, adjusts and destroys corresponding clusters and, consequently, agents.

To allow effective management of faults, the agents execute their functions in the following order. First, the platform agents performs initial mapping of an incoming application to the underlying platform. It also creates a cluster agent to monitor the region (the set of cores) where the application is mapped. Then, the monitoring starts.

The cell agents that belong to the application region monitor the state of the cells. When a high temperature or a fault is detected, the corresponding cell agent promotes this situation to the cluster agent. The cluster agent applies DVFS in order to manage the occurred inadequacy and returns control to the cell agent. The cell agent monitors the state again and promotes the monitored information to the cluster agent. This process is repeated until the situation is stabilized or frequency and voltage reach their minimum values. If the high

temperature or the fault remain in some cell, the platform agent takes over the control. It remaps either the whole application to a new region or reallocates a particular task from a faulty cell to a free one within the platform. The task reallocation procedure requires the shape adjustment of the corresponding cluster. Upon task reallocation, the local cell agent that detected inadequacy can initiate local reconfiguration of the cell. The local reconfiguration can be repeated several times until a stable state of the cell is achieved. If so, the platform agent reallocates the task back in order to restore original shape and, consequently, performance of the region. Thus, the agents can effectively manage inadequacies occurring in the NoC.

In the next section, we present the development of the hierarchical agent-based monitoring system with reconfiguration procedures in Event-B. We assume a two-dimensional NoC as the underlying many-core interconnect platform. We note however that our approach can be applied for any NoC topology.

Formal Design of an Agent-Based Management System for NoC Platforms in Event-B

The main function of the hierarchical agent-based management system lies in providing uninterruptable computations for the applications running on the NoC platform. It is achieved by specifying the three-level structure of an agent-based system that monitors the state of NoC and triggers different fault-tolerance mechanisms.

Following our development method presented in the previous section, as well as the general description of NoC management systems presented in section "Agent-based Monitoring system for NoC platforms," we start with specifying the abstract functionality of the platform agent. We then refine the platform agent to tolerate application faults by executing reallocation procedures at inter-regional level. In order for the system to tolerate faults at local (regional) levels, we incrementally introduce cluster and cell agents considering fault-tolerance mechanisms and promotion of data to higher-level agents. After adequately speficying the required functionality and communication between the agents, we derive the hierarchy of agents through decomposition. Finally, we discuss how each layer can individually be further refined, as well as we discuss the assumptions for the developed system.

Before we proceed with the formal development in Event-B, we show the summary of the used symbols in Table 1.

The Abstract Platform Agent: Basic Functionality

We first introduce a set of applications and the underlying platform. Since we do not consider specific applications, the set of applications is defined as a so- called *deferred* set. The deferred set has no elements in it when introduced but can be instantiated later on. Although this set contains no elements, it needs to be finite because the number of applications is finite.

To simplify the system development, we assume a 2D NoC mesh with the square shape to be an underlying platform. The square shape denotes a square matrix whose number of rows is the same as the number of columns. Hence, we need only

Table 1. Summary of the used symbols

Symbol	Description
$\emptyset$	The empty set
$\mathbb{P}(S)$	The power set of set S
$\mathbb{P}1(S)$	$\mathbb{P}(S) \setminus \{\emptyset\}$
card(S)	Cardinality (i.e., the number of elements) of set S
partition(S,A,B)	Enumerated set comprehension such that $S = A \cup B$ and $A \cap B = \emptyset$
finite(S)	Specifies that the set S is finite
bool(E)	Evaluates expression E and returns TRUE, if E is TRUE and FALSE, otherwise
n..m	An interval, i.e., the set of numbers starting from n and ending in m
$x \mapsto y$	An ordered pair
X×Y	Cartesian product of X and Y, i.e., the set of all possible ordered pairs where the first entry belongs to X and the second entry belongs to Y.
dom(f) $\subseteq$ S	The domain of a relation f
ran(f) $\subseteq$ T	The range of a relation f
f $\in$ S $\nrightarrow$ T	A partial function from set S to set T
f $\in$ S $\rightarrow$ T	A total function (dom(f) = S) from set S to set T
f $\in$ S $\rightarrowtail$ T	A partial injective (one-to-one) function from set S to set T
f $\in$ S $\twoheadrightarrow$ T	A partial surjective (ran(f) = T) function from set S to set T
f $\rhd$ R	Range restriction of the relation f by the set R
f $\rhd$ R	Range subtraction from the relation f the set R
f $\lhd$ R	Domain subtraction from the relation f the set R
f $\lhd$- O	Relational override of the relation f with the set O
f[g]	Relational image obtained from relation f by set g
f~	Inverse of the relation f, i.e., if $f \in S \twoheadrightarrow T$, $f\sim \in T \twoheadrightarrow S$

one constant for rows and columns, namely IPnum. The lowest number of cells the NoC can have is four (IPnum equals to 2) which describes a matrix of size 2x2. The matrix itself is represented as a set of all pairs (Cartesian product) formed from two enumerated sets ranging from one to constant IPnum. This matrix is finite because the number of cores in the NoC can only be finite. We then specify a set of applications that the system works with (Listing 1).

There are two types of applications in the system: the pending applications that are waiting for resources to be allocated to and the running applications that have already been mapped to the NoC platform and, hence, running their computations.

Pending applications represent a subset of all applications in the system. They request a particular number of resources to be allocated to in order to run their computations. Thus, we denote them as a partial function that maps applications to a number of resources they are requesting. Abstractly, an application can request a number of resources whose maximum value equals to IPnum*IPnum (i.e., the whole NoC platform):

$$\text{pending\_apps} \in \text{APPLICATIONS} \nrightarrow 1..(\text{IPnum}*\text{IPnum})$$

Running applications are the applications allocated to some cores in NoC and running their computations. They represent a subset of all applications as well:

$$\text{running\_apps} \subseteq \text{APPLICATIONS}$$

Since running applications are mapped to the NoC, the platform agent should store this dependency taking into account free regions, i.e., the regions where no application is mapped. This dependency is defined as a partial surjective function (Tab. 3) that maps resources of NoC to the running applications:

$$\text{mapping} \in \text{NoC} \nrightarrow\!\!\!\!\rightarrow \text{running\_apps}$$

Listing 1. Specifying a set of applications and a NoC

```
context NoC platform and applications
constants IPnum NoC
sets APPLICATIONS
axioms
finite(APPLICATIONS)
IPnum ∈ ℕ1
IPnum ≥ 2
NoC = 1..IPnum×1..IPnum
finite(NoC)
end
```

Each running application is not a pending one because it has been already mapped to the NoC platform and, hence, is not waiting for resources anymore. This relationship is defined by the invariance property below:

$$\forall a . a \in \text{ran(mapping)} \Rightarrow a \notin \text{dom(pending\_apps)}$$

Finally, an application can request resources whose number is not available at the moment of request. However, if there exists a region whose cardinality, i.e. the number of cores in the region, is greater or equal to the number of resources requested, then the platform agent can map the application onto this region. This situation is supported by the following invariant:

$$\forall a . a \in \text{dom(pending\_apps)} \Rightarrow$$

$$(\exists \text{reg} . \text{reg} \in \mathbb{P}1(\text{NoC}) \wedge$$

$$\text{card(reg)} \geq \text{pending\_apps}(a) \wedge$$

$$\text{mapping[reg]} = \varnothing \Rightarrow$$

$$\text{card(dom(mapping))} \leq$$

$$\text{card(dom(mapping))} + \text{card(reg)})$$

In addition to application management, we introduce application faults. Since a fault can occur in any running application, the platform agent simply tags an application where the fault has occurred. We specify this dependency as a total function (Tab. 3) that returns the result of the Boolean type where "FALSE" stands for the absence of faults and "TRUE" corresponds to their presence:

$$App\_Fault \in running\_apps \to BOOL$$

Initially, there is no application pending for resources nor running computations and, consequently, mapped onto the NoC platform.

The functioning of the system begins when an application requests a particular number of resources in order to execute its computations on them. This application belongs neither to the pending applications nor to the running ones. However, under this request the platform agent stores the mapping between the application and the number of resources it requests. This number ranges from one to IPnum*IPnum at this step (Listing 2).

If the requested number of resources is available, i.e. there exists a region in the NoC whose cardinality is at least as the number of requested resources, the platform agent maps the application onto this region. The mapping is reflected by moving this application from the set of pending applications to the set of running ones. Moreover, the platform agent stores the dependency between this region and the application that has been mapped to it (Listing 3).

In case there are not enough free resources to map an application onto, the platform agent waits until the required number of resources will be available. This occurs when one of the running applications completes its computations. This application is removed from the set of running applications and, hence, frees resources (Listing 4).

The platform agent also manages reallocation of either the whole application or the task from a faulty cell. In our system, this decision is made non-deterministically. However, a more precise order of reallocation can be instantiated depending on particular requirements.

Listing 2. Modelling a resources request

```
event Request_resources
any app res_num
where
app ∈ APPLICATIONS ∧ app ∉ dom(pending_apps) ∧
app ∉ ran(mapping) ∧ res_num ∈ 1..IPnum*IPnum
then
pending_apps := pending_apps ∪ {app↦res_num}
end
```

Listing 3. Mapping an application onto resources

```
event Resources_found
any app region
where
app ∈ dom(pending_apps) ∧ region ∈ P1(NoC) ∧
card(region) ≥ pending_apps(app) ∧
mapping[region] = ∅
then
running_apps := running_apps ∪ {app} ||
mapping := mapping ∪ (region×{app}) ||
pending_apps := {app} ⩤ pending_apps ||
App_Fault := App_Fault ∪ {app↦FALSE}
end
```

Listing 4. Freeing resources

```
event Computations_over
any app
where app ∈ ran(mapping)
then
running_apps := running_apps ∖ {app} ||
mapping := mapping ⩥ {app} ||
App_Fault := {app} ⩤ App_Fault
end
```

During the remapping of a whole application the platform agent moves a running application from the set of running ones to the set of pending ones, thus, freeing the resources. Then, it initiates the finding and mapping procedures as if this application is a pending one (Listing 5).

The remapping of a whole application keeps the shape of the application region the same, which preserves performance. However, this procedure may be very time consuming, especially in a large-scale NoC platform, due to remapping of all the application data to another region. Moreover, it initiates another searching procedure for finding an appropriate region that corresponds to the application needs. Therefore, the reallocation of a task from one cell should be taken into account, since finding a single free cell is much easier than finding a region and produces less delays.

In case of reallocating a task from a cell, the platform agent moves the task from one cell to another cell, therefore, substituting one with another. These cells must be different and the newly chosen cell must not be assigned to any application. Furthermore, the cell whose task to be reallocated should not be the only cell in the region that belongs to the application. Otherwise, it would be the remapping of the whole application (Listing 6).

Notice that whenever the computations complete or reallocation is invoked, the platform agent removes the fault tag from the application. This is because the freeing procedure completely removes information about the application while the reallocation procedures aim at eliminating faults so that the computations can continue without interruption.

To reallocate applications or tasks, the platform agent needs to know whether an application has a fault. The fault occurrence is a random process that has a non-deterministic behaviour. Hence, we model it with the event App_Fault_Monitor as shown in Listing. 7.

At this point, we have derived the abstract specification of the platform agent. The excerpt of this specification is shown in Figure 5.

Now we can refine the abstract platform agent considering more specific functions of the platform agent as well as fault management at the platform level.

The First Refinement: Platform Level Reconfiguration

At this refinement step, we introduce a function for mapping applications to the NoC platform in a regional manner. For simplicity, we assume that applications are to be allocated to rectangular regions that consist of at most two cores in a column (two rows) and a number of cores in these two

Listing 5. Reallocating the whole application

```
event Reallocate_app
any app
where
app ∈ ran(mapping) ∧
card(dom(mapping ▷ {app})) ∈ 1..IPnum∗IPnum ∥
App_Fault(app) = TRUE
then
running_apps := running_apps ∖ {app} ∥
mapping := mapping ▷ {app} ∥
App_Fault := {app} ◁ App_Fault ∥
pending_apps :=
pending_apps ∪ {app↦card(dom(mapping ▷ {app}))}
end
```

Listing 6. Reallocating a task from core with coordinates (k,l) to core with coordinates (x,y)

```
event Reallocate_task
any app x y k l
where
app ∈ ran(mapping) ∧
(x↦y) ∈ NoC ∖ dom(mapping) ∧
k↦l ∈ dom(mapping ▷ {app}) ∧
({k↦l} ◁ mapping) ▷ {app} ≠ ∅
then
mapping := {k↦l} ◁ (mapping ∪ {(x↦y)↦app}) ∥
App_Fault := App_Fault ◁- {app↦FALSE}
end
```

Listing 7. Monitoring faults of applications

```
event App_Fault_Monitor
any app f
where
app ∈ ran(mapping) ∧ f ∈ BOOL
then
App_Fault := App_Fault ◁- {app↦f}
end
```

rows (a number of columns). Then, the function for mapping applications to regions takes three arguments – the number of requested resources, current row and current column – and returns a region of the rectangular shape. The number of cores in this rectangular region equals to the multiplication result of two rows and a number of columns such that it conforms to the number of requested resources (Listing 8). Taking into account some other specific topology of the NoC platform than the 2D mesh, this function may differ.

Figure 5. The excerpt of the abstract platform agent

```
machine Abstract platform agent
sees NoC platform and applications
variables running_apps, pending_apps, mapping, App_Fault
invariants
  ...
  ∀a . a∈ran(mapping) ⇒ a∉dom(pending_apps)
  ∀a . a∈dom(pending_apps) ⇒
   (∃reg . reg∈ℙ1(NoC) ∧ card(reg) ≥ pending_apps(a) ∧
   mapping[reg] = ∅ ⇒
    card(dom(mapping)) ≤ card(dom(mapping))+card(reg))

events
  event INITIALISATION
  event Request_resources
  event Resources_found
  event Computations_over
  event App_Fault_Monitor
  event Reallocate_app
  event Reallocate_task
```

Listing 8. Introducing a function for concrete mapping

```
context Mapping function
extends NoC platform and applications
constants mapfun_hor
axioms
mapfun_hor ∈ 1..(2∗IPnum)×NoC ⇸ ℙ1(NoC)
partition(dom(mapfun_hor), {1↦(x↦y) | x↦y∈NoC},
{z↦(x↦y) | z∈2..2∗IPnum ∧
x∈1..(IPnum−1) ∧ y∈1..(IPnum+1−(z+1)÷2)})
∀x,y . (x↦y)∈NoC ⇒ mapfun_hor(1↦(x↦y))=x..x×y..y
∀x,y,z . x∈1..(IPnum−1) ∧
y∈1..(IPnum+1−(z+1)÷2) ∧ z∈2..2∗IPnum ⇒
mapfun_hor(z↦(x↦y))=x..(x+1)×y..(y−1+(z+1)÷2)
end
```

Together with this function, we introduce two variables modelling the coordinates in the NoC platform. One (r) shall be the index for rows and the other one (c) shall be the index for columns:

$r \in 1..IPnum, c \in 1..IPnum$

Considering a horizontal rectangular region, we refine the amount of resources an application can request. In this refinement this amount ranges between one and 2*IPnum. This limitation is stated as the invariant below:

$\forall a . a \in dom(pending\_apps) \Rightarrow$

$pending\_apps(a) \in 1..2*IPnum$

In this refinement step, we specify reallocation procedures performed by the platform agent in case of faults in a more concrete manner. In particular, when a task is reallocated from a faulty cell the platform agent stores the reallocation trace. The trace is represented as a partially injective (one-to-one, Tab. 3) function that accumulates the coordinates of the faulty cells and their substitutions. The platform agent can only reallocate a task to a free cell if that cell is not assigned to any application and does not belong to trace storing faulty cells. However, if a task has been reallocated there exists a cell that substitutes the one whose task has been reallocated and, thus, belongs to the region of a running application. This invariance property is supported by the following type predicate:

$Platform\_Cell\_Trace \in$

$NoC \setminus dom(mapping) \rightarrowtail dom(mapping)$

In addition, the platform agent marks a cell whose task has been reallocated in order to identify it and return its task when the local reconfiguration procedure is completed. Considering that the reallocation procedure can take place for any cell, the marking is a total function. The function returns "TRUE," if a task of a cell has been reallocated and "FALSE" otherwise:

$Reallocated \in NoC \rightarrow BOOL$

Consequently, if a cell has a trace, then it is also marked and vice versa. This dependency is specified by the following invariance property:

$\forall x . x \in NoC \Rightarrow$

$(x \in dom(Platform\_Cell\_Trace) \Leftrightarrow Reallocated(x) = TRUE)$

Those cells whose tasks are neither reallocated nor traced either belong to the regions where the running applications are allocated or they are just free cells in the platform. Thus, any cell used in computations has no trace neither has its task reallocated. This is stated using the invariant below:

$$\forall a \, . \, a \in \mathrm{ran(mapping)} \Rightarrow$$

$$\mathrm{Reallocated[dom(mapping \rhd \{a\})]} = \{FALSE\}$$

Initially, indices are set to the leftmost and uppermost cell (r:=1 and c:=1) and there are no faults in the NoC platform. Moreover, cells have no trace nor the tasks are reallocated since there are no applications running in the platform. From this point and until the end of the development we present only newly introduced formal structures in existing events, and complete new events of the refinements.

Since applications are to be mapped to rectangular regions, the resource requests are refined accordingly. Particularly, an application can request 1 resource at least and 2*IPnum resources at most (Listing 9).where "..." represents structures that appear in the refined specification in the same form as in the more abstract one, i.e., they are not changed in the refinement.

To search for the resources in a more detailed manner, we introduce two new events. One of them searches for resources in a row by changing the index of columns. The other one searches for resources in columns by increasing the value of rows. Consequently, the platform agent can always find an appropriate region if such a region exists. Both events are convergent, i.e. they must stop eventually. This non-divergence is ensured by the variant that is defined as a natural number. The value of this number must decrease every time when the value of the coordinates on the NoC platform increases. The searching procedure always starts from the uppermost and leftmost location of the 2D NoC mesh and ends in the bottommost rightmost region:

$$\mathrm{IPnum*IPnum-(r-1)*IPnum-c}$$

The searching procedure is executed while the mapping function returns a non-empty set. Specifically, it searches for a region where no application is mapped (the returning value is the empty set). If an application requests 2 resources the greatest value of the column index equals to IPnum. Since n÷m is an integer division, the substitution of pending(app) with 2 produces IPnum+1-(2+1)÷2=IPnum. If an application requests 2*IPnum resources, i.e. the greatest possible number, the value of the column index equals to 1. In other words, the substitution of pending(app) with 2*IPnum produces IPnum+1-(2*IPnum+1)÷2=1 (Listing 10).

If there are no resources available in a row the platform agent increases the current value of the row index and sets the value of the column index to one, i.e. starts searching from the leftmost cell in a row. The greatest value a row index can have equals to IPnum-1 or IPnum depending on the number of requested resources (Listing 11).

In the abstract specification the platform agent did not perform any functions if the required number of resources had not been found. In this refinement, if the searching procedure fails the platform agent sets the indexes to the initial val-

Listing 9. Refining a resources request

```
event Request_resources refines Request_resources
where ... ∧ res_num ∈ 1..2*IPnum
then ...
end
```

Listing 10. Searching for resources in rows

```
convergent event Find_resources_in_row
any app
where
app ∈ dom(pending_apps) ∧
pending_apps(app)↦(r↦c) ∈ dom(mapfun_hor) ∧
mapping[mapfun_hor(
pending_apps(app)↦(r↦c))]≠∅ ∧
(pending_apps(app) = 1 ⇒ c < IPnum) ∧
(pending_apps(app) ∈ 2..2*IPnum ⇒
c < IPnum+1−(pending_apps(app)+1)÷2)
pending_apps(app)↦(r↦c)
then
c := c+1
end
```

ues because a running application can complete its computation and release resources. Therefore, starting a new searching procedure enables the platform agent to find an appropriate region (Listing 12). Please note that this event disables itself, which follows the refinement rules.

If the searching procedure is successful, the platform agent maps an application to the found region. In addition, it resets the indices because while the searching procedure is in progress some of the running applications can complete their computations freeing resources. Moreover, the platform agent initializes parameters that belong to the application being mapped. For the application that starts its computations there are no faults occurred, no cells reallocated and, hence, no traces stored.

At the previous step, an application was mapped to a region that was chosen randomly. In this refinement, the region is chosen following a specific procedure. Consequently, the local variable representing the region in the abstract specification (i.e., region in the event Resources_found) disappears now. Instead, we provide the concrete value for this variable. This value is returned by the mapping function specified in the context of this refinement (region = mapfun_hor(pending_apps(app)$\mapsto$ ($r \mapsto c$))). As a result, all the occurrences of the disappearing local variable are substituted with the result of the function (Listing 13, where X = mapfun_hor(pending_apps(app)$\mapsto$($r \mapsto c$)) is the region returned by the mapping function).

When a running application completes its computations, the platform agent removes all information about faults, reallocated cells and traces belonging to this application. In addition, it initialises the indices so that execution of a new searching procedure for an application requesting resources may be successful (Listing 14).

The presence of a fault can initiate the reallocation procedures performed by the platform agent. We notice that at this level the platform agent does not know the exact location of the faulty cell since the cell agents are not yet introduced. Hence, the

Listing 11. Searching for resources in columns

```
convergent event Find_resources_in_col
any app
where
app∈dom(pending_apps) ∧
pending_apps(app)↦(r↦c) ∈ dom(mapfun_hor) ∧
mapping[mapfun_hor(
pending_apps(app)↦(r↦c))]≠∅ ∧
(pending_apps(app)=1 ⇒ c=IPnum ∧ r<IPnum) ∧
(pending_apps(app) ∈ 2..2∗IPnum ⇒
c=IPnum+1−(pending_apps(app)+1)÷2 ∧ r<IPnum−1)
then
r := r+1 ∥ c := 1
end
```

Listing 12. Resetting indices if resources are not found

```
event Resources_not_found
any app
where
app∈dom(pending_apps) ∧
pending_apps(app)↦(r↦c) ∈ dom(mapfun_hor) ∧
mapping[mapfun_hor(pending_apps(app)↦(r↦c))]≠∅ ∧
(pending_apps(app)=1 ⇒ c = IPnum ∧ r = IPnum) ∧
(pending_apps(app) ∈ 2..2∗IPnum ⇒
r=IPnum−1 ∧ c=IPnum+1−(pending_apps(app)+1)÷2)
then
r,c := 1,1
end
```

Listing 13. Concrete application mapping to the found resources

```
event Resources_found refines Resources_found
any app
where ... ∧
pending_apps(app)↦(r↦c) ∈ dom(mapfun_hor) ∧
mapping[X] ≠ ∅ ∧(card(X) ≥ pending_apps(app)) ∧
Reallocated[X] = {FALSE} ∧
(pending_apps(app) ∈ 2..2∗IPnum ⇒
r ∈ 1..(IPnum−1) ∧
c∈1..(IPnum+1−(pending_apps(app)+1)÷2))
then ... ∥ r,c := 1,1 ∥ mapping := mapping ∪ (X×{app})
end
```

Listing 14. Freeing resources refined

```
event Computations_over refines Computations_over
where ...
then ... ∥
Reallocated := Reallocated ◁-
(Platform_Cell_Trace~[dom(mapping▷{app})]×{FALSE}) ∥
Platform_Cell_Trace :=
Platform_Cell_Trace ▷ (dom(mapping ▷ {app}))
end
```

platform agent performs the task reallocation from the faulty cell non-deterministically. This non-determinism will be eliminated when more details about cluster and cell agents are introduced in later refinements.

The occurrence of a fault in the region that belongs to a running application triggers reallocation procedures at the global (platform) level. The platform agent either remaps the whole application or reallocates a task from the faulty cell.

The application remapping is performed in the same manner as at the abstract level. The information connected to the application being remapped (the reallocated cells and the traces of these cells) is removed. Furthermore, the platform agent assigns a particular value to the coordinate indexes wherefrom to start searching for resources in a new region. It prevents the platform agent from mapping the application to the same region where the fault has occurred (Listing 15). In Listing 15, $Z = \text{dom}(\text{mapping} \triangleright \{\text{app}\}) -$ application region.

The task reallocation event is split into two cases. The first one is the task reallocation from a faulty core to a free one. In this case, the platform agent searches for a free cell in the NoC platform and if such a cell exists, the platform agent moves the task there. Additionally, the platform agent stores the trace between the coordinates of the faulty cell and its substitution and marks the faulty cell. Since the task has been reallocated (i.e., the faulty cell is excluded from the application region), the platform agent assumes that the fault is not present anymore in the region that belongs to the running application (Listing 16). Notice however that the task reallocation from a faulty cell takes place only if there is an application fault (App_Fault(app) = TRUE).

The other case considers moving the task back when the faulty cell is repaired. To return a task to its original cell, the platform agent uses the information about the cell tracing and marking. This procedure is performed if the faulty cell has been repaired through the local reconfiguration at the hardware level. The returning of tasks helps

the application to restore the performance of its computations (Listing 17). We notice that the reallocation of a task back can take place only if there is no application fault (App_Fault(app) = FALSE). Furthermore, due to that, the platform agent does not remove the fault tag from the application when returning the task to its previous location.

The refined model of the platform agent includes functions, variables and events for searching resources in the NoC platform in a more concrete manner. In addition, this refinement contains procedures for application remapping and task

Listing 15. Refining the reallocation of the whole application

```
event Reallocate_app refines Reallocate_app
any x y
where ... ∧
 card(Z) ∈ 1..2∗IPnum ∧
card(Z↦(x↦y))∈dom(mapfun_hor) ∧
mapping[mapfun_hor(card(Z↦(x↦y)))] = ∅ ∧
(card(Z) = 1 ⇒ c<IPnum ∧ r<IPnum) ∧
(card(Z) ∈ 2..2∗IPnum ⇒ (x↦y) ∈
1..(IPnum−1)×1..(IPnum+1−card(Z)÷2))
then ... ‖ r,c := x,y ‖
Platform_Cell_Trace := Platform_Cell_Trace ▷ (Z) ‖
Reallocated :=
Reallocated ◁- (Platform_Cell_Trace∼[Z] × {FALSE})
end
```

Listing 16. Task reallocation from a faulty cell

```
event Reallocate_task refines Reallocate_task
where ... ∧ App_Fault(app) = TRUE ∧
(x↦y) ∉ ran(Platform_Cell_Trace) ∧
(k↦l) ∉ ran(Platform_Cell_Trace)
then ... ‖
Platform_Cell_Trace:=Platform_Cell_Trace∪{(k↦l)↦(x↦y)} ‖
Reallocated := Reallocated ◁- {(k↦l)↦TRUE}
end
```

Listing 17. Returning task to reconfigured cell

```
event Return_task refines Reallocate_task
where ... ∧ App_Fault(app) = FALSE ∧
(x↦y)↦(k↦l) ∈ Platform_Cell_Trace
then ... ‖
Platform_Cell_Trace := {x↦y} ◁ Platform_Cell_Trace ‖
Reallocated := Reallocated ◁- {(x↦y)↦FALSE}
end
```

reallocation upon fault occurrences. The excerpt of this refinement is shown in Figure 6. By refining the model further, we introduce variables and events that represent cluster agents.

The Second Refinement: Introducing Cluster Agents

A cluster agent monitors the state of a corresponding region where a running application is allocated. Additionally, it adjusts various regional parameters such as frequency and/or voltage. It allows the system to tolerate faults at the cluster level. The cluster agent promotes the information about the state of the region it monitors as well as the value of regional parameters to the platform agent. In case of fault occurrence in this region and the reallocation of a task by the platform agent, the topology of the region is changed. This leads to adaptation of the cluster agent to the new shape of this region. Therefore, the platform agent dynamically reconfigures the region and adapts the cluster agent to the new conditions.

The regions managed by cluster agents are represented as elements of the power set of NoC. We assume that cells that do not belong to any cluster are free and non-active so that power consumption is minimal.

In this refinement, we consider frequency and voltage as the parameters that can be adjusted. These parameters have some limits. Commonly, modern chips support voltage supply whose value ranges from 0.8V to 5.0V, i.e., it has a value of the type real. However, formal methods do not support real numbers since the set of real number is infinite and uncountable which leads to impossibility to prove properties over them. Therefore, we assume that the real decimal value representing voltage corresponds to a natural number divided by 10. That is 0.8V is equivalent to 8 in terms of the formalism, 1.2V corresponds to 12 etc.

We start this refinement with the introduction of constants that represent boundaries for frequency and voltage parameters. Frequency has only one boundary, namely the upper limit, because it ranges from zero to some maximum value. Voltage supply has two boundaries corresponding to its minimum and maximum values. In addition, we state that the voltage minimum is strictly less than its maximum (Listing 18).

To determine the exact region where an application is mapped, the cluster agent refers to the mapping function. The power set of this function *domain* represents regions, i.e. clusters, assigned to the running applications. Therefore, the functions of the cluster agent take the domain of the mapping function restricted to a particular running application as an argument.

Figure 6. The excerpt of the first refinement

```
machine Specific platform agent
refines Abstract platform agent
sees Mapping function
variables ..., r, c, Platform_Cell_Trace, Reallocated
invariants
  ...
  ∀a . a∈dom(pending_apps) ⇒ pending_apps(a)∈1..2*IPnum
  ∀x . x∈NoC ⇒
    (x∈dom(Platform_Cell_Trace) ⇔ Reallocated(x) = TRUE)
  ∀a . a∈ran(mapping) ⇒
    Reallocated[dom(mapping ▷ {a})] = {FALSE}
variant IPnum*IPnum-(r-1)*IPnum-c
events
  event INITIALISATION extends INITIALISATION
  event Request_resources refines Request_resources
  convergent event Find_resources_in_row
  convergent event Find_resources_in_col
  event Resources_found refines Resources_found
  event Resources_not_found
  event Computations_over refines Computations_over
  event App_Fault_Monitor refines App_Fault_Monitor
  event Reallocate_app refines Reallocate_app
  event Reallocate_task refines Reallocate_task
  event Return_task refines Reallocate_task
```

Listing 18. Introducing parameters for regional control

```
context Regional level parameters
extends Mapping function
constants Max_Freq Min_Volt Max_Volt
axioms
Max_Freq ∈ ℕ1
Min_Volt ∈ ℕ1
Max_Volt ∈ ℕ1
Min_Volt < Max_Volt
end
```

The faults within the clusters are represented as a function that takes a region as an argument and returns the result of type Boolean. Value "TRUE" of this result stands for the fault presence while value "FALSE" denotes the fault absence:

$$\text{Cluster\_Fault} \in \{x \mid \exists a . a \in \text{ran(mapping)} \land$$

$$x = \text{dom(mapping} \rhd \{a\})\} \rightarrow \text{BOOL}$$

One reconfiguration mechanism used in many-core platforms is dynamic frequency and voltage scaling. The cluster agent adjusts voltage supply and frequency under the circumstances of faults. Since the cluster agent adjusts frequency and voltage dynamically, we introduce new variables for them. These variables are defined as functions that take a region as an argument and return a corresponding result, i.e. the current value of frequency or voltage within their boundaries:

$$\text{Cluster\_Freq} \in \text{dom(Cluster\_Fault)} \rightarrow 0..\text{Max\_Freq}$$

$$\text{Cluster\_Voltage} \in \text{dom(Cluster\_Fault)} \rightarrow \text{Min\_Volt..Max\_Volt}$$

To allow communication between the platform and cluster agents, the following variable is required. This is the shared variable that returns "TRUE" if a new value has been read and "FALSE" otherwise:

$$\text{Cluster\_Read} \in \text{dom(Cluster\_Fault)} \rightarrow \text{BOOL}$$

The execution order of the cluster functions is deterministically specified. First, a cluster agent monitors cluster faults. If the faults are present, the cluster agent decreases frequency first. The decrease of frequency may not be enough in which case the cluster agent reduces voltage supply. To specify this order, we introduce another variable that is modified by the cluster agent only, namely Cluster_Exec:

$$\text{Cluster\_Exec} \in \text{dom(Cluster\_Fault)} \rightarrow \text{BOOL}$$

In case when the platform agent maps an application to a region in the NoC platform, it dynamically creates a cluster agent for this region and initialises regional parameters for this cluster. In other words, every running application performs its computations with some voltage supply and at some frequency. Furthermore, the cluster agent monitors faults that can occur in this region.

The presence or absence of a fault in a running application stands for the presence or absence of the fault in the region which belongs to this application. This relationship is maintained by the following invariant:

$$\forall a . a \in \text{ran(mapping)} \land$$

$$\text{Cluster\_Read(dom(mapping} \rhd \{a\})) = \text{FALSE}$$
$$\Rightarrow$$

$$\text{Cluster\_Fault(dom(mapping} \rhd \{a\})) = \text{App\_Fault(a)}$$

When a fault occurs, the platform agent tries to diminish frequency and voltage supply for the region where this fault has occurred. This adjustment allows the application located on this region to continue its computations with lower speed without interruption. However, if the decrease of these parameters for the running application reaches their minimum, then the cluster fault remains:

$$\forall a . a \in \text{ran(mapping)} \Rightarrow$$

$$(\text{Cluster\_Freq(dom(mapping} \rhd \{a\})) = 0 \land$$

Cluster_Voltage(dom(mapping $\triangleright$ {a})) = Min_Volt $\Rightarrow$ Cluster_Fault(dom(mapping $\triangleright$ {a})) = TRUE)

The Platform Agent Functionality

Initially, no cluster agents are created and, hence there is no parameter to control. When the platform agent has found the appropriate region to map an application onto, it does this and creates the cluster agent for this region. The platform agent sets up frequency and voltage supply to the maximum possible value so that the computations are running at the fastest speed. In addition, at the moment of the application mapping there are no faults in the cluster, as shown in Listing 19, where X = mapfun_hor(pending_apps(app)$\mapsto$(r$\mapsto$c)).

When the computations are completed, the platform agent destroys the cluster agent that has been monitoring the region where these computations were performed In Listing 20, X = {dom(mapping $\triangleright$ {app})}.

The introduction of the cluster agents allows us to specify application faults more specifically. In particular, the platform agent is now aware that the application fault is essentially the cluster fault. Hence, we refine the platform fault monitoring event (App_Fault_Monitor) in order to take into account this knowledge (Listing 21). We provide a specific value for the disappearing local variable, namely f = Cluster_Fault(dom(mapping $\triangleright$ {app})).

In case of remapping the whole application, the platform agent destroys the cluster agent that has been monitoring the region this application has been mapped to. Then, it creates a new cluster and, consequently, a new cluster agent to monitor the new region where the application has been remapped. In Listing 22, X = {dom(mapping $\triangleright$ {app})}.

In case of the task reallocation from a cell, there is no need to destroy the cluster agent but to adjust it to the new topology. Hence, the platform agent adapts the cluster agent to the new region

Listing 19. Creating cluster agent when application is mapped to some region

```
event Resources_found refines Resources_found
where ... ∧
then ... ||
Cluster_Freq := Cluster_Freq ∪ {X↦Max_Freq} ||
Cluster_Voltage:= Cluster_Voltage ∪ {X↦Max_Volt} ||
Cluster_Fault := Cluster_Fault ∪ {X↦FALSE} ||
Cluster_Read := Cluster_Read ∪ {X↦FALSE} ||
Cluster_Exec := Cluster_Exec ∪ {X↦FALSE}
end
```

Listing 20. Destroying the cluster agent when resources are freed

```
event Computations_over refines Computations_over
where ... ∧
then ... ||
Cluster_Freq:= X ◁ Cluster_Freq ||
Cluster_Voltage:= X ◁ Cluster_Voltage ||
Cluster_Fault := X ◁ Cluster_Fault ||
Cluster_Read := X ◁ Cluster_Read ||
Cluster_Exec := X ◁ Cluster_Exec
end
```

Listing 21. Application faults as cluster faults

```
event App_Fault_Monitor refines App_Fault_Monitor
any app
where
app ∈ ran(mapping) ∧
Cluster_Read(dom(mapping ▷ {app})) = TRUE
then
App_Fault := App_Fault ◁-
{app↦Cluster_Fault(dom(mapping ▷ {app}))} ||
Cluster_Read := Cluster_Read ◁-
{dom(mapping▷{app})↦FALSE}
end
```

Listing 22. Reallocating an application considering regional parameters

```
event Reallocate_app refines Reallocate_app
where ... ∧
Cluster_Freq(dom(mapping ▷ {app})) = 0 ∧
Cluster_Voltage(dom(mapping ▷ {app})) = Min_Volt ∧
Cluster_Read(dom(mapping ▷ {app})) = FALSE
then ... ||
Cluster_Freq := X ◁ Cluster_Freq ||
Cluster_Voltage := X ◁ Cluster_Voltage ||
Cluster_Fault := X ◁ Cluster_Fault ||
Cluster_Read := X ◁ Cluster_Read ||
Cluster_Exec := X ◁ Cluster_Exec
end
```

topology where the faulty cell is substituted with some free cell. Additionally, it runs the computations at the maximum speed In Listing 23, X = {dom(mapping $\rhd$ {app})} is the current application region and Y = dom({k$\mapsto$l} $\lhd$ mapping $\rhd$ {app}) $\cup$ {x$\mapsto$y} is a new region such that faulty cell (k$\mapsto$l) is substituted with a free cell (x$\mapsto$y) and Z = Cluster_Exec(dom(mapping$\rhd$ {app})) – the current value of Cluster_Exec.

By applying a local reconfiguration procedure, the faulty cell can be repaired. If the reconfiguration procedure has been utilized successfully, the platform agent returns the reallocated task to the reconfigured cell and adapts the cluster agent to the change of topology. This allows the computations to restore their performance. In Listing 24, X = {dom(mapping $\rhd$ {app})} is the current application region, Y = dom({k$\mapsto$l} $\lhd$ (mapping $\cup$ {(x$\mapsto$y)$\mapsto$app}) $\rhd$ {app}) is a new region such that substituting cell (k$\mapsto$l) is replaced with the reconfigured cell (x$\mapsto$y) and Z = Cluster_Exec(dom(mapping$\rhd${app})).

The Cluster Agents Functionality

While a running application performs its computations, a fault in the cluster can occur. The cluster agent detects a fault and promotes the information about it to the platform agent as shown in Listing 25.

Please notice that Cluster_Exec variable obtains the same value as the Cluster_Fault variable. This is because DVFS is only applied if a fault has occurred. However, if there is no fault the monitoring has to be repeated.

The presence of a fault triggers the cluster agent to take some actions for tolerating this fault. These actions include frequency and voltage adjustment. The cluster agent diminishes frequency that the cluster is running at. This procedure is applied repeatedly unless the value of frequency reaches its minimum, i.e., zero (Listing 26).

In case the value of frequency reaches zero, the cluster agent reduces voltage supply. Each

Listing 23. Reallocating tasks from faulty cells considering regional parameters

```
event Reallocate_task refines Reallocate_task
where ... ∧
Cluster_Voltage(dom(mapping ▷ {app})) = Min_Volt ∧
Cluster_Freq(dom(mapping ▷ {app})) = 0 ∧
Cluster_Read(dom(mapping ▷ {app})) = FALSE
then ... ||
Cluster_Freq := (X ◁ Cluster_Freq) ∪ {Y↦Max_Freq} ||
Cluster_Fault := (X ◁ Cluster_Fault) ∪ {Y↦FALSE} ||
Cluster_Read := (X ◁ Cluster_Read) ∪ {Y↦FALSE} ||
Cluster_Exec := (X ◁ Cluster_Exec) ∪ {Y↦Z} ||
Cluster_Voltage := (X ◁ Cluster_Voltage) ∪ {Y↦Max_Volt}
end
```

Listing 24. Returning the reallocated tasks considering regional parameters

```
event Return_task refines Return_task
where ... ∧
Cluster_Freq(dom(mapping▷{app})) ≠ 0 ∧
Cluster_Voltage(dom(mapping▷{app}))≠Min_Volt
then
Cluster_Fault := (X ◁ Cluster_Fault) ∪ {Y↦FALSE} ||
Cluster_Freq := (X ◁ Cluster_Freq) ∪ {Y↦Max_Freq} ||
Cluster_Voltage := (X◁Cluster_Voltage) ∪ {Y↦Max_Volt} ||
Cluster_Read := (X ◁ Cluster_Read) ∪ {Y↦FALSE} ||
Cluster_Exec := (X ◁ Cluster_Exec) ∪ {Y↦Z}
end
```

Listing 25. Monitoring cluster faults

```
event Cluster_Fault_Monitor
any app f
where
app ∈ ran(mapping) ∧ f ∈BOOL ∧
Cluster_Freq(dom(mapping ▷ {app})) ≠ 0 ∧
Cluster_Read(dom(mapping▷{app})) = FALSE ∧
Cluster_Exec(dom(mapping▷{app})) = FALSE
then
Cluster_Read := Cluster_Read ◁-
{dom(mapping▷{app})↦TRUE} ||
Cluster_Exec:=Cluster_Exec ◁-{dom(mapping▷{app})↦f} ||
Cluster_Fault := Cluster_Fault◁-{dom(mapping▷{app})↦f}
end
```

Listing 26. Diminishing frequency

```
event Diminish_Freq
any app
where
app ∈ ran(mapping) ∧
Cluster_Freq(dom(mapping ▷ {app})) > 0 ∧
Cluster_Fault(dom(mapping ▷ {app})) = TRUE ∧
Cluster_Exec(dom(mapping ▷ {app})) = TRUE
then
Cluster_Freq := Cluster_Freq ◁- {dom(mapping ▷ {app})↦
Cluster_Freq(dom(mapping ▷ {app}))−1} ||
Cluster_Exec := Cluster_Exec ◁-
{dom(mapping ▷ {app})↦FALSE})
end
```

time the value of voltage is decreased, the cluster agent restores the value of frequency to the maximum (Listing 27). Please notice that all the cluster agent events disable themselves following the refinement rules.

These operations are applied repeatedly until the regional parameters (frequency and voltage) reach their minimum values. If none of these operations helps, i.e., frequency and voltage are at the minimum, the platform agent executes fault-tolerance mechanisms at the platform level. These

mechanisms are the whole application remapping or a task reallocation from a faulty cell.

At this point, we have derived the refined specification of the platform and cluster agents and their communication interfaces. The excerpt of the second refinement is shown in Figure 7. In the next refinement, we introduce variables and events that model cell agents.

The Third Refinement: Introducing Cell Agents and Local Reconfiguration

Since neither the platform agent nor cluster agents detect faults occurring in the NoC platform, i.e., in the cells, it is necessary to implement local monitors, namely cell agents. These agents monitor the state of cells in the NoC platform at the lowest, hardware level. They inspect different parameters of cells such as local temperature and faults and promote this information to higher-level agents (the platform and cluster agents). The higher-level agents execute the necessary functions and return control to the cell agents in order to repeat monitoring. Additionally, the cell agents initiate

Listing 27. Decreasing voltage

```
event Decrease_Voltage
any app
where
app ∈ ran(mapping) ∧
Cluster_Freq(dom(mapping ▷ {app})) = 0 ∧
Cluster_Voltage(dom(mapping▷{app})) > Min_Volt ∧
Cluster_Fault(dom(mapping▷{app}))=TRUE
then
Cluster_Freq := Cluster_Freq ⊲-
{dom(mapping ▷ {app})↦Max_Freq} ‖
Cluster_Voltage :=
Cluster_Voltage ⊲- {dom(mapping ▷ {app})↦
Cluster_Voltage(dom(mapping ▷ {app}))−1}
end
```

Figure 7. The excerpt of the second refinement

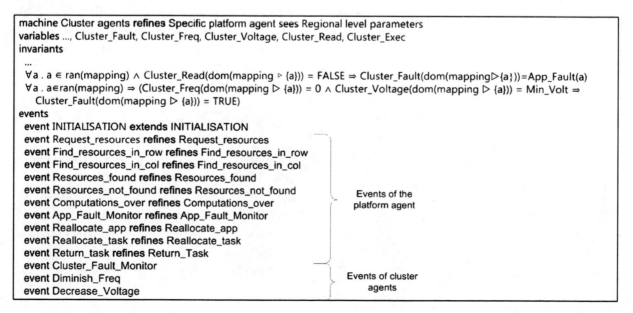

```
machine Cluster agents refines Specific platform agent sees Regional level parameters
variables ..., Cluster_Fault, Cluster_Freq, Cluster_Voltage, Cluster_Read, Cluster_Exec
invariants
   ...
   ∀a . a ∈ ran(mapping) ∧ Cluster_Read(dom(mapping ▷ {a})) = FALSE ⇒ Cluster_Fault(dom(mapping▷{a}))=App_Fault(a)
   ∀a . a∈ran(mapping) ⇒ (Cluster_Freq(dom(mapping ▷ {a})) = 0 ∧ Cluster_Voltage(dom(mapping ▷ {a})) = Min_Volt ⇒
      Cluster_Fault(dom(mapping ▷ {a})) = TRUE)
events
   event INITIALISATION extends INITIALISATION
   event Request_resources refines Request_resources
   event Find_resources_in_row refines Find_resources_in_row
   event Find_resources_in_col refines Find_resources_in_col
   event Resources_found refines Resources_found
   event Resources_not_found refines Resources_not_found
   event Computations_over refines Computations_over          Events of the
   event App_Fault_Monitor refines App_Fault_Monitor           platform agent
   event Reallocate_app refines Reallocate_app
   event Reallocate_task refines Reallocate_task
   event Return_task refines Return_Task
   event Cluster_Fault_Monitor
   event Diminish_Freq                                         Events of cluster
   event Decrease_Voltage                                          agents
```

reconfiguration at the local level. In this refinement, we introduce these cell agents.

To detect high temperature, a cell agent compares the current temperature of a cell with a specific threshold. The threshold represents the temperature exceeding which a cell may burn out. In addition to the threshold, cells have some initial (environmental) temperature at the beginning. Clearly, this temperature is strictly less than the threshold. Finally, the value of temperature cannot be infinite. Hence, there exists maximum value a cell can have. Obviously, the threshold and the initial temperature are strictly less than the maximum temperature.

We start this refinement with the introduction of the necessary constants for the initial temperature, threshold and the maximum temperature as well as their properties. Notice that we do not specify exact values; one can determine them when instantiating the model according to specific requirements of the system. Additionally, we introduce a deferred set of configurations (configuration files) for the cells. This set is finite because configuration files are stored in some memory and the volume of this memory cannot be infinite (Listing 28).

For detecting the exact location of a faulty cell, every cell in the NoC platform has a cell agent. Therefore, all variables that represent cell agents are total functions whose domain is an element of NoC (a cell).

Listing 28. Introducing local parameters

```
context Local level parameters
extends Regional level parameters
constants Temp_max Temp_Threshold Initial_Temp
sets CONFIGURATIONS
axioms
finite(CONFIGURATIONS)
Temp_max ∈ ℕ
Temp_Threshold ∈ ℕ
Initial_Temp ∈ ℕ
Initial_Temp < Temp_Threshold
Initial_Temp < Temp_max
Temp_Threshold < Temp_max
end
```

First, we define a function whose returning result corresponds to the temperature of a particular cell:

$$Cell\_Temp \in NoC \rightarrow 0..Temp\_max$$

In a similar manner, the cell agent reads the information about faults that occur in a cell. The presence or absence of faults is modelled as a function that returns "TRUE" if a fault is present and "FALSE" otherwise:

$$Cell\_Fault \in NoC \rightarrow BOOL$$

In order for a cell agent to promote the information about the current state of a cell to higher-level agents, it reads the cell state and updates a flag for this cell to value "TRUE." The cell state affects actions taken by all agents in the hierarchy so that they execute fault-tolerant mechanisms at different levels (if necessary) and reset this flag. This process is performed repeatedly and allows the agents to exchange the information about local state of cells via the following shared variable:

$$Cell\_Read \in NoC \rightarrow BOOL$$

If the temperature of a cell is high or a fault is present in a cell the corresponding cell agent can initiate the local reconfiguration procedure by sending a command to the cell. If this command equals to "TRUE" the cell starts its reconfiguration:

$$Cell\_Start\_Reconf \in NoC \rightarrow BOOL$$

When the local reconfiguration procedure is completed, the corresponding cell agent notifies the platform agent about this. This allows the platform agent to return tasks to the reconfigured cells according to the stored trace. The following variable specifies this communication between the cell platform agents:

Cell_Reconfigured $\in$ NoC $\rightarrow$ BOOL

The reconfiguration of a cell at the hardware level can be seen as the modification of the internal structure of this core as if it were a single field-programmable-gate-array chip (Altera, 2008) (Xilinx, 2006). The modification is performed via uploading a new configuration file to the core. Therefore, we introduce a total function which maps every cell in the NoC platform to some configuration, which can be modified (uploaded) when required:

CoreConfiguration $\in$ NoC $\rightarrow$ CONFIGURA-TIONS

Upon completeness of the local reconfiguration procedure, the cell agent receives feedback from its cell. This feedback is introduced by the means of CoreReconfigured variable. The variable value "TRUE" specifies the completion of the local reconfiguration:

CoreReconfigured $\in$ NoC $\rightarrow$ BOOL

The reconfiguration starts only over a cell whose task has been reallocated. In other words, if the local reconfiguration procedure has been initiated or is in progress, the task must have been reallocated from this cell to some other one and the new state of the cell has not been read yet:

$\forall x . x \in NoC \Rightarrow$

$((\text{Cell\_Reconfigured}(x) = \text{TRUE} \lor$

$\text{Cell\_Start\_Reconf}(x) = \text{TRUE}) \Rightarrow$

$\text{Reallocated}(x) = \text{TRUE} \land \text{Cell\_Read}(x) = \text{TRUE})$

Moreover, if a cell is running a task, the local reconfiguration procedure cannot start. Hence,

any cell that belongs to the application region is neither reconfigured nor is going to be:

$\forall x . x \in \text{dom(mapping)} \Rightarrow$

$(\text{Cell\_Start\_Reconf}(x) = \text{FALSE} \land$

$\text{CoreReconfigured}(x) = \text{FALSE})$

Clearly, when the local reconfiguration is completed, i.e., the corresponding cell agent receives acknowledgement from the cell, the cell agent has to notify the platform agent about this. Hence, it should communicate using the relevant data. The invariants below support this property:

$\forall x . x \in NoC \land \text{Cell\_Start\_Reconf}(x) = \text{FALSE} \Rightarrow$

$(\text{Cell\_Reconfigured}(x) = \text{FALSE} \Rightarrow$

$\text{CoreReconfigured}(x) = \text{FALSE})$

$\forall x . x \in NoC \Rightarrow$

$((\text{Cell\_Read}(x) = \text{FALSE} \land \text{Cell\_Start\_Reconf}(x) = \text{FALSE}) \Rightarrow$

$\text{Cell\_Reconfigured}(x) = \text{CoreReconfigured}(x))$

Finally, the presence of a cluster fault stands for the presence of a faulty cell and the absence of any task reallocation from the region where this fault has occurred. This relation is expressed as the following invariant:

$\forall a . a \in \text{ran(mapping)} \Rightarrow$

$(\exists x . x \in \text{dom}(\text{mapping} \rhd \{a\}) \land$

$\text{Reallocated}(x) = \text{FALSE} \land \text{Cell\_Read}(x) = \text{FALSE} \land$

(Cell_Fault(x)=TRUE ∨ Cell_Temp(x)≥Temp_ Threshold) ⇒

Cluster_Fault(dom(mapping ▷ {a})) = TRUE)

The Platform Agent Functionality

Initially, all the cells in the NoC platform have the initial temperature, no fault is detected because there are no applications running. Moreover, no state has been read, hence, the cells are not reconfigured and does not have to be reconfigured.

To map an application to a region, the platform agent needs to check if the region where the application is going to be mapped does not contain faulty cells. If this is the case, the platform agent maps an application and creates a cluster agent for the region. In Listing 29, X = mapfun_hor(pending_apps(app)↦(r↦c)) is the region returned by the mapping function.

Whenever an application completes its computations, the resources it is mapped to has to be freed. However, if the local reconfiguration procedure is in progress (the platform agent reallocated tasks), it needs to be completed before the application is removed from the platform. This is required in order to have a consistent state of the platform when a new application comes in. Similarly, if the platform agent decides to reallocate the whole application to another region, it should check if the local reconfiguration is complete. Hence, these procedures are only performed when the local reconfiguration procedure is complete. In Listing 30, X = Platform_Cell_Trace~[dom(mapping ▷ {app})] is a set of cores which have traces, i.e., the platform agent reallocated tasks from these cells.

Similarly, the platform agent needs to check if the cell, where the task is going to be reallocated, neither is being reconfigured nor has been reconfigured. This is to ensure consistency of the platform as well (Listing 31).

If reconfiguration of the cell is successful the platform agent returns the task from the substitut-

Listing 29. Mapping on an error free region

```
event Resources_found refines Resources_found
where ... ∧
Cell_Start_Reconf[X] = {FALSE} ∧
Cell_Reconfigured[X] = {FALSE} ∧
Cell_Fault[X] = {FALSE} ∧
(∀x . x∈X ⇒ Cell_Temp(x) < Temp_Threshold)
then ...
end
```

Listing 30. Freeing resources

```
event Computations_over refines Computations_over
where ... ∧ (¬ X = ∅ ⇒
(Cell_Start_Reconf[X]={FALSE} ∧
Cell_Reconfigured[X]={FALSE}))
then ...
end
```

Listing 31. Reallocating a task from a faulty cell to a free one

```
event Reallocate_task refines Reallocate_task
where ... ∧ Cell_Start_Reconf(x↦y) = FALSE ∧
Cell_Reconfigured(x↦y) = FALSE
then ...
end
```

Listing 32. Returning a task from a substituting cell to the reconfigured one according to the trace

```
event Return_task refines Return_task
where ... ∧
Cell_Start_Reconf(x↦y) = FALSE ∧
Cell_Fault(x↦y) = FALSE ∧
Cell_Temp(x↦y) < Temp_Threshold ∧
Cell_Reconfigured(x↦y) = FALSE
then ...
end
```

ing cell to the reconfigured one according to the stored trace. Additionally, this cell is marked as if it has not been reconfigured in order to allow the platform agent to start the reallocation procedure again, if necessary (Listing 32).

The Cluster Agents Functionality

The introduction of the cell agents affects the behaviour of the cluster agents. In particular, a fault within a cluster stands for the presence of a faulty cell within this cluster. Hence, we refine the monitoring of cluster faults (Cluster_Fault_Monitor event) by substituting the local variable f with the expression bool(Cell_Temp(x↦y) ≥ Temp_Threshold ∨ Cell_Fault(x↦y) = TRUE) (Listing 33). Consequently, the local variable f disappears. The primed variables in Listing 33 represent new values, X = dom(mapping ▷ {app}) and Y = bool(Cell_Temp(x↦y) ≥ Temp_Threshold ∨ Cell_Fault(x↦y) = TRUE).

A corresponding cluster agent reacts on a faulty cell by dynamically adjusting frequency and voltage. On the other hand, the cluster agent communicates with the cell agent that detected the faulty cell by returning control to the corresponding cell agent. This is performed by resetting the flag for the cell whose state has been read allowing the cell agent to read the state of this cell again. When the frequency is decreased to zero the value of this flag remains the same enabling the cluster agent to decrease voltage (Listing 34).

Every time the cluster agent decreases voltage, the flag for the cell whose state has been read is reset because the decreasing of voltage is the final operation the cluster agent can perform (Listing 35).

The Cell Agents Functionality.

After initialisation, the cell agents can start the monitoring. Specifically, each cell agent monitors the state of a cell that this agent is assigned to and promotes the information about it to the cluster and platform agents (Listing 36, where Z = bool(t≥Temp_Threshold ∨ f=TRUE). Notice that the variable Cell_Read is assigned the value "TRUE" in case of faults and the value "FALSE," otherwise, in order to allow for iterative monitoring). If the local temperature of a cell is higher than the threshold or a fault has occurred in the cell, it stands for a cluster fault where this cell belongs and denotes a corresponding application fault. In other words, any faulty cell in the region leads to

Listing 33. Monitoring cluster faults

```
event Cluster_Fault_Monitor refines Cluster_Fault_Monitor
any x y
where ... ∧
(x↦y) ∈ ran(mapping) ∧
Cell_Read(x↦y) = TRUE
then ... ||
Cluster_Fault, Cluster_Exec, Cell_Read:|
(Cluster_Exec' = Cluster_Exec ⩤ {X↦Cluster_Fault'(X)})
∧
(Cell_Read' = Cell_Read ⩤ {(x↦y)↦Cluster_Fault'(X)})
(Cluster_Fault' = Cluster_Fault ⩤ {X↦Y})
end
```

Listing 34. Diminishing frequency

```
event Diminish_Freq refines Diminish_Freq
any x y
where ... ∧ x↦y∈dom(mapping ▷ {app}) ∧
Cell_Read(x↦y)=TRUE
then ... || Cell_Read:| Cell_Read'∈NoC→BOOL ∧
(Cluster_Freq(dom(mapping ▷ {app}))−1 > 0 ⇒
Cell_Read' = Cell_Read⩤{x↦y↦FALSE}) ∧
(Cluster_Freq(dom(mapping ▷ {app})) − 1 = 0 ⇒
Cell_Read' = Cell_Read)
end
```

Listing 35. Decreasing voltage supply

```
event Decrease_Voltage refines Decrease_Voltage
any x y
where ... ∧ x↦y∈dom(mapping ▷ {app}) ∧
Cell_Read(x↦y)=TRUE
then ... || Cell_Read := Cell_Read ⩤ {x↦y↦FALSE}
end
```

Listing 36. Monitoring the state of a cell

```
event Cell_monitor
any f t x y
where t ∈ 0..Temp_max ∧ f ∈ BOOL ∧
(x↦y)∈NoC ∧ Cell_Read(x↦y) = FALSE
then Cell_Read := Cell_Read⩤{x↦y↦Z} ||
Cell_Temp := Cell_Temp ⩤ {(x↦y) ↦ t}
Cell_Fault := Cell_Fault ⩤ {(x↦y) ↦ f} ||
end
```

the cluster fault of this region and an application fault for the application mapped to this region.

Upon the successful reallocation of the cell task, the cell agent belonging to it initiates reconfiguration of this cell. If after reconfiguration of the cell the fault remains or the temperature is still high the cell agent initiates the reconfiguration procedure once again. While the cell is being reconfigured, it neither reacts on the inputs nor produces outputs. Furthermore, the frequency does not affect this cell. Therefore, several reconfiguration procedures in a row may help to cool down a cell or to upload a proper configuration file, which may lead to repairing the cell (Listing 37).

The cell reacts on the reconfiguration command and acknowledges the completion of the reconfiguration by modifying the corresponding flag. The cell agent, in its turn, updates its flags accordingly and removes the reconfiguration command (Listing 38).

Finally, all the flags including the reading flag need to be reset. This is necessary to allow the monitoring of the cell state and the iterative application of the local reconfiguration if necessary (Listing 39).

The Cells Functionality

The local reconfiguration is modelled as a couple of auxiliary events. One event specifies the modification of the cell structure by uploading a new configuration file to it (Listing 40). The other event resets the flags so that a new reconfiguration procedure can be initiated, if necessary (Listing 41).

Please note that events belonging to cells and cell agents disable themselves following the refinement rules. At this point, we have derived the functionality of all agents and defined variables for their communication. Figure 8 shows the excerpt of the third refinement.

SUMMARY OF THE DEVELOPMENT

To model the proposed system, we have used the Rodin platform. The platform generated 319 proof obligations (PO) and discharged 286 of them automatically. The remaining 33 POs

Listing 37. Initiating local reconfiguration of a cell

```
event Reconfigure_cell
any x y
where (x↦y)∈NoC ∧ Cell_Reconfigured(x↦y) = FALSE ∧
Cell_Read(x↦y) = TRUE ∧ Reallocated(x↦y) = TRUE ∧
Cell_Start_Reconf(x↦y) = FALSE
then Cell_Start_Reconf := Cell_Start_Reconf◁-
{x↦y↦TRUE}
end
```

Listing 38. Modelling response of the reaction of the cell agent to reconfiguration

```
event Fin_cell_reconfiguration
any x y
where (x↦y) ∈ NoC ∧ CoreReconfigured(x↦y) = TRUE ∧
Cell_Start_Reconf(x↦y) = TRUE
then
Cell_Start_Reconf := Cell_Start_Reconf◁-{(x↦y)↦FALSE} ‖
Cell_Reconfigured := Cell_Reconfigured ◁- {(x↦y)↦TRUE}
end
```

Listing 39. Modelling response of the reaction of the cell agent to reconfiguration

```
event Cell_reconfigured
any x y
where (x↦y) ∈ NoC ∧ Cell_Reconfigured(x↦y) = TRUE
CoreReconfigured(x↦y) = FALSE ∧
Cell_Start_Reconf(x↦y) = FALSE
then
Cell_Read := Cell_Read ◁- {(x↦y)↦FALSE} ‖
Cell_Reconfigured := Cell_Reconfigured◁-
{(x↦y)↦FALSE}
end
```

required interactive assistance. The interactive proofs included case distinction technique, which is rather difficult to be invoked automatically. Nevertheless, the proof statistics (Tab. 2) shows that the system preserves the postulated properties and, consequently, behaves according to the described requirements.

Listing 40. Modelling local reconfiguration

```
event Reconfiguring_cell
any k l cfg
where (k↦l)∈NoC ∧ cfg ∈ CONFIGURATIONS ∧
¬ cfg = CoreConfiguration(k↦l) ∧
Cell_Start_Reconf(k↦l) = TRUE ∧
CoreReconfigured(k↦l) = FALSE
then
CoreReconfigured := CoreReconfigured ◁- {(k↦l)↦TRUE} ∥
CoreConfiguration := CoreConfiguration ◁- {(k↦l)↦cfg}
end
```

Listing 41. Finalizing local reconfiguration of a cell

```
event Reconfiguration_done
any x y
where (x↦y) ∈ NoC ∧ CoreReconfigured(x↦y) = TRUE ∧
Cell_Start_Reconf(x↦y) = FALSE
then
CoreReconfigured(x↦y) := FALSE
end
```

Please notice that the presented system has originally been published in (Ostroumov & Tsiopoulos, Formal Development of Hierarchical Agent-Based Monitoring Systems for Dynamically Reconfigurable NoC Platforms, 2012). However, the system specification in this paper has been updated according to the requirements that have to be fulfilled by the decomposition and code generation. Moreover, the updates enhanced the structure and clarity of the system specification.

Decomposing the Platform, Cluster and Cell Agents

After specifying the functionality and communication of the agents, the model is decomposed into three sub-models using the corresponding plug-in of the Rodin platform. These sub-models

Figure 8. The excerpt of the third refinement

```
machine Cell agents refines Cluster agents sees Local level parameters
variables ..., Cell_Temp, Cell_Fault, Cell_Start_Reconf, Cell_Read, Cell_Reconfigured, CoreReconfigured, CoreConfiguration
invariants ...
    ∀a . a∈ran(mapping) ⇒ (∃x · x∈dom(mapping ▷ {a}) ∧ Reallocated(x) = FALSE ∧ Cell_Read(x) = FALSE ∧
    (Cell_Fault(x) = TRUE ∨ Cell_Temp(x) ≥ Temp_Threshold) ⇒ Cluster_Fault(dom(mapping ▷ {a})) = TRUE)
    ∀x . x∈dom(mapping) ⇒ (Cell_Start_Reconf(x) = FALSE ∧ CoreReconfigured(x) = FALSE)
    ∀x . x∈NoC ⇒ ((Cell_Reconfigured(x) = TRUE ∨ Cell_Start_Reconf(x) = TRUE) ⇒ Reallocated(x) = TRUE ∧ Cell_Read(x) = TRUE)
    ∀x . x∈NoC ∧ Cell_Start_Reconf(x) = FALSE ⇒ (Cell_Reconfigured(x) = FALSE ⇒ CoreReconfigured(x) = FALSE)
    ∀x . x∈NoC ⇒ ((Cell_Read(x) = FALSE ∧ Cell_Start_Reconf(x) = FALSE) ⇒ Cell_Reconfigured(x) = CoreReconfigured(x))
events
    event INITIALISATION extends INITIALISATION
    event Request_resources refines Request_resources
    event Find_resources_in_row refines Find_resources_in_row
    event Find_resources_in_col refines Find_resources_in_col
    event Resources_found refines Resources_found                          Events
    event Resources_not_found refines Resources_not_found                  of the
    event Computations_over refines Computations_over                      platform
    event App_Fault_Monitor refines App_Fault_Monitor                      agent
    event Reallocate_app refines Reallocate_app
    event Reallocate_task refines Reallocate_task
    event Return_task refines Return_Task
    event Cluster_Fault_Monitor refines Cluster_Fault_Monitor              Events
    event Diminish_Freq refines Diminish_Freq                              of
    event Decrease_Voltage refines Decrease_Voltage                        cluster
                                                                           agents
    event Cell_monitor                                                     Events
    event Reconfigure_a_cell                                               of cell
    event Fin_cell_reconfiguration                                         agents
    event Cell_reconfigured
    event Reconfiguring_cell                                              Events
    event Reconfiguration_done                                            of cells
```

Table 2. Proof statistics

Model	Number of PO	Automatically discharged	Interactively discharged
Contexts	3	2	1
Abstract model *(Applications and platform)*	32	32	0
1st refinement *(Concrete Platform agent)*	83	72	11
2nd refinement *(Cluster agents)*	89	78	11
3rd refinement *(Cell agents)*	112	102	10
Total	319	286	33

represent the platform agent, cluster agents and cell agents, respectively. The agents communicate via shared variables (Abrial, 2009) and store necessary information using the variables in the boxes as illustrated in Figure 9.

A very important shared variable is mapping since each agent deals with cells in the platform

and regions over these cells and applications mapped to these regions. Since a fault presence should be identified at every level of the hierarchy, Cell_Fault, Cell_Temp and Cluster_Fault are shared variables. The variables Cluster_Freq and Cluster_Voltage, which the cluster agent uses to adjust frequency and voltage, are also shared since

Figure 9. The overall diagram of the agents, their data and communication

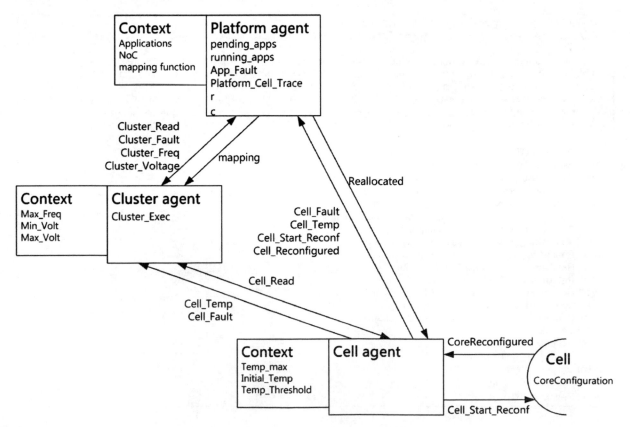

the platform agent dynamically creates, adjusts and destroys clusters. Furthermore, it initiates the reallocation and reconfiguration procedures after these parameters reach their minimum values. Additionally, the agents manipulate the variables showing that the state of the system has been read, namely Cluster_Read and Cell_Read. Therefore, these variables are shared as well. When the platform agent reallocates a task from a faulty cell, it informs the corresponding cell agent through the shared variable Reallocated. Finally, the variables Cell_Start_Reconf and Cell_Reconfigured that reflect the current state of the reconfiguration procedure are also shared because the cell agent monitors the local reconfiguration. In addition, it notifies the platform agent when the procedure is completed so that the platform agent can return the task to the reconfigured cell.

The decomposition is performed rather automatically through the interface provided by the plug-in. This interface allows one to express which variables are shared. It also allows one to distribute events that modify the shared variables to the corresponding models.

DISCUSSION

During the system development in the Event-B formalism, we assumed a 2D mesh topology as the underlying interconnect platform. However, if one would like to use, for instance, a torus 2D mesh topology, the platform agent would be specified in such a manner that the definition of the mapping function (and the corresponding events) would take into account rectangular regions with one-hop outermost connections. In addition, we assumed that non-allocated free cores can be shutdown when they are inactive in order to reduce power consumption.

As a fault-tolerance mechanism at the cluster level, we considered frequency and voltage as parameters that can be decreased in order to lower the temperature of cells and tolerate faults. However, when the temperature of cells becomes lower than the threshold or faults do not occur anymore, one can add event(s) for stepwise increasing regional frequency and voltage supply, so that the overall performance of computations is restored. If a fault occurred once again, the diminishing of the values of regional parameters would be repeated.

This work is one of the first steps in the direction of formal designing agent-based management systems for dynamic reconfiguration of many-core platforms. The continuation of it can be found in (Ostroumov S., Tsiopoulos, Plosila, & Sere, 2013), where we proposed algorithms for more efficient utilization of cores using task graphs (Hu & Marculescu, 2003). Moreover, we derived an implementation for the cell agents by generating a VHDL description from the formal model (Ostroumov & Tsiopoulos, 2011).

RELATED WORK

The work that influenced our research is the hierarchical agent monitoring system proposed by (Guang, Yang, Plosila, Latif, & Tenhunen, 2010), which allows effective monitoring of the state of the NoC platform. The authors present a structured framework for designing such a system. However, the framework only describes the main definitions of the hierarchical agent-based system. Furthermore, the framework is not formally verified. In comparison, we integrate reconfiguration techniques into agents functionality in addition to monitoring and formally verify the proposed system following the refinement-based and correct-by-construction approach while discharging proof obligations.

Guang et al. (Guang, Jafri, Yang, Plosila, & Tenhunen, 2012) have also proposed to incorporate reconfiguration procedures at coarse-grained (system) and fine-grained (local) levels for tolerating permanent and transient faults in many-core (thousand-core) Systems-On-Chip. They have suggested a two-level architecture where the system agent manages the whole platform and the local agent monitors the local component such as a router. The system also uses a portion of spare cores that are utilized if some processor fails. However, the authors do not show where these spare cores are located nor describe the algorithm for utilizing these spare cores. Moreover, the authors only consider the faults of the routers in which case the local agent executes reconfiguration by replacing a broken wire with a spare one. In addition, the functionality of the system agent includes many activities that may lead to a failure state of the agent itself, although the system agent is designed with higher reliability.

In contrast to (Guang, Jafri, Yang, Plosila, & Tenhunen, 2012), we focus on faults that can occur in the cells (cores) rather than in the routers. We believe that the probability of fault occurrence in cores is higher than that of the router since cores execute intensive computations. Moreover, we adopt a three-level architecture where reconfiguration procedures are incorporated into different levels of the hierarchy such that the platform can be dynamically adapted and healed, if necessary. In particular, the platform agent can remap the entire application or reallocate a task of a particular cell within the platform. The cluster agent can apply DVFS to its region in order to tolerate faults. These agents are dynamically created and destroyed when an application is mapped to and released from the platform, respectively. Finally, a cell agent can initiate local reconfiguration aiming at recovering the functionality of the core. In our opinion, the proposed three-level agent-based management system provides better scalability and structure for complex NoC platforms. Moreover, the functionality of the agents in this architecture is more balanced enhancing dependability of the system without overloading the platform (system) agent. Furthermore, our approach has been developed following the refinement-based and correct-by-construction approach allowing formal verification by discharging proof obligations.

A fault-tolerant reconfigurable NoC has been proposed by Motamedi et al. (Motamedi, Ionnides, Rümmeli, & Schagaev, 2009). The authors consider application specific architecture for avionic systems. In particular, they use a star network topology as the main active formation where the so called cockpit switch is placed in the center of the topology. The redundancy is achieved by placing redundant links in the system. When a fault is detected in the network, the topology is switched (reconfigured) from the star formation to the ring one. In addition, the authors utilize an Embedded Reliable Reduced Instruction Processor (ERRIC) as a computational unit, which has been specially designed for permanent faults. Additionally, the authors show the prototyping results where the overhead of using their approach is marginal while the required level of fault-tolerance is achieved. Although ERRIC is used as a computational unit, it has a reduced instruction set which may not be applicable to application domains other than avionics. Moreover, ERRIC is implemented on Field-Programmable-Gate-Array (FPGA) or Application-Specific-Integrated-Circuit (ASIC), where a fault may also occur and, hence, this processing unit is not available any more.

In comparison, instead of reconfiguring the NoC topology, we consider a topologically fixed NoC platform that is not application specific. We note however that the approach we propose in this paper can be applied to any topology and any type of routing schemes since it does not depend on the underlying platform. Our approach allows for executing applications without interruption and recovering the functionality of the platform by applying dynamic task reallocation and local cell reconfiguration procedures, respectively. The local reconfiguration, which is executed on the processing unit instead of the topology, recovers the operational mode of the former. Nevertheless, redundant routers (and/or links) can complement our approach.

Another formal approach to specifying agent-based systems is presented by (Andres, Molinero, & Nuez, 2008). This approach allows one to describe an agent-based system in terms of communication cellules that are organised into a hierarchy. The authors focus on a mathematical framework for describing such a generic hierarchical agent-based system. However, as the authors state, this approach is difficult to apply for complex systems such as NoC platforms. Furthermore, this approach does not support reconfiguration procedures, neither provides verification means for proving the correctness of the system being modelled. Instead, the approach supports simulation of a system.

The integration of Z notation and X-machines presented in (Ali & Zafar, 2010) enables modelling of agent-based system behaviour and supports data modelling as well as property analysis. The authors focus on developing specifications using X-machines and proving their properties using the Z notation. However, the authors do not consider hierarchical scheme of an agent-based system within their framework, which may lead to increased complexity in its application to large-scale NoC platforms.

In another Event-B approach presented in (Lanoix, 2008) the authors refer to a so-called platoon problem where several vehicles are moving one after another simultaneously. The authors consider the vehicles as a situated multi-agent system where agents exchange the information at one level, i.e., a system with a flat architecture. Hence, this approach may not be applicable for hierarchical agent-based systems nor provide scalability for such complex systems. Moreover, the authors do not consider faults that may occur in the system and, consequently, there are no reconfiguration procedures integrated.

ADAM is an informally developed framework for run-time agent-based distributed application mapping to NoC platforms proposed in (Al Faruque, Krist, & Henkel, 2008). The focus is on monitoring NoC traffic and avoiding congestions. This approach allows one to develop distributed agent-based systems tolerant to single point of fail-

ure, possibly manifesting in centralised schemes. This approach can possibly be adopted for our proposed system in order to overcome single point of failure of the platform agent assuming that a fault can also occur in it.

CONCLUSION

We presented an agent-based management system for many-core platforms such as NoC. The system follows the centralised three-level scheme where the platform agent, cluster and cell agents communicate with each other enabling efficient monitoring. This scheme is extended by adding reallocation and reconfiguration procedures for tolerating spontaneous faults that occur in the system. The reallocation procedure permits the computations to continue without interruption while the reconfiguration procedure makes it possible to restore performance of the computations. Furthermore, we have presented the formal development of the system within the Event-B framework following the refinement approach by discharging proof obligations. This shows functional correctness of the proposed system w.r.t. postulated properties.

We focused on some specific parameters for monitoring and manipulating, namely temperature, faults, frequency and voltage supply. However, an obvious direction of the future work is to explore and integrate into the proposed approach other characteristics of many-core platforms such as traffic, activity of routers, etc.

Another future direction is the instantiation of the presented system with application specific characteristics and the evaluation of efficiency based on different case studies. Finally, we will also consider various NoC topologies other than 2D mesh in our future work.

ACKNOWLEDGMENT

The authors would like to thank Adj. Prof. Pasi Liljeberg and Dr. Masoud Daneshtalab for fruitful discussions on the topic of this chapter. The authors would also like to thank the reviewers for their valuable feedback.

REFERENCES

Abrial, J.-R. (2009). *Event Model Decomposition.* Academic Press.

Abrial, J.-R. (2010). *Modelling in Event-B: System and Software Engineering.* Cambrige, UK: Cambrige University Press. doi:10.1017/CBO9781139195881

Al Faruque, M., Krist, R., & Henkel, J. (2008). ADAM: Run-time agent-based distributed application mapping for on-chip communication. In *Proceedings of Design Automation Conference* (pp. 760 - 765). Anaheim, CA: IEEE.

Ali, G., & Zafar, N. A. (2010). Modelling Agent-Based Systems Using X-Machine and Z Notation. In *Proceedings of International Communication Software and Networks* (pp. 249–253). Singapore: IEEE. doi:10.1109/ICCSN.2010.76

Altera. (2008). Retrieved from http://www.altera.com/literature/wp/wp-01055-fpga-run-time-reconfiguration.pdf

Andres, C., Molinero, C., & Nuez, M. (2008). A formal methodology to specify hierarchical agent-based systems. In *Proceedings of Signal Image Technology and Internet Based Systems* (pp. 169–176). Bali: IEEE. doi:10.1109/SITIS.2008.70

Benini, L., & De Micheli, G. (2002). Networks on chips: A new SoC paradigm. *Computer*, 70–78. doi:10.1109/2.976921

Butler, M. (2009). Decomposition Structures for Event-B. In *Proceedings of International Conference on Integrated Formal Methods* (pp. 20 - 38). Berlin: Springer-Verlag.

Event-B and the Rodin Platform. (2008). Retrieved from http://www.event-b.org/

Guang, L., Jafri, S., Yang, B., Plosila, J., & Tenhunen, H. (2012). Embedding Fault-Tolerance with Dual-level Agents in Many-Core Systems. In *Proceedings of MEDIAN Workshop* (pp. 41-44). MEDIAN.

Guang, L., Yang, B., Plosila, J., Latif, K., & Tenhunen, H. (2010). Hierarchical power monitoring on NoC - A case study for hierarchical agent monitoring design approach. In *Proceedings of NORCHIP* (pp. 1 - 6). Tampere, Finland: IEEE.

Hu, J., & Marculescu, R. (2003). Energy-aware mapping for tile-based NoC architectures under performance constraints. In *Proceedings of Design Automation Conference* (pp. 233-239). IEEE.

Kamali, M., Petre, L., Sere, K., & Daneshtalab, M. (2011). Formal Modelling of Multicast Communication in 3D NoCs. In *Proceedings of 14th Euromicro Conference on Digital System Design (DSD 2011)* (pp. 634-642). Oulu, Finland: IEEE.

Khatib, I. A., Bertozzi, D., Poletti, F., Benini, L., Jantsch, A., Bechara, M., & Jonsson, S. (2006). MPSoC ECG biochip: A multiprocessor system-on-chip for real-time human heart monitoring and analysis. In *Proceedings of Conference on Computing Frontiers* (pp. 21-28). New York: ACM.

Lanoix, A. (2008). Event-B Specification of a Situated Multi-Agent System: Study of a Platoon of Vehicles. In *Proceedings of International Symposium on Theoretical Aspects of Software Engineering* (pp. 297 - 304). Nanjing: IEEE.

Métayer, C., Abrial, J.-R., & Voisin, L. (2005, May 31). *Deliverables.* Retrieved from http://rodin.cs.ncl.ac.uk/deliverables/D7.pdf

Motamedi, K., Ionnides, N., Rümmeli, M., & Schagaev, I. (2009). Reconfigurable Network on Chip Architecture for Aerospace Applications. In *Proceedings of 30th IFAC Workshop on Real-Time Programming and 4th International Workshop on Real-Time Software* (pp. 131-136). IFAC.

Ostroumov, S., & Tsiopoulos, L. (2011). VHDL Code Generation from Formal Event-B Models. In *Proceedings of Digital System Design (DSD), 2011 14th Euromicro Conference* (pp. 127-134). Oulu, Finland: IEEE.

Ostroumov, S., & Tsiopoulos, L. (2012). Formal Development of Hierarchical Agent-Based Monitoring Systems for Dynamically Reconfigurable NoC Platforms. *International Journal of Embedded and Real-Time Communication Systems, 3*(2), 40–72. doi:10.4018/jertcs.2012040103

Ostroumov, S., Tsiopoulos, L., Plosila, J., & Sere, K. (2013). Formal Approach to Agent-Based Dynamic Reconfiguration in Networks-On-Chip. *Journal of Systems Architecture, 59*(9), 709–728. doi:10.1016/j.sysarc.2013.06.001

Rantala, P., Isoaho, J., & Tenhunen, H. (2007). Novel Agent-Based Management for Fault-Tolerance in Network-on-Chip. In *Proceedings of Euromicro Conference on Digital System Design Architectures, Methods and Tools* (pp. 551 - 555). Lubeck: IEEE.

Robinson, K. (2010, October 10). *System Modelling & Desing using Event-B*. Retrieved from http://wiki.event-b.org/images/archive/20101010115803!SM%26D-KAR.pdf

RODIN. (2011). Retrieved from http://sourceforge.net/projects/rodin-b-sharp/

Waldén, M. (1998). Layering Distributed Algorithms within the B-Method. In Proceedings of Second International {B} Conference (pp. 243-260). Springer-Verlag.

Xilinx. (2006). Retrieved from http://www.xilinx.com/support/documentation/application_notes/xapp441.pdf

Yin, A., Guang, L., Liljeberg, P., Nigussie, E., Isoaho, J., & Tenhunen, H. (2009). Hierarchical Agent Based NoC with Dynamic Online Services. In Proceedings of Industrial Electronics and Applications (pp. 434 - 439). Taichung: IEEE.

KEY TERMS AND DEFINITIONS

Agent: Software that enables specific functionality.

Agent-Based Management System: A system consisting of a high number of agents that are organized into some architecture or hierarchy and perform management (monitoring and orchestrating) activities.

Dynamic Reconfiguration: Modification of the system (platform) structure, architecture, spatial mapping etc.

Event-B: Mathematical framework for specifying and verifying systems by proofs.

Event-B Decomposition: A process of transforming a monolithic specification into a set of interdependent specifications.

NoC (Network-on-Chip): An on-chip interconnect that adopts a network-based communication scheme in order to allow scalability and high performance for many-core platforms.

Refinement: A verifiable process of the formal model transformation. Refinement gradually introduces implementation details into formal models such that implementations can be derived.

Section 4
Technologies for Network-On-Chip

Chapter 14
System–Level Analysis of MPSoCs with a Hardware Scheduler

Diandian Zhang
RWTH Aachen University, Germany

Gerd Ascheid
RWTH Aachen University, Germany

Jeronimo Castrillon
Dresden University of Technology, Germany

Rainer Leupers
RWTH Aachen University, Germany

Stefan Schürmans
RWTH Aachen University, Germany

Bart Vanthournout
Synopsys Inc., Belgium

ABSTRACT

Efficient runtime resource management in heterogeneous Multi-Processor Systems-on-Chip (MPSoCs) for achieving high performance and energy efficiency is one key challenge for system designers. In the past years, several IP blocks have been proposed that implement system-wide runtime task and resource management. As the processor count continues to increase, it is important to analyze the scalability of runtime managers at the system-level for different communication architectures. In this chapter, the authors analyze the scalability of an Application-Specific Instruction-Set Processor (ASIP) for runtime management called OSIP on two platform paradigms: shared and distributed memory. For the former, a generic bus is used as interconnect. For distributed memory, a Network-on-Chip (NoC) is used. The effects of OSIP and the communication architecture are jointly investigated from the system point of view, based on a broad case study with real applications (an H.264 video decoder and a digital receiver for wireless communications) and a synthetic benchmark application.

INTRODUCTION

Heterogeneous Multi-Processor Systems-on-Chip (MPSoCs) are nowadays widely used in the embedded domain, such as in wireless communication and multimedia applications, since they can provide efficient trade-offs between the computational power, the energy consumption and the flexibility of the system. One big challenge that comes with heterogeneous MPSoCs is system programming. This becomes even more critical, when taking runtime task scheduling and mapping

DOI: 10.4018/978-1-4666-6034-2.ch014

into consideration. In large-scale systems, runtime scheduling and mapping are highly demanding from the performance and energy perspective, since it is very difficult to consider different dynamic effects at design time. At the same time, applications are becoming more dynamic in nature, making previous static scheduling approaches obsolete. For these reasons, even MPSoCs for deeply embedded applications employ some kind of runtime scheduling.

The past decade has seen a myriad of techniques for runtime task and resource management, from extensions and optimizations of traditional Operating Systems (OS) to dedicated hardwired solutions for heterogeneous MPSoCs. Today, even commercial platforms such as the Texas Instruments KeyStone II provide hardware support for task management (Biscondi et al. 2012). In such systems, from the software perspective, a programmer uses a high-level Application Programming Interface (API), which internally calls hardware primitives that enable efficient, system-wide application scheduling. In this chapter we analyze MPSoCs with this kind of support. As hardware scheduler, we analyze the so-called OS Instruction-set Processor (OSIP), a custom processor optimized for task management in heterogeneous MPSoCs (Castrillon et al., 2009). OSIP provides a balance between the efficiency of hardwired runtime managers and the flexibility of pure software solutions.

Typically runtime managers are analyzed and benchmarked either in isolation or within an entirely customized MPSoC. This makes it difficult to assess the overhead of the runtime manager when integrated in a different platform, or to understand how platform-wide design decisions affect the overall performance of an application running on the MPSoC. In this chapter we provide a thorough characterization of the performance of OSIP on different types of systems. Two main on-chip interconnect paradigms are covered in the analysis:

Bus-based systems, e.g., AMBA buses (ARM Ltd., 2013) or CoreConnect (IBM Corporation, 2013), and larger-scale systems with a Network-on-Chip (NoC) as interconnect (Benini & De Micheli, 2002; Jantsch & Tenhunen, 2003). These two interconnect paradigms account for most of the systems designed today. For the analysis, typical architectural features of bus-based systems (e.g., cache subsystem) and NoC-based systems (e.g., peripherals for Direct Memory Access (DMA)) are considered in this chapter.

The joint characterization of OSIP and the different communication architectures is carried on by varying the number of Processing Elements (PEs) and analyzing the application performance as well as the runtime management overhead. The latter is compared to an off-the-shelf RISC processor and to an ideal manager (i.e., a manager that processes scheduling and mapping requests in zero time). For benchmarking, the H.264 video decoder and a Multiple-Input Multiple-Output (MIMO) digital receiver are used. Additionally, a synthetic benchmark complements the analysis by providing an idea of the limits of OSIP.

The rest of this chapter is organized as follows. After a brief survey of runtime managers, an introduction to OSIP and OSIP-based MPSoCs is given. Next, the impact on application performance of real-life communication architectures is demonstrated by means of examples for bus-based and NoC-based MPSoCs. The baseline systems introduced in the example are then enhanced with architectural features that improve the system performance. These interconnect paradigms and features are analyzed in a comprehensive case-study. Furthermore, the joint effects between OSIP and the communication architecture are extensively investigated, using the above-mentioned H.264 video decoding and MIMO receiver application and an additional generic synthetic application. Finally, conclusions are drawn and an outline for future work is provided.

RUNTIME MANAGEMENT

Various approaches have been proposed in academia and industry for runtime management. Generally, these approaches can be categorized into two groups: software and hardware solutions. In a software solution, an Operating System (OS) runs on one of the processors (typically a RISC) and dynamically distributes the workload to the other Processing Elements (PEs). This is the case of the TI OMAP (Texas Instruments, Inc., 2013), the Cell Broadband Engine (Kistler, Perrone, & Petrini, 2006) and the Qualcomm Snapdragon processors (Qualcomm, 2013). Software approaches are very flexible, but have low efficiency, especially when it comes to small tasks. The main reason for this is the high OS overhead in terms of power, memory footprint and performance.

This problem has been tackled by moving OS functionality to hardware accelerators, both for single-core platforms (Nakano, Utama, Itabashi, Shiomi, & Imai, 1995; Kohout, Ganesh, & Jacob, 2003; Murtaza, Khan, Rafique, Bajwa, & Zaman, 2006, Nordström & Asplund, 2007) and for multi-core platforms (Park, Hong, & Chae, 2008; Seidel, 2006; Limberg, et al., 2009; Lippett, 2004; Nácul, Regazzoni, & Lajolo, 2007; Pan & Wells, 2008). In the following, hardware solutions for MPSoCs are further discussed.

The hardware OS kernel – HOSK introduced by Park, Hong and Chae (2008) is a coprocessor that performs scheduling (fair and priority based) on a homogeneous cluster of simplified RISC processors. It features a low multi-threading overhead (less than 1% for 1-kcycle-tasks). However, a dedicated context controller should be included into the RISC processor to exchange context data between the processor and HOSK. This impedes its integration into traditional component-based design with off-the-shelf processors. Furthermore, to our best knowledge, no programming model exists for the HOSK-based MPSoCs.

In the work of Seidel (2006) and Limberg, et al. (2009), a hardware scheduler called Core-Manager is used to detect task dependencies at runtime and schedule tasks. A programming model is provided along with CoreManager, following the synchronous data flow (SDF) model. Hence, this solution is only applicable to a limited set of applications. High scheduling efficiency has been reported for CoreManager (60 cycles to schedule a task in average), which however is at the cost of high area overhead.

The approach of SystemWeaver (Lippett, 2004) focuses on the issue of task scheduling and mapping on heterogeneous MPSoCs, supported with a programming model. It has a slightly higher flexibility than HOSK and CoreManager by allowing the user to compose different basic scheduling primitives so as to implement complex scheduling decisions. However, its flexibility is still rather limited and the usage is rather difficult due to its design complexity.

Hardware supported scheduling has also been considered in dynamically Reconfigurable Systems-on-Chip (RSoCs) to improve the utilization of the reconfigurable components of the system. In the work presented by Pan and Wells (2008), a hardware unit is in charge of task scheduling for system reconfiguration and execution. Two types of tasks are distinguished: configuration tasks for reconfigurable logic cells and application software tasks. A priority-based policy is applied for the scheduling in this system.

The need for efficient task management has also been identified in industry. Paulin et al. (2006) proposed the MultiFlex architecture, containing a HW accelerator for task mapping. A similar runtime manager can be found in the P2012 platform (STMicroelectronics and CEA, 2010). As mentioned above, Texas Instruments recently included hardware support for scheduling in the KeyStone architecture (Biscondi et al, 2012).

In general, these pure hardware solutions provide high efficiency in dynamic scheduling in MPSoCs. However, they suffer from problems inherent to hardware such as low scalability, extensibility and usability, fixed scheduling properties and difficulties in system integration. These reasons motivated the use of dedicated processors for task management such as OSIP (Castrillon et al., 2009) and an ASIP-based CoreManager (Arnold, Noethen, & Fettweis, 2012). In this chapter we build from the work presented by Castrillon et al. (2009) and Zhang et al. (2011). Note that the experiment results presented in this chapter differ slightly from the ones presented by Zhang et al. (2011). This is due to applying a faster memory configuration in the current systems.

The dynamic task management can be performed either using a centralized or distributed approach. While a centralized approach is more common in the bus-based systems, both approaches can be found in the NoC-based systems. The examples of the centralized approach are given by Chou and Marculescu in 2008, de Souza Carvalho, Calazans and Moraes in 2010 and Khajekarimi and Hashemi in 2012, and the distributed examples are presented by Al Faruque, Krist and Henkel in 2008, Shabbir, Kumar, Mesman and Corporaal in 2011, Castilhos, Mandelli, Madalozzo and Moraes in 2013 and Garibotti et al. in 2013.

Both centralized and distributed approaches have advantages and disadvantages. A big advantage of using a central approach is the straightforward implementation and programming of the system. And a central manager in general has a global overview of the system, which enables making proper scheduling and mapping decisions. Its main drawback is the limited scalability, which can be critical in large-scale systems with tens or hundreds of cores. In comparison to the centralized approaches, distributed approaches have a unique advantage in the scalability. They are more suitable for large systems. However, the task management with the distributed approaches can be sub-optimal due to lacking global system information. In fact, even in the distributed approaches, many approaches (e.g. Al Faruque, Krist and Henkel, 2008 and Castilhos, Mandelli, Madalozzo and Moraes, 2013) include both global and local managers. With these two different types of managers, the resource management is still performed in a kind of centralized way for the

global system and local PE domains, respectively. Therefore, whether the systems are bus-based or NoC-based, efficient runtime system schedulers are needed.

OSIP-BASED MPSOCS

In OSIP-based MPSoCs, a customized processor called OSIP is used as a programmable hardware scheduler to perform runtime task scheduling, mapping and synchronization. It features special instructions to efficiently support scheduling and mapping algorithms, including instructions for the comparison of task descriptors and queue operations. In order to support different PE classes and application requirements (e.g. hard/soft real time, best effort), the scheduling and mapping are performed in a hierarchical way, similar to the concept introduced by Goyal, Guo, and Vin (1996). In this section we describe the architecture of OSIP, its hardware and software interfaces and introduce the system-level integration on bus-based as well as NoC-based MPSoCs.

The OSIP Architecture

As mentioned before, to balance the efficiency of hardwired solutions and the flexibility of software runtime managers, OSIP was designed as an Application-Specific Instruction-set Processor (ASIP) (Ienne & Leupers, 2006). OSIP is the result of extending the LTRISC processor that comes with Synopsys Processor Designer starter kit (Synopsys, Inc., 2013b). An overview of the resulting architecture is shown in Figure 1. In addition to the actual core, OSIP has a slave register interface that serves to communicate with the processors in the MPSoC. The master interface to the right of Figure 1 is composed of a set of interrupt ports that serve to trigger task execution on the processors. The core itself is a load-store architecture with a 6-stage pipeline: pre-fetch (PFE), fetch (FE), decode (DC), execute (EX),

Figure 1. OSIP Architecture

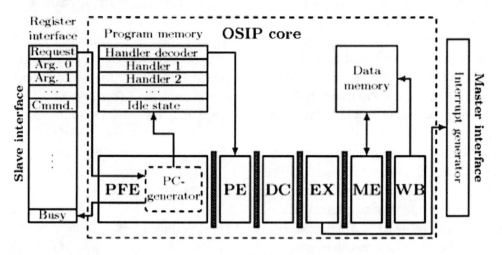

memory access (ME) and write-back (WB). The last five stages serve the original purpose as in the baseline LTRISC processor. The additional PFE stage is used to compute the Program Counter (PC) upon arrival of a system event, e.g., task creation or task synchronization.

When operational, OSIP receives requests through the register interface. Depending on the request, the *handler decoder* selects the appropriate routine, e.g., task creation or task synchronization. While processing a request, OSIP may generate interrupt control signals at the execute stage in order to control system-wide task execution. To accelerate the execution of the most critical requests, OSIP includes custom instructions. Among others, these instructions provide support for quickly updating linked lists of task descriptors, for comparing task descriptors according to configurable scheduling policies and for quickly computing addresses to a tightly coupled memory. For further details the reader is referred to the work of Castrillon et al. (2009).

OSIP Software Integration

The PEs in an OSIP-based MPSoC interact with OSIP by sending low-level commands to its register interface. There are around 50 different commands with a compact binary encoding for the command *opcode* and its arguments. Clearly, directly writing these commands in user code is a cumbersome task. To simplify the process of MPSoC programming, OSIP-based systems are delivered with software layers, as shown on the left of Figure 2. In addition to standard micro-kernels of the PEs for individual task scheduling, a software API layer is inserted to support system-wide scheduling and mapping. It provides an abstraction of the actual low level communication primitives for multi-task management, e.g. task creation, task suspension or task deletion. These primitives are translated into commands and off-loaded to OSIP for dynamic task scheduling and mapping. An example of creating a task is given in Figure 3, in which the low-level primitives of creating the task in the memory and inserting the task into a given task queue at OSIP are abstracted by the API *CreateTask()*.

Figure 2. OSIP software layers

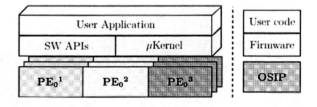

Conceptually, OSIP allows to group PEs into several processing classes for performing runtime mapping decisions. These classes are user-defined and can be for example determined according to PE types such as RISC, DSP, hardware accelerator etc. In this way, a task can be mapped onto any PE in the corresponding class, which improves the utilization of PEs and greatly reduces the programming effort when extending the system. The superscripts of PEs in Figure 2 represent the different processing classes and the subscripts represent the individual PEs inside each class.

Apart from the software APIs, a firmware containing low level functions is provided to optimally exploit OSIP's dedicated hardware features (see right-hand side of Figure 2). It comprises several basic scheduling algorithms such as *round-robin*, *FIFO*, *priority-based* and *fair queue*. Since OSIP is provided with a C-compiler, the user has the full flexibility to implement his own code to extend or add new scheduling features on top of the provided firmware. Two exemplary firmware function signatures are given in Figure 4. The first example inserts a task into a task queue based on a given insert policy. The second example is to solve the possible task dependencies in a given task queue. OSIP_DT is a specific data structure describing the tasks and queues in OSIP.

OSIP Hardware Integration

As shown in Figure 1, the main hardware interfaces of OSIP are the register interface and the interrupt generation interface. These interfaces are relevant for integrating OSIP into the bus-based and NoC-based systems analyzed in this chapter. The main features of the two interfaces are described in the following.

Register Interface

This interface enables OSIP to be accessed as a standard memory-mapped I/O, which eases the integration of OSIP. Information is exchanged between the PEs and OSIP through this interface. Typical information includes the commands and arguments generated from the PEs to OSIP for multi-task management, the status (busy/idle) of OSIP and its return value. To synchronize the tasks, hardware spinlock registers are also implemented in this interface, following the *compare-and-swap* approach.

Since OSIP is a shared hardware resource, a special semaphore register is used in the register interface to guarantee that only one processor can access OSIP. This is necessary for avoiding accidental mixing the commands and arguments

Figure 3. An API example for creating a task

```
TaskParameters params;
InitTask(&params, taskFuncPtr, prio, queueID, ...);
CreateTask(&params);
```

Figure 4. OSIP firmware examples

```
typedef OSIP_DT DT;
InsertTask(DT *pQueue, DT *pTask, POLICY insertPol);
SyncTask(DT *pTask, DT *pQueue, TYPE syncType);
```

from different PEs. Only after successful acquisition of this semaphore, a PE is allowed to send commands. The other PEs will poll the semaphore until it is freed by the owning PE.

Interrupt Interface

Upon receiving a command from a PE, a corresponding handler is executed in OSIP. Depending on the commands and system state, interrupts might, but need not be generated from OSIP to the PEs through dedicated interrupt lines in order to control platform-wide task execution. Typically, an interrupt from the OSIP triggers a new task execution on a corresponding PE. The current implementation of OSIP supports up to 32 interrupt lines.

System-Level Integration

As mentioned in the introduction, we analyze the performance of OSIP in a system context. Since the communication architecture has become one of the dominant factors for performance in modern MPSoCs, we focus our analysis on the on-chip interconnect. More specifically, we target the main interconnect paradigms, namely, bus-based and NoC-based. These two paradigms find application in different MPSoCs. Buses are common on small-scale platforms whereas the NoCs are found in large-scale platforms. As stated by Wentzlaff et al. in 2009, beyond 8 – 16 processor cores, buses do not scale well and NoCs are a preferred communication architecture. This section discusses the basic concept for integrating OSIP into such systems. Detailed architectural optimizations and the impact on performance are the matter of later sections in this chapter.

OSIP in Bus-Based MPSoCs

OSIP was originally designed for easy integration in bus-based systems thanks to the memory-mapped register interface. A typical bus-based

MPSoC is illustrated in Figure 5, which includes PEs of various types, a memory subsystem, peripherals and OSIP itself. From OSIP's point of view, PEs are grouped into several classes. As shown in the figure, OSIP is connected to the interrupt lines of all PEs.

In a bus-based system, the communication between the PEs and OSIP (including polling) is implemented via the bus. Naturally, due to sharing, the specific bus architecture has an impact on the overall performance (e.g., AHB and multi-layer AHB, as will be discussed later). Similarly, the communication among PEs is implemented over shared memory. The specific memory architecture is therefore relevant for the system performance (e.g., the impact of caches as discussed later in this chapter).

OSIP in NoC-Based MPSoCs

For larger systems, the overhead due to bus contention motivated the introduction of NoCs. In order to integrate OSIP in NoC-based MPSoCs, the HW interfaces must be adapted in order for OSIP not to become the system's bottleneck. Both the slave and the master interfaces pose integration problems:

- **Register Interface:** In the original OSIP design, communication heavily relied on polling. After issuing certain commands,

Figure 5. OSIP in bus-based MPSoCs

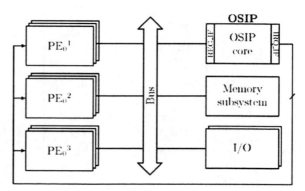

a PE would poll for results. More importantly, PEs would poll for OSIP spinlocks and semaphores in order to get access to shared resources and OSIP itself. Polling through the NoC would introduce a high communication overhead, thereby reducing system efficiency.

- **Interrupt Interface:** Direct interrupt lines from OSIP to PEs in a large system would lead to long wires across the chip. Such long connections are typically undesired in modern chips. The long wire delay would potentially reduce the system clock frequency or introduce unpredictability when transferring the signal.

To overcome these interfacing problems, a subsystem for the NoC integration was designed, depicted in Figure 6. The subsystem includes small hardware proxies for the different PEs. These proxies implement the original communication protocol on the OSIP side, i.e., polling OSIP's register interface and catching interrupt signals. On the side of the NoC, the proxies communicate with the PEs using a protocol that is better suited for NoC-platforms, based on messages. Note that it is not mandatory to have a separate proxy for each PE. A proxy can be shared between several PEs, and it can be a hardwired state machine or

a programmable processor. What kind of proxies to choose and implement in a concrete system largely depends on the PE types communicating with OSIP.

The OSIP subsystem of Figure 6 can be then integrated in the NoC-based MPSoCs. A sample system is illustrated in Figure 7, which consists of several PE subsystems, memory subsystems and the OSIP subsystem. As with bus-based systems, different NoC configurations, e.g. different link data widths and virtual channel configurations of the routers, have an impact on the system-level performance of NoC-based MPSoCs. Hardware support for communication (e.g., DMA) is also crucial for performance. As will be discussed later in this chapter, the system behavior will be jointly affected by these communication-related design decisions and the OSIP scheduling efficiency.

COMMUNICATION OVERHEAD: OSIP BASELINE SYSTEMS

Before analyzing different platform improvements for bus-based and NoC-based systems in upcoming sections, this section quantifies the cost of communication on baseline systems. This demonstrates that even an optimized controller such as OSIP can behave inefficiently if no further

Figure 6. OSIP subsystem for NoC integration

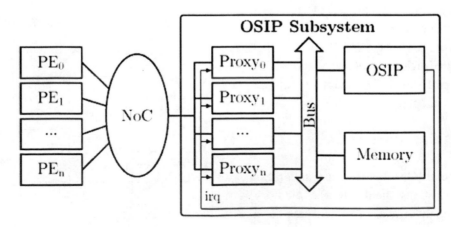

Figure 7. OSIP in NoC-based MPSoCs

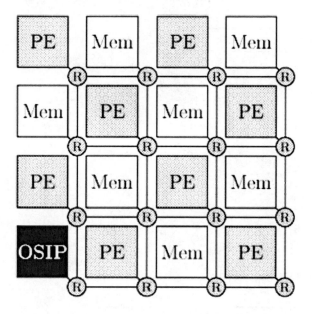

system-level considerations are made. Therefore, this section serves as a motivational example for the remainder of this chapter.

Bus-Based MPSoCs

For the analysis of bus-based MPSoCs we use the system described by Castrillon et al. (2009) as reference. It consists in a virtual platform modeled in SystemC using Synopsys Platform Architect (Synopsys, Inc., 2013a). In the work of Castrillon et al., (2009), the effects of OSIP were analyzed in isolation for which an ideal communication architecture was employed. The platform contains several ARM926EJS instruction-accurate processor models, a cycle-accurate model of OSIP, a shared memory and several peripherals (input stream, virtual LCD and UART). The components are connected with an ideal bus with no contention (zero communication overhead). The clock frequency is 200 MHz.

As done by Castrillon et al., (2009), we use a parallel implementation of the H.264 video decoding application in the system. Based on this application, we compare the performance of

realistic bus-based systems and systems with an ideal bus. The frames-per-second (fps) achieved by the decoder is used as comparison metric. Figure 8 shows the performance of the baseline bus-based system with an accurate AMBA AHB bus and the system with an ideal bus for a varying number of PEs (1 to 11). A big performance gap between both systems can be observed, which increases with the number of PEs. This increasing performance gap is not only due to the latency of the real bus, but mostly due to the bus contention, which introduces a high communication overhead. In fact, starting from the 6-ARM-system with the AHB bus, adding more processors does not improve the system performance. In this case, the benefit gained from the parallel execution of tasks is lost by the additional communication overhead. On average, by replacing the ideal bus with the real AHB bus, the frame rate is slowed down by a factor of 2.2.

Figure 9 provides details of the communication overhead in the baseline bus-based system by separating the execution time of individual processors into three parts:

- **Active Time (t_A):** The time that the PEs spend on executing the instructions. If an instruction accesses a shared component (e.g., the shared memory) through the bus, the time spent on the bus is excluded from t_A.
- **Idle Time (t_I):** The stall time, during which the PEs lie in a low-power state, waiting for interrupts from OSIP to get new tasks.
- **Communication Time (t_C):** The time spent on the communication from the PEs to OSIP, the shared memory and I/O.

For all systems, the communication time is the main contributor to the total execution time. In most of the cases, around 50% of the total execution time is spent on the communication, which is clearly the system bottleneck. In contrast, the active time continuously decreases with the in-

Figure 8. Frame rate comparison: ideal bus vs. AHB

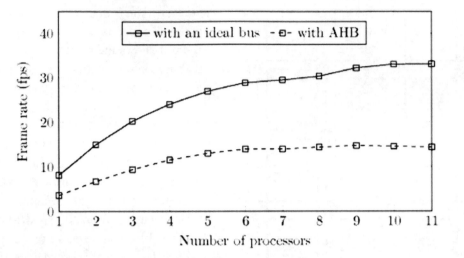

Figure 9. Time components from processor's view in AHB systems

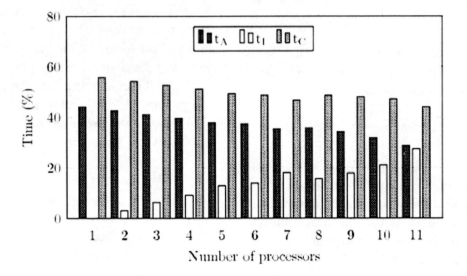

creasing number of PEs, since each PE gets less tasks assigned with more PEs used. So they enter more frequently the idle state, which is illustrated by the trend of increasing idle time in the figure.

The impact of the communication architecture on the system performance is also reflected in the state of OSIP. Naturally, if more PEs are integrated into the system, requests to OSIP are more frequently generated. This should keep OSIP more often in the busy state. However, as shown in Figure 10, this effect is much stronger in systems with the ideal bus, in which the busy

time of OSIP increases steadily with more PEs. As a comparison, for the systems with the real AHB bus, OSIP stays mostly at the idle state ($t_{OSIP\text{-}busy}$ below 8%), because most of time is spent on the communication. In fact, for systems with a large number of PEs, the percentage of busy time of OSIP remains nearly unchanged. This indicates that the tasks are not assigned fast enough to the PEs due to the communication overhead, which results in the performance loss.

NoC-Based MPSoCs

For the analysis of NoC-based MPSoCs we use an abstract virtual platform with the mesh-like NoC used by Schürmans, et al. (2013) (see Figure 13). The NoC is modeled in SystemC in a cycle-accurate way, which has the same timing behavior as the actual RTL implementation. The MPSoC is modeled using Synopsys Platform Architect (Synopsys, Inc., 2013a). PEs are modeled as Virtual Processing Units (VPUs) (Kempf, Ascheid, & Leupers, 2011) of the Synopsys MCO technology, which are abstract modules that can mimic the timing behavior of different processor types (e.g., RISCs, ASIPs and hardware accelerators). The system is driven by a clock frequency of 300 MHz.

Since the parallel implementation of the H.264 video decoder heavily relies on communication over shared memory, it is not well suited for NoC-based systems. For the analysis of the communication overhead, we therefore selected a second benchmark from the wireless communication domain, namely a 2×2 Multiple-Input Multiple-Output (MIMO) Orthogonal Frequency-Division Multiplexing (OFDM) digital receiver. Applications in this domain are typically specified as so-called dataflow graphs which are suitable for distributed memory architectures.

For receiver, we use a doubly iterative design with a similar specification as those for the 4G LTE communication standard (3GPP, 2013). A block diagram of the receiver is shown in Figure 11. For this study we focus on the digital part of the receiver (inside the dashed box). In order to achieve low Bit-Error-Rate (BER) of the system, four iterations between the channel decoding and the de-mapping are executed. The throughput constraint of the receiver is *150 Mbit/s*, corresponding to the LTE standard, while the latency constraint is defined as *2.5 ms,* considering the algorithmic complexity of the whole transceiver including the transmitter and the receiver.

NoC Baseline System and Application Mapping

In order to evaluate the NoC-based system performance and the OSIP efficiency, it is sufficient to annotate the task execution time to the code executed in the VPUs. The execution timing information is obtained from the hardwired solutions for MIMO-OFDM receivers in the literature. Instead of detailing the algorithmic and hardware

Figure 10. Influence of communication architecture on OSIP

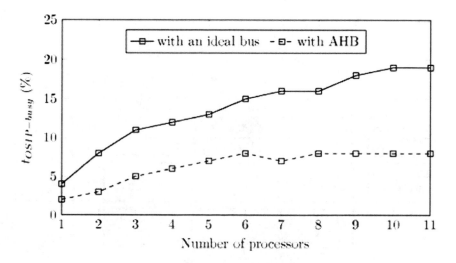

Figure 11. Block diagram of MIMO-OFDM receiver.

RF Frontend
FFT
Channel Estimation
Pre-Processing
De-mapping
Channel Decoding

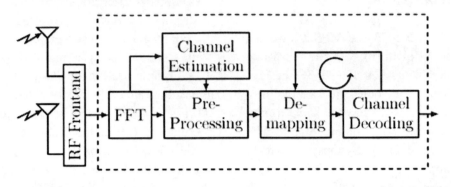

implementation of these blocks, we list the referenced implementations in Figure 12.

With the timing information, the number of the required VPUs is determined, which can fulfill the timing constraints if not considering the communication overhead. In total, we use 10 processing nodes, 3 memory nodes, 1 node for the OSIP subsystem and leave 2 nodes unused. Figure 13 shows the result of mapping the MIMO-OFDM receiver onto the 2D mesh. FFT processing is mapped to the nodes on the left-hand side, two VPUs are used for channel estimation, one for the pre-processing, one for the de-mapping (i.e., a cluster of sphere decoders) and four for the channel decoder (TD). The OSIP subsystem is placed at the center of the NoC and the memory nodes are placed closely to the processing nodes, to which they have intensive data communications.

Figure 12. Functional kernels of the digital receiver

Kernel	Description / Implemented Algorithm	Reference
FFT	Converting signals from time domain to frequency domain / 2048-point FFT	Boyapati & Kumar, 2010
Channel Estimation	Estimating the distortion in the wireless channel / Matching pursuit	Maechler, Greisen, Felber, & Burg, 2010
Pre-Processing	Decomposition of estimated channel matrix / Sorted QRD	Hwang & Chen, 2008
	Matrix-Vector multiplication / $Q^H Y$	Estimated from vector processor by Minwegen, Auras, Deidersen, & Ascheid, 2012
De-Mapping	Detecting of constellation points of the signals / Sphere Decoding (SD)	Witte et al., 2010
Channel Decoding	Recovering the message based on the code structure / Turbo Decoding (TD)	May, Ilnseher, Wehn, & Raab, 2010

Figure 13. Node assignment for the MIMO-OFDM receiver in the NoC-based MPSoC

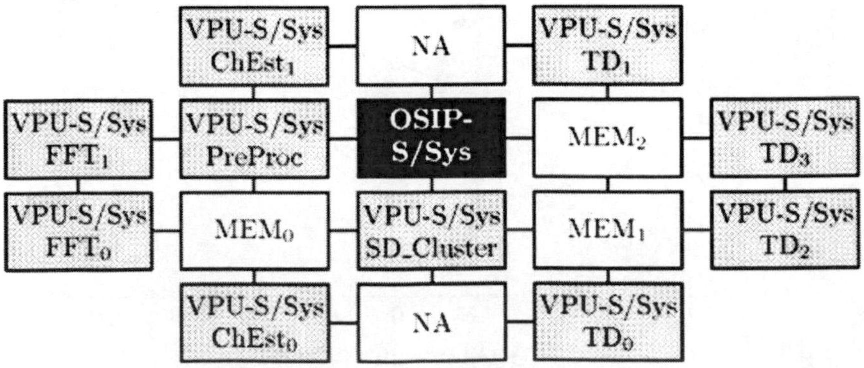

VPU-S/Sys ChEst1
NA
VPU-S/Sys TD1
VPU-S/Sys FFT1
VPU-S/Sys PreProc
OSIP-S/Sys
MEM2
VPU-S/Sys TD3
VPU-S/Sys FFT0
MEM0
VPU-S/Sys SD_Cluster
MEM1
VPU-S/Sys TD2
VPU-S/Sys ChEst0
NA
VPU-S/Sys TD0

Each node in the network includes a network interface and a router with four virtual channels on each port. Each channel is implemented as an 8-position FIFO buffer. The X/Y-routing is used in the NoC, following the wormhole scheme. The arbitration between the packets is priority-based. A data packet consists of a number of flits, which are the basic flow control units for the NoC. In our system, the flit size corresponds to the physical link data width between the routers. The maximum size of a data packet is *4 kB*.

A detailed view of the OSIP subsystem is given in Figure 14. An ARM processor is used to program the system, and an adapter is used to create and interpolate the packets for OSIP. For this case study, for each VPU a hardware proxy is added to interact with OSIP. Since the VPUs in the current system are all based on hardwired components, the number of the required operation types performed by the proxies is low. In general, the proxies create three commands to OSIP to *request a task*, *fetch a task* and *synchronize tasks*. The task request and task synchronization are directly issued by the request packets from the corresponding VPU. In contrast, a proxy fetches a task only when it receives an interrupt signal from OSIP. Upon the interrupt, it sends the task

Figure 14. OSIP subsystem

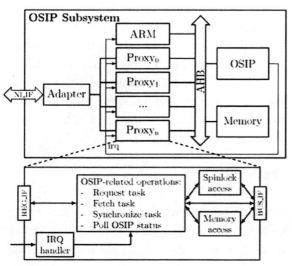

information to the VPU master subsystem via the adapter. In this system, the task information only includes the data information, i.e. the node coordinates of the source and destination data and their addresses as well as their sizes at these nodes. In addition to the command-related operations to OSIP, the proxies also poll the information from the OSIP register interface, which includes the OSIP status and the spinlock status.

Impact of the Communication

The impact of the communication for this system is illustrated in Figure 15. It compares the obtained latency and throughput of the receiver on an ideal NoC and on the real baseline NoC. The ideal NoC has an unlimited communication bandwidth with a link data width (DW) towards infinity, and the real baseline NoC has an initial data width of 32 bit, corresponding to the data width used in the bus-based system. As the figure shows, the latency of the receiver increases by a factor of 4.4, if using the baseline NoC instead of the ideal NoC. Similarly, the throughput decreases from 150 Mbit/s to a moderate 50 Mbit/s. Note that 150 Mbit/s is the maximum achievable throughput due to the maximum input data rate received at the antennas.

The analysis in this section for both bus-based and NoC-based systems shows that even with an optimized processor such as OSIP, a low system-level performance is achieved if no optimizations are performed on the system interconnect. The following sections will introduce general optimizations to both bus-based and NoC-based MPSoCs and measure their impact on OSIP-based systems.

INTERCONNECT ENHANCEMENTS AND PERFORMANCE ANALYSIS

There are a myriad of techniques to improve the performance of both bus-based and NoC-based systems. In this section we introduce basic techniques and evaluate their performance on the

Figure 15. Comparison of MIMO receiver on NoC-based MPSoCs

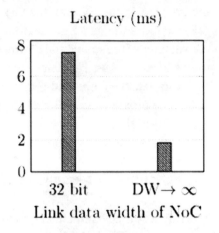
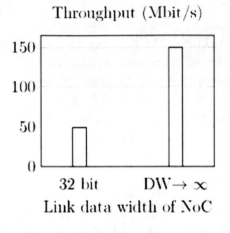

motivational examples presented above (H.264 decoder for bus-based systems and MIMO receiver for NoC-based systems). The effects of different implementation of OSIPs on the system-level performance is analyzed later in this chapter.

Enhancements: Bus-Based MPSoCs

In the following we introduce three enhancements that are common to bus-based systems with communication over shared-memory.

Multi-Layer AHB

The advantages of a multi-layer AHB (ARM Ltd., 2013) against an AHB bus is that parallel accesses from the masters to the slaves of the bus are possible, as long as the addressed slaves are different. This fits well to the characteristics of the data communication in OSIP-based systems, in which three types of independent data communications exist:

- **Communication between PEs and OSIP:** OSIP is accessed by PEs through the register interface as a standard peripheral connected to the bus. Information for system-wide scheduling, mapping and synchronization is exchanged through this interface. This communication type can be considered as a control-centric communication.

- **Communication between PEs and Shared Memory:** Based on this communication, data are exchanged between different tasks running on different PEs. This communication type is data-centric.

- **Communication between PEs and I/O:** This communication serves for reading an input data stream and outputting results, which is also data-centric. Normally this traffic contributes only to a minor part of the total communication.

By using a multi-layer AHB, the communication overhead due to the unnecessary bus contention caused in an AHB bus can be reduced, especially when the control-centric communication and the data-centric communication are well balanced. In the current simulation platform, each processor is connected to a bus layer, since all processors are treated as equal in this system.

Cache System

Compared to a multi-layer AHB, which mainly focuses on improving the bus contention, a cache focuses on reducing bus accesses by exploiting data locality, which reduces the communication cost due to the bus latency. More importantly, the reduced accesses to the bus potentially reduce bus contention.

However, in multi-processor systems cache coherence becomes of vital importance. This problem has been thoroughly studied in the literature (Tomasevic & Milutinovic, 1994a, 1994b; Eggers & Katz, 1988; Loghi, Poncino, & Benini, 2006). In this chapter we analyze a cache system based on the write-broadcast approach.

As shown in Figure 16, the cache system contains two basic modules: a local cache module for each individual PE and a global Cache Coherence Management Unit (CCMU) ensuring the consistency between the local cache modules. The local cache module consists of a cache memory and a cache controller. Besides maintaining the cache memory, the local cache controller acts as a bridge to the CCMU with additional ports, such that information exchange is possible across all cache modules.

When reading data, the cache system behaves like a normal single-cache system. The difference occurs when data are written into the memory, in which case the CCMU is involved. Whenever a cache controller captures a write request, it first sends the request to the bus to store the data into the shared memory. In addition, it forwards the request to the CCMU, which broadcasts the request to all cache modules. Upon receiving the broadcast from the CCMU, each cache module checks whether a local data copy at the required write address exists in the cache und updates the data correspondingly if it is the case. At the end, each cache controller sends a confirmation to the CCMU, which in turn generates a response over the initial cache module back to the PE to complete the write procedure. In this way, CCMU guarantees the consistency of the data stored in the caches and shared memory.

Note that the cache system mainly improves the data access to the shared memory. For reading the register interface of OSIP, the cache mechanism does not help, because mostly the PEs poll the register interface of OSIP for status information (e.g., to request a spinlock). Due to the *compare-and-swap* control mechanism for the spinlocks, reading the spinlock information must exactly take place at the register interface to ensure that only one PE acquires the spinlock. In the implementation, all read requests to the OSIP register interface are excluded from the cache modules.

Write Buffers

Write buffers are used to improve the data writing. In fact, write buffers do not reduce the number of bus accesses, nor the bus contention, but they enable the parallelization of the communication

Figure 16. Cache system

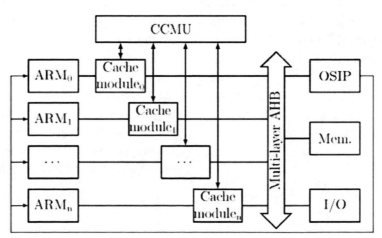

and computation. The PEs offload the data transfer to the write buffers and then continue with the task execution, such that the write to the bus is transparent to the PEs.

The implementation of the write buffers follows a similar principle as introduced by Sloss, Symes, and Wright (2004). In this approach, hardware FIFO queues are inserted between the cache modules and the bus. A write buffer confirms the write request of the PE before the request truly reaches the bus. Whenever there is a write request to the bus, it will be first appended to the FIFO queue, if the queue is not full, and a response to the PE will be sent immediately. Otherwise, the write buffer will wait, until at least one entry in the queue becomes free. In order to ensure the correctness of reading data, any read access to the bus will be delayed, until all requests in the write buffer have been processed. In other words, a read access empties the local write buffer.

Similar to caches, multiple write buffers also have the consistency problem. Suppose that PE_A writes a word to the shared memory, which shall be used by PE_B. It can happen that the word from PE_A still lies in the write buffer, when PE_B reads the shared memory and thus fetches the outdated value. However, in OSIP-based systems this race condition is prevented by the spinlock registers in the OSIP register interface. In the example, there is a data dependency between the tasks on PE_A and PE_B. A spinlock will be occupied by PE_A, such that PE_B will not be allowed to fetch the data. After PE_A finishes the task, it frees the spinlock by writing a dedicated value. Since the write buffer is based on a FIFO queue, the write request to free the spinlock always follows the write requests before it. So, at the time when the spinlock register is written, the value in the memory has already been updated. Therefore, the race condition is prevented.

Performance Analysis: Bus-Based MPSoCs

The aforementioned enhancements for bus-based systems are evaluated using the same H.264 video decoding mentioned in the motivational example. The setup of the evaluation system remains the same as introduced above, with the communication architecture enhanced by a multi-layer AHB, a cache system and write buffers. We use the following notation for the different versions of the communication architectures:

CA0: AHB bus (unoptimized communication).
CA1: multi-layer AHB bus.
CA2: multi-layer AHB bus with cache.
CA3: multi-layer AHB bus with cache and write buffers.
CA4: ideal bus (no communication overhead).

In the following discussions, the cache size and the write buffer size are set to 16 kB and 4 words respectively. The local cache module is implemented as a four-way set associative cache with a Least Frequently Used (LFU) replacement policy.

Performance Comparison

Figure 17 gives an overview of the effect of the different optimizations on the system performance by comparing the frame rate of H.264. The performance improvement is high: From CA0 to CA3 in the 11-PE-system, the frame rate is increased by about 10 fps, which is a factor of 1.7.

For this application, the improvement is mainly due to the cache system and the write buffers. In the 11-ARM-system, 84.4% of the improvement is gained with these two optimization steps. In contrast, using a multi-layer AHB only shows a slight speed-up due to the dominating communication between the PEs and the shared memory.

However, for balanced communications from the PEs to the slave components, more benefit is expected from the multi-layer bus.

Naturally, the performance gap between the system with the ideal bus (CA4) and the systems with realistic communication architectures (CA0-CA3) grows with the number of PEs, as shown in the figure. In the realistic systems, the bus contention increases with the number of PEs, which largely reduces the benefits obtained from the task processing parallelism.

A detailed analysis of the communication overhead at different optimization levels is given in Figure 18, based on the 11-ARM-system. It can be observed that the communication time (t_C) is reduced significantly by a factor of 2.2 from CA0 to CA3. At the same time, the idle time of the PEs (t_I) is decreased by a factor of 1.7, which is an important side effect of the optimization for the communication architecture. By optimizing the communication, the execution time for the tasks is reduced, which in turn leads to shortened stall time of PEs for dependent tasks.

In comparison to t_C and t_I, the active time (t_A) of processors is only slightly changed in systems from CA0 to CA3. This is because each PE gets the same amount of tasks assigned independent on the communication architecture used. In the

Figure 17. Frame rate at different optimization levels

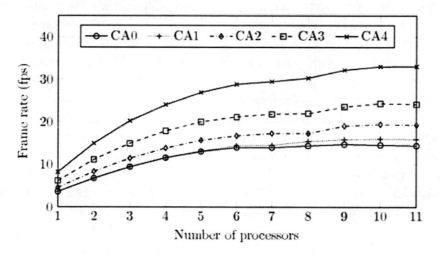

Figure 18. Time components at different optimization levels (11-ARM-system)

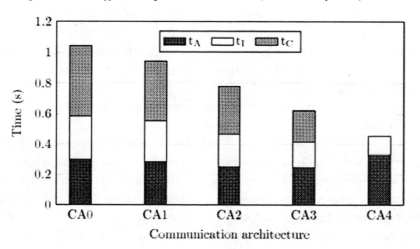

case of CA4, the active time increases since the PEs spend more time in the software API calls for polling the interface registers of OSIP due to the zero-latency in the bus. In this sense, the communication overhead is partially shifted to the execution of software APIs.

Hardware Analysis

The performance improvement achieved from the optimizations for the communication architecture is however at the cost of the additional hardware, i.e. the full multi-layer AHB and the cache system with write buffers.

Currently the area cost of the cache system with write buffers is estimated by the gate-level synthesis with Synopsys Design Compiler using a 90 nm standard cell library. The supply voltage is 1.0 V and the temperature is 25°C. A memory compiler is used to estimate the memory area used for caches. The synthesis result shows that the area is dominated by the memories. In the largest system with 11 PEs, around 90% of the area of the cache system with write buffers are contributed by the memories, if a cache line consist of four words, each stored in a separate memory bank. In total, 1.3 MGate are reported for the area. The maximum achievable clock frequency is 210 MHz.

Enhancements: NoC-Based MPSoCs

To the baseline NoC design, we also present three performance enhancements to increase the communication latency of the NoC and, more importantly, to overlap task dispatching with computation in a pipelined fashion.

Interconnect Analysis

Naturally, the format of the communication has a high impact on the performance of NoC-based MPSoCs. This covers several parameters, such as the communication protocol, buffering scheme, link data width, number of virtual channels and packet size among others. To optimize all these different parameters is outside the scope of this work. We therefore fixed most of the NoC parameters to typical values (4 virtual channels, 8-flit buffers and 4 kB packet size), and focus on a single free parameter, namely the link data width (i.e. flit size) of the NoC. This parameter allows to easily match the average interconnect latency with the average response time of OSIP.

DMA Support

It is common practice to have DMA support for communication on distributed-memory architectures. Good examples can be found in the Cell processor (Kistler, Perrone, & Petrini, 2006) and the KeyStone II processor (Biscondi et al., 2012). A DMA offloads the processors from transferring the data to other nodes of the system so that communication and computation overlap.

As the PEs in the currently nodes are ASIC-based, they do not control the DMA directly. Therefore, an additional controller is added in the master subsystem for this purpose, shown in Figure 19. The DMA contains a separate read and write channel, corresponding to the bi-directional port of the network interface (NI). Based on the write and read signals as well as the source and destination address at the local memory provided by the controller, the DMA performs the data transfer between the local memory and the remote memory over the NoC. In order to get aligned data accesses to the network interface and the local memory respectively, the controller is also responsible for creating and extracting the data packets.

Pipelined Execution

DMAs allow to *pre-fetch* data so that once the PE is ready, the computation can immediately start. This reduces the communication overhead. Different from the distributed systems, in the OSIP-based systems, another type of communication overhead is introduced by accessing the central

Figure 19. VPU master subsystem

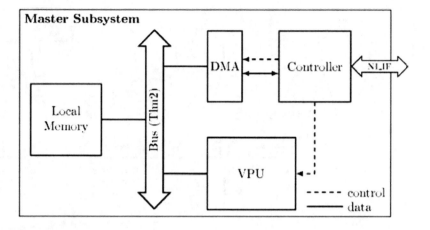

controller – OSIP. This communication overhead is not only caused by the traffic in the network, but more due to the fact, that there is a latency when OSIP processes the PE requests. Therefore, it will help in improving the system performance if the OSIP latency can be hidden, e.g., by means of pre-fetching the tasks during the execution of the PEs. In fact, in this system, pre-fetching tasks is obligatory for the pre-fetch of data. Without task, the data information including the addresses and the data size is unknown to the DMA.

Considering the ASIC-based PEs of the system, the execution pattern of the VPU master subsystems for each task can be simplified into five steps:

1. Requesting a task from OSIP to get input data information, including remote data address and data size (*req(task)*);
2. Using the DMA to load input data from a remote memory to local (*R2L(task)*);
3. Executing the task (*exec(task)*);
4. Using the DMA to transfer the output data from the local memory to remote (*L2R(task)*);
5. Sending synchronization information to OSIP, acknowledging the finish of the task and requesting OSIP to solve possible task dependencies (*sync(task)*).

To hide the communication overhead of both the data transfer and accessing OSIP, the pipeline concept is applied. In the current implementation, step 1) and 2) are grouped into first pipeline stage, step 3) is in the second stage and step 4) and 5) are grouped into the third stage. The pipelining is only applied for the de-mapping (SD_cluster) and channel decoding (TD) subsystems due to the frequent data and task information exchange and iterations between them. Deeper pipelining would be also possible by assigning each step to a separate stage. Note that in the non-pipelined execution of the system (i.e., DMA is not applied in the system), the controller transfers the data in step 2) and 4).

An exemplary execution pattern of the non-pipelined and pipelined control flow of the master subsystem is illustrated in Figure 20 und Figure 21, respectively. As shown in the example of pipelined control flow, while executing task$_B$, task$_C$ is pre-fetched from OSIP and the input data are prepared, while the output data of task$_A$ are sent and a task synchronization is issued to OSIP.

The control flow in the master subsystem can be implemented in different ways. As the PEs are ASICs, the control flow is currently modelled like a state machine in SystemC. Since OSIP is in principle a slave component of the system, the *request-response* communication from the master

Figure 20. Non-pipelined control of task execution

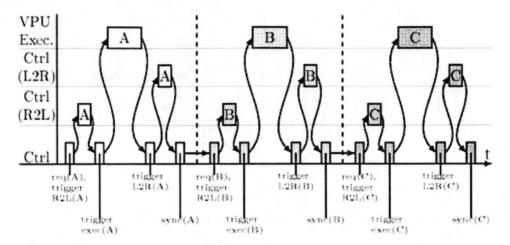

Figure 21. Pipelined control of task execution

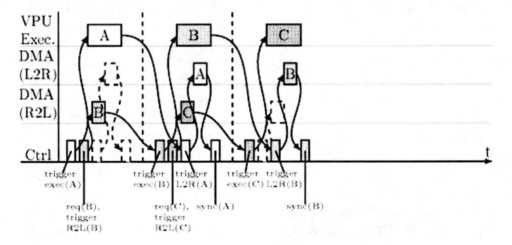

subsystem to the OSIP subsystem is intuitive. The controller will not send the next request to OSIP, until it gets the response from OSIP. This results in a simple control mechanism. An exception exists, if the controller does not get a valid response from OSIP. For example, if the controller requests for a new task from OSIP, but there are currently no ready tasks existing for that PE, the OSIP subsystem will not be able to provide valid task information to the controller. In this case, this request information should be not be removed from the control flow, but buffered both in the master subsystem by the controller and in the OSIP subsystem by the OSIP adapter. Later

when a new task becomes available for the PE, the buffered information is needed to establish the packet for sending the task from OSIP to the PE and to identify the packet at the PE side.

Performance Analysis: NoC-Based MPSoCs

The performance improvement of the NoC-based system with the above-mentioned communication enhancements is illustrated in Figure 22.

By varying the link data width, a better application performance can be obtained. For the MIMO receiver, both the latency and the through-

Figure 22. Improvement of system performance with optimizations in the NoC-based MPSoC

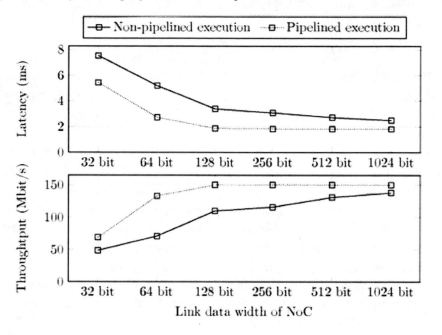

put get significantly improved. Considering the non-pipelined communication, the latency is reduced by a factor of 3 by increasing the data width from 32 bit to 1024 bit, and the throughput is increased by a factor of 2.8. However, in the non-pipelined communication, even at the largest link data width the achieved throughput is still below the required throughput, and the latency just matches the requirement. Moreover, it can be observed that the performance first increases very fast from 32-bit to 128-bit data width. Then, beyond 128 bit the performance increment is clearly slowed down. This results from the fact that in the small configurations, the communication bandwidth is the major factor which influences the system performance. In contrast, in the large NoC configurations, the scheduling efficiency of OSIP has an increasing impact. As for a given application, the latency of task scheduling and mapping by OSIP is almost constant, this latency is the reason for the low performance increment in the large network configurations.

Of course, higher performance can be achieved by further increasing the link data with. This is however impractical due to the high area overhead and also the achievable clock frequency. To be mentioned, with place & route, the 128-bit-based NoC achieves only slightly higher clock frequency than the system clock frequency of 300 MHz. Therefore, instead of having further large NoC configurations, pipelining the communication and computation is a better choice. Shown by the dotted lines in the figure, by hiding the latency of the data transfer of DMAs and the requests to OSIP during the task execution, more performance can be gained than simply increasing the link data width. Already at the data width of 128 bit, both the latency and the throughput constraint are fulfilled based on the pipelining concept. And the performance at the 64-bit configuration with the pipelining is comparable with the one at the 1024-bit configuration without pipelining. The performance saturation beyond 128-bit is caused by the maximum input data rate at the antennas. As the chosen ASIC implementations for the VPUs can only achieve slightly higher throughput than the input data rate, beyond the 128-bit configuration, the main factor influencing the system performance is data processing speed of the ASICs.

OSIP-CENTRIC SYSTEM LEVEL ANALYSIS

The performance analysis above has solely focused on the impact of the communication architecture on the system performance. In OSIP-based systems, the performance is also affected by other factors, most notably, the efficiency of OSIP for task management. When designing OSIP-based systems, both the communication architecture and OSIP have to be considered together. In this section, the joint effects of the communication architecture and OSIP are discussed based on the H.264 video decoder, the MIMO receiver and a synthetic benchmark.

To analyze the joint effects, two other implementations of the same OSIP functionality have been made for comparison:

- **UT-OSIP:** It is a pure untimed SystemC implementation, which implements the OSIP functionality within zero simulation time, hence can be regarded as a hypothetical ASIC implementation with extremely high scheduling efficiency.

- **LT-OSIP:** It corresponds to the functionality of OSIP running on the LTRISC processor. Recall that this processor was the starting point for the OSIP design. It therefore represents a slower scheduler and serves to assess the quality of the extensions built in OSIP.

H.264 Decoding: Bus-Based MPSoC

In this section we analyze the impact of different implementations of OSIP on the performance of the H.264 video decoder application when running on bus-based MPSoCs. For the sake of simplicity, we focus on the variations CA0, CA3 and CA4. The achieved frame rates are shown in Figure 23. Naturally, the system performance with LT-OSIP is lower than that with UT-OSIP and OSIP, as LT-OSIP is not an efficient scheduler. However, the performance gap between the systems with a fast scheduler (UT-OSIP/OSIP) and a slow scheduler (LT-OSIP) is gradually narrowed from CA4 over CA3 to CA0. Considering the largest system with 11 PEs applying an ideal communication architecture (CA4), replacing LT-OSIP with UT-OSIP/OSIP causes an improvement of the frame rate by 19.6 fps/17.4 fps. As a comparison, in the system with an unoptimized communication architecture (CA0), the improvement lies at 5.8 fps/5.5fps,

Figure 23. Joint impact of OSIP and the bus-based communication architecture

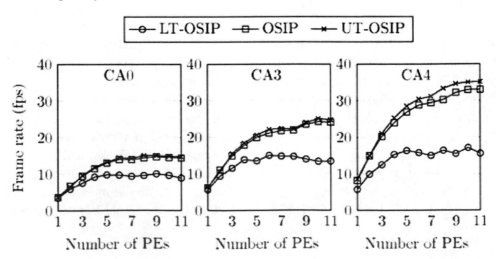

which is a factor of 3.4/3.2 less than with CA4. This means, the possible speedup with a fast scheduler is largely limited by the unoptimized communication architecture. In other words, the efficiency of a fast scheduler will be wasted to a large extent, if the communication architecture is not properly designed.

Once the communication architecture is highly optimized, the system performance largely benefits from a fast scheduler. In this case, a slow scheduler quickly becomes the bottleneck of the system. This can be reflected in the small improvement from CA3 to CA4 observed in LT-OSIP-based systems. Even with an ideal communication architecture the performance improvement is moderate. In contrast, OSIP-based systems can still manage to achieve a significant speedup.

Furthermore, while a slight performance difference still exists between OSIP- and UT-OSIP-based systems at CA4, this difference becomes only marginally in the systems with a realistic communication architecture, even if it is highly optimized. This shows that OSIP optimizations apply not only to ideal but also to realistic systems.

MIMO Receiver: NoC-Based MPSoC

A similar analysis is done for the MIMO-OFDM receiver system, in which the efficiency of the different OSIP implementations and optimizations for the NoC-based communication are jointly investigated.

Figure 24 shows the latency and throughput results for the different optimization approaches for improving the communications on the NoC. Naturally the best performance results are obtained with the UT-OSIP. Especially in the non-pipelined communication configurations, the UT-OSIP has a clear advantage against the other two OSIP implementations. In these configurations, the scheduling latency of OSIP and LT-OSIP is simply added to the communication latency, resulting in an additional communication overhead. However, in the NoC with low communication bandwidths (from 32-bit to 128-bit link data width), in which the data communication is the main contributor to the overall communication overhead, the advantage of using UT-OSIP is not very large. In these cases, the system performance with OSIP

Figure 24. Joint impact of OSIP and the NoC-based communication architecture

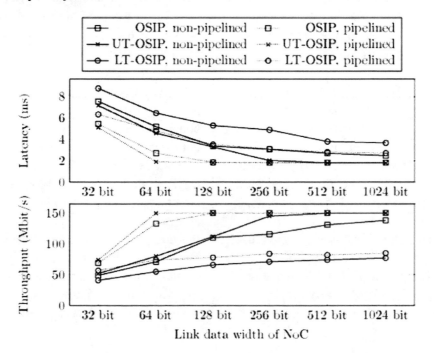

is very close to that based on UT-OSIP. For the very low communication bandwidth based on 32 bit, even LT-OSIP results in a comparable performance as the other two OSIP implementations. By increasing the communication bandwidth, a fast OSIP implementation gains more advantage. Starting from the 256-bit link data width, UT-OSIP performs much better than OSIP. If comparing LT-OSIP and UT-OSIP, a big performance difference already occurs at 64 bit. This shows that in this application, without pipelining the communication and the computation, OSIP can efficiently support a link data width up to 128 bit, but LT-OSIP is only suitable for the NoC with a link data width of 32 bit.

After pipelining the communication and the computation, the advantage of UT-OSIP against OSIP is greatly reduced. Shown in the figure, beyond the 128-bit configuration, both UT-OSIP and OSIP result in the same system performance. In these configurations, the latency of OSIP is completely hidden by the task execution, and the data processing speed of the application processors is the major factor influencing the system performance. The biggest performance difference based on these two OSIP implementations can be found at the 64-bit configuration. In this configuration, the data transfer can be hidden by the task execution, but adding the additional OSIP latency to the data transfer exceeds the task execution time. However, the performance loss of using OSIP instead of UT-OSIP is relatively small. If choosing an even smaller configuration of 32 bit, for both UT-OSIP- and OSIP-based systems, the communication and the computation cannot totally overlap and the data transfer plays a more important role than the scheduling efficiency. Therefore, their performance difference becomes very small again.

Pipelining the LT-OSIP-based systems also increases the system performance. However, in comparison to the two fast OSIP implementations, the high latency of LT-OSIP largely prevents the total communication overhead from being hidden

by the task execution. This results in a much lower system performance even with the pipelining than for the other two implementations, especially in the systems with a high communication bandwidth for the data transfer. As shown in the figure, starting from the link data width of 128 bit, the performance increment becomes very small. In these systems, the overhead caused by the data transfer is not critical, but accessing LT-OSIP dominates the communication time, which does not match the task execution speed.

Synthetic Benchmarking: Bus-Based MPSoC

The performance of OSIP-based systems greatly depends on both the scheduler efficiency and the communication architecture. Both are however in turn influenced by the task sizes (execution time of tasks on PEs). Different task sizes and also different numbers of tasks will cause different loads to the scheduler and the communication architecture. The two previous sections have analyzed specific applications, so that it is difficult to assess the performance for other applications. In this section we therefore introduce a synthetic benchmark that helps to investigate the limits of OSIP in terms of number of tasks, task size, number of PEs and amount of data traffic. Note that this synthetic analysis is only carried out for the bus-based systems. A synthetic analysis for the NoC-based systems is much more complex, since a large number of factors need to be taken into consideration. Apart from the above-mentioned configurable communication bandwidths and pipelined execution, other important factors that can have a big impact on the system performance could be e.g. the location of the OSIP subsystem and the memory subsystems, the packet priorities and virtual channel configurations. Also the data communication protocols, whether using a master-to-master communication or a master-to-slave communication, are important for the analysis as well. This complexity makes a synthetic analysis for the NoC-based OSIP systems have little practical use.

As illustrated in Figure 25, the synthetic application consists of three major tasks: *data producing (task$_p$)*, *task generation (task$_{tg}$)* and *data consuming (task$_c$)*. The first two tasks are mapped onto one ARM processor as a producer PE (*PPE*) and the last one is mapped onto the other ARM processors as a class of consumer PEs (*CPEs*). The *PPE* first produces a data block into the shared memory and then generates a set of *tasks$_c$* for consuming the data. OSIP will schedule the *CPEs* to execute the *tasks$_c$*, which read and sum up the data from the shared memory and output the result to the I/O. A system with *n* consumer processors is referred to as *n-CPE-system*.

The following parameters are configurable in the synthetic application to influence the load of OSIP:

- **N_CPES** $\in$ **{1, 3, 5, 7, 9, 11}**: Number of *CPEs*. Increasing N_CPEs will potentially create more requests to OSIP and more traffic, hence increases the load to OSIP and the communication architecture.

- **N_TASKS** $\in$ **{11, 88, 165}**: Number of generated *tasks$_c$*. These numbers are chosen as multiples of 11, such that in an 11-CPE-system the tasks could be distributed to the *CPEs* in a balanced way.

- **N_ACCESSES** $\in$ **{1, 6, 11}**: Frequency of accessing the same data. The size of the data block is fixed to 500 words, each 32 bits. The data access is configured in such a way that each word in the data block is read for multiple times, which is specified by N_ACCESSES. The function of this parameter is three-fold. First, changing N_ACCESSES will change the task size. An access frequency of 1, 6 and 11 represents a task size of 2.5 Kcycle, 15 Kcycle and 27.5 Kcycle, respectively. Note that these cycle numbers are obtained by running *task$_c$* on the processor, with data located in the local memory. Certainly if the data are located in the shared memory, the cycle numbers will vary due to the communication overhead in the bus. However, they show qualitatively the different task sizes. Second, a larger N_ACCESSES indicates

Figure 25. Configuration of the synthetic benchmark application

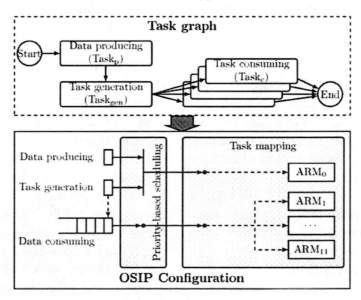

more data traffic. Third, the effectiveness of the cache can be easily modified by changing this parameter.

In addition, for including the cache coherence effect, the flow "data producing – data consuming" is iterated. In each iteration, the values in the data block are updated. Furthermore, the scheduling algorithm applied for the task queue is priority-based in order to stress the scheduler.

For evaluating OSIP-based systems, three representative application scenarios are derived from the benchmark application by configuring the above-mentioned parameters to produce different workloads to the scheduler.

- **Best Case Scenario (N_ACCESSES = 11, N_TASKS = 11):** In this scenario, the size of $task_c$ is set to maximum within the scope of the benchmark parameters and the number of $tasks_c$ is set to minimum. For this type of applications, it takes a long time for a PE to finish one task before it requires a new task from OSIP. Therefore, the frequency of accessing OSIP is low. Furthermore, it takes OSIP relatively little time to handle a request (to find the best candidate task from the task queue in this case), because the task queue is short. Therefore, low workloads are generated to OSIP in this scenario.
- **Worst Case Scenario (N_ACCESSES = 1, N_TASKS = 165):** This scenario is exactly the opposite case to the one above, in which the task size is set to minimum and the number of $tasks_c$ is set to maximum. In this case, the PEs are able to finish the tasks within a very short time, hence request very frequently new tasks from OSIP. The scheduling performed in OSIP also becomes much more intensive due to a much larger task queue. In general, high workloads are generated to OSIP in the case.

- **Average Case Scenario (N_ACCESSES = 6, N_TASKS= 88):** In this scenario, both the task size and the number of tasks are set to medium. The load of OSIP is supposed to be at a moderate level in this case.

Based on these scenarios, the behavior of the targeted evaluation systems is discussed in detail, considering the different schedulers and number of *CPEs*.

Best Case Scenario

The upper part of Figure 26 compares the execution time of the benchmark application in the best case scenario. In general, UT-OSIP- and OSIP-based systems have a very similar performance profile, while LT-OSIP-based systems have a reduced performance characteristic.

Similar to the observation made with the H.264 application, using an unoptimized communication architecture leads to small performance differences between the systems with different schedulers. This is due to the fact that the communication dominates the execution time of the application, which is the bottleneck of the system. By optimizing the communication architecture, the bottleneck starts to move to the other parts of the system.

With an optimized communication architecture, the effect of the different schedulers is still relatively small in small systems. In this case, the dominating factor is the execution time of the tasks. As the number of *CPEs* increases, LT-OSIP begins to reach its limitations in terms of scheduling tasks. Starting from a 7-CPE-system, the performance gap between the LT-OSIP-based systems and the UT-OSIP-/OSIP-based systems becomes large. In this case, a slow scheduler is not anymore able to efficiently distribute the various tasks to the PEs. This is also the reason for the limited performance improvement from CA0 to CA4 in LT-OSIP-based systems with a large number of *CPEs*, while in UT-OSIP/OSIP-based systems the execution time is significantly reduced.

Figure 26. Best case scenario (N_ACCESSES = 11, N_TASKS = 11)

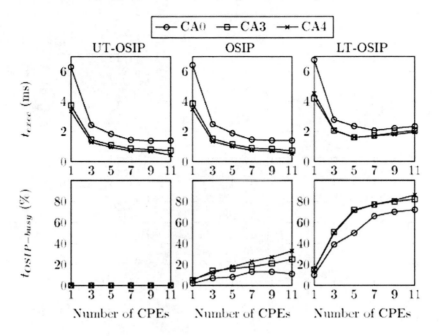

The scheduling efficiency can be well reflected by the busy state of the scheduler, which is illustrated in the lower part of Figure 26. Naturally, UT-OSIP is never busy, since it can perform any kind of scheduling within zero time. In comparison to UT-OSIP, OSIP and LT-OSIP in general become busier with increasing number of *CPEs*. Also further optimization in the communication architecture makes both schedulers more frequently enter the busy state, because the task execution time is shortened, which in turn increases the request frequency of *CPEs* to the scheduler. However, in comparison to LT-OSIP, OSIP is much less stressed. In the 11-CPE-system with a realistic but highly optimized communication architecture (CA3), the busy time of OSIP is around 23%, while LT-OSIP is busy with the scheduling for more than 80% of the time.

Worst Case Scenario

The execution time of the applications in Figure 27 demonstrates more peculiar system behaviors in the worst case scenario compared to the best case

scenario presented above. In this scenario, the LT-OSIP becomes impractical. While it is still able to reduce the execution time from the 1-CPE-system to the 3-CPE-system, the system performance becomes disastrous when more *CPEs* are used. Even an ideal communication architecture is not able to really improve the system performance. In this case, not only the task scheduling is very inefficient, the time spent on the scheduling algorithm itself contributes to the major part of the total execution time of the application.

In contrast, the scheduling by UT-OSIP cannot be the bottleneck of the system. Therefore, significant improvement can be achieved by optimizing the communication architecture. However, it can also be observed that applying more than *3 CPEs* cannot help to continue improving the system performance even at CA4, in which the communication architecture cannot be the system bottleneck, either.

The main reason lies in the low task parallelism caused by synchronizations. In the application, the *PPE* and *CPEs* compete for the same spinlock at the OSIP register interface to access the protected

Figure 27. Worst case scenario (N_ACCESSES = 1, N_TASKS = 165)

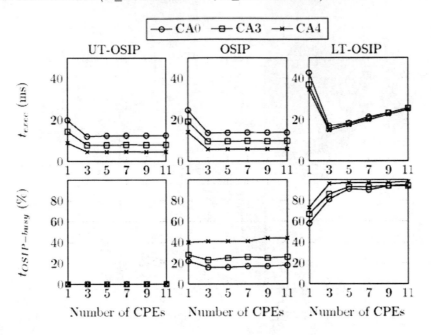

resources. As there are only a single *PPE* but multiple *CPEs*, the *PPE* is often blocked by the *CPEs*, when trying to create new tasks to OSIP. This becomes especially critical if the task size is small. In this case, the *CPEs* finish the tasks fast, which increases the possibility that the *PPE* competes with several *CPEs* at the same time. So the competition between them happens very frequently. It leads to the fact that the *PPE* is not able to prepare the tasks fast enough, which are needed by the *CPEs*. In other words, the task parallelism cannot match the number of *CPEs*. More details of this and an improvement of this synchronization is presented by Zhang et al. (2013).

In a wider sense, the synchronization can also be regarded as one form of the communication, which appears to be the bottleneck of the system in the worst case scenario. This is also the reason that in this case, OSIP-based systems still perform as well as UT-OSIP-based systems. In fact, mostly OSIP stays at an idle state in the system with a realistic communication architecture, as shown in the figure. Even in the largest system with the ideal bus, OSIP stays at busy state for 43% of the time.

Average Case Scenario

Often task partitioning is done in such a way that the task sizes are neither extremely large nor very small. Too large tasks would potentially reduce the parallelism degree of an application, while too small tasks would cause too much communication and synchronization overhead. Therefore, an average case scenario with moderate task sizes and number of tasks is studied.

The analysis results in Figure 28 show high similarity to those in the best case scenario. Only in the systems with an unoptimized communication architecture, the LT-OSIP-based systems still have comparable performance to the other two. In other cases, the UT-OSIP- and OSIP-based systems, which have similar performance, are much better than the LT-OSIP-based systems.

For OSIP and UT-OSIP, with an unoptimized communication architecture (CA0), adding more than 5 *CPEs* even worsens the system performance. In this case, very high bus contention occurs, which in fact decreases the parallel task execution.

Figure 28. Average case scenario (N_ACCESSES = 6, N_TASKS = 88)

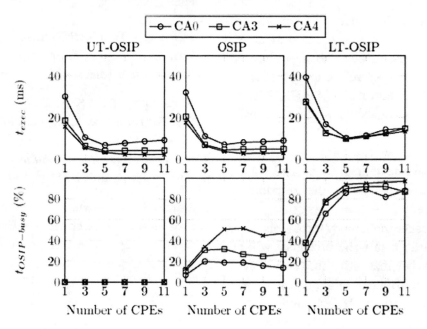

By comparing the average case scenario and worst case scenario for OSIP-based systems, an interesting phenomenon can be observed in the systems using CA3 and CA4. In the average case scenarios, OSIP enters more frequently the busy state. The reason has been implicitly given above. In the worst case scenario, the task parallelism is in fact not high due to the synchronization effect. Therefore, the size of the task queue in OSIP in the worst case is on the contrary smaller than in the average case. This leads to a higher load of OSIP in the average case scenario. Still, in comparison to the task size, the scheduling overhead of OSIP is minor.

The behavior of OSIP-based systems is influenced by several factors including the scheduling efficiency, the communication and synchronization overhead and the task sizes. The bottleneck of the system can shift, depending on the different application scenarios. The case study above shows that OSIP has a high efficiency in performing task scheduling and mapping at runtime for realistic systems. Furthermore, in order to make OSIP-based systems work in an optimal way, the com-

munication architecture and the synchronization overhead need to be carefully considered during system design.

CONCLUSION AND OUTLOOK

In this work, we investigated the performance behavior of OSIP-based systems from a system point of view, considering both the OSIP efficiency and the communication overhead. We studied the most common interconnect paradigms, namely, bus and NoC. For bus-based systems, we analyzed the impact of including a multi-layer bus, a cache system and write buffers. For NoC-based systems, the effects of the communication bandwidth, the use of DMAs and the pre-fetching of tasks were investigated. The results showed a significant performance improvement by reducing the communication overhead.

The analysis, based on real-life applications and a synthetic benchmark, gives a broad view of the joint effects caused by OSIP and the communication architecture. The results show a very high efficiency of OSIP for dynamic scheduling and mapping in MPSoCs. However, in order to fully utilize the OSIP-approach, not only an efficient OSIP processor, but also an optimized communication architecture is required. In addition, the synchronization overhead in OSIP-based systems should also be well considered to make the systems work in a proper way. More generally, designing MPSoCs with a hardware scheduler must take special care of the balance between the efficiency of the scheduler and the communication architecture.

In future, cache coherency protocols with less area-overhead than write-broadcast will be analyzed. Furthermore, sharing the bus layers of the multi-layer AHB among the processors will also be considered to trade off area against system performance.

For the NoC-based MPSoCs, systems with programmable processors instead of ASIC subsystems will be targeted. The flexibility of the processors will introduce more complexity in the control flow of the system, for which more complex OSIP proxies are most probably requested. Instead of the current state machine based proxies, small RISC processors can be considered for the proxies to cope with the increased flexibility.

Furthermore, in current systems, the NoC size is still relatively small. In future many-core systems with hundreds of cores, applying a single OSIP processor for the task and resource management of the whole system is obviously impractical. However, the efficiency and the flexibility of OSIP motivate the future work of studying hierarchical solutions with multiple OSIP instances for the management in large-scale systems.

REFERENCES

3GPP. (2013). *3GPP Release 8*. Retrieved December 06, 2013, from http://www.3gpp.org/specifications/releases/72-release-8

Al Faruque, M. A., Krist, R., & Henkel, J. (2008). ADAM: Run-time Agent-based Distributed Application Mapping for on-chip Communication. In *Proceedings of Design Automation Conference* (pp. 760-765). doi: 10.1145/1391469.1391664

Arnold, O., Noethen, B., & Fettweis, G. (2012). Instruction Set Architecture Extensions for a Dynamic Task Scheduling Unit. In *Proceedings of IEEE Computer Society Annual Symposium on VLSI* (pp. 249-254). IEEE. doi: 10.1109/ISVLSI.2012.51

Benini, L., & De Micheli, G. (2002). Networks on Chips: A New SoC Paradigm. *IEEE Computer*, 35(1), 70–78. doi:10.1109/2.976921

Biscondi, E., Flanagan, T., Fruth, F., Lin, Z., & Moerman, F. (2012). *Maximizing Multicore Efficiency with Navigator Runtime* (White Paper). Retrieved December 8, 2013, from www.ti.com/lit/wp/spry190/spry190.pdf

Castilhos, G., Mandelli, M., Madalozzo, G., & Moraes, F. (2013). Distributed Resource Management in NoC-based MPSoCs with Dynamic Cluster Sizes. In *Proceedings of IEEE Computer Society Annual Symposium on VLSI* (pp. 153-158). IEEE. doi: 10.1109/ISVLSI.2013.6654651

Castrillon, J., Zhang, D., Kempf, T., Vanthournout, B., Leupers, R., & Ascheid, G. (2009). Task Management in MPSoCs: An ASIP Approach. In *Proceedings of International Conference on Computer-Aided Design* (pp. 587-594). New York: ACM. doi: 10.1145/1687399.1687508

Chou, C., & Marculescu, R. (2008). User-Aware Dynamic Task Allocation in Networks-on-Chip. In *Proceedings of Design* (pp. 1232–1237). Automation and Test in Europe.

Corporation, I. B. M. (2013). *CoreConnect Bus Architecture*. Retrieved December 15, 2013, from https://www-01.ibm.com/chips/techlib/techlib. nsf/productfamilies/CoreConnect_Bus_Architecture

de Souza Carvalho, E. L., Calazans, N. L. V., & Moraes, F. G. (2010). Dynamic Task Mapping for MPSoCs. *Design & Test of Computers*, *27*(5), 26–35. doi:10.1109/MDT.2010.106

Eggers, S. J., & Katz, R. H. (1988). *Evaluating the Performance of Four Snooping Cache Coherency Protocols* (Tech. Rep. UCB/CSD-88-478). University of California, Berkeley, EECS Department.

Garibotti, R., Ost, L., Busseuil, R., & Kourouma, M. Adeniyi-Jones, Sassatelli, G., & Robert, M. (2013). Simultaneous Multithreading Support in Embedded Distributed Memory MPSoCs. In *Proceedings of Design Automation Conference (DAC)* (pp. 1-7). DAC. doi: 10.1145/2463209.2488836

Goyal, P., Guo, X., & Vin, H. M. (1996). A Hierarchical CPU Scheduler for Multimedia Operating Systems. In *Proceedings of Symposium on Operating Systems Design and Implementation* (pp. 107-121). New York: ACM. doi: 10.1145/238721.238766

Ienne, P., & Leupers, R. (Eds.). (2006). *Customizable Embedded Processors: Design Technologies and Applications (Systems on Silicon)*. San Francisco, CA: Morgan Kaufmann Publishers.

Jantsch, A., & Tenhunen, H. (Eds.). (2003). *Network on Chips*. Boston, MA: Kluwer Academic Publishers.

Kempf, T., Ascheid, G., & Leupers, R. (2011). *Multiprocessor Systems on Chip: Design Space Exploration*. Springer. doi:10.1007/978-1-4419-8153-0

Khajekarimi, E., & Hashemi, M. R. (2012). Communication and Congestion Aware Run-Time Task Mapping on Heterogeneous MPSoCs. In *Proceedings of International Symposium on Computer Architecture and Digital Systems* (pp. 127-132). doi: 10.1109/CADS.2012.6316432

Kistler, M., Perrone, M., & Petrini, F. (2006). Cell Multiprocessor Communication Network: Built for speed. *IEEE Micro*, *26*(3), 10–23. doi:10.1109/MM.2006.49

Kohout, P., Ganesh, B., & Jacob, B. (2003). Hardware Support for Real-Time Operating Systems. In *Proceedings of IEEE/ACM/IFIP International Conference on Hardware/Software Codesign and System Synthesis* (pp. 45-51). New York: ACM. doi: 10.1145/944645.944656

Limberg, T., Winter, M., Bimberg, M., Klemm, R., Tavares, M., & Ahlendorf, H. … Fettweis, G. (2009). A Heterogeneous MPSoC with Hardware Supported Dynamic Task Scheduling for Software Defined Radio. In *Proceedings of DAC/ISSCC Student Design Contest*. doi: 10.1.1.150.6363

Lippett, M. (2004). An IP Core Based Approach to the On-Chip Management of Heterogeneous SoCs. In *Proceedings of IP Based SoC Design Forum & Exhibition*. Academic Press.

Loghi, M., Poncino, M., & Benini, L. (2006). Cache Coherence Tradeoffs in Shared-Memory MPSoCs. *ACM Transactions on Embedded Computing Systems*, *5*(2), 383–407. doi:10.1145/1151074.1151081

Ltd, A. R. M. (2013a). *AMBA On-Chip Connectivity*. Retrieved December 15, 2013, from http://www.arm.com/products/system-ip/amba/amba-open-specifications.php

Murtaza, Z., Khan, S. A., Rafique, A., Bajwa, K. B., & Zaman, U. (2006). Silicon Real Time Operating System for Embedded DSPs. In *Proceedings of International Conference on Emerging Technologies* (pp. 188-191). doi: 10.1109/ICET.2006.336032

Nácul, A. C., Regazzoni, F., & Lajolo, M. (2007). Hardware Scheduling Support in SMP Architectures. In *Proceedings of the Conference on Design, Automation and Test in Europe* (pp. 642-647). San Jose, CA: EDA Consortium. doi: 10.1109/DATE.2007.364666

Nakano, T., Utama, A., Itabashi, M., Shiomi, A., & Imai, M. (1995). Hardware Implementation of a Real-Time Operating System. In *Proceedings of TRON Project International Symposium* (pp. 34-42). doi: 10.1109/TRON.1995.494740

Nordström, S., & Asplund, L. (2007). Configurable Hardware/Software Support for Single Processor Real-Time Kernels. In *Proceedings of International Symposium on System-on-Chip* (pp. 1-4). doi: 10.1109/ISSOC.2007.4427426

Pan, Z., & Wells, B. E. (2008). Hardware Supported Task Scheduling on Dynamically Reconfigurable SoC Architectures. *IEEE Transactions on Very Large Scale Integration (VLSI). Systems*, 16(11), 1465–1474. doi: doi:10.1109/TVLSI.2008.2000974

Park, S., Hong, D.-S., & Chae, S.-I. (2008). A Hardware Operating System Kernel for Multi-Processor Systems. *IEICE Electronics Express*, 5(9), 296–302. doi:10.1587/elex.5.296

Paulin, P., Pilkington, C., Langevin, M., Bensoudane, E., Lyonnard, D., & Benny, O. et al. (2006). Parallel Programming Models for a Multiprocessor SoC Platform Applied to Networking and Multimedia. *IEEE Transactions on Very Large Scale Integration (VLSI). Systems*, 14(7), 667–680. doi: doi:10.1109/TVLSI.2006.878259

Qualcomm, Inc. (2013). *Snapdragon*. Retrieved December 16, 2013, from http://www.qualcomm.com/snapdragon

Schürmans, S., Zhang, D., Auras, D., Leupers, R., Ascheid, G., Chen, X., & Wang, L. (2013). Creation of ESL Power Models for Communication Architectures using Automatic Calibration. In *Proceedings of Design Automation Conference (DAC)* (pp. 1-6). doi: 10.1145/2463209.2488804

Seidel, H. (2006). *A Task-level Programmable Processor*. Duisburg, Germany: WiKu-Verlag.

Shabbir, A., Kumar, A., Mesman, B., & Corporaal, H. (2011). Distributed Resource Management for Concurrent Execution of Multimedia Applications on MPSoC Platforms. In *Proceedings of International Conference on Embedded Computer Systems* (pp. 132-139). doi: 10.1109/SAMOS.2011.6045454

Sloss, A., Symes, D., & Wright, C. (2004). *ARM System Developer's Guide: Designing and Optimizing System Software*. San Francisco, CA: Morgan Kaufmann Publishers.

STMicroelectronics & CEA. (2010). *Platform 2012: A Many-core Programmable Accelerator for Ultra-Efficient Embedded Computing in Nanometer Technology* (White Paper). Retrieved December 10, 2013, from http://www.2parma.eu/index.php/documents/publications.html

Synopsys, Inc. (2013a). *Synopsys Platform Architect*. Retrieved December 06, 2013, from http://www.synopsys.com/Systems/ArchitectureDesign/Pages/PlatformArchitect.aspx

Synopsys, Inc. (2013b). *Synopsys Processor Designer*. Retrieved December 06, 2013, from http://www.synopsys.com/Systems/BlockDesign/ProcessorDev/Pages/default.aspx

Texas Instruments, Inc. (2013). *OMAP*. Retrieved January 25, 2013, from http://focus.ti.com/docs/prod/folders/print/omap3530.html

Tomasevic, M., & Milutinovic, V. (1994a). Hardware Approaches to Cache Coherence in Shared-Memory Multiprocessors, Part 1. *IEEE Micro*, *14*(5), 52–59. doi:10.1109/MM.1994.363067

Tomasevic, M., & Milutinovic, V. (1994b). Hardware Approaches to Cache Coherence in Shared-Memory Multiprocessors, Part 2. *IEEE Micro*, *14*(6), 61–66. doi:10.1109/40.331392

Wentzlaff, D., Griffin, P., Hoffmann, H., Bao, L., Edwards, B., & Ramey, C. et al. (2007). On-Chip Interconnection Architecture of the Tile Processor. *IEEE Micro*, *27*(5), 15–31. doi:10.1109/MM.2007.4378780

Zhang, D., Lu, L., Castrillon, J., Kempf, T., Ascheid, G., Leupers, R., & Vanthournout, B. (2013). Efficient Implementation of Application-Aware Spinlock Control in MPSoCs. *International Journal of Embedded and Real-Time Communication Systems*, *4*(1), 64–84. doi:10.4018/jertcs.2013010104

Zhang, D., Zhang, H., Castrillon, J., Kempf, T., Vanthournout, B., Ascheid, G., & Leupers, R. (2011). Optimized Communication Architecture of MPSoCs with a Hardware Scheduler: A System-Level Analysis. *International Journal of Embedded and Real-Time Communication Systems*, *2*(3), 1–20. doi:10.4018/jertcs.2011070101

KEY TERMS AND DEFINITIONS

Application-Specific Instruction-Set Processor (ASIP): An ASIP is a special instruction-set processor tailored for certain application domains. Typically it features customized instructions and/or special memory architectures.

Cache Coherency: It means that on update of a data item in a cache, its copies in shared resources and other caches are updated before they are accessed by other processors. In multiprocessor systems, data inconsistency can happen when updating the shared data, if no additional protection mechanisms are applied.

Communication Architecture: A communication architecture is a hardware subsystem which connects the system components. Typical communication architectures are based on busses or Networks-on-Chip (NoCs).

Direct Memory Access (DMA): It is a mechanism for directly transferring data without involving processors by means of using specific hardware blocks. This can largely reduce the load the processors.

Dynamic Scheduling and Mapping: It refers to making task scheduling and mapping decisions at runtime, typically depending on the current system states and/or scenarios.

Multi-Processor System-on-Chip (MPSoC): It refers to a System-on-Chip, in which multiple processor elements exist. MPSoCs can be homogeneous or heterogeneous. In a homogeneous MPSoC, all processing elements are of the same type, while in a heterogeneous MPSoC, the types of the processing elements are different. A processing element can be a general-purpose processor, a customized processor, a dedicated hardware IP, or reconfigurable hardware such as FPGA, etc.

Network-on-Chip (NoC): NoC is a communication paradigm for Systems-on-Chip, in which the data communication between the system components (such as processing elements, memories, etc.) is done through a network, typically constructed using routers. A major advantage of a NoC against the traditional busses is its scalability.

Resource Management: It refers to the management of the utilization of hardware resources in the system. Usually it is performed for achieving certain design or optimization goals, e.g., maximizing system performance or minimizing energy consumption, etc. The resource management typically considers the load of the processing elements, the communication and memory bandwidth and/or even reconfigurable hardware blocks.

Chapter 15
Efficiency Analysis of Approaches for Temperature Management and Task Mapping in Networks-on-Chip

Tim Wegner
University of Rostock, Germany

Martin Gag
University of Rostock, Germany

Dirk Timmermann
University of Rostock, Germany

ABSTRACT

With the progress of deep submicron technology, power consumption and temperature-related issues have become dominant factors for chip design. Therefore, very large-scale integrated systems like Systems-on-Chip (SoCs) are exposed to an increasing thermal stress. On the one hand, this necessitates effective mechanisms for thermal management and task mapping. On the other hand, application of according thermal-aware approaches is accompanied by disturbance of system integrity and degradation of system performance. In this chapter, a method to predict and proactively manage the on-chip temperature distribution of systems based on Networks-on-Chip (NoCs) is proposed. Thereby, traditional reactive approaches for thermal management and task mapping can be replaced. This results in shorter response times for the application of management measures and therefore in a reduction of temperature and thermal imbalances and causes less impairment of system performance. The systematic analysis of simulations conducted for NoC sizes up to 4x4 proves that under certain conditions the proactive approach is able to mitigate the negative impact of thermal management on system performance while still improving the on-chip temperature profile. Similar effects can be observed for proactive thermal-aware task mapping at system runtime allowing for the consideration of prospective thermal conditions during the mapping process.

DOI: 10.4018/978-1-4666-6034-2.ch015

INTRODUCTION

The emergence of nanotechnology is accompanied by cumulative power densities and switching activities per unit area. Therefore, increasingly complex and highly integrated systems like SoCs have to contend with well-known challenges. Amongst others, this concerns heat dissipation, leading to high circuit temperatures and possibly strongly unbalanced on-chip temperature distributions (see ΔT [°C] in Table 2 for examples regarding on-chip temperature variations). In the light of a growing number of transistors per chip, which are increasingly susceptible to environmental influences and deterioration, this issue is topical more than ever. As a consequence, thermal stress and physical effects exponentially depending on temperature (JEDEC, 2009) threaten the integrity of Integrated Circuits (ICs) and have major influence on operability, lifetime and performance.

The relationship between temperature and deterioration is illustrated by the Arrhenius model (Srinivasan & Adve, 2003) describing the influence of temperature on the velocity of chemical reactions. This model originates from the Van't Hoff rule saying that chemical reactions take place twice as fast when temperature is increased by 10 K. As a rule of thumb, this also can be interpreted as a bisection of lifetime of ICs with every 10 K temperature increase. For this reason, monitoring and control of on-chip temperature distribution are important tasks to secure system functionality and ensure high performance.

Typically, monitoring of on-chip temperature is performed by collecting temperature-related data (e.g. by using integrated diodes). In order to react to undesirable temperatures this data has to be transferred to a component responsible for data evaluation and determination of appropriate reactions (i.e. thermal management). Then instructions are sent to the concerned components. For NoC-based systems, commonly the NoC infrastructure is used for this communication. Despite the importance of thermal management the utilization of the communication infrastructure represents an intrusion into the system and the induced traffic curtails the availability of the network for regular communication.

A particular drawback of reactive approaches is the comparatively long response time of thermal management. This time is in the order of several milliseconds and is caused by the time that passes between the occurrence of a switching activity and a corresponding heat propagation (see end of section *Reactive Management* for an example calculation) as well as the transmission delay (see D_p in Table 2 for an uncongested NoC), which is induced when using the NoC for monitoring purposes. Since two transmissions (i.e. reporting temperature and sending instructions) are necessary, an already congested NoC additionally exacerbates thermal management. This results in possibly delayed reactions to undesired temperatures and unnecessary long periods of reduced performance.

Hence, a method to predict the on-chip temperature profile is proposed. The prediction is based on a model that is realized as part of a Thermal Management Unit (TMU), instead of only reverting to physical sensors. This model uses switching activities to calculate the temperature profile. By means of the made predictions, the TMU is able to immediately initiate execution of instructions for thermal management. Such a TMU can be implemented in software running on a core of the SoC or it is an inherent part of a core implemented in hardware. Thereby, the response time for thermal management is shortened by avoiding the waiting period between the occurrence of a switching activity and subsequent heat propagation.

Another field of application for the prediction model is thermal-aware task mapping at system runtime. Basically, this approach relies on the prediction model to estimate the impact of a task placement on the development of the temperature profile of a NoC-based system. For this purpose the model reverts to certain task properties (i.e. task runtime, expected computation effort, generated communication traffic). The estimated impact of a particular task placement can then be considered for future mapping decisions. Compared to static design-time mapping algorithms this offers the advantage of including the system status (e.g. availability of system components, condition of the communication infrastructure) into mapping decisions. This is especially beneficial for systems exhibiting unknown application profiles or high application dynamics. Furthermore, the proactive thermal-aware strategy is expected to achieve additional improvements regarding the on-chip temperature profile compared to conventional runtime mapping algorithms.

Prerequisites for the targeted use cases are that predictions can be accomplished rather fast and without generating too much additional heat induced by the logic necessary for the activity counters (this applies only to thermal management) and the calculation process for the model. To ascertain to which extent the proactive thermal-aware approaches influence system performance and on-chip temperature distribution, they are compared to setups reverting to reactive strategies. The proactive management is additionally compared to a setup without any thermal management. Briefly summarized, the contributions of this article are the following:

- Introduction of a software-based temperature model for ICs
- Application for proactive temperature management of NoC-based systems
- Application for proactive thermal-aware task mapping at runtime of NoC-based systems
- Evaluation of proactive management and task mapping as well as comparison to conventional reactive methods

The remainder of this chapter is organized as follows. In section *Related Work* an overview over existing work regarding modeling of on-chip temperature is given. Furthermore, strategies for reactive and proactive management and task mapping are introduced. In section *Simulation Environment* the basic environment for the simulation of approaches for thermal-aware management and task mapping of NoC-based systems is presented. Specific extensions to this environment facilitating thermal-aware system management (both reactive and proactive) are illustrated in chapter *Thermal-aware Management*. In chapter *Thermal-aware Task Mapping* the simulation environment is extended by an application model in order to enable thermal-aware algorithms for runtime task mapping. In section *Experiments and Results* experiments focusing on the impact of proactive and reactive management and task mapping on different system parameters (i.e. performance, temperature and effort) are conducted. Finally, *Future Research Directions* and a *Conclusion* are given.

RELATED WORK

Numerous investigations have already been conducted in the field of modeling thermal behavior (Skadron, Stan, W., Velusamy, Sankaranarayanan, & Tarjan, 2003), (Shang, Peh, Kumar, & Jha, 2004), (Liu, Calimera, Nannarelli, Macii, & Poncino, 2010), (Tockhorn, Cornelius, Saemrow, & Timmermann, 2010) of ICs by exploiting the equivalence of electrical and thermal energy flows (Krum, 2000), since this approach implicates some worthwhile consequences regarding effort for thermal management. In (Skadron, Stan, W., Velusamy, Sankaranarayanan, & Tarjan, 2003), electrical RC-circuits are used to model the thermal behavior of an entire chip. Variability of modeling granularity allows for a trade-off between modeling accuracy and speed. Temperature of the functional blocks is computed by using values for average power dissipation. In (Shang, Peh, Kumar, & Jha, 2004) this approach is tailored to the simulation of the thermal behavior of on-chip networks. For this purpose, the model of equivalent RC-circuits is extended by the integration of heat spreading angles. Temperature estimation is performed by capturing the network traffic, using these statistics for estimation of power consumption and computing the temperature profile. The creation of SPICE netlists consisting of RC-circuits in order to model on-chip thermal properties is proposed in (Liu, Calimera, Nannarelli, Macii, & Poncino, 2010), (Tockhorn, Cornelius, Saemrow, & Timmermann, 2010).

Research related to reactive management strategies for on-chip networks is available abundantly. A general concept of an event-based runtime monitoring service for NoC components using hardware probes is proposed in (Ciordas, Basten, Rădulescu, Goossens, & Meerbergen, 2005). In (Ciordas, Goossens, Radulescu, & Basten, 2006) this concept is examined with focus on the integration into an existing NoC and the arising implications. In (Guang, Nigussie, Rantala, & Isoaho, 2010) a hierarchical agent framework to realize monitoring services on parallel SoC systems in order to provide for fault tolerance is proposed. An approach specified to reactive monitoring and control of temperature in NoC-based systems is provided by (Martinez & Atienza, 2009). Sensors monitor the component temperatures and use the NoC infrastructure to report temperature to a central TMU.

Proactive thermal management can be defined as predicting temperature at runtime and taking appropriate actions instead of monitoring temperature and reacting to undesired changes. Assuming this, investigations in this field are available more sparsely. In (Coskun, Rosing, & Gross, 2008) autoregressive moving average (ARMA) modeling is used to predict temperature of SoCs by regressing previous measurements from thermal sensors. Predictions are employed for thread allocation in order to balance on-chip temperature distribution. An approach using a thermal model based on equivalent RC-circuits in order to apply reactive and proactive measures

for thermal management is introduced in (Shang, Peh, Kumar, & Jha, 2004).

Besides the given works on proactive and reactive thermal management, there exist several design-time thermal-aware solutions based on task and application mapping. In (Hung, Addo-Quaye, Theocharides, Xie, Vijaykrishnan, & Irwin, 2004) an algorithm for IP virtualization and placement for NoC-based architectures is presented. This approach aims at balancing the temperature profile of a system and reducing hot spots during the design process. This is achieved by IP virtualization and placement applying genetic algorithms, which are based on evolutionary principles of natural selection. Due to the nature of the process of IP placement and mapping, this approach can be considered rather inflexible. Similarly, (Zanini, Atienza, & G., 2009) focuses on reducing hot spots and balancing the temperature profile of Multi-Processor Systems-on-Chip (MPSoCs). For this purpose, the thermal behavior of MPSoCs is modeled as a control theory problem. This facilitates the design of a frequency controller for thermal management (i.e. Dynamic Frequency Scaling (DFS)) without the necessity of knowing the thermal profile (i.e. the workload profile) of the chip at design-time. The approach is verified by application to the Sun Niagara 8-core MPSoC. Although this work has to be attributed to design-time thermal-aware solutions, it yields a method for runtime thermal management. In (Coskun, Rosing, Whisnant, & Gross, 2008) Integer Linear Programming (ILP) is utilized for task scheduling for MPSoCs in order to minimize on-chip hot spots and temperature imbalances. This approach adjusts the workload distribution at design time to optimize temporal and spatial temperature distributions provided that the applied set of tasks and its characteristics is known. Therefore, the presented approach can be considered rather static, though an extension facilitating the possibility to predict the impact of (conceivable) workload variations at runtime is presented.

Approaches dynamically applying task mapping at system runtime are presented in (Hölzenspies, 2008) and(Singh, 2010) for example. Generally, such approaches are optimized for certain aspects. Amongst others, this concerns load balancing of the system components (Coskun, Rosing, Whisnant, & Gross, 2008), minimization of transmission distance between communicating components (Kaushik, 2011) and balancing of the temperature profile (Kursun, 2006).

This chapter mainly focuses on three issues. Firstly, thermal models eventually depend on offline profiling for the extraction of values for power consumption. This makes these models more suitable for thermal-aware placement and mapping than for the dynamic modeling of thermal properties of ICs. Secondly, due to their nature reactive strategies for thermal-aware management and task mapping suffer from long response times since between the occurrence of a switching activity and corresponding heat propagation a significant period of time passes. This for example delays the sending of control instructions or leads to inappropriate mapping decisions from a temperature related point of view. Thirdly, in many cases existing proactive approaches are not tailored to management of NoC routers and links (e.g. measures like software-based thread allocation are deployed) or they partially still rely on external tools for profiling (Shang, Peh, Kumar, & Jha, 2004).

For these reasons the main contributions of this chapter are as follows. A simulation environment for NoC-based systems is extended by a thermal model, which is based on equivalent RC-circuits. Therefore, this model neither relies on thermal sensors nor depends on any external tools. Most important, it allows for simultaneous system simulation and thermal modeling. The model is used in conjunction with DFS and task relocation in order to allow for simulation of proactive thermal management. To determine the impact on system performance and on-chip temperature distribution this setup is compared to

an analog reactive implementation and a reference system without thermal management. In order to assess the applicability of the model for proactive thermal-aware task mapping an appropriate algorithm is compared to two algorithms focusing on communication distance and load balance and to a reactive thermal-aware implementation.

SIMULATION ENVIRONMENT

The simulation environment developed for the evaluation of reactive and proactive approaches for temperature management and thermal-aware task mapping allows for functional simulation of NoCs based on a 2D mesh topology, wormhole packet switching and XY routing. Amongst others, parameters like NoC size, link width and simulation duration can be specified. The system components, which are connected by the NoC, are represented by Intellectual Property (IP) cores. The IP cores are individually configurable concerning generation frequency, length and destination address of packets. For more detailed information regarding system parameters like packet size, flit size and operating frequencies refer to Table 1. Furthermore, the sample period T_S determining the rate of capturing statistics can be set.

By default IP core data generation is executed randomly based on fixed load factors assigned to the cores resulting in uniform traffic patterns. The load factors serve to emulate unspecific tasks running on the cores. This implies that the overall number of running tasks directly depends on the NoC size (e.g. a 4x4 NoC comprises 16 tasks) and that there are no specified task parameters (e.g. task runtime, overall data amount or scheduling policies). This mode of data generation is used in conjunction with both reactive and proactive temperature management. Although a more sophisticated model for data generation will be introduced subsequently, this model was not available at the time of management implementation. Hence, there is no possibility to draw conclusions regarding the impact of thermal management on task runtime or overall application runtime.

In contrast, for the evaluation of thermal-aware mapping strategies a task-based model for data generation is applied. Basically, this model uses task graphs (Dick, Rhodes, & Wolf, 1998) in order to simulate the execution of applications by the system. A single task graph consists of several tasks with individual values for properties like runtime, processing effort, dependencies to other

tasks and generated communication traffic. This not only allows for simulation of realistic load and traffic patterns, but is also the foundation for the evaluation of various mapping algorithms regarding application performance and runtime.

To preserve consistency the thermal model was developed by using the SystemC Analog Mixed Signal (AMS) library (Accellera, SystemC Analog Mixed Signal Extensions 1.0, 2010), since the simulation environment itself is based on the SystemC (Accellera, Core SystemC Language 2.2, 2006) and SystemC Transaction-Level Modeling (TLM) (Accellera, SystemC Transaction-Level Modeling Library 2.0.1, 2009) libraries. By deploying the AMS library, the dualism of electrical and thermal energy flows can be exploited for modeling. This allows for simultaneous system simulation and thermal modeling, while preserving system integrity (i.e. independence from external tools for power tracing). For modeling, the NoC infrastructure is mapped on a regular grid of RC-tiles (Tockhorn, Cornelius, Saemrow, & Timmermann, 2010). A single RC-tile consists of four horizontal and one vertical electrical resistance modeling the resistance for heat flow in the respective direction (see Figure 1). Furthermore, every tile contains an electrical capacitance representing the amount of thermal energy this tile is able to absorb before heat starts to spread to adjacent tiles. Every component of the NoC (i.e.

Table 1. Design parameters of the applied simulation environment

Parameter	Value
NoC topology	2D mesh
Switching	Wormhole
Routing	XY
Router arbitration	Round-robin
Min. router delay	4 cycles/header flit 2 cycles/data flit
Router buffer size	8 flits
Flit size	64 bit
Packet header size	1 flit
Packet size	64-2000 flits
IP core output rate	0.5 flits/cycle
Link capacity	64 bits/cycle
Period / frequency	1 ns/1 GHz
Init. chip temperature	60.0 °C
Ambient temperature	45.0 °C

Table 2. Router delay D_R, delay of packet delivery D_P, net data throughput $Data_{Net}$, the number of delivered packets P_{Trans}, average temperature T_{Avg}, the peak temperature difference ΔT and the maximum temperature T_{Max} for reference systems of different size

Reference	2x2	3x3	4x4
D_R [cycles]	5	5	6
D_P [cycles]	28	32	38
$Data_{Net}$ [bit/cycle]	28	61	113
P_{Trans} [# in million]	50.4	109.8	201.7
T_{Avg} [°C]	58.5	65.5	71.9
ΔT [°C]	11.4	22.8	30.4
T_{Max} [°C]	67.6	84.8	96.7

link, router and IP core) is assigned a current source (black dots in Figure 1), which corresponds to the source of heat for that component. The current I_{Inj} representing the heat generated by a component is commensurate to the observed activity $Act_{Observed}$ and is calculated according to Equation (1). In addition to the actual chip the thermal model also accounts for a complete chip package including heat spreader and sink (see Figure 2) (Skadron, Stan, W., Velusamy, Sankaranarayanan, & Tarjan, 2003). By varying the number of RC-tiles per component simulation speed can be traded off against accuracy.

The general flow of the parallel functional and thermal simulation is depicted in Figure 3. The simulation is divided into three phases. During starting phase the simulation parameters are

gathered and analyzed. The setup phase consists of 4 steps. First the NoC topology is set up ① by analyzing simulation parameters and configuring the IP. Subsequently, the equivalent RC-network is established ② according to the specified geometry and modeling parameters. For this purpose an Electrical Linear Network (ELN) belonging to the AMS library is used. After the ELN model setup the model's internal sample period is calculated and propagated ③. Finally, the resulting Differential Equation System (DES) is composed and checked for solvability ④. In the simulation phase first the temperature model is initialized ⑤. After this, the functional simulation and the simulation of the thermal behavior of the system are conducted simultaneously ⑥. Every time the specified sample period T_s expires, the simulation

Figure 1. Mapping of a 2x2 NoC to a regular grid of RC-tiles

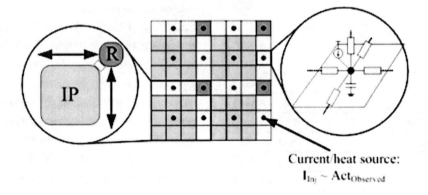

Current heat source:
$I_{Inj} \sim Act_{Observed}$

Figure 2. Modeling of a complete chip including heat emission to ambience through a heat spreader and a heat sink

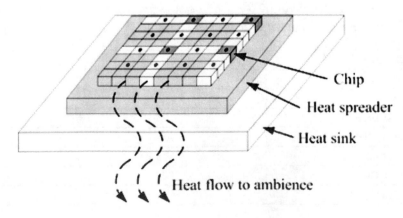
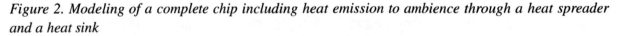

Chip

Heat spreader

Heat sink

Heat flow to ambience

is stalled, NoC component activity statistics are passed to the thermal model for current calculation and temperature output of the thermal model is delivered to the functional part of the simulation (e.g. for thermal management). The temperature output originates from the periodic solving of the DES. During simulation the output of the thermal model is updated every clock cycle. The electrical current I is fed into the RC-network, is calculated by Equation (1).

$$I_{Inj} \sim \frac{\sum Trans_{0 \rightarrow 1} \times E_{Trans}}{T_S} + P_{Static} \tag{1}$$

$Trans_{0 \rightarrow 1}$ is the number of bit transitions from 0 to 1 for a particular NoC component (note that for the sake of simplicity energy consuming transitions from 1 to 0 are omitted, since their consideration requires knowledge of the gate level circuit design), E_{Trans} is the energy a single transition consumes, T_S is the sample period and P_{Static} is the value for static power consumption for active components. A more simple but also more inaccurate approach to comprise switching activity is to replace $Trans_{0 \rightarrow 1}$ by the number of monitored flits. In this case E_{Trans} needs to be replaced by an average value for a single flit. For routers E_{Trans} is set to 1.5 pJ due to energy consumption of 0.096 nJ, caused by a 64 bit wide flit crossing a router (Ye, 2004). Concerning routers P_{Static} for input and output modules as well as FIFO buffers has to be considered (Cornelius, 2011). The value of 20 pJ for E_{Trans} of an IP core is unreferenced and only serves to reflect the proportion of IP core to router accounting for the variability of heat generation depending on IP core activity. P_{Static} for an IP core is estimated to be about 100 mW based upon power dissipation of an IBM PowerPC 405 (IBM, 2006) being suitable for NoC integration. For NoC links E_{Trans} is set to 11.62 fJ, assuming a wire length of at most 200 µm, random traffic patterns and a transition rate of 50 % (Gag, Wegner, & Timmermann, 2010).

Figure 3. Flow of simultaneous functional simulation and thermal modeling

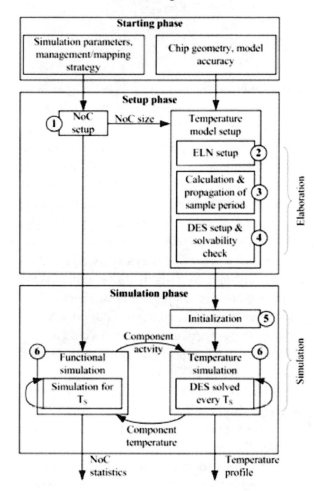

Thermal-Aware Management

Irrespective of the method of applying measures for Dynamic Thermal Management (DTM) to micro- or nanoelectronic systems these measures themselves have not changed significantly and mainly focus on the reduction of dynamic power consumption. This is due to the fact, that on the one hand the biggest part of the consumed power is dissipated as heat and on the other hand this power can be influenced rather directly. As the commonly known equation for dynamic power consumption (see Equation (2)) shows, P_{dyn} is calculated from the switching activity factor α, the frequency f, the supply voltage V_S and the switched capacitance C_{Load}.

$$P_{dyn} = \alpha \times C_{Load} \times f \times V_S^2 \qquad (2)$$

in order to dynamically control power consumption and eventually reduce heat generation, typically α, V_S and f have to be modified. This has resulted in widely spread DTM techniques like DFS, Dynamic Voltage Scaling (DVS), clock gating, task migration or relocation and various other approaches (Skadron, Stan, W., Velusamy, Sankaranarayanan, & Tarjan, 2003). In order to be able to adequately apply DTM measures, sensors are needed in order to monitor particular system parameters and to reliably draw conclusions regarding on-chip temperature distribution. The most obvious and straightforward approach is to monitor the temperature of a certain chip area or component and to report violations of predefined rules to initiate a DTM technique.

Transferred to DTM for regular NoC-based systems, this results in numerous components (i.e. routers, IP cores) that have to be monitored in parallel. Since monitoring every component with a dedicated sensor would be disproportionate to costs, a part of the NoC for which a single sensor or probe P is responsible for (see Figure 4) is defined. Since this chapter focuses on the impact of DTM

communication traffic on system performance the applied DTM techniques are limited to DFS and task relocation. For the latter a task running on an IP core is migrated to another if required.

Reactive Approach

The scheme of the simulation environment for reactive temperature management of NoC-based MPSoCs is depicted in Figure 5. The management is represented by an implementation of a monitoring and control system deploying event-based monitoring for NoCs (Ciordas, Basten, Rădulescu, Goossens, & Meerbergen, 2005).

Regarding the monitoring system a probe P is attached to every NoC-tile. The probe is responsible for managing its associated tile. This includes the IP core, the router and the two links from north to south as well as from east to west (see Figure 4). The general communication flow for reactive thermal management is depicted in Figure 6. The probe constantly monitors temperature of all components of its associated tile. In case an event is detected at a sensor, the probe generates a packet containing an event message and the involved components. For reactive temperature management such an event is a temperature change ΔT exceeding a predefined threshold T_{Thresh} for one or more components. The packet is sent

Figure 4. Components of a NoC-tile a probe P is responsible for; every monitoring point represents a thermal sensor or activity counter respectively

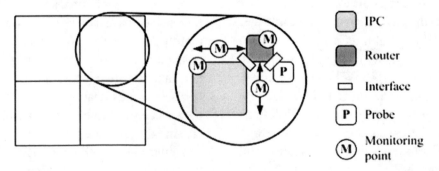

to the Thermal Management Unit (TMU) via the NoC infrastructure where it is analyzed and, if necessary, appropriate instructions are sent back to the affected components.

In order to enable this flow, the TMU and the monitoring system are closely intertwined with the simulation environment, since both are implemented as a part of the functional simulation. This allows for consideration of the impact of management traffic on overall system performance. The monitoring system is supplied with temperature values by the "Actual profile," which in turn is updated by the functional part of the simulation and represents the current temperature profile of the system.

The overall scheme of the reactive TMU is illustrated in Figure 7. As long as there are no events the TMU stays in idle mode. All arriving event messages are processed in the receiver in the sequence of their arrival. In case a message arrives, it is analyzed regarding type (i.e. link, router, IP core), position and temperature value of the involved components. Thereupon, the TMU's internal status profile of the concerned system components is updated and an appropriate reaction (i.e. DFS or task relocation) is determined if necessary. Reaction policies follow specified values for step size of DFS as well as maximum and minimum frequency boundaries. Furthermore, a temperature limit T_{Bound} and a limit for

Figure 5. Scheme of simulation environment for reactive thermal-aware management (legend: —> simulated data transmission, ⋯> internal communication)

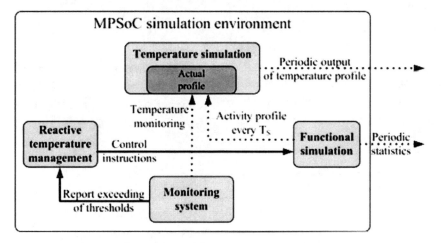

Figure 6. Exemplary communication between a probe and the central TMU

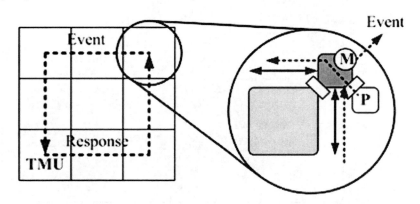

temperature variation ΔT_{Max} between IP cores are defined both triggering IP core task relocation. T_{Bound} serves to reduce hot spots by relocating a task to the IP core with the lowest temperature in case a maximal allowed temperature is exceeded. ΔT_{Max} is used to balance the thermal profile by relocating a task to the IP core with the biggest temperature variation compared to the affected IP core (i.e. in case the biggest temperature variation exceeds ΔT_{Max} the current task is relocated to the IP core with the lowest temperature). ΔT_{Max} is only used as long as T_{Bound} is not exceeded for that IP core. The TMU itself is not excluded from this process. Thus, every IP core is a potential TMU, since replacing a whole IP core by a TMU would induce unacceptable overhead. Hence, the TMU can be regarded as being implemented in software. While the TMU is in idle mode, the concerned IP core switches to normal operation mode and regularly sends and receives data packets.

The main disadvantage of this reactive approach is that response times to temperature changes are comparatively long. Equation (3) shows how the response time Δt_{Re} is composed of the time Δt_{Spread}, that passes between a switching activity occurs (i.e. energy is consumed) and a corresponding heat propagation is detectable, and twice the time Δt_{Trans}, that is necessary for the transmission of the monitoring and DTM instruction data. As it can be seen, Δt_{Spread} is a function of the thermal time constant τ_{th}. Basically, this constant describes how long it takes until an object is heated up or annealed. For reasonable sizes of NoC components this constant is in the range of milliseconds leading to comparatively long response times. According to Equation (4) for a cube consisting of silicon with an edge length of 500 μm τ_{th} would be 4.375 ms (for more information regarding calculation of τ_{th} please refer to (Skadron, Stan, W., Velusamy, Sankaranarayanan, & Tarjan, 2003)) (R_{th} is the thermal resistance and C_{th} is the thermal capacitance of a body).

$$\Delta t_{Re} = \Delta t_{Spread}\left(\tau_{th}\right) + 2 \times \Delta t_{Trans} \qquad (3)$$

$$\tau_{th} = R_{th} \times C_{th} \qquad (4)$$

On the one hand this leads to delayed reactions to critical temperature levels and unnecessarily long periods of applied DTM mechanisms, since recovery of normal operation mode depends on

Figure 7. Reactive approach for Thermal Management Unit (TMU)

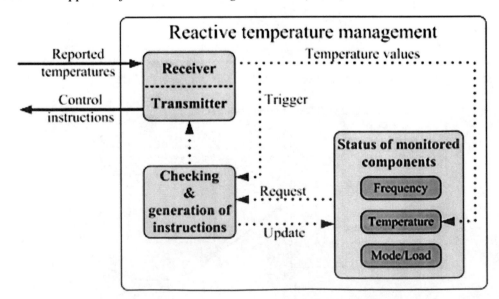

decrease in reported temperature. On the other hand the need to report decrease in temperature as well causes additional monitoring traffic possibly detracting regular traffic and Δt_{Re}.

Proactive Approach

Proactive thermal management does not solely rely on sensors for temperature monitoring. Instead, simple activity counters are deployed, which monitor the switching activity of all components of their associated NoC-tiles. This results in one counter for the router, one for the IP core and one for every link (see Figure 4). To consider temperature influences outside the model (e.g. a hot day) and to facilitate mechanisms for model calibration and correction, thermal sensors are still needed of course, but their time of operation can be reduced clearly. Thereby, analog measurements and resulting extensive Analog-to-Digital (AD) conversions, additionally delaying transmission and consuming power, can be minimized. It is assumed that the additional power consumption induced by the activity counters is negligibly small compared to the monitored components (and to that of the temperature sensors including AD converters). For now, no calibration or correction techniques are considered, since firstly for simulation all relevant parameters are defined (e.g. ambient temperature and initial chip temperature) and secondly the main objective is to identify the impact of a pure proactive temperature management on system performance and on-chip temperature distribution (i.e. a correction by means of thermal sensors would distort the results). Of course, for future work as well as application to real scenarios these mechanisms have to be incorporated.

The general scheme of the simulation environment for proactive thermal management is illustrated in Figure 8. Identically to the reactive approach, the TMU and the monitoring system are directly integrated into the simulation environment. The main difference to the reactive implementation is the availability of two temperature profiles. The "Predicted profile" is calculated from the activities reported by the monitoring system (activities are monitored during functional simulation) and represents the temperature profile of the system from the perspective of the proactive TMU. The TMU uses this profile in order to determine appropriate control instructions, if necessary. The "Actual profile" is identical to the reactive implementation, represents the current temperature profile and serves for the purpose of comparison.

Every time new data (i.e. a new flit) arrives at one of the monitored components the corresponding counter is increased. This can be done by counting the actual bit switches caused by the arrival of the flit or by just counting the arriving flit. The former approach results in more accurate power estimations but is also comparatively complex (i.e. time and energy consuming), while the latter approach is more inaccurate and lightweight. As soon as the monitored activity for one or more of the observed components exceeds a predefined threshold Act_{Thresh}, the activity value of the concerned component and the period, in which this activity was observed, are sent to the TMU via the NoC for further processing. Unlike for reactive management, in this case no thermal sensor is needed, since the corresponding data can be transmitted directly to the TMU (e.g. no AD conversion is necessary).

At the TMU (see Figure 9 for detailed scheme of the proactive TMU) the values for activity and observed period are used to inject a current into the equivalent RC-network (according to Equation (1)) representing the NoC. In parallel the TMU periodically updates its internal status information of the monitored components. For this purpose the "Prediction profile" is checked for predicted temperature violations. This corresponds to event-based temperature monitoring executed by the probes of the reactive approach using an identical threshold T_{Thresh}. In case violations are detected, according measures are determined. The measures follow the same policies described for reactive management, again including the TMU itself for possible relocation and therefore turning every IP core into a potential TMU. Once measures are scheduled, instruction packets are transmitted to the targeted system component. As long as no instruction packets are generated, the IP core, in which the TMU is currently located, performs normal operation (i.e. sending and receiving of data packets).

The main challenge to enable a TMU to model the thermal profile of a practical NoC is to provide the TMU with activity statistics of all network

components. For now it is assumed that this task can be accomplished by exploiting hardware counters for capturing activity statistics, which reside in the NoC components and the IP cores and use the NoC infrastructure to transmit these statistics to the TMU.

Another possibility would be to use dedicated control wires for transportation to the TMU, which may already be available in the components of the NoC infrastructure or to exploit structures and mechanisms integrated for the purpose of testability (i.e. Built-In-Self-Test, Design for

Figure 8. Scheme of simulation environment for proactive thermal-aware management (legend: —> simulated data transmission, ⋯> internal communication)

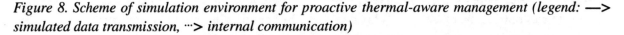
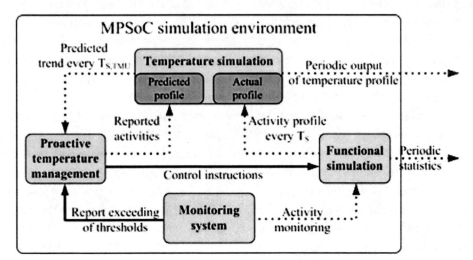

Figure 9. Proactive approach for Thermal Management Unit (TMU)

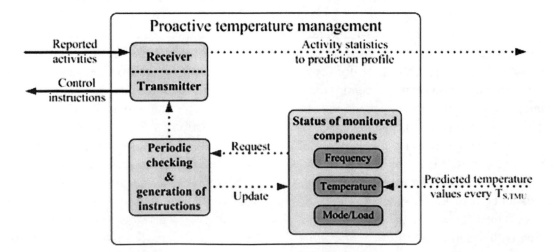

Test) (Durupt, Bertrand, Beroulle, & Robach, 2006). To determine IP core activity an approximation using typical load factors or reverting to integrated IP core elements may also be sufficient. This especially applies to third-party IP cores, where access to the hardware is constricted. Admittedly, transmission of activity statistics to the TMU using the NoC slightly diminishes the performance advance of proactive management.

$$\Delta t_{Pro} = \Delta t_{Compute} + 2 \times \Delta t_{Trans} \qquad (5)$$

Anyway, the most important advantage is the reduction of response time Δt_{Res} for thermal management. As it can be seen in Equation (5) Δt_{Spread} is now replaced by $\Delta t_{Compute}$, which is the time that is necessary for calculating the impact of newly arrived activity statistics and updating the temperature profile. More detailed RC-networks for modeling the on-chip temperature distribution will produce more accurate temperature predictions, but calculation effort and therefore $\Delta t_{Compute}$ will also rise. In order to guarantee that Δt_{Res} is shorter for proactive thermal management an appropriate temperature model must be chosen (i.e. $\Delta t_{Compute} < \Delta t_{Spread}$ and temperature predictions are sufficiently accurate). Furthermore, shortening $\Delta t_{Compute}$ contributes to the reduction of additional heat generation, since the core activity required for computation is curtailed.

Due to the fact that only activities are monitored, the amount of monitoring traffic can be reduced, since a decrease in temperature is automatically calculated inside the TMU instead of being reported by a probe. This minimizes overall traffic load and reduces additional heating caused by monitoring packets crossing the NoC. However, this is only true if the frequency of reporting the switching activity is lower than that of reporting temperature changes. The assumption that Δt_{Re} is shorter for proactive management and that this saving of time is actually relevant is supported by the fact that Δt_{Trans} is in the range

of few nanoseconds (see Table 2 for packet delay D_p of regular data packets) and therefore is negligible for further considerations. Additionally, Δt_{Spread} of reactive management is in the range of milliseconds (see example in section *Reactive Approach*) for reasonable sizes of NoC-tiles, so that the time saving per notification of activity (instead of waiting for a temperature change) may also be in the range of milliseconds (if $\Delta t_{Compute}$ is kept accordingly short). As it can be seen in (Wegner, Cornelius, Gag, Tockhorn, & Uhrmacher, 2010) this assumption is justified, since temperature modeling of a 2x2 NoC over 1 ms (using the same modeling accuracy as here) takes roughly 8.5 s. This results in 8.5 ms for every μs for four identical NoC-tiles. For comparison, τ_{th} for this example amounts to 6.3 ms. Due to a reduced Δt_{Re} and lowered traffic load, proactive management additionally implies two possible advantages compared to reactive approaches, provided that temperature can be influenced positively. Either, thermal stress and peak temperatures are reduced, when applying identical adjustment measures, leading to increased reliability and lifetime. Or, to achieve identical results, for the proactive approach less effort regarding adjustment measures has to be put in, resulting in lower detraction of overall system performance.

THERMAL-AWARE TASK MAPPING

The concept introduced in this section realizes an approach for proactive thermal-aware task mapping at system runtime. This concept employs the introduced temperature model, which is integrated in an existing simulation environment. However, in order to provide for the basis of a proactive mapping algorithm the environment is extended by the modeling of task graphs (TGs). The TGs represent applications executed by the simulated system and are modeled using Task Graphs For Free (TGFF) (Dick, Rhodes, & Wolf, 1998). By using this model the simulation of realistic load and traffic patterns is facilitated, since the TGs consist of single tasks exhibiting individual properties like computation time and effort, generated traffic volume and inter-task dependencies.

The overall scheme of the simulation environment is depicted in Figure 10. The TGs supplied for simulation are kept in the task pool until they are mapped by the applied mapping algorithm. Since some tasks feature periodic processing, simulation finishes either after a specified time or after a certain amount of tasks is completed. Once the algorithm gets permission, always a complete TG is mapped. This is due to strong inter-task dependencies, otherwise potentially causing deadlocks by permanently waiting tasks. Before a certain mapped task is finally executed in the functional part, the mapping algorithm turns over mapping decision as well as task properties to the "Predicted profile." Properties include processing time, amount of processed data, involved IP cores and expected traffic volume. This information is used to predict the thermal impact of the mapping decision. The estimated trend of the "Predicted profile" is forwarded to the mapping algorithm every $T_{S.Mapping}$ so that the next pending task can be placed on the presumably coolest IP core. Analogous to proactive temperature management, the activity profile generated during functional

simulation is fed to the "Actual profile" every T_S. This profile represents the current temperature distribution.

In order to estimate the impact of proactive thermal-aware mapping on system performance and temperature distribution, three established mapping algorithms are consulted for comparison. This concerns a thermal-aware reactive approach reverting to the "Actual profile" for taking mapping decisions. A similar approach is used in (Kursun, 2006) with the difference that mapping is executed statically at design-time.

Furthermore, a load balancing algorithm is deployed. This approach is also known as "First Free" algorithm (Coskun, Rosing, & Gross, 2008) and maps pending tasks to the IP core exhibiting the lowest number of assigned tasks (properties of already assigned tasks are not considered). On the one hand, the distributed mapping raises the probability that tasks can be processed in parallel. The uniform core load possibly influences the temperature distribution positively. On the other hand, the effort for inter-task communication may be increased due to extended communication paths generating additional heat.

Finally, a distance-oriented algorithm also known as "Nearest Neighbor" is applied (Kaushik, 2011) minimizing the communication distance

Figure 10. Scheme of simulation environment for proactive thermal-aware task mapping for NoC-based MPSoCs

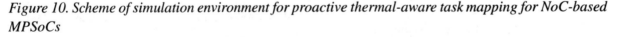
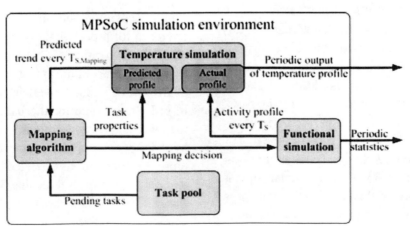

between interdependent tasks. This potentially results in task concentration leading to the formation of hot spots. This especially is the case for TGs exhibiting tasks, which have to be processed sequentially.

The introduced algorithms are suitable for mapping focusing either on temperature or performance to a different extent. This strongly depends on the composition of the TGs, which have to be mapped (i.e. number of tasks, computation effort and traffic volume). Therefore, in addition to different mapping algorithms it is essential to include different TG types. For this work, this concerns the following three types:

- **TG1:** Huge number of tasks; short computation time, low computation effort and traffic volume per task
- **TG2:** Small number of tasks; long computation time, high computation effort and traffic volume per task
- **TG3:** Mixture from TG1 and TG2

TG1 represents applications, which are separable into a big number of subtasks. These subtasks require much less effort for computation, processing and communication. Applications of this type strongly benefit from parallelization. TG2 emulates applications consisting of few but complex subtasks. The subtasks exhibit high computational and temporal effort. In most cases parallelization of applications of this type is not advisable due to the huge amount of data that has to be exchanged between the subtasks causing high traffic volume. TG3 is a mixture of the previous types.

EXPERIMENTS AND RESULTS

In this section the impact of the proactive approaches for thermal-aware management and task mapping on temperature and performance of NoC-based systems is investigated.

Investigations regarding impact on performance focus on net data throughput $Data_{Net}$, router delay D_R (i.e. time a flit needs to cross a router), the number of delivered packets P_{Trans} and the average time $t_{R,DFS}$ a single router operated in lowered frequency. All values refer to regular data (i.e. traffic for thermal management is excluded). For calculation of the temperature profile the switching activities caused by monitoring messages and management instructions are considered. While this is true only for thermal management, the evaluation of proactive mapping focuses on the number of completed tasks and the average idle time of the IP cores. This is due to the fact that parameters like $Data_{Net}$ or P_{Trans} do not give sufficient information about the velocity of the completion of single tasks or TGs.

Concerning management effort the average number of task relocations for every IP core as well as the number of monitoring messages and management instructions is analyzed. Since for proactive thermal-aware mapping no monitoring and control messages are generated, this point of analysis is omitted for mapping.

In order to also account for the impact on the on-chip temperature distribution the average temperature T_{Avg}, the peak temperature difference ΔT and the maximum on-chip temperature T_{Max} are analyzed for both thermal-aware management and mapping.

Investigations concerning thermal-aware management include simulations for NoC sizes of 2x2, 3x3 and 4x4 using the standard configuration C1.1 (see Table 3). Moreover, different configurations varying T_{Thresh}, T_{Bound}, ΔT_{Max} and Act_{Thresh} are applied to a 4x4 NoC (see Table 3) to investigate their influence on the observed parameters for both proactive and reactive thermal management for a comparatively large NoC. For the purpose of comparability, a reference system without any thermal management is simulated for all concerned NoC sizes. The results are depicted in Table 2.

For the evaluation of thermal-aware task mapping simulations are limited to a NoC size of 4x4, since smaller NoCs unnecessarily restrict the capabilities of the mapping algorithms. In order to provide for comprehensive examination, the task graph types TG1, TG2 and TG3 are simulated for the proactive and the reactive thermal-aware task mapping as well as the "Nearest Neighbor" and the "First Free" algorithms.

Since currently simulation of more practical periods of time (e.g. a couple of minutes) turns out to be very time consuming, for now all simulation runs are restricted to 1 s. The initial chip temperature and the ambient temperature are set to 60 °C (Shang, Peh, Kumar, & Jha, 2004) and 45 °C (Skadron, Stan, W., Velusamy, Sankaranarayanan, & Tarjan, 2003).

Performance

Thermal-Aware Management

The results for D_R and $t_{R,DFS}$ for NoC sizes from 2x2 up to 4x4 using configuration C1.1 are depicted in Table 4. Generally, both reactive and proactive management decrease overall performance for all NoC sizes with degradation growing with larger NoC sizes when using the standard configuration C1.1. This can be attributed to increased traffic for management due to a higher number of com-

ponents requiring management. This potentially causes longer transmission times for data packets and extends periods in which the NoC components operate with reduced performance (i.e. $t_{R,DFS}$ rises and consequently D_R increases).

When compared to the reactive approach, proactive management using configuration C1.1 is able to achieve improvements only for bigger NoC sizes. For D_R reactive and proactive management significantly increase delay. For $t_{R,DFS}$ results are acceptable with routers operating with reduced frequency only one fifth of the overall simulation time (1 s). For larger NoCs this value rises to almost 60 % (proactive) and nearly 100 % (reactive) and results in strongly delayed flit transmission.

To investigate if a modification of management parameters results in better performance and proactive management is able to outperform the reactive approach, the simulation is repeated for a 4x4 NoC varying management parameters (see Figure 11). For D_R the best results can be achieved for both approaches when increasing ΔT_{Max} (C3.2 and C3.3), while an excessive increase of T_{Thresh} has a negative impact for both approaches (because return to full performance after cooling down is delayed) and an increase of Act_{Thresh} naturally only affects proactive management negatively (i.e. delayed detection of temperature rises and

Table 3. Configurations for simulation varying the threshold for detection of temperature changes T_{Thresh}, the temperature limit of IP cores T_{Bound}, the temperature variation between IP cores ΔT_{Max} and the threshold for the detection of switching activity Act_{Tresh} (C2.1 = C3.1 = C4.1 = C1.1)

Config	T_{Thresh} [°C]	T_{Bound} [°C]	ΔT_{Max} [°C]	Act_{Thresh} [flits]
C1.1	1	64	1	100 000
C1.2	2	64	1	100 000
C1.3	4	64	1	100 000
C2.2	1	66	1	100 000
C2.3	1	68	1	100 000
C3.2	1	64	2	100 000
C3.3	1	64	4	100 000
C4.2	1	64	1	50 000
C4.3	1	64	1	200 000

Table 4. Average values for router delay D_R and time routers operate with reduced frequency $t_{R,DFS}$ for NoC sizes from 2x2 to 4x4 for proactive and reactive management and the reference system

Config C1.1	D_R [cycles]			$t_{R,DFS}$ [ms]		
	2x2	3x3	4x4	2x2	3x3	4x4
Proactive	7	13	17	365	187	548
Reactive	7	12	23	134	206	957
Reference	5	5	6	/	/	/

therefore longer periods of low performance). This leads to the conclusion that reducing the amount of task relocations (i.e. increasing ΔT_{Max} and also T_{Bound}) for the purpose of balancing the on-chip temperature profile contributes most to the reduction of D_R. Though, differences between C3.2 and C3.3 are insignificant. Concerning D_R proactive management is not able to outperform its reactive counterpart at least for the applied management settings.

Regarding $t_{R,DFS}$, C2.3 exhibits the best results for proactive management while C3.2 (i.e. a raise of ΔT_{Max}) achieves the best result for the reactive approach. Absolute improvements compared to C1.1 are considerable and amount to few hundreds of ms. It can be assumed that the advancement of proactive management increases with growing NoC sizes, since this behavior is already indicated in Table 4. Regarding the other configurations a behavior similar to D_R can be observed. Theoretically, an increase of T_{Thresh} should result in fewer monitoring packets for reactive management and therefore also in fewer instruction packets congesting the NoC, while for proactive management naturally only the latter are reduced. At first glance a higher T_{Thresh} seems to have a counterproductive effect, since the performance decrease is exacerbated especially for proactive management (for the reactive approach no increase is possible). Actually, the increased amount of regular data crossing the NoC leads to this ostensive exacerbation (compare reduced penalties for P_{Trans} in Figure 12 for C1.2 and C1.3), especially as the remaining management instruc-

tions still decrease performance and the routers' operating frequency has a major influence on the delay (i.e. D_R).

The conducted simulations as well as the resulting considerations suggest that the behavior of D_R and $t_{R,DFS}$ only provides conclusions of limited expressiveness regarding the impact of different configurations on system performance. Nevertheless, an increase of T_{Thresh} is not advisable to minimize D_R and $t_{R,DFS}$. This is also true for increasing Act_{Thresh} (i.e. reduction of monitoring messages) yielding aggravations for $t_{R,DFS}$ and slightly impairing D_R. Thus, in case router performance is decisive, a small T_{Thresh} is advisable for both reactive and proactive management in order to avoid unnecessary decrease of performance. Of course, this negatively affects T_{Avg} and ΔT.

Regarding $Data_{Net}$ and P_{Trans} the reactive approach is able to avoid penalties almost completely for the considered NoC sizes using configuration C1.1, while proactive management partly exhibits clear cutbacks (see Table 5). Basically, this is also true for the simulations conducted for a 4x4 NoC using different configurations (see Figure 12). Exceptions are C1.3 (i.e. proactive management reduces penalties) as well as configurations C3.2 and C3.3 (i.e. reactive management suffers from reductions). The apparent advantage of the reactive approach results from multiple aspects. Reactive management is hardly able to reduce thermal stress at least for the included NoC sizes and configurations (see Table 6 as well as Figures 14, 15 and 16). Therefore, a performance penalty would be inacceptable anyway. Another reason may be the design of the TMU. As stated before, the TMU behaves like a regular IP core (i.e. receiving and sending data packets), as long as no management has to be performed. Apparently, this case occurs more infrequently for proactive management than for reactive management, since incoming packets containing activity statistics need to be processed.

Another cause may be the disproportionately high number of monitoring packets crossing the NoC (see Table 7). This increases traffic load and potentially raises the probability of congestions. Therefore, a higher value for Act_{Thresh} may be advisable in order to relieve the NoC (i.e. abet $Data_{Net}$ and P_{Trans}), reduce temperature (monitor-

Figure 11. The average values of router delay D_R and the average time a router operated with reduced frequency $t_{R,DFS}$ for configurations C1.1 to C4.3 for a 4x4 NoC

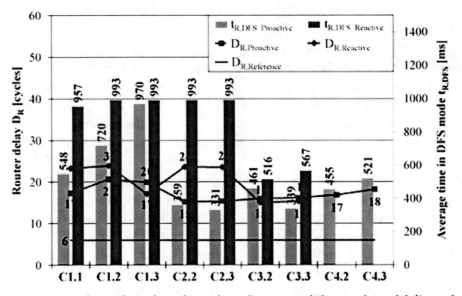

Figure 12. The average values of net data throughput $Data_{Net}$ and the number of delivered data packets P_{Trans} for configurations C1.1 to C4.3 for a 4x4 NoC

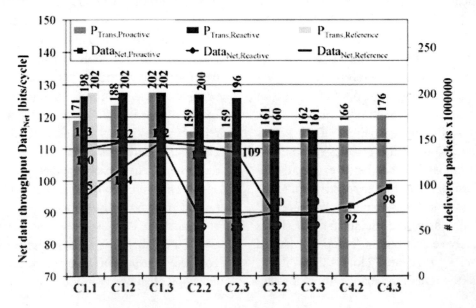

ing and instruction packets are also incorporated into temperature calculation), eliminate redundant management effort and thereby exploit the advantages of proactive management more effectively.

The impact of different management configurations on proactive and reactive management is depicted in Figure 12. Naturally, for reactive management the number of transmitted probe packets Mon_{Trans} can be reduced by raising T_{Thresh}. For proactive management this even has a negative effect (see Figure 19), since a higher T_{Thresh} only delays the forwarding of TMU management instructions, leaving the NoC components in full performance mode longer (i.e. more switching

activity can be detected and reported). Nevertheless, best results for both approaches regarding $Data_{Net}$ and P_{Trans} are achieved using a high value for T_{Thresh} (i.e. C1.3) yielding identical results. For proactive management Mon_{Trans} can be reduced by raising Act_{Thresh} (C4.3), while this has no effect on reactive management of course. In this case improvements induced by proactive management for $Data_{Net}$ and P_{Trans} are noticeable, but still do not come up to C1.2 and C1.3. Varying T_{Bound} (C2.2 and C2.3) as well as ΔT_{Max} (C3.2 and C3.3) obtains no acceptable results for proactive management.

To conclude, in order to reduce penalties for $Data_{Net}$ and P_{Trans} a reduction of monitoring pack-

Table 5. Results for average values of net data throughput $Data_{Net}$ and the number of delivered data packets P_{Trans} for NoC sizes from 2x2 to 4x4 for proactive management, reactive management and the reference system

Config C1.1	$Data_{Net}$ [Bit/Cycle]			P_{Trans} [# in Billion]		
	2x2	3x3	4x4	2x2	3x3	4x4
Proactive	27	55	95	49	99.7	171.3
Reactive	28	61	110	49.7	109.2	197.8
Reference	28	61	113	50.4	109.8	201.7

Table 6. Results for average temperature T_{Avg}, peak temperature difference ΔT and the maximum temperature T_{Max} for NoC sizes from 2x2 to 4x4 for proactive management, reactive management and the reference system

Config C1.1	T_{Avg} [°C]			ΔT [°C]			T_{Max} [°C]		
	2x2	3x3	4x4	2x2	3x3	4x4	2x2	3x3	4x4
Proactive	58.2	64.2	69.3	12.2	22.3	27.0	68.5	83.2	92.0
Reactive	58.3	65.4	71.5	12.9	23.1	31.5	69.3	85.0	97.9
Reference	58.5	65.5	71.9	11.4	22.8	30.4	67.6	84.8	96.7

Table 7. Results for the number of task relocations as well as the number of transmitted monitoring and instruction messages for thermal management for NoC sizes from 2x2 to 4x4 for proactive management and reactive management (1 flit per message)

Config C1.1	# Task Relocations			# Mon_{Trans}			# $Instr_{Trans}$		
	2x2	3x3	4x4	2x2	3x3	4x4	2x2	3x3	4x4
Proactive	2	14	38	17250	48496	105408	124	273	274
Reactive	14	62	85	179	345	366	179	220	222

ets is the best approach. Increasing T_{Thresh} works best for both reactive and proactive management. While for reactive management Mon_{Trans} can be significantly reduced, for proactive management Mon_{Trans} is even increased to due to delayed initiation of management instructions. This may cause undesirable side effects like increased traffic load and congestions. Therefore, for proactive management a combined increase of Act_{Thresh} and T_{Thresh} is recommendable eliminating side effects and yielding most considerable improvements compared to reactive management.

Thermal-Aware Task Mapping

The applied mapping strategies revert to different criteria in order to place pending tasks. This naturally results in deviations regarding system performance. Analysis focuses on the number of completed tasks as well as the average idle time of the IP cores. Idle time denotes the time no task is available for processing in the internal queue of an IP core. Therefore, it is a measurement for the ability of a mapping algorithm to constantly provide all IP cores with tasks and to avoid unnecessary periods of inactivity. Figure 13

depicts the results for the three TG types using the introduced mapping strategies. Note that all values are normalized to proactive thermal-aware task mapping.

In principle the number of completed tasks directly corresponds to the prevailing temperature conditions. This means that a lower average temperature automatically indicates a reduced number of completed tasks, since a simple task rearrangement would only influence the location of heat propagation but not the amount. Furthermore, a reduction of the number of completed tasks is always accompanied by an increased idle time. Compared to nearest neighbor and load balanced algorithms the proactive approach reduces the number of completed tasks only for TG1. This also is reflected in the corresponding idle time. For this TG type reactive thermal-aware mapping clearly falls behind. For TG types TG2 and TG3 performance can be preserved almost completely for all mapping strategies.

Poor results for TG1 caused by proactive and reactive thermal-aware mapping originate from a disadvantageous placement from a performance-oriented point of view. Thereby, communication of interdependent tasks is slowed down, leading to

Figure 13. Results for the number of completed application tasks and the average IP core idle time in a 4x4 NoC using different mapping strategies and application types (values normalized to proactive strategy)

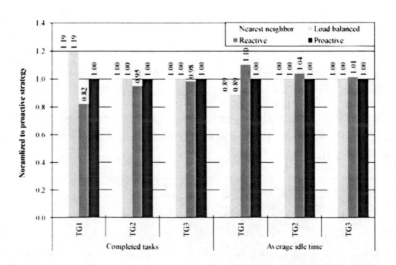

Figure 14. The average temperature T_{Avg} for configurations C1.1 to C4.3 for a 4x4 NoC

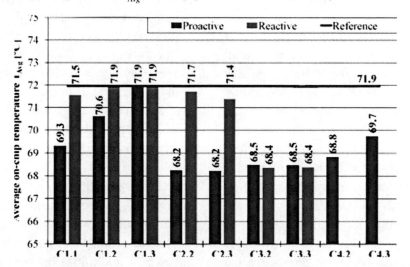

Figure 15. Peak temperature difference ΔT for configurations C1.1 to C4.3 for a 4x4 NoC

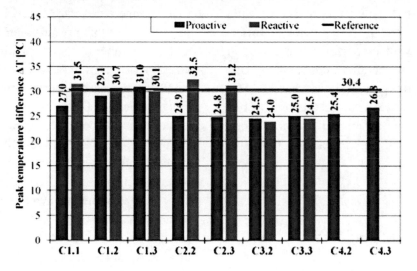

Figure 16. The maximum temperature T_{Max} for configurations C1.1 to C4.3 for a 4x4 NoC

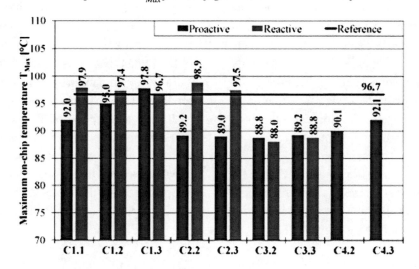

delayed task completion. This in turn reduces the overall number of completed tasks. Furthermore, IP cores possibly have to wait for completion of tasks of the same TG being processed by other IP cores in order to proceed. This causes avoidable idle time. Against this background a temperature reduction at the expense of system performance is not desirable, since in contrast to thermal-aware management performance penalties are not permitted. The existing tasks rather have to be mapped as efficient as possible without causing any penalties.

For these reasons the assessment of the applicability of different mapping strategies for varying TGs cannot be conducted from a pure performance-focused or temperature-oriented point of view. Instead, analysis has to consider temperature related parameters (i.e. maximum temperature T_{Max} and peak temperature difference ΔT_{Max}) as well as performance issues in order to draw comprehensive conclusions. Due to the nature of the prediction mechanism TGs like TG2 are expected to reveal the best results for proactive thermal-aware mapping, since long processing times and high execution effort abet accurate predictions.

Temperature

Thermal-Aware Management

The impact of proactive and reactive management on the average on-chip temperature T_{Avg}, the peak temperature difference ΔT and the maximum temperature T_{Max} is shown in Table 6. As it can be seen, both approaches are not capable of significantly reducing T_{Avg} as well as ΔT and T_{Max} for configuration C1.1 with growing NoC size. In some cases even impairments appear. This indicates that both approaches need further adjustment.

The impact of configurations from C1.1 to C4.3 on T_{Avg}, ΔT and T_{Max} are illustrated in Figure 14, 15 and 16 respectively. As it can be seen, both approaches achieve the best results for all three parameters when applying a higher ΔT_{Max} (C3.2

and C3.3). For proactive management similar results can be achieved by applying a higher T_{Bound} (C2.2 and C2.3). In contrast, raising T_{Thresh} or Act_{Thresh} acts counterproductive, since localization of potential hot spots and execution of appropriate countermeasures is delayed for reactive as well as for proactive management. The only exception is C4.2 for proactive management representing a decreased Act_{Thresh} and therefore causing improvements relative to C1.1. These characteristics indicate that especially a higher ΔT_{Max} (and partially a higher T_{Bound}) contributes to the uniformity of on-chip temperature distribution. Due to the fact that these parameters mainly influence how and when task relocations between the IP cores are executed, improvements are apparently limited to the temperature of IP cores. However, Figure 18 shows that this presumption is wrong, since a higher number of task relocations seems to be directly correlated to higher temperature values.

When directly comparing temperature reductions, proactive management almost completely outperforms the reactive approach at least for the used configurations. Only for C3.2 and C3.3 slight advantages of the reactive approach can be observed. Besides increased values for ΔT_{Max} and T_{Bound} proactive management benefits most from reduced values for Act_{Thresh} (C4.2). In contrast, the reactive approach only achieves improvements for increased values of ΔT_{Max}. Otherwise, reactive management even causes impairments concerning T_{Max} and ΔT. This indicates that at least for the analyzed configurations the reactive approach is unable to balance the overall temperature profile and to reduce hot spots. A reduced value for T_{Avg} is always accompanied by significant performance penalties (see performance analysis above).

In case a reduction of T_{Avg} is desired, a trade-off between performance and temperature has to be taken. However, a reduction of temperature imbalances may be achieved without forfeiting performance to that extent. Nevertheless, this requires appropriate configurations.

Thermal-Aware Task Mapping

Analogous to temperature management, the most critical factor for the quality of a mapping strategy is its impact on the temperature profile of the underlying system. In order to evaluate to which extent the proactive thermal-aware mapping provides improvements compared to commonly used approaches, again the average on-chip temperature T_{Avg}, the peak temperature difference ΔT and the maximum temperature T_{Max} are analyzed. Figure 17 illustrates the results for the introduced mapping strategies applying TG types TG1, TG2 and TG3.

Unexpectedly, proactive task mapping and the reactive thermal-aware approach achieve the most distinctive reductions of T_{Avg} for TG1. This indicates that a temperature-based mapping is not employable for applications consisting of a large number of simple subtasks. This evaluation is based on the fact that a reduction of T_{Avg} can only be achieved in conjunction with performance penalties (see Figure 13). However, a reduction of performance is prohibited, since theoretically varying task placements do not influence the overall activity. Therefore, the total quantity of generated heat cannot be reduced without impairing performance. The observed behavior can be explained by the composition of TG1. In detail, the subtasks exhibit short execution times as well as low processing effort obviously preventing accurate prediction of thermal behavior. Finally, this results in mapping decisions positively influencing T_{Avg} but at the same time preventing efficient and quick task eradication. Except for reactive thermal-aware mapping none of the strategies exhibits this behavior for TG2 and TG3.

Regarding T_{Max} the proactive approach performs like expected and achieves significant improvements for TG2. Solely the load balanced approach outperforms the proactive strategy. This indicates that for TGs comprising few but complex tasks (i.e. high processing effort and data volume, long execution times) both proactive thermal-aware as well as load balanced mapping

are capable of balancing the thermal profile of the underlying system. However, further improvements for the proactive strategy require more sophisticated abilities for long-term predictions. For TG1 the strategies consulted for comparison outperform the proactive approach. High values for T_{Max} indicate task congestions, which are caused by insufficient prognoses resulting from the task properties. Best results are achieved by the load balanced strategy again naturally preventing such congestions. Partly, this is also true for TG3. Compared to TG1 the proactive approach is able to conduct more accurate prognoses leading to better results. This can be attributed to mixed task pool containing both TGs of type TG1 (i.e. poor predictability) and TG2 (i.e. good predictability).

The tendency of the results for T_{Max} can be transferred to ΔT as well. Especially for tasks showing long execution times as well as high effort for processing and communication (i.e. TG2), the proactive approach shows its strengths. For these proactive mapping is capable of arranging tasks such that the on-chip temperature profile is more balanced.

Management Effort

Thermal-Aware Task Management

The effort conducted for thermal management is not only crucial for temperature distribution but also impacts system performance, since for example task relocations necessitate reorganization or rescheduling (in this work this is not considered) and DFS obviously slows down performance. Therefore, management measures should be limited to a necessary minimum. Three parameters indicating the actual management effort are Mon_{Trans} (the number of transmitted monitoring messages), $Instr_{Trans}$ (the number of transmitted management instructions) and the number of task relocations between the IP cores. The results for NoC sizes from 2x2 to 4x4 using configuration C1.1 are depicted in Table 7. As

Figure 17. Results for average temperature T_{Avg}, maximum on-chip temperature T_{Max} and peak temperature difference ΔT in a 4x4 NoC using different mapping strategies and application types

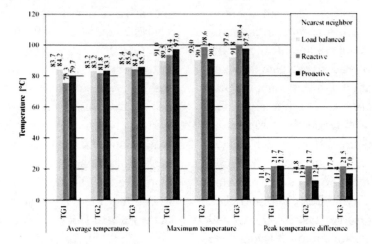

Figure 18. Number of task relocations for configurations C1.1 to C4.3 for a 4x4 NoC

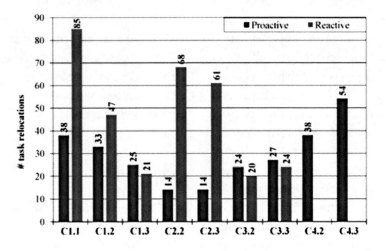

Figure 19. Number of monitoring and instruction messages for configurations C1.1 to C4.3 for a 4x4 NoC

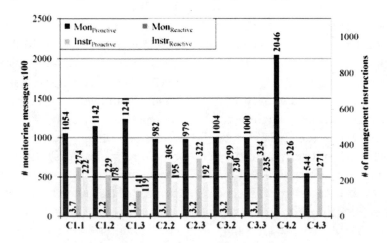

expected, for both approaches all three parameters increase with growing NoC sizes with proactive management generally indicating a higher management effort. This becomes most apparent for Mon_{Trans} revealing a disproportionate high amount of monitoring packets.

The impact of configurations from C1.1 to C4.3 on the number of task relocations as well as the number of monitoring and instruction messages is illustrated in Figures 18 and 19 respectively. As it can be seen in Figure 18 an increased T_{Thresh} (C1.2 and C1.3) significantly contributes to the reduction of the number of task relocations with proactive management outperforming reactive management for C1.1 and C1.2. This originates from a deferred initiation of management mechanisms. As seen before, this diminishes the impairment of performance (see Figure 12). Furthermore, since every relocation is connected with a certain amount of management effort (e.g. scheduling, recovery of execution), a reduced number of task relocations certainly relieves system management (this is currently not considered in our performance analysis). Of course, this antagonizes the reduction of temperature imbalances. Increasing T_{Bound} and ΔT_{Max} has moderate impact on the number of task relocations. This way T_{Avg} and ΔT can be influenced positively, but performance penalties still have to be accepted. Configuration C4.3 is attended by reduced performance penalties (see Figure 12) but also yields slightly worse results for all temperature related parameters.

Regarding a reduction of the number of monitoring messages, due to their nature reactive and proactive management benefit most from an increase of T_{Thresh} (C1.3) and Act_{Thresh} (C4.3) (see Figure 19). In order to further relieve the NoC by limiting the number of instruction messages to a minimum, raising T_{Thresh} (C1.3) is the best choice for both approaches, since initiation of instructions is directly linked to this parameter. An increase of Act_{Thresh} for proactive management has a positive effect, since the TMU's internal temperature model of the NoC is updated more infrequently leading to a reduced number of management instructions. All other parameter combinations still have no significant impact.

To conclude, a trade-off between the desired results for temperature and the admissible penalties regarding performance and effort for system management has to be carefully chosen. Furthermore, cross effects have to be considered complicating explicit suggestions which set of parameters to choose.

Thermal-Aware Task Mapping

The intended design of thermal-aware proactive task mapping renders the consideration of effort for mapping dispensable. This is due to the fact that mapping will be executed as a part of the operating system directly accessing the task pool and centrally coordinating task placement. Therefore, no monitoring or control messages are expected to be generated.

The simulations conducted in this section show that parameters for thermal management, system performance, temperature distribution and NoC size are closely intertwined. Results clarify that management parameters have to be individually adapted to different NoC sizes in order to guarantee a certain level of system performance while maintaining positive influence on temperature. This applies to both reactive and proactive management. Furthermore, results are strongly sensitive to the amount of monitoring data (i.e. temperature values for reactive and activity statistics for proactive management), since this traffic naturally constitutes the main part of overall management traffic and therefore contributes most to NoC congestion leading to delays for both temperature management and regular traffic.

Modification of management parameters facilitates adaptation of thermal management to the performance needs of the underlying system. Generally, results show that for both reactive and proactive management positive effects on temperature distribution can be traded off against performance. In case temperatures and an impairment of delay (e.g. D_R) are required to be as low as possible, parameters for task relocation need to be relaxed. In contrast, if high data throughput ($Data_{Net}$, P_{Trans}) is preferred, thresholds for the detection of temperature violations have to be increased.

However, in order to exploit the potentially shorter response times of proactive management (see Equation 5) especially Act_{Thresh} needs to be adjusted precisely. As shown in Figure 19 a low value (configuration C4.2) results in a disproportionately large amount of packets containing activity statistics when compared to reactive management. This leads to comparatively moderate results not revealing the full capabilities of proactive management especially for larger NoCs, which is also reflected in direct comparisons of reactive and proactive management for particular configurations. It is important to mention that the results at hand are achieved with settings reducing the number of monitoring packets for proactive management as much as possible. Nevertheless, this number still is significantly higher than for any results of reactive management (see Figure 19). Therefore, it stands to reason that a further increase of Act_{Thresh} for proactive management would yield a more equitable comparison and possibly better results.

Regarding proactive thermal-aware task mapping analyses show that this approach is suitable for particular applications. Typically, such applications are of complex nature forming specific TG types. These TGs consist of subtasks exhibiting long execution times, high computation effort and high traffic volume (originates from communication between interdependent tasks) at best. In case the named prerequisites are fulfilled, the proactive thermal-aware strategy is capable of reducing the maximum on-chip temperature as well as temperature differences compared to commonly used strategies. At the same time overall system performance is not impaired noteworthy. Consequently, hot spots and temperature imbalances can be reduced without accepting performance penalties by placing single tasks anticipatory. Nevertheless, this cannot be translated to arbitrary application types. Especially tasks featuring short execution times and low effort for processing and communication exacerbate reliable temperature predictions. This in turn prevents appropriate task placements. Therefore, the ability of proactive thermal-aware task mapping to improve the temperature profile is strongly bound to the applications processed by the underlying system.

FUTURE RESEARCH DIRECTIONS

The introduced temperature model, as well as the resulting concepts for proactive thermal-aware management and task mapping, provides the basis for further research. Additionally, results for the conducted analysis indicate possible margins for improving the concepts.

Regarding the temperature model, modifications improving the accuracy of low resolution models or reducing the time effort for high resolution models are desirable. Thereby, a more precise and simultaneously more efficient runtime modeling of the thermal behavior of NoC-based systems would be possible. Additionally, it has to be determined to which extent this influences the concepts for management and task mapping.

Since the improvements of the thermal behavior induced by proactive management are inevitably connected with performance penalties, in this field modifications are desirable as well. A possible approach is to dynamically adjust the management parameters at system runtime. Thereby, requirements of the currently processed applications can be considered more flexibly. High priority real time applications could be executed using full performance, while low priority applications could be processed with curtailed performance to favor temperature reductions. Similarly, the concept of runtime task mapping could be extended by a dynamic component choosing the appropriate mapping strategy according to the current situation.

Another aspect not covered exhaustively is the long-term effect of reduced temperatures and balanced temperature profiles on the underlying system. In detail, this concerns the impact of the realized concepts on the lifetime of ICs and therefore on the reliability of NoC-based systems. In this context more sophisticated long-term simulations of representative systems are required.

CONCLUSION

In this chapter proactive approaches for temperature management and thermal-aware task mapping for NoC-based systems are proposed. For this purpose, the NoC infrastructure is mapped on a network of RC-tiles. Thereby, the dualism of electrical and thermal energy flows can be exploited in order to model the thermal behavior of NoC-based systems.

The RC-model is used to simulate the proactive thermal management of corresponding systems executed by a central Thermal Management Unit (TMU). The TMU uses the temperature model to predict the temperature distribution based on activity statistics and triggers appropriate measures, instead of relying on temperature values, which are transmitted by thermal sensors. This contributes to the reduction of response times for thermal management, since on the one hand only activities are reported (i.e. delay between switching activity and an according heat propagation is avoided) and on the other hand decrease in temperature is simply calculated instead of reverting to more time-consuming temperature reports. Nevertheless, thermal sensors are still needed for model calibration and periodical correction of predicted values.

Comparisons between proactive thermal management and an equivalent reactive implementation show, that with the applied settings improvements regarding performance as well as temperature distribution are feasible. However, this always requires trade-offs with the proactive approach showing significantly more room for modifications. To utilize the full potential of proactive management further modifications of the management parameters especially concerning the frequency of capturing activity statistics have to be evaluated. Nevertheless, results for different configurations show that in order to achieve practical advancements for on-chip temperature distribution, a decrease of performance has to be accepted in either case. Furthermore, management parameters have to be individually adapted to particular NoC sizes to sustain sufficient system performance while applying thermal management.

Furthermore, the applicability of the thermal model for proactive thermal-aware task mapping is evaluated with respect to the impact on temperature distribution and system performance. The proactive approach uses the RC-model to predict the impact of a task placement on the overall temperature profile before the actual mapping is executed. In order to estimate the impact as accurate as possible execution time, effort for processing and communication as well as all involved IP cores are included in the prognosis. This prognosis enables the proactive task mapping to place successive tasks on the presumably coolest IP core.

Comparison of this approach with three conventional strategies (i.e. reactive thermal-aware, load balanced and distance oriented task mapping) reveals three main findings. Firstly, the average on-chip temperature cannot be influenced positively by simply rearranging tasks. This can only be achieved at the expense of system performance (i.e. reduced number of tasks). Secondly, proactive thermal-aware task mapping is capable of reducing the maximum on-chip temperature as well as peak temperature differences without impairing system performance. This leads to a balanced temperature profile and reduces thermal stress. Thirdly, the achievable improvements strongly depend on the applications executed by the underlying system.

REFERENCES

Accellera. (2006). Retrieved January 19, 2012, from Core SystemC Language 2.2: http://www.accellera.org/downloads/standards/systemc/

Accellera. (2009). *SystemC Transaction-Level Modeling Library 2.0.1*. Retrieved January 19, 2012, from http://www.accellera.org/downloads/standards/systemc/

Accellera. (2010). *SystemC Analog Mixed Signal Extensions 1.0*. Retrieved January 19, 2012, from http://www.accellera.org/downloads/standards/systemc/

Ciordas, C., Basten, T., Rădulescu, A., Goossens, K., & Meerbergen, J. V. (2005). An event-based monitoring service for networks on chip. [ACM Press.]. *ACM Transactions on Design Automation of Electronic Systems*, *10*(4), 702–723. doi:10.1145/1109118.1109126

Ciordas, C., Goossens, K., Radulescu, A., & Basten, T. (2006). NoC Monitoring: Impact on the Design Flow. *IEEE International Symposium on Circuits and Systems* (pp. 1981-1984). IEEE.

Cornelius, C. (2011). Design of complex integrated systems based on networks-on-chip - Trading off performance, power and reliability. *Dissertation (in press)*.

Coskun, A. K., Rosing, T. S., & Gross, K. C. (2008). Proactive temperature balancing for low cost thermal management in MPSoCs. *2008 IEEE-ACM International Conference on Computer Aided Design* (pp. 250-257). IEEE.

Coskun, A. K., Rosing, T. S., Whisnant, K. A., & Gross, K. C. (2008). Temperature-Aware MPSoC Scheduling for Reducing Hot Spots and Gradients. *Proc. of 13th Asia and South Pacific Design Automation Conference (ASP-DAC 2008)* (pp. 49-54). IEEE Computer Society Press.

Dick, R. P., Rhodes, D. L., & Wolf, W. (1998). TGFF: Task Graphs For Free.

Durupt, J., Bertrand, F., Beroulle, V., & Robach, C. (2006). A DFT Architecture for Asynchronous Networks-on-Chip. *Eleventh IEEE European Test Symposium ETS06* (pp. 219-224). IEEE.

Gag, M., Wegner, T., & Timmermann, D. (2010). System Level Power Estimation of System-on-Chip Interconnects in Consideration of Transition Activity and Crosstalk. *Proceedings of the 20th international conference on Integrated circuit and system design: power and timing modeling, optimization and simulation* (pp. 21-30). Springer.

Guang, L., Nigussie, E., Rantala, P., & Isoaho, J. (2010). Hierarchical Agent Monitoring Design Approach towards Self-Aware Parallel Systems-on-Chip. *ACM Transactions on Embedded Computing Systems*, *9*(3), 1–24. doi:10.1145/1698772.1698783

Hölzenspies, P. H. (2008). Run-time spatial mapping of streaming applications to a heterogeneous multi-processor system-on-chip (MPSoC). *Proceedings of the conference on Design, automation and test in Europe*, (pp. 212-217).

Hung, W., Addo-Quaye, C., Theocharides, T., Xie, Y., Vijaykrishnan, N., & Irwin, M. J. (2004). Thermal-Aware IP Virtualization and Placement for Networks-on-Chip Architecture. *IEEE International Conference on Computer Design (ICCD'04)* (pp. 430-437). ACM Press.

IBM. (2006). *IBM PowerPC 405 CPU Core Product Overview*. Retrieved January 19, 2012, from https://www-01.ibm.com/chips/techlib/techlib.nsf/techdocs/3D7489A3704570C0872571DD0065934E/$file/PPC405_Product_Overview_20060902.pdf

JEDEC. (2009, March). Failure Mechanisms and Models for Semiconductor Devices. *JEDEC publication JEP122F* .

Kaushik, S. S. (2011). Preprocessing-based run-time mapping of applications on NoC-based MP-SoCs. *IEEE Computer Society Annual Symposium on VLSI*, (pp. 337-338).

Krum, A. (2000). Thermal Management. In F. Keith (Ed.), *The CRC Handbook of Thermal Engineering* (pp. 1–92). Boca Raton: CRC Press.

Kursun, E. C.-Y. (2006). Investigating the effects of task scheduling on thermal behavior. *In 3rd Workshop on Temperature-Aware Computer Systems*, (pp. 1-12).

Liu, W., Calimera, A., Nannarelli, A., Macii, E., & Poncino, M. (2010). On-chip Thermal Modeling Based on SPICE Simulation. [Springer.]. *PATMOS, 5953*, 66–75.

Martinez, E., & Atienza, D. (2009). Inducing Thermal-Awareness in Multicore Systems Using Networks-on-Chip. *2009 IEEE Computer Society Annual Symposium on VLSI* (pp. 187-192). IEEE.

Shang, L., Peh, L.-S., Kumar, A., & Jha, N. K. (2004). Thermal Modeling, Characterization and Management of On-Chip Networks. *37th International Symposium on Microarchitecture MICRO3704* (pp. 67-78). IEEE.

Singh, A. K. (2010). *Run-time mapping of multiple communicating tasks on MPSoC platforms* (pp. 1019–1026). Procedia Computer Science.

Skadron, K., Stan, M., W., H., Velusamy, S., Sankaranarayanan, K., & Tarjan, D. (2003). Temperature-Aware Microarchitecture: Exended Discussion and Results. *Computer Engineering, (CS-2003-08)* .

Srinivasan, J., & Adve, S. (2003). *RAMP: A Model for Reliability Aware MicroProcessor Design.* IBM Research Report, RC23048.

Tockhorn, A., Cornelius, C., Saemrow, H., & Timmermann, D. (2010). Modeling temperature distribution in Networks-on-Chip using RC-circuits. *13th IEEE Symposium on Design and Diagnostics of Electronic Circuits and Systems* (pp. 229-232). IEEE.

Wegner, T., Cornelius, C., Gag, M., Tockhorn, A., & Uhrmacher, A. (2010). Simulation of Thermal Behavior for Networks-on-Chip. *28th NORCHIP Conference* (pp. 1-4). IEEE Xplore.

Ye, T. (2004). Packetization and routing analysis of on-chip multiprocessor networks. [Elsevier North-Holland, Inc.]. *Journal of Systems Architecture, 50*(2-3), 81–104. doi:10.1016/j.sysarc.2003.07.005

Zanini, F., Atienza, D., & G., D. M. (2009). A Control Theory Approach for Thermal Balancing of MPSoC. *Proc. of 14th Asia and South Pacific Design Automation Conference (ASP-DAC 2009)* (pp. 37-42). IEEE.

KEY TERMS AND DEFINITONS

Network-on-Chip: Communication architecture for connecting components of Many-core systems and Multi-Processor Systems-on-Chip.

Network-on-Chip Performance: Parameters of a NoC-based system referring to data throughput, router and packet transfer delay as well as network utilization.

On-Chip Temperature Distribution: Thermal properties of a NoC-based system including average component temperature, hot spots and temperature imbalances.

Proactive Task Mapping: Deployment of the RC-based temperature model to predict the impact of task placements on the temperature profile of NoC-based systems and place successive tasks accordingly.

Proactive Temperature Management: Deployment of the RC-based temperature model to predict thermal behavior of NoC-based systems and initiate measures in advance.

RC-Based Temperature Modeling: Exploitation of the equivalency of the flow of electrical and thermal energy to model thermal behavior of integrated circuits.

Thermal-Aware System Management: Deployment of the RC-based temperature model to improve the temperature profile of NoC-based systems.

Thermal-Aware Task Mapping: Deployment of the RC-based temperature model to perform temperature oriented runtime task mapping.

Chapter 16
Wearout and Variation Tolerant Source Synchronous Communication for GALS Network–on–Chip Design

Alessandro Strano
Intel Mobile Communications, Germany

Federico Silla
Universitat Politècnica de València, Spain

Carles Hernández
Barcelona Supercomputing Center, Spain

Davide Bertozzi
Università degli studi di Ferrara, Italy

ABSTRACT

In the context of multi-IP chips making use of internal communication paths other than the traditional buses, source synchronous links for use in multi-synchronous Networks-on-Chip (NoCs) are becoming the most vulnerable points for correct network operation and therefore need to be safeguarded against intra-link delay variations and signal misalignments. The intricacy of matching link net attributes during placement and routing and the growing role of process parameter variations in nanoscale silicon technologies, as well as the deterioration due to the ageing of the chip, are the root causes for this. This chapter addresses the challenge of designing a timing variation and layout mismatch tolerant link for synchronizer-based GALS NoCs by implementing a self-calibration mechanism. A timing variation detector senses the misalignment, due to process variation and wearout, between data lines with themselves and with the transmitter clock routed with data in source synchronous links. Then, a suitable delayed replica of the transmitter clock is selected for safe sampling of misaligned data. This chapter proves the robustness of the link in isolation with respect to a detector-less link, also addressing integration issues with the downstream synchronizer and switch architecture, proving the benefits in a realistic experimental setting for cost-effective NoCs.

INTRODUCTION

The Globally Asynchronous Locally Synchronous (GALS) paradigm for chip synchronization structures the system into voltage and frequency islands working at differentiated and decoupled voltages and frequencies. In this context, multi-synchronous designs, making use of synchronizers and source-synchronous communication for clock domain crossing, turn out to be a more flexible

DOI: 10.4018/978-1-4666-6034-2.ch016

and readily available alternative to embody the GALS concept into industry-relevant designs rather than clockless handshaking.

In source synchronous links, data is routed together with a strobe signal (it might be the transmitter clock itself) which enables safe sampling at the receiver side regardless of the clock phase and/or frequency ratio between the communicating clock domains. However, synchronization interfaces now become the true weak-point of the system that needs to be safeguarded against timing failures. In fact, source synchronous links are generally designed under the assumption that there will be no or very limited routing skew between data lines and the transmitter clock, an assumption that might be easily impaired in nanoscale CMOS processes. The reason for this is twofold. On one hand, significant length, resistance and load deviations among different wires of a link should be expected even when advanced bus routing features of place-and-route tools are used. On the other hand, process parameter variations impact various device characteristics, such as effective gate length, oxide thickness and transistor threshold voltages, which may in turn lead to significant variations in power consumption and to timing violations. Furthermore, the use of VLSI chips along time also causes that the initial parameters found at the chip manufacturing stage change, thus being another source for variations in power consumption and timing. All these effects impair the functionality of the link by reducing the data stability window and by causing the uncertainty on the precise sampling time over the clock period.

Although reducing delay variations between wires of a GALS NoC link would be the ideal requirement for advanced NoC implementation styles, it is virtually impossible to completely predict nominal delay and routing skew deviations occurring in the manufacturing process. Predicting wearout effects is also a complicated task. Therefore, NoC link design should consider self-calibration, which is the approach taken by this manuscript. Instead of relying on the worst-case characterization of design parameters, self

calibrating systems determine autonomously the boundary of correct behavior, and set design parameters accordingly.

In this work we apply the self-calibration design principle to source synchronous links for use in synchronizer-based GALS NoCs. The basic idea is to use a variability detector at the receiver side of a source synchronous link with the capability of sensing the offset between data wires with each other and with the transmitter clock routed along with them. Based on misalignment quantification, the circuit selects the earliest delayed replica of the transmitter clock which can safely sample input data in the synchronization interface. The detector is meant for use during system reset, during which each GALS link is supposed to perform a self-calibration procedure. Repeated at every system bootstrap, the procedure can ensure robustness against wearout effects.

RELATED WORK

Many recent works analyze the impact of process variations on the performance of integrated circuits, providing data on how parameter variations impact the maximum design frequency (Bowman, 2002) or variability models that characterize variations in microarchitecture (Sarangi, 2008; Bonesi, 2008). However, these studies do not consider the implications of variations in the interconnect infrastructure. Although (Nicopoulos, 2010) is a step forward in this direction, this study neglects the impact of manufacturing deviations on NoC links. Unfortunately, this impact is not negligible (Mondal, 2007; Hernández, 2010; Hassan, 2009). On one hand, although there are examples of repeater-less NoC self-calibrating links (Jose, 2005), they typically undergo repeater insertion. Therefore, they suffer from Lgate variations and dopant fluctuations in the transistors building up repeater stages, and also suffer from the variability introduced by the chemical metal planarization process (Mondal, 2007).

In addition to process variation, the effect of wearout in circuit timing behavior cannot be neglected anymore. According to (Bernstein, 2006) negative-bias temperature instability (NBTI) and hot carrier injection (HCI) are two key mechanisms affecting the timing of transistors. NBTI and HCI effects cause a shift in transistor threshold voltage which leads to an increase in transistor delay. Both effects can be treated as an additional source of timing variability affecting circuit operation.

All these sources of instability, when applied to NoC links, cause that links in the network feature different delays, despite that they were initially designed to be identical (Hernández, 2010). Thus, some links will not be able to switch at the intended frequency, thus reducing overall performance. This will be the case when the NoC operating speed is constrained by link delay (Gilabert, 2009). A more destructive effect concerns the delay differentiation between the individual wires of a NoC link. In the case of GALS links, bridging mesochronous or fully asynchronous clock domains, this delay differentiation increases the probability of sampling failure at the receiver end (Loi 2008; Ludovici, 2010).

Architecture and circuit designers have to deal both with the random and with the systematic components of link delay variability. The systematic component can be partially addressed by making all wires undergo the same physical routing conditions, like in (Kakoee, 2010). Unfortunately, this approach retains sensitivity to random variations. One common choice is to add worst-case guard bands to critical paths (Borkar, 2004). However, this comes at the expense of a high reduction in performance. A more refined technique is using statistical delay calculation tools in order to reduce design margins and improve design speed (Synopsys, 2007). This probabilistic framework avoids the pessimistic timing estimation introduced by the classical static timing analysis (Orshansky, 2002). Other kinds of solutions are based on tolerating infrequent run-time timing violations, where delay failures are tolerated at the cost of performance (Ernst, 2003; Murali, 2006). In this

regard, there are several proposals that can deal with timing failures by tolerating the presence of faulty wires. These proposals are based on performing flit serialization (Changlin, 2012) or splited flit transmissions (Vitkovskiy, 2012). Some other fault-tolerant proposals focus on increasing interconnect yield by using redundant hardware. In this sense, the authors of (Grecu, 2006) propose the use of spare wires to tolerate a bounded number of faults without decreasing communication performance.

On the other hand, there exist several proposals that perform post-silicon detection and compensation. In these techniques the circuit is designed for the typical case and variability compensation is performed post-silicon at some cost (performance, power) (Bonesi, 2008; Paci, 2009). An effective technique to design differently than worst-case is to use self-calibrating links. The work in (Worm, 2005) dynamically scales down the voltage swing, while ensuring data integrity. In (Merdadoni, 2008) the voltage swing is adapted to the link delay budget during a calibration phase. The Stars system uses non intrusive monitors to measure the delay across the longest path in a device and adjust clock frequency accordingly (Diamos, 2007). Finally, in (Höppner, 2010) data is serialized into several parallel high speed lanes and the delay deviation of each lane is compensated by means of a digitally controlled delay cell.

Source synchronous communication has recently emerged as a potential alternative to clockless handshaking for clock domain crossing in GALS NoCs (Panades, 2008). In contrast to well-consolidated application domains such as off-chip memory sub-systems and networking as well as I/O (Collins, 1999), its application to an on-chip setting is challenged by the uncertainty of the manufacturing process discussed so far. In particular, there is consensus among industrial designers that naively implementing source synchronous links in an on-chip setting is very likely to lead to timing violations. Typical solutions to mitigate this concern include configurable delay elements on data and/or clock wires (Yu, 2006), adaptive delay tuning (Elrabaa, 2006), clock recovery schemes (Kihara, 2003) (Moore, 2000), multiple registers on the data bus (Dally, 1998), or alternating sample edges (Tran, 2009). In all cases, proposed schemes are either overly costly since inspired by off-chip networking, or assume timing constraints that in real-life are unpredictable until tape-out and testing, or assume matched delays between similar components, which is hard to achieve in a variability-dominated environment.

Recently, the use of resonant clocks to implement source synchronous NoCs has been proposed in (Mandal, 2013). However, the use of resonant clocks challenges the timing analysis of circuits as its use is not currently supported by regular design flow tools. With respect to previous research, the work in this manuscript applies, for the first time, the self-calibration design principle to source synchronous links for use in GALS NoCs. Differently than shared busses, flow control issues need to be more closely accounted for.

TARGET GALS PLATFORM AND MOTIVATION

Without lack of generality, the synchronizer-based GALS network addressed in this work consists of implementing the on-chip network and the networked IP cores as disjoint clock domains, and therefore to place circuitry to reliably and efficiently move data across asynchronous clock boundaries between NoC switches and connected network interfaces. Figure 1 depicts our target GALS platform.

Unfortunately, the network ends up spanning the entire chip and might be difficult to clock due to the growing chip sizes, clock rates, wire delays and parameter variations. We find that mesochronous synchronization can relieve the burden of chip-wide clock tree distribution while requiring simpler and more compact synchronization interfaces than dual-clock FIFOs. Hierarchical clock tree synthesis is an effective way of inferring mesochronous links, as already experimented in (Panades, 2008).

Figure 1. GALS target platform of this work

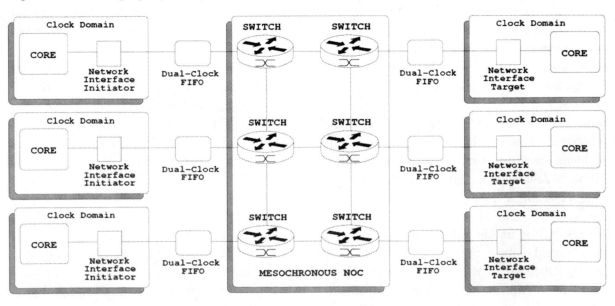

In our architecture, the upstream links of both dual-clock FIFOs and mesochronous synchronizers are source synchronous: data is routed together with the transmitter clock, which is used at the receiver end to safely sample input data in the receiver (mesochronous or asynchronous) clock domain. Minimizing routing skew between data wires and transmitter clock is a challenging task for state-of-the-art physical routing tools (Kakoee, 2010), as current support of physical routing tools for bundled routing proves not always capable of matching wire characteristics (e.g., resistance, length, delay), especially over long links. Moreover, the reduced signal slope over long links causes a larger delay of downstream logic cells, thus resulting in a delayed generation of the sampling control signal.

Even assuming ideal routing, the onset of significant random process variation effects makes it very difficult to meet the routing skew constraints. As a consequence, a sampling failure may occur at the receiver side of a source synchronous link. This scenario calls for an augmented GALS link architecture for robustness to layout mismatches and process variations

ARCHITECTURE OF THE VARIATION DETECTOR

This section presents a novel architecture circuit (the *variation detector*) guaranteeing the reliability of NoC source synchronous interfaces under high circuit and wire delay variability. Implicitly, the circuit copes with mismatches of layout characteristics as well (such as unmatched link net attributes).

The variation detector we propose should be placed in front of the regular synchronizer (such as dual-clock FIFOs or input latches of mesochronous synchronizers) and aims at restoring the required alignment between data lines and the transmitter clock for safe sampling. The operating principle is the following. At first, the amount of misalignment between the wires of the source synchronous link (data plus clock) is sensed. Then, the receiver synchronizer is fed with a clock signal version that guarantees correct sampling. The clock signal is a delayed replica of the transmitter clock. A schematic of the proposed architecture is illustrated in Figure 2. It is composed of a parametric number of brute force synchronizers sampling the output

Figure 2. Variation detector architecture

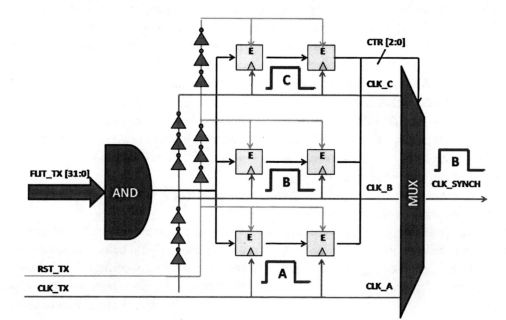

of an AND block. The AND gate receives the incoming data as input generating a high value as soon as all the data input bits are high. The source synchronous clock signal (routed along with the data) crosses a set of delay chains introducing an incremental amount of delay with respect to the nominal source clock. The delayed clocks feed 2-flop (or more) brute force synchronizers that sample data at different time instants within the clock period. Therefore, at least one of them will be able to sample input data during the stability window of these latter.

In order to preserve the synchronism property between reset and clock signal, the reset is bundled with the data as well and it crosses the delay chains together with the clock. The number of delay chains depends on the number of brute force synchronizers in order to guarantee to every synchronizer block a clock/reset with a different and equally interspaced offset. Moreover, it depends on the amount of expected routing skew between data and clock and between data lines with themselves: the shorter the data stability window, the higher the number of brute force synchronizers needed to correctly sample the data at least once.

Finally, the outputs of the brute force synchronizers drive the multiplexer control signals and select through it the delayed clock signal replica required to feed the receiving mesochronous synchronizer with a safe strobe signal. The selection of the safe clock signal takes place during the NoC reset phase and it represents a key step in the proposed architecture. It is detailed in the following section ("*Operating principle*"). However, it is worth recalling that once a suitable strobe signal is selected during the reset process (namely, at every system bootstrap), then the selection stays the same throughout the entire use cycle of the system, thus not generating an overhead for normal system operation. Used over time, the proposed architecture allows the network to deal also with wearout effects that might prolong gate and wire delays in an unpredictable way.

Two implementation issues should be observed. On one hand, by increasing the number of flip flops in the brute force synchronizers, the only implication would be the prolonging of the reset phase. This course of action might be taken so to be able to resolve metastability in future technology nodes, where the resolution time constant of synchronizers is expected to degrade (Beer, 2010). On the other hand, the logical AND gate introduces a delay that actually induces further routing skew between data and clock wires. However, this effect is mitigated by practical considerations.

First, the AND gate could be synthesized for maximum performance, thus minimizing delay at the cost of area. Second, when the data stability window is short, design guidelines in the "*Design guidelines*" section will indicate that a higher number of brute force synchronizers is needed. Therefore, the selected clock signal replica will typically sample data close to the middle of the stability window. Finally, it is possible to engineer the delay lines in such a way to mask the AND gate delay. In fact, a clock signal *clk_A'* might be derived by delaying *clk_A* by an amount of time equal to the AND gate delay. Then, *clk_A* would be sent to the multiplexer, while *clk_A'* would feed the first brute force synchronizer stage. The same would then be done for the later stages. In practice, for the sake of clock signal selection, the clock signals would be rigidly delayed by the same delay of the AND gate. This latter solution has been experimented in this work and makes the variability detector operate correctly with 5 synchronization stages at 1 GHz with only 50% data stability window (with respect to the clock period) in 65nm CMOS process.

Operating Principle

The architecture of the variation detector provides the safe clock signal before the downstream synchronizer starts its operation. To achieve this result, the proposed architecture comes into play during the NoC reset phase. Furthermore, to guarantee the synchronization between the incoming data and the source synchronous clock signal, the circuit must detect the exact arrival time of every data bit in the GALS link. This is possible by sensing the incoming bits when they are switching from a high logic value to a low logic value or vice versa. As a result, the proposed architecture can properly sense routing skews when it receives two consecutive data patterns so that every data bit of the first transaction is negated with respect to the data bit of the other one. These two transactions can be purposely generated from the output buffer of the upstream switch during the reset process. To meet this goal, we designed an output buffer in the xpipesLite architecture (Stergiou, 2005) capable of driving, during the reset phase, a sequence of low-logic value bits (00..0) followed by a sequence of high-logic value bits (11..1). To note that different patterns could be implemented in order to better cope with the crosstalk effect. In fact, the AND block can be always set in compliance with the expected patterns (i.e. the high/low bits of the first/second pattern can be negated at the AND input).

Figure 3 reports the post-layout reset phase waveforms of a variation detector composed of 3 delay lines and 3 brute force synchronizers, like the one in Figure 2. We assume in this figure that input data is sampled on *positive* clock edges. As represented in the figure, once the first sequence of zeros is driven at the variation detector input port, the output of the AND module is set to a low logic value (*and_out*). In the meantime, the source synchronous clock signal (*clk_tx*) is propagated through the delay lines and it generates 3 replicated clock signals with a different offset (*clk_A, clk_B, clk_C*). Until the incoming sequence has low value, the brute force synchronizers sample the low AND output and they set to zero the control signal driving the multiplexer (see *ctr_A, ctr_B, ctr_C* in Figure 3). In this configuration the multiplexer output is permanently at a low logic value by default and the downstream mesochronous synchronizer does not receive any clock signal.

Then, the transmitter starts to drive the sequence of ones and the data bits (*flit_tx*) start to switch at the variation detector input port. Finally, when the incoming data is stable and every data bit has switched to the high logic value, the AND output assumes a high logic value. Although the AND output feeds all the brute force synchronizers, only some of them are able to sample the high value at the AND output. In fact, the brute force synchronizers driven by clock signals with an early positive edge will not reveal the high value; on the contrary, the brute force synchronizers driven

Figure 3. Variation detector reset phase

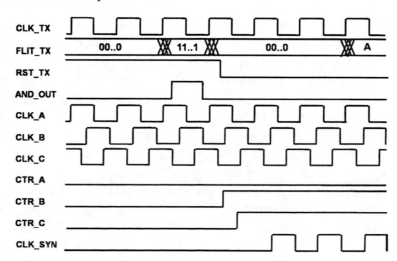

by the clock signal with a late positive edge will sample it and will set the multiplexer control signal. As a result, the control signal collects all the information about the result of the AND output sampling and this latter information is exploited to distinguish between the safe clock signal and the potentially unsafe clock signals (early clock signals). In Figure 3, the *ctr_B* and the *ctr_C* signals are set to the high logic value (i.e. *CTR[2:0]=110*) and this means that the *clk_B* and *clk_C* are safe: the positive edges of these latter clock signals occur after all data bits have switched to the high logic value (i.e., the data is stable).

Since it is the first of the safe clock signals to be selected to cross the multiplexer (i.e. the first high bit of *CTR[2:0]* enables the respective input port of the multiplexer), in our example, the *clk_B* is the clock signal driving the receiver synchronizer (*clk_synch*). It should be noted that a source synchronous synchronizer in a similar scenario could not properly work if driven by the nominal source clock signal (*clk_tx*). In fact, as shown in Figure 3, the positive edge of the source clock signal occurs close to the switching window of input data, hence resulting in a failure of the hold constraint.

The sequence of 1s is driven by the transmitter during the last clock cycle of the reset phase. As result, once the control signal in the variation detector is correctly set, the reset switches to the low value and the variation detector circuit interrupts its operation by freezing the safe clock signal at the multiplexer output. Since the detector circuit works only during the reset phase, the power consumption overhead is marginal over the system lifetime.

As a general comment, the architecture of the variability detector is quite simple and its operation in clear in principle. The main challenge consists however of assessing its timing robustness with respect to layout mismatches and to process variation-induced delay variability of link wires and of detector logic gates as well. An extensive validation effort of the architecture follows hereafter.

Delay Variability Robustness

In order to test the capabilities of our design, we set up a Verilog testbench able to drive the data lines and the transmitter clock signal to a mesochronous synchronizer in the receiver clock domain. The mesochronous synchronizer is the same as in (Ludovici, 2010), however the considerations that follow are synchronizer architecture independent. Therefore, a dual-clock FIFO in place of the mesochronous synchronizer would not make any difference for the results that follow.

Variability detector and companion synchronizer have been synthesized, placed and routed with a commercial toolflow on a 65nm STMicroelectronics technology.

Misalignment between Data and Clock

In order to study the robustness of the communication when the ideal alignment between data and transmitter clock is not respected, we inject an increasing negative/positive skew into the clock signal wire. Therefore, we compare the performance of a baseline source synchronous architecture with the one achieved by a source synchronous architecture augmented with the proposed variation detector.

Figure 4(a) reports the timing margin derived from a post-layout analysis of a source synchronous mesochronous synchronizer synthesized at 1 GHz without the variation detector. Setup and hold times have been experimentally measured by driving the mesochronous synchronizer under test with a transmitter clock affected by an increasing routing skew and by monitoring the relative waveforms. *FF_Time* refers to the minimum values required by the technology library for correct sampling.

First of all, we observe that the setup time increases and the hold time decreases linearly with the increase of the clock skew. When the positive clock skew reaches 20% of the clock period, the hold time violates the minimum timing margin and a failure is detected in the mesochronous

Figure 4. Timing margin of the baseline mesochronous synchronizer (a) and the mesochronous synchronizer with a variation detector in front of it (b)

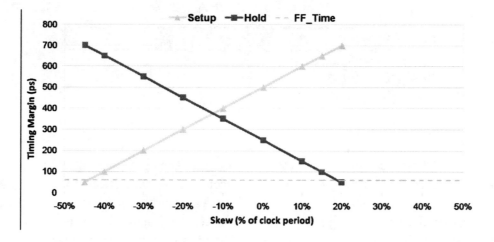

synchronizer. On the contrary, a setup violation is verified when injecting a negative clock skew higher than 45%. As a conclusion, the baseline mesochronous synchronizer is able to support a skew between clock and data within a range of -45% and +20% of the clock period.

Overall, 35% of the clock period is unsafe for sampling and a positive clock edge, loosely synchronous with the data, can only occur during the remaining 65% of the clock period. This experiment was carried out considering an incoming data stable for 75% of the clock period at the mesochronous synchronizer input port (see Figure 5). This accounts for the delay variability between data lines and for the non-null settling time usually found in post-layout link switching.

In order to perform a fair comparison, a similar experiment, shown in Figure 4(b), was performed for a mesochronous synchronizer with the variation detector (composed of 5 delay lines and 5 brute force synchronizers) connected in front of it. We can notice a significant difference of the results due to the insertion of the variation detector. The setup margin is strictly symmetric with respect to the hold margin. Moreover, the timing margin is composed of the periodic repetitions of the same triangular-shaped trend.

As before, at the increase of the clock skew corresponds a decrease of the hold margin although, in this case, as soon as the hold time degradation becomes significant and dangerous for correct circuit operation the detector re-establishes safe margins. In conclusion, the proposed architecture guarantees a safe communication in a mesochronous link in every clock delay scenario as opposed to the baseline detector-less GALS link architecture.

Misalignment between Wires of the Data Link

Although it is essential to guarantee the in-phase property between the data wires and the transmitter clock signal, the size of the timing window during which data is stable at the input port of the receiver synchronizer is also an important parameter for reliable communication. In particular, we measured timing margins of the baseline source synchronous architecture vs. the detector-augmented architecture when varying this latter parameter.

Since the considered timing window is longer when the data bits arrive simultaneously, this parameter is tightly dependent on the amount of process variations affecting the upstream link. Therefore, we injected an increasing amount of

Figure 5. Settling time of the incoming data

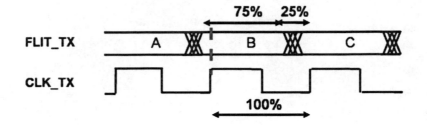

delay variability in the wires carrying the data bits and we monitored the behavior of the synchronizer under test.

Figure 6 reports the timing margin derived from a post-layout analysis of a mesochronous synchronizer synthesized at 1GHz. The timing window when input data is stable is identified by the two red lines crossing the y-axis in -50% and +50%. Y-axis reports the position of the data sampling edge inside (i.e., as a percentage of) the data stability window. 0% means sampling in the middle of the window, while sampling close to the upper bound (50%) incurs setup time violations; on the contrary, a sampling close to the lower bound (-50%) can incur hold time violations.

The figure also compares the results in the presence of a different amount of delay variability in the data link wires, i.e. different sizes of the data stability window. The analyzed timing windows range from 95% to the 50% of the clock period size. As expected, in the configuration without the variation detector (*TimeBase 95%*, *TimeBase 75%*, *TimeBase 50%*), at the increase of the positive/negative clock skew, the data sampling time moves closer to the lower/higher bound of the window until an hold/setup violation occurs. Moreover, the higher the delay variability in the data wires the lower the probability of sampling in the safe timing window.

A similar experiment was carried out by reporting the timing margin of a mesochronous synchronizer augmented with the variation detector (*TimeDet 95%*, *TimeDet 75%*, *TimeDet 50%*). As result, the data sampling occurs inside the safe timing window in every experimented scenario. We can notice how the sampling time moves closer to the unsafe lower/upper bound as the timing window size shrinks. In every case, the proposed architecture does not incur timing violations even when the data is stable for only 50% of the clock cycle.

Timing Variability Affecting the Detector

While the robustness to link delay variability is already implicit in the above experimental results, the robustness of the detector to timing variations in its own logic cells still has to be demonstrated. Timing variability, in the form of process variations, is caused by threshold voltage variation due to Gaussian Random Dopant Fluctuations (RDF). Briefly, the concept behind RDF is the following: as technology scales down, the number of dopant atoms that fit into the transistor channel area is getting smaller and smaller. For 65nm and later technologies, this number is in the range of a few tens atoms. Thus, with these numbers, a few atoms more or less considerably matter. This variation in the number of dopant atoms in the transistor channel is what is known as Random Dopant Fluctuations, which are expected to be the major source of unpredictability affecting future VLSI circuits.

For this purpose, we characterized the circuit behavior by modeling the random variations for technologies ranging from 65nm to 16nm, based on the methodology presented in (Hernández,

Figure 6. Timing margin in the presence of process variation in the data link wires

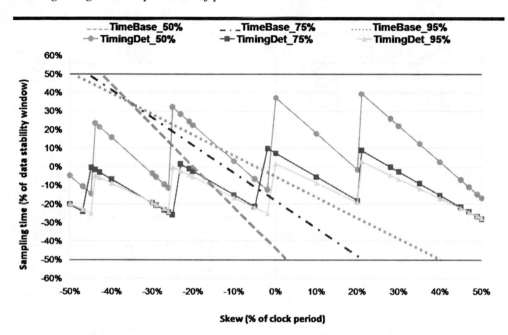

2010). Therefore, this section shows the effect of process variations affecting the detector on the correct operation of this latter.

In Figure 7, we measured the timing margin of the proposed architecture in the presence of best, nominal and worst random process variation conditions. We considered a pessimistic timing window, when the data is unstable at the synchronizer input port for 50% of the clock period. The timing margins are derived from a post-layout analysis performed at 1GHz.

As reported in the figure, the variation detector proves robust to the random delay variability achieving similar margins in the different process variation scenarios. Similar tests have been performed by applying the variability model considering even the wider variances of parameter spread foreseen by the ITRS for the 45nm, 32nm, 22nm and 16nm technology nodes, but are omitted for lack of space. Again, the result proved the relative insensitivity by construction of our detector to the amount of injected delay variability.

ARCHITECTURE INTEGRATION

Since a mesochronous synchronizer is designed to support a given amount of skew between two mesochronous domains, it is relevant to analyze how link delay variability affects the skew tolerance of the synchronizer and whether the integration of the variation detector into the synchronizer degrades its final skew robustness figures. At this point, the actual synchronizer architecture comes into play. This manuscript builds on the latest advances in synchronizer design for cost-effective on-chip networks.

In fact, there are two main schemes to instantiate a mesochronous synchronizer in a source synchronous link. The basic scheme consists of placing the synchronizer in front of the downstream switch input port, but as an external block to it. A link taking this approach follows a *Loosely Coupled Synchronization* scheme (Ludovici, 2009) and the synchronizer used in this context might be similar to the one graphically illustrated in Figure 8.

Figure 7. Random variations affecting the delay of detector logic cells

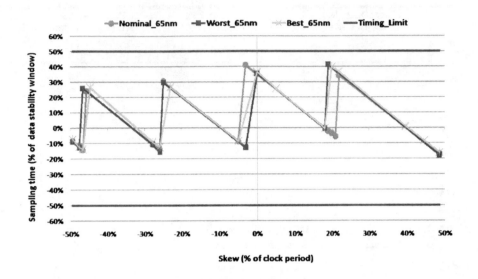

The circuit receives as its inputs a bundle of NoC wires representing the regular NoC link, carrying data and/or flow control commands, and a copy of the clock signal of the sender. The circuit is composed by a front-end and a back-end. The front-end is driven by the incoming clock signal, and of parallel latches in a rotating fashion, based on a counter. The back-end of the circuit leverages the local clock of the downstream switch, and samples data from one of the latches in the front-end thanks to multiplexing logic which is also based on a counter. The rationale is to temporarily store incoming information in one of the front-end latches, using the incoming clock wire to avoid any timing problem related to the clock phase offset.

Once the information stored in the latch is stable, it can be read by the target clock domain and sampled by a regular flip-flop. In practice, such a sampling is performed by the input buffer of the downstream switch. Anyway, the loosely coupled design introduces a severe overhead in the communication. A key principle for reducing the latency, area and power overhead of switches with loosely coupled synchronizers is to co-design the synchronizer with the downstream switch input port. The co-optimization consists of using the buffering resources of the switch input stage not only for retiming and flow control as in the synchronous architecture, but also for synchronization purposes. This would allow to completely remove the switch input buffer and to replace it with the synchronizer itself and the ultimate consequence is that the mesochronous synchronizer becomes the actual switch input stage.

The basic architecture of the new switch input port is illustrated in Figure 9 and denoted *Tightly Coupled Synchronizer* (Ludovici, 2009). In this case, the ultimate sampling of the data flowing out of the synchronizer multiplexer and going through the switch internal logic will occur at the switch output port. On the contrary, in the loosely coupled synchronizer, the synchronized data is directly sampled by a regular flip-flop. As a consequence, the two architectures are affected by different timing constraints and exploit a distinct phase offset budget.

In particular, the skew tolerance of the two architecture schemes depends on the relative alignment of data arrival time at latch outputs, multiplexer selection window and sampling edge in the receiver clock domain.

Figure 8. Loosely coupled synchronizer architecture

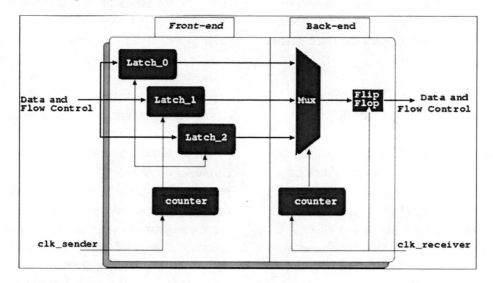

Figure 9. Tightly coupled synchronizer architecture

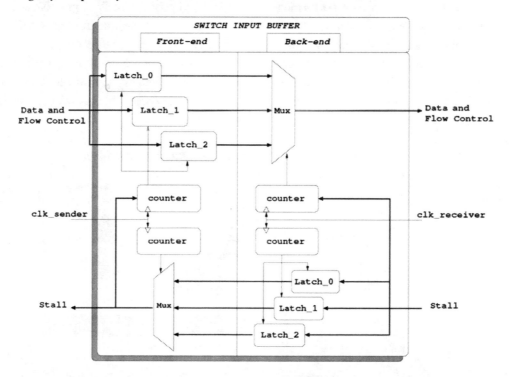

Anyway, due to the worst case timing path between the tightly coupled synchronizer output and the switch output port, a switch with tight coupling of the synchronizer with the switch incurs in a degradation of the timing margin (which is actually the skew tolerance of the synchronizer).

This is a clear trade-off that is at the disposal of the designer, who selects the suitable design point based on his requirements.

The variability detector is integrated with the synchronizer based on the scheme in Figure 10. This scheme refers to a loose coupling with the

NoC, but the same holds also for a tight coupling. The output of the variability detector feeds a counter which in turn generates the latch enable signal for the synchronizer.

Impact of Variability Detector on Synchronizer Timing Margins

In order to evaluate the skew tolerance in the case of link delay variations in the source synchronous link, we injected an increasing negative/positive skew into the transmitted clock signal wire, thus inducing a routing skew with the data wires. A the same time, we applied a different clock phase offset between the transmitter clock (*clock_TX*) and the receiver clock (*clock_RX*), to assess evolution of the timing margin. This experiment is performed for both the detector-augmented and the detector-less synchronization architecture where a tightly coupled and a loosely coupled synchronizer are alternatively instantiated. Ultimately, we aim at assessing to which extent the detector impacts

the skew tolerance of the two synchronization schemes in a noisy environment.

In Figure 11 the results for the Loosely Coupled Synchronizer are reported. The y-axis reports negative and positive values of the skew, expressed as percentage of the clock period, while the x-axis reports the misalignment of the transmitted clock signal wire with respect to data wires.

As expected, the baseline detector-less synchronizer is not able to work in every variability scenario. It is able to properly absorb a skew between two mesochronous domains only when the skew between the transmitted clock and data is within a range of -38% and +3% of the clock period. In this latter range the architecture supports 100% of negative skew and between 60% and 90% of positive skew. A positive skew means that the clock at the transmitter is delayed with respect to the one at the receiver.

On the contrary, by applying the proposed variation detector to the synchronizer, the architecture guarantees an acceptable skew tolerance in every

Figure 10. Integration of the variation detector with the loosely coupled synchronizer

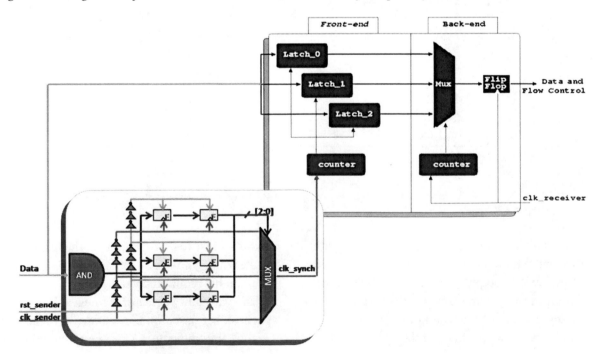

Figure 11. Skew tolerance as a function of the misalignment between data and transmitter clock in the loosely coupled synchronizer augmented with the variability detector

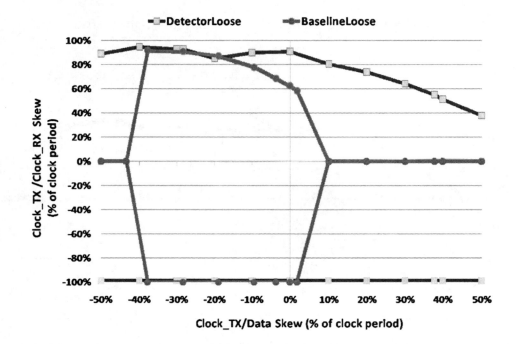

process variation scenario. In particular it achieves similar performance to the baseline architecture when the skew between the transmitted clock and data is within a range of -38% and +3% while it supports 100% of negative skew and between 40% and 90% of positive skew in the other scenarios. As a result, in those cases where the detector-less architecture works fine, the detector-augmented one does not do a worse job, while at the same time prolonging the feasibility space.

Finally, Figure 12 reports the results for the Tightly Coupled Synchronizer. As expected, the tightly coupled synchronizer is not able to work in every variability scenario as the loosely coupled counterpart. In particular, the tightly coupled synchronizer is able to properly absorb a skew between two mesochronous domains only when the skew between the transmitted clock and data is within a range of -28% and +3% of the clock period. In this latter range the tightly coupled architecture supports 100% of negative skew but only 40% and 60% of positive skew. As expected,

its timing margin allows to absorb a lower positive skew and to work in a shorter range than the loosely coupled counterpart. Anyway, by applying the proposed variation detector to the synchronizer, it is still guaranteed a skew tolerance in every process variation scenario although the performance for positive skews are clearly lower due to the severe timing constraints. Especially, similar performance to the stand-alone tightly coupled architecture are achieved when the skew between the transmitted clock and data is within a range of -28% and +3% while it supports still 100% of negative skew and until 50% of positive skew in the other scenarios.

In practice, without the variation detector, skew tolerance of the mesochronous synchronizer at the receiver end is seriously impacted both in the loosely coupled and in the tightly coupled scenario by even small misalignments between data and transmitter clock. Vice versa, the variation detector enables a smoother degradation of the skew tolerance and its extension to operating

Figure 12. Skew tolerance as a function of the misalignment between data and transmitter clock in the tightly coupled synchronizer augmented with the variability detector

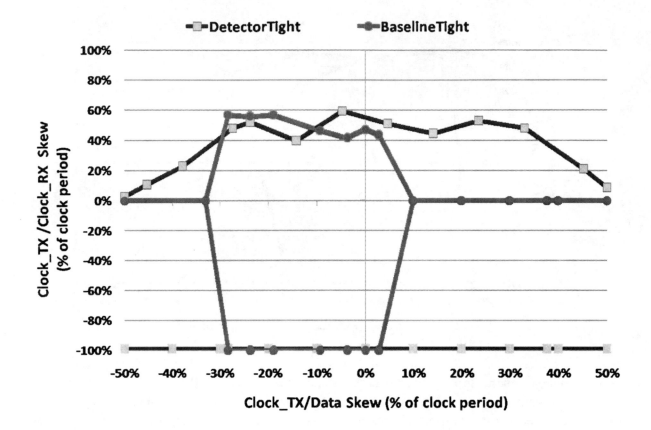

conditions featuring large routing skew between data and clock. Ultimately, having the variability detector can help better cope with the layout constraints and manufacturing process uncertainties, while marginally impacting skew tolerance in the best cases.

Area Overhead

According to our post-synthesis area figures, our variability detector consumes approximately one third of the area of the compact loosely coupled mesochronous synchronizer in (Ludovici, 2009) that we have used throughout this manuscript for the experiments (Figure 8). The detector with 5 brute-force synchronizers consumes 440 um2 as opposed to the 1150 um2 of the mesochronous

synchronizer. When considering a 5x5 switch commonly found in a 2D mesh and synthesized for max performance, the variation detector takes less than 1% of total switch area.

DESIGN GUIDELINES

As a final contribution, the variation detector can be designed to match well-defined operating conditions at the minimum implementation cost. Indeed, the proposed architecture can be parameterized as a function of the target performance and of the timing margins expected for the source synchronous interface. Next, design equations follow for this purpose.

In a high variability environment, the alignment between the sampling edge of the clock signal and the data can be unpredictable. As a result, sampling can occur regardless of the stability or settling timing window of data. In order to ensure the reliability of the source-synchronous communication, the detector should always guarantee a safe clock version with a sampling edge falling in the data stability window. To meet this requirement, a minimum number of clock versions should be designed to sweep the entire clock period in such a way that stable data can be safely sampled at least once regardless of its position within the clock period.

As a consequence, to increase the reliability of the proposed architecture and to minimize the number of required stages, the number of clock versions and the delay introduced by each delay line must be constrained. In particular, the length of the clock period to be swept and the length of the stability data window are the two main parameters to be taken into account to properly design a reliable variation detector. Then, we can define the safe timing window (*Swd*) where the output of the AND block is stable at a high value (i.e. all the bits of the incoming data arrived at the receiver end) and the metastability risk is avoided, as follows:

$$Swd = Tstability - Tsetup - Thold_F \qquad (1)$$

where *Tstability* represents both the timing window where the AND output is high and the window where the incoming data is stable. Moreover, *Tsetup* and *Thold* correspond to the minimum timing margins required by the technology library for correct sampling.

Intuitively, to ensure that at least a clock version will safely sample, we can suppose to split the clock period into equal parts of *Swd* length and generate a rising clock edge in each sub-period. The following formula can be exploited

to determine the minimum amount of variation detector stages:

$$\frac{Tclk}{Swd} < NumStage \qquad (2)$$

It should be observed that *Tclk* is the source clock period and the delay chains are supposed to be composed by the same number of delay elements to guarantee an equal offset between the falling edge of the clock versions. Clearly, *NumStage* represents the minimum number of stages for a reliable communication but a greater number of stages can be adopted without affecting the correct operating principle of the system.

In the baseline architecture, described in the *"Operating Principle"* section, the multiplexer selects the first safe clock between the available versions. However, it is possible to enable a circuit variant exploiting the multiplexer control signals to further increase the robustness of the variation detector. In fact, an unlucky condition can occur. A clock version can violate the hold/setup time constraint of its brute-force stage. As a consequence, the resulting output can fall into a metastability state and it can erroneously set to 1 the multiplexer control bit. Since this clock version will be the first to set the multiplexer, it will be allowed to cross the multiplexer. As a result it will provoke potential metastability concerns in the source synchronous interface since its rising edge will be premature for the safe sampling of the data. In this case, the real safe clock version was not considered since it was the second one to set the multiplexer.

Anyway, we can force a *Swd* length constraint more severe than in Formula 1 to avoid the metastability concerns. In fact, we can require to have two safe clock versions for every *Swd* scenario and as a result, we can set the multiplexer to allow always the second clock version instead of the first one to cross its logic. In this case, the clock version violating the brute-force hold/setup time constraint will be not considered since it will be the first to set the multiplexer. To conclude, Formula 1 should be extended as follows for increased robustness to metastability concerns:

$$Swd = \frac{Tstability - Tsetup - Thold}{2} \quad (3)$$

$$Swd > Max[Tsetup; Thold] \quad (4)$$

Note that metastability robustness is guaranteed when both conditions (Formula 3 and 4) are respected. The violation of condition in Formula 4 can bring two consecutive clock versions to violate the brute-force hold/setup time and set erroneously the multiplexer (the worst case scenario). In practice, we have to avoid that sampling occurs at such a fine granularity that two consecutive sampling edges occur in the metastability window.

In order to quantify the number of required brute-force and delay line stages for real-life design problems, Table 1 shows some of the possible scenarios as a function of target frequency and data instability timing window. It should be observed that in Table 1 Formula 3 was applied and a *Tsetup* and *Thold* of 100ps was considered.

Interestingly, the *Swd > Max[Tsetup; Thold]* condition is not verified when working at 1.25Ghz with 500ps of settling time. In fact, this can be considered an infeasible scenario where the safe sampling data window lasts just 12% of the clock period. As a final remark, throughout the manuscript safe timing margins with a 5 stage variation detector working at 1Ghz were achieved since the *Tsetup/Thold* constraints of the used technology library were clearly lower than the pessimistic constraints considered in Table 1.

CONCLUSION

This chapter has presented a novel timing variation detector architecture for use in source synchronous links in front of the receiver synchronizer (either a mesochronous synchronizer or a dual-clock FIFO). The detector was in particular conceived to deal with layout mismatches and with link delay variability (either in the form of process variations or in the form of ageing variations) and builds on the principle of self-calibration, i.e., adaptation of the transmitter clock signal phase to in-situ actual operating conditions. We have proved the ability of detector-augmented GALS links to cope with large signal misalignments and small data stability windows, unlike baseline GALS links. Above all, the variation detector has proven robust to the delay variability of its logic cells. Overall, analyzing the comprehensive metric of skew tolerance for a mesochronous synchronizer has revealed that the insertion of the variation detector has advantages in most cases and for the main synchronization architectures of practical interest (loosely vs. tightly coupled with the NoC). Even for non-variability-dominated links, it can better cope with the layout constraints of the link.

Table 1. Number of stages of the variability detector based on the frequency and the settling time requirements

Settling Time	Target Frequency				
	1.25Ghz	1Ghz	750Mhz	500Mhz	250Mhz
500ps	X	7	5	4	3
300ps	6	4	4	3	3
100ps	4	3	3	3	3

REFERENCES

Asenov, A., Kaya, S., & Davies, J. H. (2002). Intrinsic threshold voltage fluctuations in decanano MOSFETs due to local oxide thickness variations. *IEEE Transactions on Electron Devices*, *49*(1), 112–119. doi:10.1109/16.974757

Beer, S., Ginosar, R., Priel, M., Dobkin, R., & Kolodny, A. (2010). The Devolution of Synchronizers. In *Proceedings of IEEE Symposium on Asynchronous Circuits and Systems* (ASYNC), (pp. 94-103). IEEE.

Bernstein, J. B., Gurfinkel, M., Li, X., Walters, J., Shapira, Y., & Talmor, M. (2006). Electronic circuit reliability modeling. *Microelectronics and Reliability*, *46*(12), 1957–1979. doi:10.1016/j.microrel.2005.12.004

Bonesi, S., Bertozzi, D., Benini, L., & Macii, E. (2008). Process Variation Tolerant Pipeline Design Through a Placement-Aware Multiple Voltage Island Design Style. In Design, Automation and Test in Europe, (pp. 967-972). Academic Press.

Borkar, S., Karnik, T., & Vivek, D. (2004). Design and reliability challenges in nanometer technologies. In *Proceedings of the 41st Design Automation Conference*. Academic Press.

Bowman, K. A., Duvall, S. G., & Meindl, J. D. (2002). Impact of die-to-die and within-die parameter fluctuations on the maximum clock frequency distribution for gigascale integration. *IEEE Journal of Solid-State Circuits*, *37*(2), 183–190. doi:10.1109/4.982424

Changlin, C., Ye, L., & Cotofana, S. D. (2012). A Novel Flit Serialization Strategy to Utilize Partially Faulty Links in Networks-on-Chip. In *Proceedings of Sixth IEEE/ACM International Symposium on Networks on Chip* (NoCS), (pp. 124-131). IEEE/ACM. doi: 10.1109/NOCS.2012.22

Collins, H.A., & Nikel, R.E. (1999). *DDR-SDRAM, high-speed, source synchronous interfaces create design challenges*. EDN article, September 1999.

Dally, W. J., & Poulton, J. W. (1998). *Digital Systems Engineering*. Cambridge University Press.

Diamos, G., Yalamanchili, S., & Duato, J. (2007). STARS: A system for tuning and actively reconfiguring SoCs link. In *Proceedings of DAC Workshop on Diagnostic services in NoCs*. DAC.

Elrabaa, M. E. S. (2006). An all-digital clock frequency capturing circuitry for NRZ data communications. In *Proceedings of IEEE Int. Conf. on Electronics, Circuits and Systems*, (pp. 106-109). IEEE.

Ernst, D. Kim, Das, S., Pant, S., Rao, R., Pham, … Mudge, T. (2003). Razor: A low-power pipeline based on circuit-level timing speculation. In *Proceedings of 36th Annual IEEE/ACM International Symposium on Microarchitecture*, (pp. 7-18). IEEE/ACM.

Faiz-ul-Hassan. Cheng, B., Vanderbauwhede, W., & Rodriguez, F. (2009). Impact of device variability in the communication structures for future synchronous SoC designs. In *Proceedings of International Symposium on System-on-Chip*, (pp. 68-72). Academic Press.

Gilabert, F., Ludovici, D., Medardoni, S., Bertozzi, D., Benini, L., & Gaydadjiev, G. N. (2009). Designing Regular Network-on-Chip Topologies under Technology, Architecture and Software Constraints. In *Proceedings of International Conference on Complex, Intelligent and Software Intensive Systems*, (pp. 681-687). Academic Press.

Grecu, C., Ivanov, A., Saleh, R., & Pande, P. P. (2006). NoC interconnect yield improvement using crosspoint redundancy. In *Defect and Fault Tolerance in VLSI Systems* (pp. 457–465). Academic Press.

Hernández, C., Roca, A., Silla, F., Flich, J., & Duato, J. (2010). Improving the Performance of GALS-Based NoCs in the Presence of Process Variation. In *Proceedings of Fourth ACM/IEEE International Symposium on Networks-on-Chip* (NOCS), (pp. 35-42). ACM/IEEE.

Höppner, S., Walter, D., Eisenreich, H., & Schüffny, R. (2010). Efficient compensation of delay variations in high-speed network-on-chip data links. In *Proceedings of International Symposium on System on Chip* (SoC), (pp. 55-58). Academic Press.

ITRS. (2007). *International Technology Roadmap for Semiconductors*. Retrieved from http://www.itrs.net/Links/2007ITRS/Home2007.htm

Jose, A. P., Patounakis, G., & Shepard, K. L. (2005). Near speed-of-light on-chip interconnects using pulsed current-mode signalling. *Digest of Technical Papers Symposium on VLSI Circuits*, 108- 111.

Kakoee, M. R., Loi, I., & Benini, L. (2010). A New Physical Routing Approach for Robust Bundled Signaling on NoC Links. In *Proceedings of GLSVLSI10*. GLSVLS.

Kenyon, C., Kornfeld, A., Kuhn, K., Liu, M., Maheshwari, A., Shih, W., … Zawadzki, K. (2008). Managing Process Variation in Intel's 45nm CMOS Technology. *Intel Technology Journal*.

Kihara, M., Ono, S., & Eskelinen, P. (2003). *Digital Clocks for Synchronization and Communications*. Artech House, Inc.

Loi, I., Angiolini, F., & Benini, L. (2008). Developing Mesochronous Synchronizers to Enable 3D NoCs. In Design, Automation and Test in Europe, (pp. 1414-1419). Academic Press.

Ludovici, D., Strano, A., Bertozzi, D., Benini, L., & Gaydadjiev, G. N. (2009). Comparing tightly and loosely coupled mesochronous synchronizers in a NoC switch architecture. In *Proceedings of 3rd ACM/IEEE International Symposium on Networks-on-Chip*, (pp. 244-249). ACM/IEEE.

Ludovici, D., Strano, A., Gaydadjiev, G. N., Benini, L., & Bertozzi, D. (2010). Design space exploration of a mesochronous link for cost-effective and flexible GALS NOCs. In Design, Automation & Test in Europe, (pp. 679-684). Academic Press.

Mandal, A., Khatri, S. P., & Mahapatra, R. N. (2013). Exploring topologies for source-synchronous ring-based Network-on-Chip. In *Proceedings of Design* (pp. 1026–1031). Automation & Test in Europe Conference & Exhibition.

Merdadoni, S., Lajolo, M., & Bertozzi, D. (2008). Variation tolerant NoC design by means of self-calibrating links. In Proceedings of Design, Automation and Test in Europe, (pp. 1402-1407). Academic Press.

Mondal, M., & Ragheb, T. Wu, Aziz, A., & Massoud, Y. (2007). Provisioning On-Chip Networks under Buffered RC Interconnect Delay Variations. In *Proceedings of 8th International Symposium on Quality Electronic Design*, (pp. 873-878). Academic Press.

Moore, S. W., Taylor, G. S., et al. (2000). Self Calibrating clocks for globally asynchronous locally synchronous systems. In *Proceedings of IEEE Int. Conf. on Computer Design*, (pp. 73-78). IEEE.

Murali, S., Tamhankar, R., Angiolini, F., Pulling, A., Atienza, D., Benini, L., & De Micheli, G. (2006). Comparison of a Timing-Error Tolerant Scheme with a Traditional Re-transmission Mechanism for Networks on Chips. In *Proceedings of International Symposium on System-on-Chip*, (pp. 1-4). Academic Press.

418

Nicopoulos, C., Srinivasan, S., & Yanamandra, A., Park, Narayanan, V., Das, C.R., & Irwin, M.J. (2010). On the Effects of Process Variation in Network-on-Chip Architectures. *IEEE Transactions on Dependable and Secure Computing, 7*(3), 240–254. doi:10.1109/TDSC.2008.59

Orshansky, M., & Keutzer, K. (2002). A general probabilistic framework for worst case timing analysis. In *Proceedings 39th Design Automation Conference*, (pp. 556- 561). Academic Press.

Paci, G., Bertozzi, D., & Benini, L. (2009). Effectiveness of adaptive supply voltage and body bias as post-silicon variability compensation techniques for full-swing and low-swing on-chip communication channels. In Proceedings of Design, Automation & Test in Europe, (pp. 1404-1409). Academic Press.

Panades, I. M., Clermidy, F., Vivet, P., & Greiner, A. (2008). Physical Implementation of the DSPIN Network-on-Chip in the FAUST Architecture. In *Proceedings of Int. Symp. On Networks-on-Chip*, (pp. 139-148). Academic Press.

Sarangi, S. R., Greskamp, B., Teodorescu, R., Nakano, J., Tiwari, A., & Torrellas, J. (2008). VARIUS: A Model of Process Variation and Resulting Timing Errors for Microarchitects. *IEEE Transactions on Semiconductor Manufacturing, 21*(1), 3–13. doi:10.1109/TSM.2007.913186

Stergiou, S., Angiolini, F., Carta, S., Raffo, L., Bertozzi, D., & De Micheli, G. (2005). Xpipes Lite: a synthesis oriented design library for networks on chips. In Proceedings, Design, Automation and Test in Europe, (pp. 1188-1193). Academic Press.

Synopsys. (2007). *PrimeTime VX Application Note - Implementation Methodology with Variation-Aware Timing Analysis, Version 1.0*. Author.

Tran, A. T., Truong, D. N., & Baas, M. B. (2009). A low-cost high-speed source-synchronous interconnection technique for GALS Chip multiprocessors. In *Proceedings of ISCAS*, (pp. 996-999). ISCAS.

Vitkovskiy, A., Soteriou, V., & Nicopoulos, C. (2012). A Dynamically Adjusting Gracefully Degrading Link-Level Fault-Tolerant Mechanism for NoCs. *IEEE Transactions on Computer-Aided Design of Integrated Circuits and Systems, 31*(8), 1235–1248. doi:10.1109/TCAD.2012.2188801

Worm, F., Ienne, P., Thiran, P., & De Micheli, G. (2005). A robust self-calibrating transmission scheme for on-chip networks. *IEEE Transactions on Very Large Scale Integration (VLSI). Systems, 13*(1), 126–139.

Yu, Z., & Baas, B. M. (2006). Implementing tile-based chip multiprocessors with GALS clocking style. In *Proceedings of ICCD*, (pp. 174-179). ICCD.

KEY TERMS AND DEFINITIONS

Delay Variations: Delay uncertainties caused by deviations introduced in the chip manufacturing process.

GALS (Globally Asynchronous Locally Synchronous): Architecture with several synchronous clock domains clocked at different frequencies.

NoCs: Architecture used to connect the CPU, memory, and peripherals within the chip.

Process Variations: Deviations introduced in the manufacturing process of the integrated circuits.

Self-Calibration: The ability of a hardware module to find its optimal operating point.

Source Synchronous Links: Links routing the clock signal together with data to enable safe sampling at the receiver side.

Synchronizers: Hardware modules that enable the interconnection of different clock domain circuits.

Wearout: Degradation caused by system aging that leads to an increase in the failure rate.

Chapter 17
Silicon Validation of GALS Methods and Architectures in a State-of-the-Art CMOS Process

Milos Krstic
IHP, Germany

Christoph Heer
Intel Mobile Communications, Germany

Xin Fan
IHP, Germany

Birgit Sanders
Intel Mobile Communications, Germany

Eckhard Grass
IHP, Germany

Alessandro Strano
Intel Mobile Communications, Germany

Luca Benini
University of Bologna, Italy

Gabriele Miorandi
University of Ferrara, Italy

M. R. Kakoee
University of Bologna, Italy

Alberto Ghiribaldi
University of Ferrara, Italy

Davide Bertozzi
University of Ferrara, Italy

ABSTRACT

The GALS methodology has been discussed for many years, but only a few relevant implementations in silicon have been done. This chapter describes the implementation and test of the Moonrake Chip – complex GALS demonstrator implemented in 40 nm CMOS technology. Two novel types of GALS interface circuits are validated: point-to-point pausible clocking GALS interfaces and GALS NoC interconnects. Point-to-point GALS interfaces are integrated within a complex OFDM baseband transmitter block, and for NoC switches special test structures are defined. This chapter discloses the full structure of the respective interfaces, the complete GALS system, as well as the design flow utilized to implement them on the chip. Moreover, the full set of measurement results are presented, including area, power, and EMI results. Significant benefits and robustness of our applied GALS methodology are shown. Finally, some outlook and vision of the future role of GALS are outlined.

DOI: 10.4018/978-1-4666-6034-2.ch017

INTRODUCTION

Globally Asynchronous Locally Synchronous (GALS) technology has been proposed many years ago as an alternative to the traditional synchronous paradigm for chip synchronization (Krstic, 2006). Although significant potential was reported by the academia, the GALS methodology has never taken off in the industry. However, the growing challenges, imposed by the unrelenting pace of technology scaling to the nanoscale regime, urge for an efficient and safe system-level integration methodology. Consequently, this section provides the overview of the recent implementation of a chip, named Moonrake, in advanced 40 nm CMOS process, aiming at the assessment of GALS technology for nanoscale designs.

The intention of this implementation was to evaluate GALS vs. standard synchronous technology on the same die, by fabricating synchronous and GALS counterparts of the same baseline designs, both in the point-to-point as well as in the network-on-chip (NoC) scenarios for on-chip communication. The two scenarios are very different, hence motivating the different choice of baseline designs for their analysis. In point-to-point communication, once an optimized GALS interface is selected, the focus is on the implications of redesigning an entire system around these links. With this respect, a state-of-the-art multi-million gate synchronous system, an OFDM baseband transmitter developed for a 60 GHz transceiver with a gigabit throughput as presented by Krstic in 2008 has been taken, and re-implemented it with GALS methodology, using the optimized interfaces for pausible (stoppable) clocking as defined by Fan in 2009. One major goal was to explore Electromagnetic Interference (EMI) and switching noise properties of GALS designs and special algorithms and circuits for noise reduction based on the GALS methodology, initially analyzed by Fan in 2010. Within the chip, the switching noise (and correspondingly EMI) is caused by simultaneous switching activity of the digital circuits and it can lead to various problems including ground bounce, power integrity, IR drop, substrate noise etc.

For on-chip networking applications, the communication landscape is more heterogeneous since it results from the interconnection of domains with different synchronization assumptions. Therefore, the focus was on the provision of flexible and cost-effective interfaces for arbitrary composability. In this direction, the novel synchronization interfaces presented by Strano (2010) and Ludovici (2010), aiming at low-area/power/latency overhead while preserving timing robustness, were integrated into NoC test structures exposing (and comparing) a range of flexible GALS solutions.

The main intentions of this chapter are as follows:

- The GALS partitioning criteria for a state-of-the-art OFDM transmitter is presented, highlighting the optimized asynchronous link crossing scheme and the partitioning granularity and strategy at the system level.
- The design flow followed for different GALS systems is illustrated: from pausible clocking to mesochronous synchronization to mixed-timing systems. Compatibility with mainstream standard cell libraries and design toolflows is discussed.
- The feasibility of GALS NoCs linking sub-systems with heterogeneous timing assumptions by means of area/power/latency optimized interfaces while preserving timing margins has been demonstrated.
- Synchronous and GALS counterparts of the same baseline designs (the OFDM transmitter and a NoC sub-set), implemented in the same demonstrator chip, have been compared in terms of area, pointing out counterintuitive benefits of the GALS design style.
- Finally, the test and measurement results of Moonrake chip are presented and analyzed, with the focus on EMI and power measure-

ments showing the benefits of GALS for complex system integration. Additionally, NoC test structures getting the clock from the external world provided an excellent result: frequencies from 25 to 265 MHz were swept, while at the same time varying the clock phase offset from 0 to 360 degrees. This means that the synchronization mechanisms, considered by themselves, can be ported to the 40 nm technology and prove functional in such an environment.

BACKGROUND: GALS SYSTEMS AND DEMONSTRATORS

To validate the GALS methodology, the theoretical and simulation based approach is not sufficient. It is indeed necessary to validate the methods also on silicon and to evaluate the potential benefits of GALS in praxis including the measurements.

In past years several GALS chip implementations were reported, where the methodology validation was one of the main goals. Many of them were focused on point to point GALS architectures (several designs from ETHZ (outlined in Krstic, 2007), Zeke from Sutherland in 1999, WLAN baseband processor described from Krstic in 2007, while more recent implementations have also explored the GALS NoC concept (NEXUS chip in Lines, 2004), recent SpiNNaker chip in Painkras (2012) and Plana (2007), and three implementations from LETI, namely FAUST (Vivet 2006), ALPIN (Beigne 2008), and MAGALI (Lemaire 2010) chip).

Table 1 illustrates an overview of basic features of the GALS demonstrators in the last few years. The complexity of the silicon validator plays important role, since it can reflect the real system integration challenges, which may be hidden in simpler implementations. Additionally, to have the overview of the actual problems in

implementation, it is important the relevance of the technology process to the state-of-the-art. The Magali chip is the most complex one, followed by the Moonrake chip. The improvement of this latter upon state-of-the-art is with respect to the most aggressive manufacturing process, higher industrial relevance of its "galsified" designs, and the maturity of implemented synchronization interfaces. Above all, both the baseline synchronous and the GALS counterparts of the same designs are now available on the same chip, thus paving the way for benchmarking in a truly homogeneous experimental setting at the architecture and at the technology level.

MOONRAKE CHIP

Chip Architecture

As anticipated, the Moonrake architecture is based on the parallel implementation of the synchronous and the GALS variants of the same baseline designs. In order to reduce chip area, the same pad frame for both was re-used, and therefore data input and output pins were in most cases multiplexed. The chip architecture is pictorially illustrated in Figure 1. Depending on the functional mode, input data will be processed by the synchronous or the GALS OFDM transmitter. In addition to the global dataflow, there is also a BIST logic used for testing of both synchronous and GALS transmitters, PLLs that trigger the synchronous Transmitter and the NoC parts, and JTAG interfaces used for programming the PLLs, setting the parameters of local clock generators, programming NoC traffic generators and controlling the different modes of operations. It should be observed that the OFDM sub-system and the NoC test structures are independent from each other, except for the sharing of some input pins where possible: 37% of NoC pins are shared with the OFDM designs.

Table 1. Recent GALS demonstrators

GALS Chip Demonstrators		Technology Node	Year of Fabrication	Size (mm²)	Power (mW)
Moonrake chip (Krstic 2012)		40 nm	2010	9	220-250
	Main features	GALS demonstrator, based on pausible clocking that serves the function of an OFDM transmitter for 60 GHz communication with the processing throughput of 1.6 -2.5 Gbps. It contains parallel synchronous and GALS implementation and NoC test structures.			
SpiNNaker chip (Plana 2007, Painkras 2012)		130 nm	2011	102	1000
	Main features	The chip, emulating neural system, contains 18 ARM9 processor cores, each running at around 200 MHz. These cores must all communicate 166 MHz mobile DDR SDRAMs. The chip uses two distinct NoCs based on Chain, a delay-insensitive (DI) communication technology.			
Magali chip (Lemaire 2010)		65 nm	2008	32	500
	Main features	Digital baseband circuit for software-defined-radio and cognitive-radio applications that features less than 50 μs for full reconfiguration. Developed for 4G mobile phones, is based on mesh asynchronous network-on-chip infrastructure delivering 2.2GB/s/link. The chip includes 23 integrated signal processors and processing with an ARM1176 processor. The circuit exhibits less than for up to 40 GOPS performance. MAGALI has been tested on a 3GPP-LTE application, delivering 50Mbits/s on a 4×2 MIMO scheme.			
Alpin chip (Beigne 2008)		65 nm	2008	11.5	20-81
	Main features	Aiming at reducing both dynamic and static power consumption. Based on GALS methodology, each synchronous island is an independent frequency and power domain. The proposed DVFS architecture is based on the association of pausible clocking and supply voltage management unit.			
Faust (Vivet 2006)		130 nm	2006	0.28–0.18	3.69 – 5.89
	Main features	FAUST is a stream-oriented multiapplication SoC platform for telecommunications addressing IEEE 802.11a and MC-CDMA standards. The Chip comprises the physical implementation of both asynchronous QDI ANoC Network on Chip and multi-synchronous DSPIN.			
Nexus (Lines 2004)		180, 150, 130 nm	2004	1.75 (in 130 nm)	Not provided
	Main features	Nexus has a crossbar that interconnects locally synchronous modules on the same chip. Asynchronous QDI channels carry the data across the chip to and from the central crossbar. The crossbar includes routing and arbitration circuitry to resolve contention on output ports. All parts of the system are safely flow controlled.			

Implementation of OFDM GALS Transmitters

When galsifying functional blocks in the signal processing domain, it is very frequent to end up exposing point-to-point asynchronous links between partitions, thus conditioning area, power and performance metrics of the entire design. For this reason, the Moonrake framework presents a pilot partitioning experience of an industry-relevant design, where the system-level GALS partitioning methodology is as important as the design of efficient signaling schemes at the link level.

Starting Point: The Synchronous Transmitter

The foundation of this implementation was the OFDM transceiver system described by Krstic in 2008, developed for 60 GHz communications and offering gigabit throughput. The OFDM transmitter is a fairly complex structure that includes, among the other blocks, 256-point FFT and six different interleaver units. This circuit was originally implemented in Xilinx FPGA, with the estimated complexity of 7.8 M gates. The design is dominated by the various memory structures (64

Figure 1. Block diagram of the Moonrake chip architecture

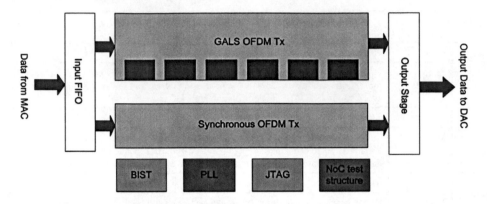

RAM blocks, and 43 ROMs), occupying around 70% of the total area.

Pausible Clocking Interfaces

Interface circuits are of key importance for the performance and reliability of GALS design. For years, the pausible clocking GALS scheme has been argued in respect of the throughput drop caused by unexpected clock stretching and the synchronization failure due to clock distribution delay. In the Moonrake chip design, an optimized interface circuit for pausible clocking scheme, as shown in Figure 2 (Fan, 2009), has been applied. A feedback loop was introduced in the ring oscillator to minimize the clock acknowledge latency for asynchronous arbitration and reduce the performance drop by intensive arbitration. The configuration of the programmable and/ or fixed delay lines was optimized taking into account both the throughput and the reliability requirements. A pair of mutually exclusive latches was further inserted on the data path in order to isolate the data sampling in the receiver from the data capture in the datalink. By this means, the safe timing region provided for the clock tree distribution without introducing synchronization failures was maximized.

The delay lines were implemented in principle by the chains of inverters with even number. The length of each inverter chain was configured by a control unit so as to adjust the clock frequency of the ring oscillator. Since the ring oscillators were running independently from each other, the active edges on each clock domain is asynchronous and random distribution with regards to the others. It leads to a dramatic spread of switching activity over time in different GALS islands compared to the synchronous design. Further, due to the delay sensitivity of inverter chains on the PVT variations, the working frequency of a ring oscillator is drifting from cycle to cycle around the target speed. Therefore, it intrinsically presents a fine-grained frequency modulation on the GALS clock, and eventually results in an efficient spread of peak power at clock harmonics over spectrum. Both of the factors contribute to the EMI attenuation by the GALS design based on the pausible clocking scheme since the switching activity is distributed in time and not bound to the global clock frequency.

GALS Partitioning Methodology

The main methodology for partitioning was the area/power equalization between the GALS blocks, thus aiming at easier timing closure and optimized supply noise profile. It is conceivable to expect that such balanced partitioning would result in the set of the compact blocks that can be much easily optimized by the CAD tools. As a result it is expected to achieve relaxed timing closure, the clock trees of significantly reduced

Figure 2. An optimized interface circuit for pausible clocking scheme

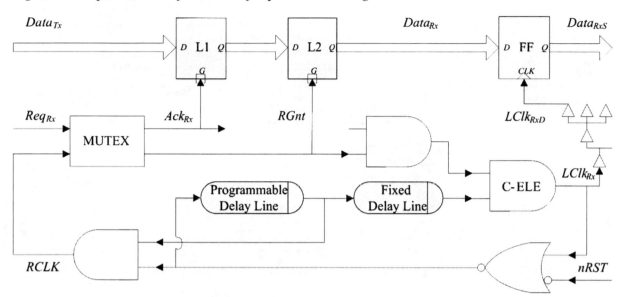

complexity, and reduced gate insertion penalty introduced during the timing optimization by the back-end tools. Furthermore, such balancing is a key part of our EMI reduction strategy that is based on timing distribution of switching activities of the locally synchronous blocks. With averaged blocks in terms of power/area this strategy could be consequently utilized leading to suppressed current peaks and reduced EMI in frequency domain. As stated above, the GALS concept that is used here is based on pausible clocking using the optimized GALS interface circuits that can offer improved performance under increased communication load and improved safety of the generated clocks. Although the optimized GALS interfaces could reduce the performance losses introduced by the concurrent arbitration of the parallel asynchronous links up to 33% (see work from Fan in 2009), extensive arbitration might still result in reduced throughput. Therefore, an additional goal of the partitioning process was to reduce the number of communication links between the mutually-asynchronous blocks in order to preserve performance. As a result, the complete transmitter was partitioned into 6 independent GALS blocks (Figure 3), mutually connected

over asynchronous links. The partitions are very balanced in respect to power and area, as can be seen from Table 2. The most complex block is only 15% larger than the average and the most power hungry block consumes only 11% more than the average value. It can be observed that the complete set of front-end processing (symbol mapping, scrambler, encoder, mapper, pilot insertion etc.) is integrated into a single block; the three pairs of interleavers build up GALS blocks 2 – 4, and finally the 256-point FFT was partitioned into the last two GALS blocks (5-6). We ended up implementing 16 different asynchronous links between the locally synchronous blocks (and consequently 32 asynchronous ports - OPC (output port controllers) and IPC (input port controllers) at figure 3), without any significant performance degradation.

Design Flow Concept

Being a GALS design, the Moonrake chip was implemented using the mixed synchronous-asynchronous design flow. The block diagram of the proposed modified GALS design flow is shown in Figure 4. The main flow additions, compared to the previously used GALS flows, such as proposed

Figure 3. GALS partitioning of the transmitter

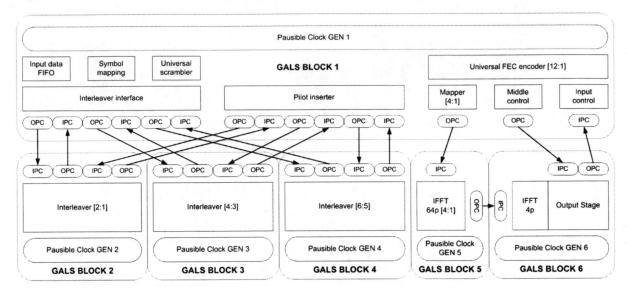

Table 2. Power/area partitioning in Moonrake chip

	GALS Block 1								
	Input controller	Symbol mapping	Scrambl.	Middle control	FEC encoder [12:1]	Output interface	Pilot insertion	Mapping [4:1]	*Total*
Power	0.1%	0.5%	0.0%	7.0%	0.09%	0.1%	3.1%	0.08%	I. 10.97%
Area	0.1%	1.0%	0.0%	12.8%	0.06%	0.1%	5.1%	0.14%	II. 19.3%
	GALS Block 2			GALS Block 3			GALS Block 4		
	Inter-leaver 1	Inter-leaver 2	*Total*	Inter-leaver 3	Inter-leaver 4	III. Total	Inter-leaver 5	Inter-leaver 6	IV. Total
Power	8.7%	8.7%	*17.4%*	8.7%	8.7%	V. 17.4%	8.7%	8.7%	VI. 17.4%
Area	8.9%	8.9%	*17.8%*	8.9%	8.9%	VII. 17.8%	8.9%	8.9%	VIII. 17.8%
	GALS Block 5					GALS Block 6			Post-synth OFDM TX
	FFT_64P 1	FFT_64P 2	FFT_64P 3	FFT_64P 4	IX. Total	FFT_4P	Out Stage	*Total*	
Power	4.9%	4.3%	4.3%	4.3%	X. 17.8%	11.3%	7.2%	*18.5%*	240mW
Area	2.7%	2.4%	2.4%	2.4%	XI. 9.9%	10.3%	6.7%	*17%*	2.2mm²

by Gurkaynak in 2003, are partitioning methodology and co-simulation flow. The asynchronous components, such as port controllers, have been modeled as Signal Transition Graphs (STGs). At the behavioral level it is always an open issue how to effectively co-simulate asynchronous event driven components and synchronous clocked RTL components. HDL languages, such as VHDL, are not the perfect choice for the modeling of asynchronous components since timed simulator cannot visualize correctly events and modeling of asynchronous channels must be done at the very low level. The proposed solution is to use a co-simulation tool which enables heterogeneous system simulation (in this context synchronous-asynchronous). One example of such tool is

AsipIDE, developed by University of Manchester for the purposes of the GALAXY project (Janin in 2010). This XML-based IP packaging format (named ASIP) is mainly compatible to the one proposed SPIRIT consortium. It can be used as a binding level for VHDL/Verilog synchronous design instances and STG asynchronous components. This allows GALS system verification at the early RTL stage, before going into synthesis level, which was a missing step in previous GALS implementations.

The port controllers and pausible clock generators have been and synthesized in hazard-free combinational logic using the tool Petrify (Cortadella in 1997) and separately laid out as hard-macros, in order to fully control the timing. For the implementation, an extended standard cell library should be used including additionally developed asynchronous components - various C-element types and mutual exclusion elements. After the layout, the components should be characterized and on the top level back-annotated including respective delay corresponding to the physical properties of the modules. The top level system implementation is relatively straightforward, starting with the floorplaning of the memories and hard macros. The best approach is to generate top level constraints in such a way that also global asynchronous paths are covered. This can be achieved by the careful definition of minimum and maximum delays of the asynchronous paths and generation of the virtual clocks between the GALS controllers. This approach is usually difficult for pure asynchronous designs due to the number of paths that have to be constrained. However, since in many GALS systems (like the one considered here) the number of asynchronous paths is limited, the constraining could be successfully applied. In this way, the standard tools for timing closure and STA can be used. This was a very important achievement that enables effective usage of the timing optimization, placement and routing capabilities offered by the standard front-end and back-end tools, thus exploiting the relaxation of the global timing introduced by the balanced partitioning.

Figure 4. GALS design flow

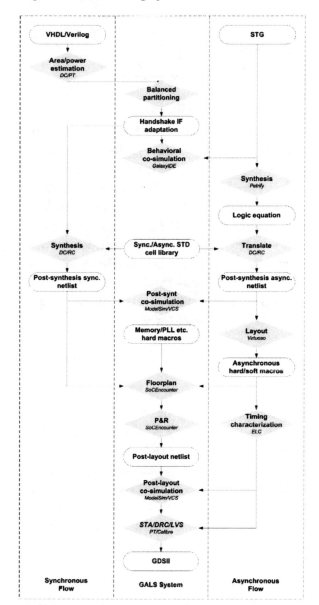

Overhead of GALS Infrastructure

In general, introducing GALS is coupled with adding the local clock generators, port controller interfaces and additional interface data registers. The cost, in terms of area and power for a typical GALS interface is provided in Table 3. This represents usually quite limited overhead to the design. In case of the Moonrake transmitter system, the complete GALS infrastructure for all six different asynchronous wrappers adds ad-

Table 3. Power dissipation and silicon area of GALS interface

	Local Clock Gen		I/O Port Cntr		Input Data Reg	
Power (*mW*)	4.05	1.77%	0.19	0.08%	1.80	0.79%
Area (*μm²*)	30K	1.25%	640	0.03%	5.7K	0.26%

ditionally 2.7% in power and 1.6% in area to the overall GALS TX design. The dominating factor within this overhead are local clock generators. This shows also a benefit of pausible clocking scheme in comparison to the frequently used FIFO approach for GALS. In particular, proposed double-latching synchronization structure, from Figure 2, in Moonrake chip results in 740 extra latches. If FIFO was used as an interface this number of memory elements will enable less than 1 word FIFO between the domains!

Test Structures for GALS NOCs

Other application domains currently call for a larger communication parallelism, which can be efficiently provided by on-chip interconnection networks in a structured and scalable way. In these cases, applying pausible clocking to multi-port components like a NoC switch brings significant performance penalty. In contrast, it is important to engineer synchronization interfaces able to marginally impact NoC quality metrics while at the same time enabling arbitrary composition of timing-heterogeneous sub-systems. Moreover, timing the NoC itself is an additional problem that point-to-point links do not have to address. Therefore, providing efficient interfaces for truly heterogeneous and cost-effective GALS partitioning is currently more critical than the assessment of system-level partitioning schemes. For this reason, the Moonrake chip aims at validating the new synchronization interfaces for synchronizer-based GALS NoCs presented by Strano (2010) and Ludovici (2010). Their features are as follows:

- Besides enabling the traditional loose coupling between NoC building blocks and the synchronizers (i.e., synchronizers are placed in front of the switch/network interface they serve), they enable a tight coupling design style (i.e., synchronizer and switch input buffer are merged together) and other variants for increased timing robustness (e.g., the hybrid coupling, implementing partial merging and link retiming).

- They target a heterogeneous timing landscape, made possible by dual-clock FIFOs, mesochronous synchronizers or a mix thereof. The focus is on synchronizer-based GALS NoCs and not on fully asynchronous interconnect fabrics, since these latter feature a wider gap with respect to traditional synchronous architectures and design tools. In fact, they are typically available only through hard macros as outlined by Thonnart in 2010.

For the sake of validating these interfaces and of their comparison with traditional synchronous ones, the Moonrake chip implements simple test structures. The baseline structure consists of a 2-ary 1-mesh topology with 4 cores (two traffic generators and two memory cores) ensuring bi-directional communication across the simple topology. Each traffic generator is programmable and connected to a JTAG interface for I/O. See Figure 5 for a pictorial overview.

The decision was to replicate sub-systems validating the same concepts in order to reflect different operating speeds. In practice, some sub-systems were selected for clocking from an external low-speed clock source through the JTAG interface, while other sub-systems with similar functionality were selected for higher-speed clocking from the PLL. The lower speed of clocking from an external source is associated not only with the inherent limitations of a possible test setup,

Figure 5. Block diagram of the NoC section of the testchip

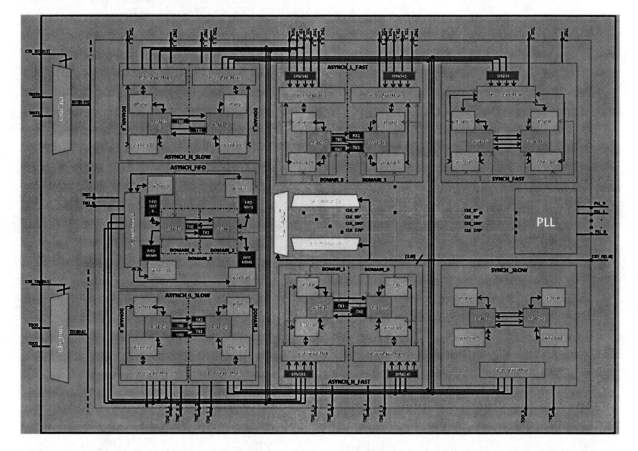

but also with the limited driving capability of I/O pins. In both cases (internal vs external clocking), the sub-systems testing mesochronous interfaces were made capable of tolerating an increasing amount of clock skew between transmitter and receiver clock domains.

Figure 5 shows the block diagram of the NoC section of the testchip. Each sub-system replicates the same baseline network-on-chip template, a 2-ary 1-mesh topology, where every switch of the network is connected to 2 cores (a memory core and a tester block). The cores are connected to JTAG interfaces in order to be programmable by external input pins.

In particular, the testchip compares a fully-synchronous NoC sub-system with two meso-chronous NoC sub-systems (a first one composed by loosely-coupled and a second one by tightly-

coupled mesochronous synchronizers, with some performance optimizations) and a GALS NoC sub-system integrating also dual-clock FIFO interfaces. As anticipated above, in order to have a backup option for the experimental characterization in case of unexpected manufacturing issues for the most complex test structures, every sub-system is designed twice to work at low as well as high frequency. The sub-systems designed to work at low frequencies (i.e. frequency lower than 260MHz) are fed by an external clock injected through the input pins of the JTAG interface. On the contrary, the sub-systems designed to work at high frequency receive the clock by a PLL (Phase Lock Loop) integrated into the testchip itself.

The PLL is able to generate a clock signal with a maximum speed of 400 MHz, together with different phase shifted replicas, hence enabling to test mesochronous synchronization under 0%, 25%, 50% and 75% clock phase offset. For the mixed-timing sub-system, the PLL clocks the network mesochronous domains, while cores and memories are clocked externally with a slow 25 MHz clock. The externally injected clock starts from a baseline speed of 25 MHz.

To demonstrate the feasibility of different composability options of timing assumptions, we implemented four different versions of the baseline sub-system:

- Synchronous reference design, shown in Figure 6. This sub-system is demonstrating a fully synchronous NoC communication. The sub-system is composed by 1 JTAG controller, 2 switches, 2 memory cores and 2 tester blocks and is clocked by a unique external clock.

- Loosely-coupled mesochronous synchronization. This sub-system has two clock domains having the same clock frequency but unknown phase offset (mesochronous synchronization). This platform uses loosely-coupled (i.e., not embedded into the switches) mesochronous synchronizers (RX and TX) placed in the bidirectional link next to the switches. The RX module synchronizes the data while the 1-bit

TX module synchronizes the flow control signal. This is the scheme of Figure 7. Mesochronous synchronization could be a promising option to reduce the complexity and enable feasibility of the top-level clock tree through relaxation of the global skew constraint (Panades 2008).

- Mesochronous synchronization with hybrid coupling. Same as before, but only the 1-bit TX synchronizer is not embedded into the switch, while the most complex RX one is. As proven by Ludovici in 2010, this approach can be used to push a higher operating speed while still preserving the low area/power/latency footprint of tight coupling.

- Mixed timing. In this sub-system, the network is an independent clock domain with respect to traffic injectors and memory cores, hence separated by means of dual-clock FIFOs. In turn, the network is split into two mesochronous sub-domains, implemented with hybrid coupling. See Figure 8. Only one pair of dual-clock FIFOs was tightly coupled (with the NoC switch), while the remaining pair was not, since this would have implied their integration into the network interfaces, which was outside the scope of this paper.

Figure 6. Baseline synchronous test structure for NoCs

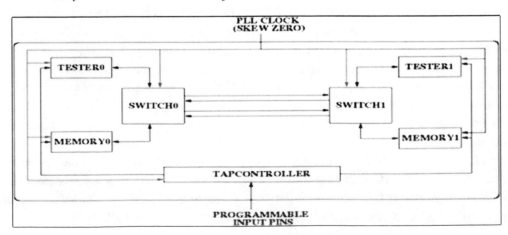

430

Figure 7. Mesochronous synchronization with loose coupling with the NoC

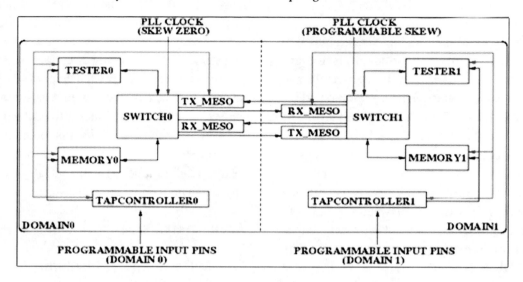

Figure 8. NoC test structure where the NoC is an independent clock domain featuring hybrid coupling

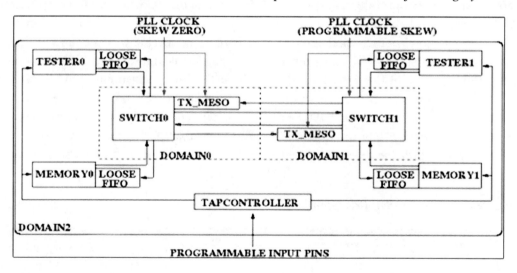

Reference synchronous architecture, as well as Loose and Hybrid variants of the Mesochronous architectures have been implemented in two variants: slow and fast. In the former case, each sub-system has been synthetized at a target frequency of 100 MHz and is driven by an external clock source. On the other hand, the fast versions have been designed to run at a maximum frequency of 400 MHz, and clock signal is provided by the on-chip PLL. In this latter case, an additional block has been implemented in order to force synchronization between the JTAG inputs, feeding configuration data, and the sub-system.

The entire design flow was based on standard cells (i.e., no full-custom designed components). Only during place-and-route we forced rigid fences for placement of external synchronizers, and used native CAD tool support for bundled routing of the source-synchronous links (for delay matching). Link length/switch separation was set to be 0.5 mm. As a result, the entire design flow was within reach of current industrial design tools.

Floorplanning Constraints

The modularity of the testchip NoC architecture allowed an easier place-and-route. In fact, every sub-system was synthesized independently and treated as a soft-macro. Anyway, global and local constraints were imposed respectively in the testchip floorplan and in each sub-system.

At a global floorplan level, as shown in Figure 9, the soft-macros of the fast sub-systems were constrained to be placed as close as possible to the PLL. In fact, the distance between the PLL and the fast designs was minimized in order to reduce the length of the clock tree branches. This strategy mitigated the unpredictability of the skew between the clock leaves and, as a result, it increased the accuracy of the sub-system intra-domain skew set by the top-level clock distribution pins.

Additionally, at a sub-system level, the source synchronous communication was constrained as well. In the hybrid-coupled sub-system, the two 1-bit TX mesochronous synchronizers were placed close to their downstream switches. That was required to ensure that an additional link delay does not nullify the expected timing margins of the TX synchronizer's output (i.e., of the flow control signal). Moreover, the links were constrained in order to have the same delay regardless of their crossing direction. As a result, the skew test experiments can be performed with an additional degree of freedom regardless of the data crossing direction. Finally, the data crossing the source synchronous links was routed together with the strobe signal to match their propagation delay (bundled routing) with industry-available physical synthesis techniques.

Area Analysis

OFDM Baseband Transmitter

The area breakdown after the chip back-end for the OFDM transmitter is provided in Table 4. It is quite noticeable that the area utilized by the synchronous transmitter (SYNC Tx) is larger than the GALS transmitter (GALS Tx) by around 5% including PLL, although the GALS version contains additional asynchronous controllers and six pausible clock generators, and the design is largely dominated by fixed size memory blocks.

Figure 9: Section of the testchip floorplan for the NoC test structures

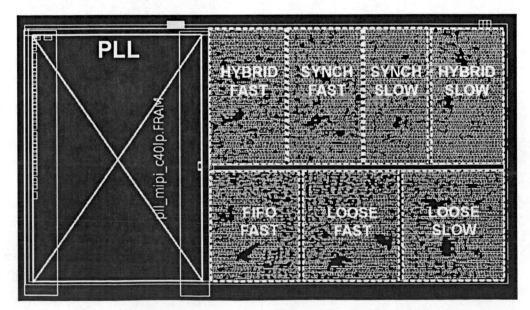

Indeed after the logic synthesis the asynchronous blocks contributed to the slight GALS overhead (1% w/o PLL), even when the synchronous transmitter was extended with scan test logic. However, during the back-end and IPO (in-place optimization), the effort for timing closure on the synchronous side was much more challenging due to the more complex global synchronous domain and chip-wide floorplan. This led to the insertion of additional buffers and to the resizing of existing cells, and resulted, together with the PLL over-head, in increased complexity of the synchronous transmitter (0.7% w/o PLL). Additionally, with the removal of the global clock constraint and with the balanced GALS block complexity, the clock network complexity was reduced, as described in detail in the following text. This also contributed to the better area results of the GALS blocks. The observed area and clock tree complexity reduction on the GALS side was a clear effect of our GALS partitioning methodology, where no block has significantly higher area or power than others. Also, it was an effect of constraining the design

using standard back-end design tools. Combining these two features together, the engineered GALS partitions proved a very good starting point for the back-end design steps.

Network on Chip Synchronization Interfaces

When it comes to the NoC test structures, the area breakdown of the seven sub-systems is illustrated in Figure 10.

Interestingly, the fast solutions present a similar area footprint with respect to the slow counter-parts. In fact, although the fast sub-systems were synthesized at a higher frequency than the slow sub-systems, both the solutions meet the target frequency constraint with a large slack and as a result the synthesis tool was able to provide a well optimized gate-level netlist in terms of area footprint in both the cases.

As far as the slow sub-systems are concerned, evolving the synchronous system to the mesochronous one with the loosely-coupled design style

Table 4. Area distribution in Moonrake chip after back-end

Total [μm²]	OFDM GALS and Synchronous Transmitter				NOC Test Structures [μm²]	Pads [μm²]
	GALS Tx [μm²]	SYNC Tx [μm²]	IO Stage, BIST [μm²]	Total [μm²]		
5406853 (100%)	2220080 (41%)	2334712 (43.2%)	91916 (1.7%)	4643900 (85.9%)	227374 (4.2%)	537075 (9.9%)

Figure 10. Area breakdown for the NoC test structures

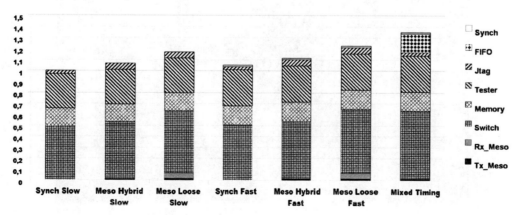

clearly comes at a significant 17% area penalty. However, the newer hybrid-coupled design style enables to contain this penalty within 7% due to the intensive buffer reuse for different purposes (synchronization, performance buffering, flow control). When considering switches in isolation (i.e., excluding testers, memories, etc.), the savings of hybrid over loose coupling amount to 40%.

The synchronous fast sub-system required 4% of additional area than the slow counterpart mainly due to the higher synthesis frequency. The fast mesochronous sub-systems followed the same area trends of their slow counterparts, +7% and +17%, respectively.

Also, dual-clock FIFOs have been demonstrated to be clearly more expensive than mesochronous synchronizers (+30% area occupancy), thus suggesting their mild use and fostering their combination with other synchronizers in real-life platform architectures.

Test Results

The Moonrake chip was taped-out in April 2010, and afterwards fabricated in 40 nm technology run. The layout photo of Moonrake chip is given on Figure 11, showing the floorplaning of GALS and synchronous transmitter, as well as position of NoC test structures. The chip was successfully tested. Main part of testing was devoted to the functional verification of the system including BIST transmitter tests in the synchronous and GALS mode, and test of various NoC modes.

The overall test activity included approximately 32 working days, where 9 days were devoted to test program generation, specification of the test board etc. The rest (23 days) were devoted to chip debugging and various functional tests and measurement. However, in the possible commercial GALS projects the time required for debugging could be dramatically reduced since significant time was spent to getting the experience with GALS testing and thorough measurements relevant only to research projects.

We have performed the power evaluation of GALS and synchronous transmitter in BIST mode for the system running with the same throughput of 1.6 Gbps. Synchronous transmitter has utilized in average 228.8mA plus 1.25mA additional current required for PLL. On the other hand, GALS transmitter required in average 215.1mA including local clock generators. This shows that in Moonrake the power (energy) benefit is up to 6% including PLL consumption. PLL power consumption is in overall picture neglectable, and without taking it into consideration the difference

Figure 11. Layout photo of Moonrake chip

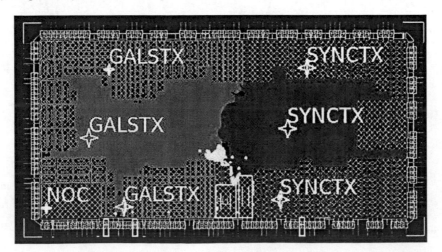

is still 5.9%. This benefit is mainly achieved due to the simplified clock network of GALS system.

In particular one complex clock tree on the synchronous side have been replaced with much smaller local clock trees. The analysis of the clock tree complexity is shown in Table 5. Instead of the single global clock tree distributed over the entire synchronous TX, 6 independent local clock trees were inserted in the GALS design. This analysis shows that in average the number of clock tree levels has been decreased by 3.6 times in average for GALS blocks and that the overall number of clock buffers is 6% lower. As a consequence, the power consumption of the clock networks has been decreased by 24% in case of GALS design. Similarly, due to the reduced number of flip-flop sinks for the local reset signals fulfilling the recovery/removal violations have been simpler ending again in 30% less reset buffers compared to the single reset tree in the synchronous case.

The test performed on the four NoC sub-systems was of special interest to validate the proposed synchronization technology in 40nm. In particular, the goal of the test was to verify the functionalities of a NoC sub-system at a time with different frequencies and skew scenarios. The test was performed under specific input patterns while performing a real time comparison between expected and received data by means of vector sets in EVCD format (i.e. standard event-based simulation traces). The EVCDs were generated in order to stimulate every possible NoC state.

The "functional test" was performed in two steps. In the first step, the slow sub-systems were analyzed ("Mesochronous Hybrid Slow," "Mesochronous Loose Slow" and "Synchronous Slow"). Since the slow sub-systems are clocked from an external low-speed clock, the PLL was no longer required and it was switched-off. The goal of this test was to verify the functionalities of a sub-system at a time in each frequency and skew scenario.

The frequency sweep was performed by increasing gradually the speed of the injected external clocks. The control input signal and the data input signal were injected in compliance with the target operating frequencies. The strobe setting at the output pins was performed similarly.

The selected lower bound for the frequency sweep was 25 MHz. Clearly, the test could also be performed below this frequency but we did not expect relevant results under this threshold. On the contrary, the upper bound of the frequency sweep was not specified a priori since the test was also meant to determine the maximum operating frequency that the slow sub-systems were able to achieve. This maximum speed is associated not only with the inherent limitations of the test setup, but also with the limited driving capability of the I/O pins. To note that the slow sub-systems were synthesized at 100 MHz although the synthesis tool met the frequency constraints with a large slack.

Additionally, a skew sweep was performed for the "Mesochronous Hybrid Slow" and "Mesochronous Loose Slow" at every operating frequency under test. In this case, the clock phase of one mesochronous domain was gradually increased until 100% misalignment of the clock periods was achieved. As a result, a functional test for a given clock phase passes when the NoC source

Table 5. Clock and reset tree complexity

	Clock Trees				Reset Trees	
	Num.	**Level**	**Buffer**	**Power**	**Num.**	**Buffer**
SYNC TX	1	27	1645	42*mW*	1	582
GALS TX	6	≤ 10	1549	32*mW*	6	405

synchronous link is able to absorb the skew. This approach allows to reduce the final effort required to define the test specification and to perform the testing. Anyway, in order to perform a safe test, the skew was not only applied to the external clock of one mesochronous domain, but it was also applied to the control/data input signal and to the output pin strobe of the same domain.

As a result, the "Synchronous Slow" sub-system passed every functional test in the range of 25 MHz and 265 MHz. Interestingly the sub-system was able to work at a frequency significantly higher than the frequency of synthesis. That can be due to two main reasons:

- A significant slack was reported after the 100 MHz synthesis. Then the failure of the sub-system is actually expected when the slack is completely absorbed.

- The designs were synthesized by means of the worst case standard cell library. The worst case library was probably assuming overly pessimistic conditions.

Concerning the synchronizer-based sub-systems, the skew and frequency sweep results of the "Mesochronous Loose Slow" sub-system are pictorially illustrated in Figure 12.

Clearly, the synchronizer-based slow sub-system was able to pass the functional tests in the range of 25 MHz and 265 MHz like the "Synchronous Slow" sub-system. Moreover, it well absorbed the applied skew tolerating up to 100% of the clock period offset. In particular, the "Mesochronous Loose Slow" sub-system was able to tolerate every injected skew up to 265 MHz. Similar test results were obtained for the "Mesochronous Hybrid Slow" design, and are therefore

Figure 12. Frequency and skew sweep in the "Mesochronous Loose Slow" subsystem

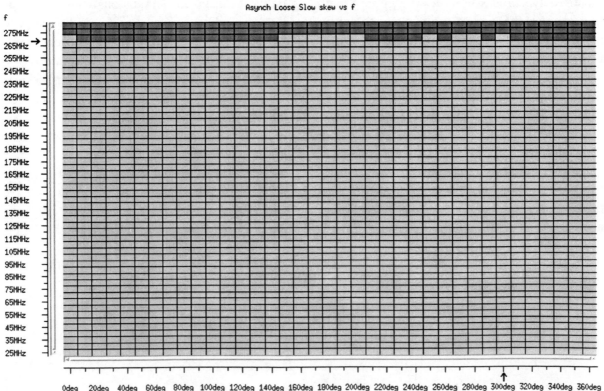

not reported here to save space. Overall, this validates the fundamental functional correctness of the designs, and provides the availability of reference designs upon which the test infrastructure could be calibrated in light of the more aggressive tests that are reported next.

In the second part of the "functional tests," the fast sub-systems were analyzed ("Mesochronous Hybrid Fast," "Mesochronous Loose Fast," "Mesochronous FIFO" and "Synchronous Fast").

Since in this case the sub-systems were clocked by the PLL, the frequency and the skew sweep was performed according to the limitations dictated by the PLL specification. In particular, the PLL provides the following 4 clock phases: $0°$, $90°$, $180°$ and $270°$ (i.e., the highest provided skew corresponds to 75% of the clock cycle). Moreover, the PLL clock frequency is generated according with the following formula:

$$PLL_{freq} = [PLL_{external\_clock}*(n+1)]/2 \qquad (1)$$

To note that the *"n"* parameter can be set through dedicated external pins.

Following the reported formula, a sweep of the $PLL_{external\_clock}$ frequency could directly generate the desired sweep of the PLL clock. Anyway, the PLL has been designed to properly work with an external clock of around 25 MHz. As a consequence, the experiments were performed with a fixed external clock (25 MHz) and the *"n"* parameter was used to modify the PLL frequency. Theoretically, 8 bits are available to set the *"n"* parameter but three of these bits were hard-wired at a low value in order to meet the maximum pin number requirements (i.e., 5 input pins are still available). As a result, the PLL can run at the maximum frequency of 400 MHz (supposing n=00011111 and $PLL_{external\_clock}$=25 MHz).

The selected PLL frequencies were 200 MHz and 400 MHz in the 4 available clock phase variants. The input pins were supposed to inject the data/control signals with a frequency of 25 MHz (going through synchronizers with the on-chip

clock frequency). The sub-systems had been synthesized at the target speed of 500 MHz.

The results for the four NoC structures were as follows:

- The "Synchronous Fast" functional test passed with the PLL frequency set at 200 MHz and 400 MHz with all the chips.
- The "Mesochronous Loose Fast" functional test passed with the PLL frequency set at 200 MHz in each of the 4 skew scenarios for 100% of the chips. The same sub-system operating at a frequency of 400 MHz passed the test with 94.4% of the chips. The sub-system was not able to work at 400 MHz in only 5.6% of chips most probably due to the effect of process variations.
- The "Mesochronous Hybrid Fast" functional test passed with the PLL frequency set at 200 MHz in each of the 4 skew scenarios for 100% of the chips. The same sub-system operating at a frequency of 400 MHz passed the test with 66.6% of the chips. The sub-system was not able to work at 400 MHz in 33.3% of the chips. Anyway, each of these latter chips passed the functional test with a phase offset of $90°$, $180°$ and $270°$, on the contrary, they did not pass the test with $0°$ of skew. The $0°$ of skew represents the fully synchronous scenario and intuitively the source synchronous interfaces should provide the best timing margins in this condition. We identified the problem to be in the reset circuit of the source-synchronous link which introduces an unpredictable misalignment between the PLL clock signal and the reset signal driven externally. Indeed the synchronization of an asynchronous reset in the source-synchronous link represents an ongoing research topic for which a circuit has recently been proposed by Verbitsky in 2011. Robust reset circuitry will be the target of our future work.

- Concerning the "Mixed-Timing" test structure, the sub-system should not just absorb the skew into the source synchronous link but also synchronize data from 25 MHz to 200 MHz and vice versa. For this purpose, at design time the dual-clock FIFOs had been implemented with 2-stage brute force synchronizers. Such a minimum number of cascaded stages was selected to minimize latency and buffering requirements of the FIFO, although we were aware of the resolution time constant degradation concern pointed out by Beer in 2010. We aimed at assessing robustness of such minimum-cost solutions in 40nm. As a result, the "Mixed-Timing" functional test passed with the PLL frequency set at 200 MHz in each of the 4 skew scenarios for 100% of the chips. On the other hand, the same sub-system operating at a frequency of 400 MHz passed the test with 61.1% of the chips. In this case, the output of the failing chips was not stable. In fact, although the sampled output was in most of the cases matching the expected one, the correct output was still not guaranteed when repeating the test several times. Interestingly, the failing tests presented few unstable output bits always located in the same position of

the output vector. On the contrary, when the "Mesochronous Hybrid Fast" and the "Mesochronous Loose Fast" sub-systems were failing the output vectors presented the anomalous bits spread homogeneously all over the output vector. As a consequence, this information led us to suppose that specific patterns bring few bits of the "Mixed" sub-system to a metastability condition that results into unpredictable states. Then we identified the root cause for this effect in the brute force synchronizers consisting only of 2 cascaded flip-flops. Such result is of high practical relevance, since it validates the theoretical concerns from Beer (2010), and advocates for further cascaded flip flops in the brute-force synchronizers at the scaling of the technology node and with the increase of the operating frequencies. This will make dual-clock FIFOs progressively more expensive and call for synchronization solutions making a conscious use of these resources.

Summing up, Figure 13 illustrates on the y-axis the percentage of chips (out of the ones which passed the continuity test) that proves functional in each test case (identified by the operating frequency, skew and architecture variant parameters).

Figure 13. Percentage of working chips in each test case

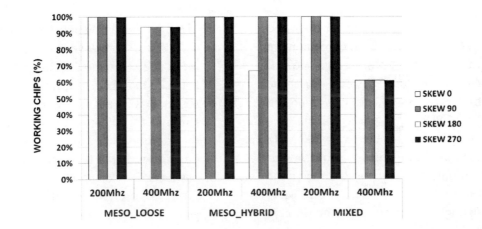

The success percentage is high. We start observing some issues at 400 MHz, as we approach the target synthesis frequency of 500 MHz. However, the failure pattern and its thorough understanding are such that improving upon such results should not be an issue in future designs.

EMI Measurement Results

Furthermore, as already described, one of the main motivations for this chip fabrication was to determine the level of simultaneous switching noise, and consequently EMI reduction using GALS techniques. Therefore, the extensive measurement of EMI have been performed by measuring on on-chip and on-board supply voltage variations in frequency domain. Figure 14a) shows the supply voltage variations measurement setup. Two test points have been used to perform the measurements. VDD_AE22 is a direct pad connection to the core supply used for the peering of supply voltage on chip. This pad is not used as a power

driver, and since is not connected to the rectifying capacitors it is actually used to sense the internal supply voltage on chip. The supply on the board is measured using the test point VDD_BOARD that is connected to the voltage supply. To have comparable results between synchronous and GALS transmitter the same BIST test has been used in both modes and the operating frequencies of both transmitter versions has been set such to achieve the same data throughput. In GALS mode all local clock generators have been set to approximately same frequency, i.e. the system was running practically in plesiochronous mode. On-chip voltage variations have been shown in frequency domain in Fig 14b). One can observe there that the first harmonic of the main clock frequency for synchronous system confronted to aggregated spectrum of first harmonics of GALS module clocks has shown the reduction by 26 dB, second by 16 dB, and third by 30 dB in favor of GALS. This result correspond to some extent to the expectations based on the simulation results

Figure 14. EMI measurement setup (a) and results for on-chip measurement (b)

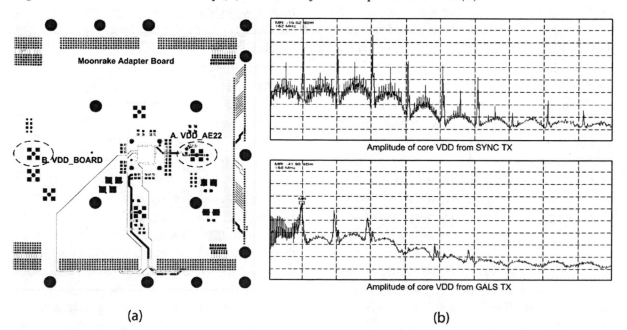

(a)   (b)

that have shown the potential for EMI reduction of 16-20 dB without taking into account clock tree simplification that was achieved as described in previous section. However, the result is probably too optimistic due to the fact that the pad used for voltage sensing is very near to the part of the die occupied by the synchronous transmitter and far away from GALS transmitter. Therefore, the additional on-board measurements are performed showing slightly reduced 19 dB EMI reduction in GALS mode for the first harmonic of the clock frequency. This shows that our utilized GALS methodology successfully reduces the noise caused by simultaneous switching in clocked digital systems, which is extremely important for several application fields such as security, automotive, high performance computing, and mixed-signal design in general.

FUTURE OF GALS

The example of Moonrake chip clearly shows that the application of GALS technology is possible using the standard CAD tools and advanced CMOS processes. It has been also demonstrated that the various benefits can be achieved including improved system integration, increased modularity of the IPs, but also area, power, and EMI improvement.

If we analyze the future trends in respect to GALS use this will certainly give different results depending on the application field and type of the GALS technique.

In this chapter the application of pausible clocking has been illustrated. It has been shown that use of this technique could lead to very good power, EMI and area results in point-to-point GALS systems. However, there are also the limiting factors which disable pervasive use of this method. The main limitations are the clock generators themselves, which have limited resolution, high jitter, and, due to pausing and arbitration, lead to non-deterministic performances. Moreover, for more complex interconnect topologies (such as bus) or for large number of ports in a single locally synchronous block, significant performance loss can be obtained. As a consequence, the main application fields of this technology can be seen in the areas where such features are not the limiting factor but it can also have positive influence. In particular, this can be interesting for mixed-signal designs where EMI and substrate noise play significant role. Moreover security applications, such as crypto-processors, could also profit from non-determinism in timing imposed by pausible clocking. For complex system interconnects it is difficult to imagine the strong application of pausible clocking due to the QoS requirements and expected deterministic clocking for many IPs.

However, for such applications, particular GALS-like methods have been already established, such as source synchronous clocking. This alternative for transferring data across clock domains was originally proposed for off-chip interconnects. In this approach, the source clock signal is sent along with the data to the destination. At the destination, the source clock is used to sample and write the input data into a FIFO queue, while the destination clock is used to read the data from the queue for processing. This method achieves high efficiency by obtaining an ideal throughput of one data word per source clock cycle with a very simple design that is also similar to the synchronous design methodology; hence it is easily compatible with common standard cell design flows. When it comes to the application of this approach to on-chip interconnection networks, the timing constraints associated with source-synchronous clocking should nonetheless be handled with care, especially as an effect of the co-design with the flow control protocol. In this chapter, novel design methods for source-synchronous communication interfaces targeting NoCs have been validated, devising special solutions for fast timing convergence (the hybrid coupling style with the NoC architecture). Overall, the achieved results enable a high level of optimism on the future use of mesochronous and/or dual-clock synchronization interfaces in real hardware platforms, to such an extent that it becomes possible to project system-level GALS architectures building on top of them. GALS system architectures will become increasingly mainstream for a number of reasons. Basically, since each core operates in its own frequency domain, it becomes possible to reduce the power dissipation, increase energy efficiency and compensate for some circuit variations on a fine-grained level. More in detail:

- GALS clocking design eliminates the need for complex and power hungry global clock trees.
- Unused cores can be effectively disconnected by power gating, and thus reducing leakage.
- When workloads distributed for cores are not identical, we can allocate different clock frequencies and supply voltages for these cores either statically or dynamically. This allows the total system to consume a lower power than if all active cores had been operating at a single frequency and supply voltage.
- More power can be saved by architecture-driven methods such as parallelizing or pipelining a serial algorithm over multiple cores.
- Computationally intensive workloads can also be spread around the chip to eliminate hot spots and balance temperature.
- GALS design flexibility supports remapping or adjusting frequencies of processors in an application that allows it to continue working well even under the impact of variations.

From these advantages in both performance and power consumption, many-core GALS style is highly desirable for designing programmable/reconfigurable DSP computational platforms. Clearly, a design methodology is required to partition a given NoC architecture into multiple voltage and frequency islands (VFIs), and to assign the supply and threshold voltages (hence the corresponding clock frequencies) to each domain such that the total energy consumption is minimized under given performance constraints. There are a couple of relevant options for GALS partitioning in this respect.

1. The former option consists of subdividing the system into disjoint physical domains, where each domain can be operated at its own independent voltage and frequency. Network components are in this case interspersed into the different domains as well, therefore the network results from the combination of baseline switching elements, each belonging to a distinct VFI. There is no specific physical domain for the network as such, therefore synchronization elements need to be inserted when connecting two network components with each other. The tested dual-clock FIFOs in the mixed-timing scenario can be used for this purpose. While the construction methodology of this architecture template is very modular, the resulting architecture has a bottleneck in the switching elements belonging to domains that at a certain point in time end up running at a slow speed. Since many communication flows will be crossing such elements at runtime, they will be all slowed down, and the network performance rapidly aligns to that of its slowest building block. To the limit, when power management techniques are implemented, a physical domain can be shutdown or placed in sleep mode, while cores in the other domains can be operational. This may require a rerouting action of the running and future communication flows crossing the VFI that needs to be shutdown while they are progressing on their way to destination. Such rerouting actions should be performed in a deadlock-free fashion, and be synchronized in time

with the power management command. Other techniques can be implemented to work around the same issue. For instance, in the context of application-specific irregular NoCs, the problem could be solved during topology synthesis, by instantiating backup VFIs that are never shutdown and that preserve connectivity under all possible power management decisions. For regular NoCs in use in general-purpose systems, the routing reconfigurability option at runtime today seems the best candidate to become mainstream.

2. The latter option is to have an always-on island containing the network-on-chip as a whole, so to preserve connectivity under all usage and management scenarios. This way, the NoC would be loosely coupled with the cores' VFIs, and each core/cluster of cores could be shutdown without any impact on global network connectivity, since the associated switch would be always on. From a synchronization viewpoint, the network would then be implemented as a large synchronous clock domain distributed throughout the entire chip. The feasibility of this solution mainly depends on physical implementation cost considerations. In fact, if the entire NoC lies in the same island, it is difficult to route the power and ground lines for the NoC across the entire chip. Another issue concerns the feasibility of the global clock tree under tight skew constraints. There is today little doubt on the fact that successful NoC design on nanoscale technologies will only be feasible under relaxed synchronization assumptions. Even though inferring a global clock tree will still be feasible for some time, it will come at a significant power cost. The growing concern associated with the reverse scaling of on-chip interconnects, and the increasingly important role of process variations will do the rest. As a workaround, the network could be inferred as a collec-

tion of mesochronous domains, instead of a global synchronous domain, yet retaining a globally synchronous perspective of the network itself. There are several methods to do that. One simple way is to go through a hierarchical clock tree synthesis process. In practice, a local clock tree is synthesized for each mesochronous domain, where the skew constraint is enforced to be as tight as in traditional synchronous designs. Then, a top-level clock tree is synthesized, connecting the leaf trees with the centralized clock source, with a very loose skew constraint. This way, many repeaters and buffers, which are used to keep signals in phase, can be removed, reducing power and thermal dissipation for the top-level clock tree. The granularity of a mesochronous domain can be as fine as a NoC switch block. The communication between neighboring switches is then mesochronous as the clock tree is not equilibrated, while the communications between switch and IP cores are fully asynchronous because they belong to different clock domains. Bisynchronous FIFOs are therefore used to connect the network switches to the network interfaces of the cores. Again, the silicon-validated mixed-timing sub-system represents a key enabler for this paradigm.

Looking forward, another option is potentially appealing for the synchronization in a multi-synchronous environment. It consists of purely asynchronous clockless handshaking, that uses multiple phases (normally two or four phases) of exchanging control signals (request and ack) for transferring data words across clock domains. Unfortunately, these asynchronous handshaking techniques typically use unconventional circuits (such as the Muller C-element) typically unavailable in generic standard cell libraries. Besides that, because the arrival times of events are arbitrary, without a reference timing signal, their activities are difficult to verify in traditional digital CAD design flows. The so-called delay-insensitive interconnection method extends clockless handshaking techniques by using coding techniques such as dual-rail or 1-of-4 to avoid the requirement of delay matching between data bits and control signals. These circuits also require specific cells that do not exist in common ASIC design libraries. For these reasons, asynchronous NoC components are still typically delivered as rigid hard macrocells.

In principle, the advantages of asynchronous interconnect design are unmistakable: average-case instead of worst-case performance, no switching power of a clock tree, easier convergence of hierarchical design flows, robustness to process/voltage/temperature variations, and efficient delivery of differentiated per-link performance. As a consequence, whether or not asynchronous NoCs become mainstream will mainly depend on whether the CAD tool challenge will be decisively addressed or not, and this will also depend on economical motivations. So far, in the best case scripting frameworks have been implemented on top of mainstream tools for synchronous design, nonetheless coming up only with "quasi-"soft macros. For instance, many logic optimizations of mainstream logic synthesis tools need to be disabled not to affect functionality of the circuit. Also, the typical RTL-level design entry is currently infeasible for the asynchronous NoC design as a whole, but limited to specific components thereof. In addition to this, there is another issue that has considerably slowed down the widespread adoption of asynchronous NoCs. The vast majority of early NoC prototypes made use of four-phase return-to-zero protocols, involving two complete round-trip channel communications per transaction, as well as delay-insensitive data encoding. These design choices have typically resulted in an overly large area and energy-per-bit overhead with respect to their synchronous counterparts, to the extent that total energy savings were reported on real applications only as an effect of the lower idle power and of the relatively low network utilization of the application itself. More recently,

the above consideration has raised the interest in single-rail bundled data asynchronous protocols, which in principle provide designs with a lower timing robustness while significantly reducing area, wire-per-link and energy-per-bit overhead. At the same time, two-phase communication protocols are gaining momentum as a preferred match for high-performance asynchronous systems. Bundled-data designs use a synchronous style data channel augmented with an extra request wire, and a single transition on the request accompanying the data bundle indicates the data is valid. The request signal has a simple one-sided timing constraint that its delay be always slightly greater than the data channel. In practice, with a bundled data approach, circuit timing must be even more carefully specified and controlled to ensure correct operation in the CAD toolflow.

Overall, while application of asynchronous design in industry conservatively goes through quasi-delay insensitive design styles, a feasibility of bundled data ones subject to the efficient specification of relative timing constraints with mainstream CAD tools would certainly extend the applicability of asynchronous interconnect technology to resource-constrained platform architectures in the embedded computing domain. In order to make this happen, tunable delay lines are currently envisioned in order to compensate for technology parameter variations during the manufacturing process.

CONCLUSION

The Moonrake chip is the first GALS demonstrator in a state-of-the-art 40 nm CMOS technology. The design includes various components for evaluating point-to-point and NoC GALS architectures in an architecture- and technology-homogeneous experimental setting. The chip is functionally tested and verified with success. For point-to-point links, the Moonrake chip has provided a pilot partitioning experience with the

GALS methodology, relying on standard CAD tools, highlighting an industry-relevant co-design example between the system layer and the link layer. Also, counterintuitive area benefits of the GALS methodology have been, to our knowledge, for the first time demonstrated, due to the lower burden for timing closure. Finally, our EMI and power measurements have shown the significant benefits of the GALS technique in respect to power and EMI in particular for highly complex digital systems requiring moderate performances.

On-chip networking calls for different considerations, hence the Moonrake chip devoted several test structures to it. The outcome is the validation of the timing robustness of a newer synchronization technology merging synchronizers with NoC building blocks. Especially the NoC structures were tested at both 200 and 400 MHz while at the same time varying the clock phase offset from 0 to 270 degrees. As a result, 91% of the performed tests completed successfully. Interestingly we identified the failures to be associated with reset synchronization, with the degradation of the resolution time constant of brute-force synchronizers and, to a smaller extent, with process variations. Therefore we conclude that the new NoC synchronization schemes are mature enough for exploitation in 40nm technology and that a refinement of design techniques for the reset circuit and the brute-force synchronizers can bridge the missing gaps. Above all, the capability of the newer technology to contain area penalty for the upgrade of synchronous networks into GALS ones has been quantified. Also, it has been proved that this technology is within-reach of industrial technology libraries without any custom extensions and that the design flow can be implemented by means of mainstream industrial tools.

REFERENCES

Beer, S., Ginosar, R., Priel, M., Dobkin, R., & Kolodny, A. (2010). The Devolution of Synchronizers. In *Proceedings of IEEE Symposium on Asynchronous Circuits and Systems* (pp. 94-103). IEEE.

Beigne, E., Clermidy, F., Durupt, J., Lhermet, H., Miermont, S., & Thonnart, Y. ... Vivet, P. (2008). An asynchronous power aware and adaptive NoC based circuit. In *Proceedings of IEEE Symposium on VLSI Circuits* (pp. 190-191). VLSI.

Cortadella, J., Kishinevsky, M., Kondratyev, A., Lavagno, L., & Yakovlev, A. (1997). Petrify: A tool for manipulating concurrent specifications and synthesis of asynchronous controllers. *IEICE Transactions on Information and Systems*, 315–325.

Fan, X., Krstić, M., & Grass, E. (2009). Analysis and Optimization of Pausible Clocking based GALS Design. In *Proceedings of International Conference on Computer Design*, (pp. 358-365). ICCD.

Fan, X., Krstic, M., Wolf, C., & Grass, E. (2010). A GALS FFT Processor with Clock Modulation for Low-EMI Applications. In *Proceedings of 21st IEEE International Conference on Application-specific Systems Architectures and Processors* (ASAP) (pp. 273 - 278). ASAP.

Gurkaynak, F., Oetiker, S., Villiger, T., Felber, N., Kaeslin, H., & Fichtner, W. (2003). On the GALS design methodology of ETH Zurich. In *Proceedings of 1st Intl. Workshop Formal Methods for Globally Asynchronous Locally Synchronous Architecture* (FMGALS). FMGALS.

Janin, L., Li, S., & Edwards, D. (2010). Integrated design environment for reconfigurable HPC. In *Proceedings of 6th international conference on Reconfigurable Computing: architectures, Tools and Applications* (ARC'10) (pp. 406-413). ARC.

Krstic, M., Fan, X., Grass, E., Benini, L., Kakoee, M. R., & Heer, C. et al. (2012). Evaluation of GALS Methods in Scaled CMOS Technology: Moonrake Chip Experience. [IJERTCS]. *International Journal of Embedded and Real-Time Communication Systems, 3*(4), 1–18. doi:10.4018/jertcs.2012100101

Krstic, M., Grass, E., Gürkaynak, F., & Vivet, P. (2007). Globally Asynchronous, Locally Synchronous Circuits: Overview and Outlook. *IEEE Design & Test of Computers, 24*(5), 430–441. doi:10.1109/MDT.2007.164

Krstic, M., Piz, M., Ehrig, M., & Grass, E. (2008). OFDM Datapath Baseband Processor for 1 Gbps Datarate. In *Proceedings of IFIP/IEEE VLSI-SoC Conference 2008* (pp. 156-159). IEEE.

Lemaire, R., & Thonnart, Y. (2010). *Magali, A Reconfigurable Digital Baseband for 4G Telecom Applications based on an Asynchronous NoC.* Paper Presented at ACM/IEEE International Symposium on Networks-on-Chip. New York, NY.

Lines, A. (2004). Asynchronous interconnect for synchronous SoC design. *IEEE Micro, 24*(1), 32–41. doi:10.1109/MM.2004.1268991

Ludovici, D., Strano, A., Gaydadjiev, G. N., Benini, L., & Bertozzi, D. (2010). Design Space Exploration of a Mesochronous Link for Cost-Effective and Flexible GALS NOCs. In *Proceedings of Design, Automation & Test in Europe Conference & Exhibition* (DATE) (pp. 679-684). DATE.

Painkras, E., Plana, L., Garside, J., Temple, S., Davidson, S., & Pepper, J. ... Furber, S. (2012). SpiNNaker: A Multi-Core System-on-Chip for Massively-Parallel Neural Net Simulation. In *Proceedings of IEEE Custom Integrated Circuits Conference* (CICC) (pp. 1-4). CICC.

Panades, I. M., Clermidy, F., Vivet, P., & Greiner, A. (2008). Physical Implementation of the DSPIN Network-on-Chip in the FAUST Architecture. In *Proceedings of International Symposium on Networks-on-Chip* (NoCs) (pp.139-148). Academic Press.

Plana, L., Furber, S., Temple, S., Khan, M., Yebin, S., Jian, W., & Shufan, Y. (2007). A GALS Infrastructure for a Massively Parallel Multiprocessor. *IEEE Design & Test of Computers, 24*(5), 454–463. doi:10.1109/MDT.2007.149

Strano, A., Ludovici, D., & Bertozzi, D. (2010). A Library of Dual-Clock FIFOs for Cost-Effective and Flexible MPSoCs Design. In *proceedings of International Conference on Embedded Computer Systems* (SAMOS) (pp. 20-27). SAMOS.

Thonnart, Y., Vivet, P., & Clermidy, F. (2010). A Fully Asynchronous Low-Power Framework for GALS NoC Integration. In Proceedings of Design, Automation & Test in Europe Conference & Exhibition (DATE) (pp. 33-38). DATE.

Verbitsky, D., Dobkin, R., & Ginosar, R. (2011). A Four-Stage Mesochronous Synchronizer with Back-Pressure and Buffering for Short and Long Range Communications. In Technical Report.

Vivet, P., Lattard, D., Clermidy, F., Beigne, E., & Bernard, C. ... Varreau, D. (2006). FAUST, an Asynchronous Network-on-Chip based Architecture for Telecom Applications. In Proceedings of Design, Automation & Test in Europe Conference & Exhibition (DATE). DATE.

ADDITIONAL READING

Badaroglu, M., Tiri, K., Van. der Plas, G., Wambacq, P., Verbauwhede, I., & Donnay, S. et al. (2006). Clock Skew Optimization Methodology for Subsrate Noise Reduction with Supply Current Folding. *IEEE Transactions on Computer-Aided Design of Integrated Circuits and Systems.* doi:10.1109/TCAD.2005.855952

Cortadella, J., Kondratyev, A., Lavagno, L., & Sotiriou, C. (2006). Desynchronization: Synthesis of Asynchronous Circuits From Synchronous Specifications, IEEE Transactions on Computer-Aided Design of Integrated Circuits and Systems, Vol. 25, No. 10 (pp. 2006)

Fan, X., Schrape, O., Marinkovic, M., Dahnert, P., Krstic, M., & Grass, E. (2013). GALS Design for Spectral Peak Attenuation of Switching Current. *ASYNC, 2013*, 83–90.

Gurkaynak, F. (2006). *GALS System Design: Side Channel Attack Secure Cryptographic Accelerators* (Vol. 168). Series in Microelectronics.

Hoppner, S., Eisenreich, H., Henker, S., & Walter, D. (2013). A Compact Clock Generator for Heterogeneous GALS MPSoCs in 65-nm CMOS Technology, IEEE Transactions on Very Large Scale Integration (VLSI). *Systems, 21*(Issue: 3), 566–570.

Horak, M., Nowick, S., Carlberg, M., & Vishkin, U. (2011). A Low-Overhead Asynchronous Interconnection Network for GALS Chip Multiprocessors. *IEEE Transactions on Computer-Aided Design of Integrated Circuits and Systems, Volume, 30*(Issue: 4), 494–507. doi:10.1109/TCAD.2011.2114970

Najvirt, R., Naqvi, S. R., & Steininger, A. (2013). *Classifying Virtual Channel Access Control Schemes for Asynchronous NoCs, Asynchronous Circuits and Systems*. ASYNC.

Nowick, S. M., & Singh, M. (2011). High-Performance Asynchronous Pipelines: An Overview. *Design & Test of Computers, IEEE, 28*(5), 8–22. doi:10.1109/MDT.2011.71

Quinton, B. R., Greenstreet, M. R., & Wilton, S. J. E. (2008). Practical Asynchronous Interconnect Network Design, Very Large Scale Integration (VLSI) Systems. *IEEE Transactions on, 16*(5), 579–588.

Rabaey, J & Chandrakasan, A & Nikolic, B (2002). Digital integrated circuits, Prentice hall Englewood Cliffs

Sheibanyrad, A., Greiner, A., & Miro-Panades, I. (2008). Multisynchronous and Fully Asynchronous NoCs for GALS Architectures. *Design & Test of Computers, IEEE, 25*(6), 572–580. doi:10.1109/MDT.2008.167

Soares, R., Calazans, N., Moraes, F., Maurine, P., & Torres, L. (2011). A Robust Architectural Approach for Cryptographic Algorithms Using GALS Pipelines. *Design & Test of Computers, 28*, 62–71. doi:10.1109/MDT.2011.69

Sparsø, J., & Furber, S. (Eds.). (2001). *Principles of Asynchronous Circuit Design – A Systems Perspective*. Kluwer Academic Publishers. doi:10.1007/978-1-4757-3385-3

Thonnart, Y., Beigne, E., & Vivet, P. (2012). *A Pseudo-Synchronous Implementation Flow for WCHB QDI Asynchronous Circuits, Asynchronous Circuits and Systems* (pp. 73–80). ASYNC.

Thonnart, Y., Vivet, P., & Clermidy, F. (2010). *A fully-asynchronous low-power framework for GALS NoC integration, Design, Automation & Test in Europe Conference & Exhibition* (pp. 33–38). DATE.

Tran, X. T., Thonnart, Y., Durupt, J., Beroulle, V., & Robach, C. (2009). Design-for-test approach of an asynchronous network-on-chip architecture and its associated test pattern generation and application. *Computers & Digital Techniques, IET, 3*(5), 487–500. doi:10.1049/iet-cdt.2008.0072

Wang, Y., & Ignjatovic, Z. (2007) On-Chip Substrate Noise Suppression Using Clock Randomization Methodology, Proceedings of IEEE International Symposium on Circuits and Systems (pp.2176,2179)

Yakovlev, A., Vivet, P., & Renaudin, M. (2013). *Advances in asynchronous logic: From principles to GALS & NoC, recent industry applications, and commercial CAD tools, Design, Automation & Test in Europe Conference & Exhibition* (pp. 1715–1724). DATE.

Zhiyi, Y., & Baas, B. (2010) A Low-Area Multi-Link Interconnect Architecture for GALS Chip Multiprocessors, IEEE Transactions on Very Large Scale Integration (VLSI) Systems, vol. 18, Issue: 5 (pp. 750-762)Bertozzi D., Flich J. (2011), Designing Network On-Chip Architectures in the Nanoscale Era, CRC Press.

KEY TERMS AND DEFINITIONS

Asynchronous: The way of circuit operation without using the clock signal for control of sequential components but some sort of handshaking between the pipelines.

EMI: Electro-Magnetic Interference, interference affecting the operation of electrical components.

GALS: Globally Asynchronous Locally Synchronous Methodology for chip system level integration where the subblocks of the system operate in synchronous fashion but mutually communicate asynchronously.

Mesochronous: Clocking scheme for digital integrated systems. In a mesochronous system, communicating nodes exhibit the same clock frequency but unknown phase offset, which makes it mandatory to implement a skew absorbing mechanism (i.e., a mesochronous synchronizer) inbetween.

Mixed Timing: Synchronization scenario in modern digital integrated systems. Integrated circuits become small scale implementations of distributed systems, in that they contain multiple sub-regions, each operating at different speeds, potentially changing over time.

Moonrake Chip: Chip designed in EU Project GALAXY used for evaluation of GALS technology .

NOC: Stands for Network-on-Chip. NoC technology applies networking theory and methods to on-chip communication and brings notable improvements over conventional bus and crossbar interconnections. The NoC interconnect fabric is modular and consists of replicated switches (routing packets to destination) and interfaces for network access.

Partitioning: Process of separation of systems into the different GALS domains.

Pausible Clocking: One type of GALS methodology based on clock stretching during the data transfer between the GALS blocks.

Synchronization: Process of aligning the data transfer from two modules using the different clock sources.

Compilation of References

3GPP. (2001). *Security Threats and Requirements (Release 4), Technical Specification Group Services and System Aspects.*

3GPP. (2013). *3GPP Release 8.* Retrieved December 06, 2013, from http://www.3gpp.org/specifications/releases/72-release-8

Abadi, M., Budiu, M., Erlingsson, U., & Ligatti, J. (2009). Control-flow integrity principles, implementations, and applications.[TISSEC].*ACM Transactions on Information and System Security, 13*(1). doi:10.1145/1609956.1609960

Abrial, J.-R. (2009). *Event Model Decomposition.* Academic Press.

Abrial, J.-R. (2010). *Modelling in Event-B: System and Software Engineering.* Cambrige, UK: Cambrige University Press. doi:10.1017/CBO9781139195881

Accellera. (2006). Retrieved January 19, 2012, from Core SystemC Language 2.2: http://www.accellera.org/downloads/standards/systemc/

Accellera. (2009). *SystemC Transaction-Level Modeling Library 2.0.1.* Retrieved January 19, 2012, from http://www.accellera.org/downloads/standards/systemc/

Accellera. (2010). *SystemC Analog Mixed Signal Extensions 1.0.* Retrieved January 19, 2012, from http://www.accellera.org/downloads/standards/systemc/

Actel. (2003). *Reliability Considerations for Automotive FPGAs.* Retrieved Dec. 19, 2013 from http://www.actel.com/documents/AutoWP.pdf

Addo-Quaye, C. (2005). Thermal-aware mapping and placement for 3-D NoC designs. In *Proceedings of IEEE International SOC Conference,* (pp. 25-28). IEEE.

Agarwal, M., & Paul, B. C. Zhang, & Mitra, S. (2007). Circuit Failure Prediction and Its Application to Transistor Aging. In *Proceedings of 25th IEEE VLSI Test Symposium* (pp. 277-286). IEEE.

Agrawal, D., Archambeault, B., Rao, J. R., & Rohatgi, P. (2003). The EM Side—Channel(s). *Lecture Notes in Computer Science, 2523,* 29–45. doi:10.1007/3-540-36400-5_4

Akos, D. (2012). *GNSS RFI/Spoofing: Detection, Localization, & Mitigation.* Paper presented at Stanford's 2012 PNT Challenges and Opportunities Symposium. Stanford, CA.

Akyildiz, I., Weilian, S., Sankarasubramaniam, Y., & Cayirci, E. (2002, August). A survey on sensor networks. *Communications Magazine, IEEE, 40*(8), 102–114. doi:10.1109/MCOM.2002.1024422

Al Faruque, M. A., Krist, R., & Henkel, J. (2008). ADAM: Run-time Agent-based Distributed Application Mapping for on-chip Communication. In *Proceedings of Design Automation Conference* (pp. 760-765). doi:10.1145/1391469.1391664

Al Faruque, M., Krist, R., & Henkel, J. (2008). ADAM: Run-time agent-based distributed application mapping for on-chip communication.In *Proceedings of Design Automation Conference* (pp. 760 - 765). Anaheim, CA: IEEE.

Alam, N., Kealy, A., & Dempster, A. G. (2013). Cooperative Inertial Navigation for GNSS-Challenged Vehicular Environments. *IEEE Transactions on Intelligent Transportation Systems, 14*(3), 1370–1379. doi:10.1109/TITS.2013.2261063

Ali, G., & Zafar, N. A. (2010). Modelling Agent-Based Systems Using X-Machine and Z Notation. In *Proceedings of International Communication Software and Networks* (pp. 249–253). Singapore: IEEE. doi:10.1109/ICCSN.2010.76

Alliance, Z. (2010, December). *ZigBee Specification.* Retrieved December 16, 2010, from http://www.zigbee.org/Standards/ZigBeeSmartEnergy/Specification.aspx

Altera. (2008). Retrieved from http://www.altera.com/literature/wp/wp-01055-fpga-run-time-reconfiguration.pdf

Alves-Foss, J., Harrison, W. S., Oman, P., & Taylor, C. (2006). The MILS Architecture for High Assurance Embedded Systems. *International Journal of Embedded Systems, 2*(3-4), 239–247. doi:10.1504/IJES.2006.014859

Amorim, A., & Galdino, J. F. (2010). I2TS02-Secrecy Rate of Adaptive Modulation Techniques in Flat-Fading Channels. *Latin America Transactions, IEEE, 8*(4), 332-339.

Anastasi, G., Bandelloni, R., Conti, M., Delmastro, F., Gregori, E., & Mainetto, G. (2003). Experimenting an Indoor Bluetooth-Based Positioning Service. In *Proceedings of the 23rd International Conference on Distributed Computing Systems Workshops* (pp. 480-483). Providence, RI: IEEE.

Anderson, J. B., & Mohan, S. (1984). Sequential coding algorithms: A survey and cost analysis. *IEEE Transactions on Communications, 32*(2), 169–176. doi:10.1109/TCOM.1984.1096023

Andres, C., Molinero, C., & Nuez, M. (2008). A formal methodology to specify hierarchical agent-based systems. In *Proceedings of Signal Image Technology and Internet Based Systems* (pp. 169–176). Bali: IEEE. doi:10.1109/SITIS.2008.70

Arnold, O., Noethen, B., & Fettweis, G. (2012). Instruction Set Architecture Extensions for a Dynamic Task Scheduling Unit. In *Proceedings of IEEE Computer Society Annual Symposium on VLSI* (pp. 249-254). IEEE. doi: 10.1109/ISVLSI.2012.51

Arockia, D., & Anbumani, P. (2013). GRGPS – Geographic Routing based on Greedy Perimeter Stateless Position for Multi- Hop Mobile Ad-hoc Networks. *International Journal of Emerging Trends & Technology in Computer Science, 2*(2).

Asenov, A., Kaya, S., & Davies, J. H. (2002). Intrinsic threshold voltage fluctuations in decanano MOSFETs due to local oxide thickness variations. *IEEE Transactions on Electron Devices, 49*(1), 112–119. doi:10.1109/16.974757

Avizienis, A., Laprie, J.-C., Randell, B., & Landwehr, C. (2004). Basic Concepts and Taxonomy of Dependable and Secure Computing. *IEEE Transactions on Dependable and Secure Computing, 1*(1), 11–33. doi:10.1109/TDSC.2004.2

Bahl, P., & Padmanabhan, V. N. (2000). Radar: An In-Building RF Based User Location and Tracking System. In *Proceedings of Infocom—Nineteenth Annual Joint Conference of the IEEE Computer and Communications Societies* (pp. 775–784). Tel-Aviv, Israel: IEEE.

Balaei, A., Motella, B., & Dempster, A. (2008). A Preventive Approach to Mitigating CW Interference in GPS Receivers. *GPS Solutions, 12*(3), 199–209. doi:10.1007/s10291-007-0082-8

Balandin, S., Gillet, M., Lavrovskaya, I., Olenev, V., Rabin, A., & Stepanov, A. (2011). Co-Modeling of Embedded Networks Using SystemC and SDL. *International Journal of Embedded and Real-Time Communication Systems, 2*(3), 24–49.

Baldini, G., & Hofher, J. (2008). *IPSC Projects based on Satellite Navigation Systems. E. C. Institute for the Protection and Security of the Citizens, 1st MENTORE event.* Italy: Ispra.

Ball, T., Bounimova, E., Cook, B., Levin, V., Lichtenberg, J., & McGarvey, C. … Ustuner, A. (2006). Thorough Static Analysis of Device Drivers. In Proceedings of EuroSys'06. Academic Press.

Ballister, P. J., Robert, M., & Reed, J. H. (2006). Impact of the use of CORBA for Inter-Component Communication in SCA Based Radio. In *Proceedings of the Software-Defined Radio Technical Conference.* Orlando, FL: Academic Press.

Bandara, U., Hasegawa, M., Inoue, M., Morikawa, H., & Aoyama, T. (2004). Design and Implementation of a Bluetooth Signal Strength Based Location Sensing System. In *Proceedings of IEEE Radio and Wireless Conference* (pp. 319 – 322). Atlanta, GA: IEEE.

Barbero, L., & Thompson, J. (2006). A fixed-complexity MIMO detector based on the complex sphere decoder. In *Proceedings of the IEEE International Workshop on Signal Processing Advances for Wireless Communications* (pp. 1- 5). New York, NY: IEEE.

Barbero, L. (2006). *Rapid prototyping of a fixed-complexity sphere decoder and its application to iterative decoding of turbo-MIMO systems*. Edinburgh, UK: University of Edinburgh.

Bard, J., & Kovarik, V. J. Jr. (2007). *Software-Defined Radio – The Software Commmunications Architecture*. John Wiley and Sons. doi:10.1002/9780470865200

Bargh, M., & Groote, R. (2008). Indoor Localization Based on Response Rate of Bluetooth Inquiries. in *Proceedings of the first ACM international workshop on Mobile entity localization and tracking in GPS-less environments* (pp. 49-54). San Francisco: ACM.

Barrenechea, M., Barbero, L., Jiménez, I., Arruti, E., & Mendicute, M. (2011). High-throughput implementation of tree-search algorithms for vector precoding. In *Proceedings of the IEEE International Conference on Acoustics, Speech and Signal Processing* (pp. 1689-1692). New York, NY: IEEE.

Barrenechea, M., Burg, A., & Mendicute, M. (2012). Low complexity vector precoding for multiuser systems. In *Proceedings of the Asilomar Conference on Signals, Systems and Computers* (pp. 453-457). New York, NY: IEEE.

Barrenechea, M., Mendicute, M., Del Ser, J., & Thompson, J. S. (2009). Wiener filter- based fixed-complexity vector precoding for the MIMO downlink channel. In *Proceedings of the IEEE Signal Processing Advances in Wireless Communications* (pp. 216–220). New York, NY: IEEE.

Barrenechea, M. (2012). *Design and implementation of multi-user MIMO precoding algorithms*. Mondragon, Spain: University of Mondragon.

Barrenechea, M., Barbero, L., Mendicute, M., & Thompson, J. (2012). Design and implementation of a low-complexity multiuser vector precoder. *International Journal of Embedded and Real-Time Communication Systems*, 3(1), 1–18. doi:10.4018/jertcs.2012010102

Barrenechea, M., Mendicute, M., & Arruti, E. (2013). Fully-pipelined Implementation of Tree-search Algorithms for Vector Precoding. *International Journal of Reconfigurable Computing*. doi:10.1155/2013/496013

Barr, M., & Massa, A. (2006). *Programming Embedded Systems: With C and GNU Development Tools* (2nd ed.). O'Reilly Media, Inc.

Barry, R. (2010). *Using the FreeRTOS™ Real Time Kernel*. Real Time Engineers Ltd. Retrieved from www.freertos.org

Barry, R. (2011). *The FreeRTOS™ Reference Manual*. Version 1.2.0. Real Time Engineers Ltd. Retrieved from www.freertos.org

Bar-Shalom, Y., Li, R. X., & Kirubarajan, T. (2001). *Estimation with Applications to Tracking and Navigation, Theory Algorithms and Software*. John Wiley & Sons. doi:10.1002/0471221279

Basili, V. R., Caldiera, G., & Rombach, H. D. (1994). The Goal Question Metric Approach. In *Encyclopedia of Software Engineering*. Wiley.

Bastide, F., Akos, D., Macabiau, C., & Roturier, B. (2003). Automatic Gain Control (AGC) as an Interference Assessment Tool. In *Proceedings of the 16th International Technical Meeting of the Satellite Division of The Institute of Navigation (ION GPS/GNSS 2003)* (pp. 2042-2053). Portland, OR: ION GPS/GNSS.

Bastide, F., Chartre, E., Macabiau, C., & Roturier, B. (2004). GPS L5 And GALILEO E5a/E5b Signal-to-Noise Density Ratio Degradation Due to DME/TACAN Signals: Simulations and Theoretical Derivation. In *Proceedings of the 2004 National Technical Meeting of The Institute of Navigation* (pp. 1049-1062). San Diego, CA: Academic Press.

Bauer, T., Eschbach, R., Groessl, M., Hussain, T., Streitferdt, D., & Kantz, F. (2009). Combining Combinatorial and Model-Based Test Approaches for Highly Configurable Safety-Critical Systems. In *Proceedings of the 2ⁿᵈ Workshop on Model-based Testing in Practice at the 5ᵗʰ European Conference on Model-Driven Architecture Foundations and Applications*. Academic Press.

Bauer, T., Stallbaum, H., Metzger, A., & Eschbach, R. (2008). Risikobasierte Ableitung und Priorisierung von Testfällen für den modellbasierten Systemtest (in German). In *Proceedings of SE'08 – Software Engineering Konferenz*. München, Germany: SE.

Bauer, T., Böhr, F., Landmann, D., Beletski, T., Eschbach, R., & Poore, J. H. (2007). From Requirements to Statistical Testing of Embedded Systems. In *Proceedings of Software Engineering for Automotive Systems - SEAS 2007, ICSE Workshops*. Minneapolis, MN: SEAS. doi:10.1109/SEAS.2007.5

Becker, C., Staamann, S., & Salomon, R. (2007). Security Analysis of the Utilization of Corba Object References as Authorization Tokens. In *Proceedings of the 10th IEEE International Symposium on Object and Component-Oriented Real-Time Computing* (ISORC'07). IEEE.

Beckwith, R. W., Vanfleet, W. M., & MacLaren, L. (2004). High Assurance Security/Safety for Deeply Embedded, Real-time Systems. In *Proceedings of the Embedded Systems Conference 2004*. Academic Press.

Beer, S., Ginosar, R., Priel, M., Dobkin, R., & Kolodny, A. (2010). The Devolution of Synchronizers. In *Proceedings of IEEE Symposium on Asynchronous Circuits and Systems* (pp. 94-103). IEEE.

Beigne, E., Clermidy, F., Durupt, J., Lhermet, H., Miermont, S., & Thonnart, Y. … Vivet, P. (2008). An asynchronous power aware and adaptive NoC based circuit. In *Proceedings of IEEE Symposium on VLSI Circuits* (pp. 190-191). VLSI.

Benini, L., & De Micheli, G. (2002). Networks on chips: A new SoC paradigm. *Computer*, 70–78. doi:10.1109/2.976921

Berghel, H., & Uecker, J. (2005). WiFi Attack Vectors. *Communications of the ACM*, *48*(8), 21–28. doi:10.1145/1076211.1076229

Bernstein, J. B., Gurfinkel, M., Li, X., Walters, J., Shapira, Y., & Talmor, M. (2006). Electronic circuit reliability modeling. *Microelectronics and Reliability*, *46*(12), 1957–1979. doi:10.1016/j.microrel.2005.12.004

Bhagawat, P., Wang, P., Uppal, M., Choi, G., Xiong, Z., Yeary, M., & Harris, A. (2008). An FPGA implementation of dirty paper precoder. In *Proceedings of the IEEE International Conference on Communication*. (pp. 2761–2766). New York, NY: IEEE.

Bhattacharya, B., & Bhattacharyya, S. S. (2001). Parameterized Dataflow Modeling for DSP Systems. *IEEE Transactions on Signal Processing*, *49*, 2408–2421. doi:10.1109/78.950795

Bilsen, G., Engels, M., Lauwereins, R., & Peperstraete, J. (1996). Cyclo-static dataflow. *IEEE Transactions on Signal Processing*, *44*(2), 397–408. doi:10.1109/78.485935

Biscondi, E., Flanagan, T., Fruth, F., Lin, Z., & Moerman, F. (2012). *Maximizing Multicore Efficiency with Navigator Runtime* (White Paper). Retrieved December 8, 2013, from www.ti.com/lit/wp/spry190/spry190.pdf

Black, D., & Donovan, J. (2004). [*From the Ground Up*. New-York: Springer Science+Buisness Media, Inc.]. *System*, C.

Blome, J., Gupta, S., Feng, S., Mahlke, S., & Bradley, D. (2006). Cost-efficient Soft Error Protection for Embedded Microprocessors. In *Proceedings of the 2006 International Conference on Compilers, Architecture and Synthesis for Embedded System*, (pp. 421-431). IEEE.

Bluegiga. (2010). *Specification of the Bluegiga System, WT41-A/WT41-N Preliminary Data Sheet, v. 7*. Retrieved 7 October 2010 from http://www.bluegiga.com

Bluetooth. (2010). *Specification of the Bluetooth System, Core Specification v2.0 + EDR, Bluetooth SIG*. Retrieved 18 August 2010 from http://www.bluetooth.org/

Boesen, M. R., Keymeulen, D., Madsen, J., Lu, T., & Chao, T.-H. (2011). Integration of the reconfigurable self-healing eDNA architecture in an embedded system. In *Proceedings of 2011 IEEE Aerospace Conference*, (pp. 1-11). IEEE.

Boesen, M. R., Madsen, J., & Keymeulen, D. (2011). Autonomous distributed self-organizing and self-healing hardware architecture - The eDNA concept. In *Proceedings of Aerospace Conference*. IEEE.

Bogdanov, A. (2008). Multiple-Differential Side-Channel Collision Attacks on AES. In *Proceedings of the Workshop on Cryptographic Hardware and Embedded Systems 2008* (CHES 2008). CHES.

Bonesi, S., Bertozzi, D., Benini, L., & Macii, E. (2008). Process Variation Tolerant Pipeline Design Through a Placement-Aware Multiple Voltage Island Design Style. In Design, Automation and Test in Europe, (pp. 967-972). Academic Press.

Bonneau, J., & Mironov, I. (2006). Cache-Collision Timing Attacks Against AES. In *Proceedings of the Workshop on Cryptographic Hardware and Embedded Systems 2006* (CHES 2006). CHES.

Borio, D., O'Driscoll, C., & Fortuny, J. (2012). GNSS Jammers: Effects and Countermeasures. In *Proceedings of the 6th ESA Workshop on Satellite Navigation Technologies and European Workshop on GNSS Signals and Signal Processing (NAVITEC 2012)* (pp. 1-7). Noordwijk, The Netherlands: NAVITEC.

Borio, D., Camoriano, L., & Lo Presti, L. (2008). Two-Pole and Multi-Pole Notch Filters: A Computationally Effective Solution for GNSS Interference Detection and Mitigation. *IEEE Systems Journal, 2*, 38–47. doi:10.1109/JSYST.2007.914780

Borio, D., Lo Presti, L., Savasta, S., & Camoriano, L. (2008). Time-frequency Excision for GNSS Application. *IEEE Systems Journal, 2*, 27–37. doi:10.1109/JSYST.2007.914914

Borkar, S., Karnik, T., & Vivek, D. (2004). Design and reliability challenges in nanometer technologies. In *Proceedings of the 41st Design Automation Conference.* Academic Press.

Bose, P., Morin, P., Stojmenovic, I., & Urrutia, J. (1999). *Routing with Guaranteed Delivery in Ad-Hoc Wireless Networks.* Paper presented at the 3rd International Workshop on Discrete Algorithms and Methods for Mobile Computing and Communications. New York, NY.

Bottou, L. (2010). Large-Scale Machine Learning with Stochastic Gradient Descent. In *Proceedings of the 19th International Conference on Computational Statistics* (COMPSTAT'2010), (pp. 177–187). Paris, France: Springer.

Bottou, L., & Le Cun, Y. (2004). Large scale online learning. *Advances in Neural Information Processing Systems, 16.*

Boutellier, J., Silven, O., & Raulet, M. (2011). Scheduling of CAL actor networks based on dynamic code analysis. In *Proceedings of the IEEE International Conference on Acoustics, Speech, and Signal Processing (ICASSP).* IEEE.

Boutellier, J., Lucarz, C., Lafond, S., Gomez, V., & Mattavelli, M. (2009). Quasi-static scheduling of CAL actor networks for Reconfigurable Video Coding. *Journal of Signal Processing Systems for Signal, Image, and Video Technology,* 1–12.

Boutellier, J., Raulet, M., & Silven, O. (2013). Automatic Hierarchical Discovery of Quasi-Static Schedules of RVC-CAL Dataflow Programs. *Journal of Signal Processing Systems for Signal, Image, and Video Technology, 71*(1), 35–40. doi:10.1007/s11265-012-0676-4

Bowman, K. A., Duvall, S. G., & Meindl, J. D. (2002). Impact of die-to-die and within-die parameter fluctuations on the maximum clock frequency distribution for gigascale integration. *IEEE Journal of Solid-State Circuits, 37*(2), 183–190. doi:10.1109/4.982424

Brawerman, A., Blough, D., & Bing, B. (2004). Securing the download of radio configuration files for software-defined radio devices. In *Proceedings of the ACM International Workshop on Mobility Management and Wireless Access.* ACM.

Brown, T. X., & Sethi, A. (2007). Potential Cognitive Radio Denial of Service Attacks and Remedies. In *Proceedings of the international symposium on advanced radio technologies 2007* (ISART 2007). ISART.

Brown, S., & Sreenan, C. J. (2013). Software Updating in Wireless Sensor Networks: A Survey adn Lacunae. *Journal of Sensor and Actuator Networks, 2*, 717–760. doi:10.3390/jsan2040717

Buck, J. (1993). *Scheduling dynamic dataflow graphs with bounded memory using the token flow model.* (PhD thesis). University of California, Berkeley, CA.

Buck, J., & Lee, E. (1993). Scheduling dynamic data-flow graphs with bounded memory using the token flow model. In *Proceedings of IEEE International Conference on Acoustics, Speech, and Signal Processing*, (pp. 429-432). IEEE.

Budinger, T. (2003). Biomonitoring with wireless communications. *Annual Review of Biomedical Engineering, 5*(1), 383–412. doi:10.1146/annurev.bioeng.5.040202.121653 PMID:14527317

Bunnell, R., & Trinidad, J. (2004). *The Challenge in Developing an SCA Compliant Security Architecture that Meets Government Security Certification Requirements.* Retrieved from http://www.omg.org/news/meetings/workshops/SBC_2004_Manual/06-2_Trinidad_etal_revised.pdf

Burg, A., Seethaler, D., & Matz, G. (2007). VLSI implementation of a lattice-reduction algorithm for multi-antenna broadcast precoding. In *Proceedings of the IEEE International Symposium on Circuits and Systems* (pp. 673 – 676). New York, NY: IEEE.

Burg, A., Wenk, M., Zellweger, M., Wegmueller, M., Felber, N., & Fichtner, W. (2004). VLSI implementation of the sphere decoding algorithm. In *Proceedings of the European Solid-State Circuits Conference* (pp. 303-306). New York, NY: IEEE.

Butler, M. (2009). Decomposition Structures for Event-B. In *Proceedings of International Conference on Integrated Formal Methods* (pp. 20 - 38). Berlin: Springer-Verlag.

Bykhteev, A. (2008). Methods and facilities for systems-on-chip design. *ChipInfo microchip manual.* Retrieved from http://www.chipinfo.ru/literature/chipnews/200304/1.html

Caceres, M. A., Penna, F., Wymeersch, H., & Garello, R. (2010). *Hybrid GNSS terrestrial cooperative positioning via distributed belief propagation.* Paper presented at IEEE Global Communication Conference. Miami, FL.

Caffery, J., & Stuber, G. (1995). Radio location in urban CDMA microcells. In *Proceedings of Sixth IEEE International Symposium on Personal, Indoor and Mobile Radio Communications, Wireless: Merging onto the Information Superhighway* (pp. 858–862). Toronto, Canada: IEEE.

Caffery, J. J., & Stuber, G. L. (1998). Overview of radiolocation in CDMA cellular systems. *IEEE Communications Magazine, 36*(4), 38–45. doi:10.1109/35.667411

Carlsson, A., Eker, J., Olsson, T., & Von Platen, C. (2010). Scalable parallelism using dataflow programming. *Ericsson Review, 2*(1), 16–21.

Castilhos, G., Mandelli, M., Madalozzo, G., & Moraes, F. (2013). Distributed Resource Management in NoC-based MPSoCs with Dynamic Cluster Sizes. In *Proceedings of IEEE Computer Society Annual Symposium on VLSI* (pp. 153-158). IEEE. doi: 10.1109/ISVLSI.2013.6654651

Castrillon, J., Zhang, D., Kempf, T., Vanthournout, B., Leupers, R., & Ascheid, G. (2009). Task Management in MPSoCs: An ASIP Approach. In *Proceedings of International Conference on Computer-Aided Design* (pp. 587-594). New York: ACM. doi: 10.1145/1687399.1687508

Cedersjo, G., & Janneck, J. W. (2012). Toward Efficient Execution of Dataflow Actors. In Proceedings of Signals, Systems and Computers (ASILOMAR) (pp. 1465 – 1469). ASILOMAR.

Cerpa, A., Wong, J. L., Kuang, L., Miodrag, P., & Estrin, D. (2005). *Statistical Model of Lossy Links in Wireless Sensor Networks.* Paper presented at the 4th International Symposium on Information Processing in Sensor Networks. Los Angeles, CA.

Chandra, V., & Aitken, R. (2008). Impact of technology and voltage scaling on the soft error susceptibility in nanoscale CMOS. In *Proceedings of IEEE International Symposium on Defect and Fault Tolerance of VLSI Systems* (pp. 114-122). IEEE.

Changlin, C., Ye, L., & Cotofana, S. D. (2012). A Novel Flit Serialization Strategy to Utilize Partially Faulty Links in Networks-on-Chip. In *Proceedings of Sixth IEEE/ACM International Symposium on Networks on Chip* (NoCS), (pp. 124-131). IEEE/ACM. doi: 10.1109/NOCS.2012.22

Chen, B. B., Hao, S., Zhang, M., Chan, M. C., & Ananda, A. I. (2009). *DEAL: Discover and Exploit Asymmetric Links in Dense Wireless Sensor Networks.* Paper presented at the 6th Annual IEEE Communications Society Conference on Sensor, Mesh and Ad Hoc Communications and Networks (SECON). Rome, Italy.

Chen, Y., Chen, R., Pei, L., Kröger, T., Chen, W., Kuusniemi, H., & Liu, J. (2010). Knowledge-based Error Detection and Correction Method of a Multi-sensor Multi-network Positioning Platform for Pedestrian Indoor Navigation. In *Proceedings of IEEE/ION PLANS 2010* (pp. 873-879). Indian Wells, CA: IEEE.

Chen, L., Pei, L., Kuusniemi, H., Chen, Y., Kröger, T., & Chen, R. (2013). Bayesian Fusion for Indoor Positioning Using Bluetooth Fingerprints. *Wireless Personal Communications*, *70*(4), 1735–1745. doi:10.1007/s11277-012-0777-1

Chen, L., Yang, L., & Chen, R. (2012). Time Delay Tracking for Positioning in DTV Networks. In *Proceedings of Ubiquitous Positioning, Indoor Navigation, and Location Based Service (UPINLBS)* (pp. 1–4). Helsinki: IEEE. doi:10.1109/UPINLBS.2012.6409784

Chen, Q., Kanhere, S., & Hassan, M. (2013). Adaptive Position Update for Geographic Routing in Mobile Ad Hoc Networks. *IEEE Transactions on Mobile Computing*, *12*(3).

Chevallier-Mames, B., Ciet, M., & Joye, M. (2004). Low-Cost Solutions for Preventing Simple Side-Channel Analysis: Side-Channel Atomicity. *IEEE Transactions on Computers*, *53*(6), 760–768. doi:10.1109/TC.2004.13

Chippa, V. K., Mohapatra, D., Raghunathan, A., Roy, K., & Chakradhar, S. T. (2010). Scalable effort hardware design: Exploiting algorithmic resilience for energy efficiency. In *Proceedings of 47th ACM/IEEE Design Automation Conference (DAC)*, (pp. 555-560). ACM/IEEE.

Chou, C., & Marculescu, R. (2008). User-Aware Dynamic Task Allocation in Networks-on-Chip. In *Proceedings of Design* (pp. 1232–1237). Automation and Test in Europe.

Chu, C.-T., Kim, S. K., Lin, Y.-A., Yu, Y., Bradski, G., Ng, A. Y., & Olukotun, K. (2007). Map-reduce for machine learning on multicore. *Advances in Neural Information Processing Systems*, *19*, 281–288.

Ciordas, C., Goossens, K., Radulescu, A., & Basten, T. (2006). NoC Monitoring: Impact on the Design Flow. *IEEE International Symposium on Circuits and Systems* (pp. 1981-1984). IEEE.

Ciordas, C., Basten, T., Rădulescu, A., Goossens, K., & Meerbergen, J. V. (2005). An event-based monitoring service for networks on chip. [ACM Press.]. *ACM Transactions on Design Automation of Electronic Systems*, *10*(4), 702–723. doi:10.1145/1109118.1109126

Clabes, J., Friedrich, J., Sweet, M., DiLullo, J., Chu, S., & Plass, D. ... Dodson, S. (2004). Design and implementation of the POWER5 microprocessor. In *Proceedings of the 41st annual Design Automation Conference*, (pp. 670--672). IEEE.

Cohen, M. B., Dwyer, M. B., & Shi, J. (2006). Coverage and adequacy in software product line testing. In *Proceedings of the ISSTA 2006 workshop on Role of software architecture for testing and analysis*. ISSTA.

Collet, J. H., Zajac, P., Psarakis, M., & Gizopoulos, D. (2011). Chip Self-Organization and Fault Tolerance in Massively Defective Multicore Arrays. *IEEE Transactions on Dependable and Secure Computing*, *8*(2), 207–217. doi:10.1109/TDSC.2009.53

Collins, H.A., & Nikel, R.E. (1999). *DDR-SDRAM, high-speed, source synchronous interfaces create design challenges*. EDN article, September 1999.

Common Criteria Evaluation and Validation Scheme Validation Report. (2008). *Green Hills Software INTEGRITY-178B Separation Kernel*. Retrieved from http://www.commoncriteriaportal.org/files/epfiles/st_vid10119-vr.pdf

Constantinescu, C. (2003). Trends and Challenges in VLSI Circuit Reliability. *IEEE Micro*, *23*(4), 14–19. doi:10.1109/MM.2003.1225959

Cornelius, C. (2011). Design of complex integrated systems based on networks-on-chip - Trading off performance, power and reliability. *Dissertation (in press)*.

Corporaal, H. (1997). *Microprocessor Architectures: from VLIW to TTA*. John Wiley & Sons.

Corporation, I. B. M. (2013). *CoreConnect Bus Architecture*. Retrieved December 15, 2013, from https://www-01.ibm.com/chips/techlib/techlib.nsf/productfamilies/CoreConnect_Bus_Architecture

Correia, M. P., & Sousa, P. J. (2010). *Segurança no Software*. FCA Editora de Informática.

Cortadella, J., Kishinevsky, M., Kondratyev, A., Lavagno, L., & Yakovlev, A. (1997). Petrify: A tool for manipulating concurrent specifications and synthesis of asynchronous controllers. *IEICE Transactions on Information and Systems*, 315–325.

Coskun, A. K., Rosing, T. S., & Gross, K. C. (2008). Proactive temperature balancing for low cost thermal management in MPSoCs. *2008 IEEEACM International Conference on ComputerAided Design* (pp. 250-257). IEEE.

Coskun, A. K., Rosing, T. S., Whisnant, K. A., & Gross, K. C. (2008). Temperature-Aware MPSoC Scheduling for Reducing Hot Spots and Gradients. *Proc. of 13th Asia and South Pacific Design Automation Conference (ASP-DAC 2008)* (pp. 49-54). IEEE Computer Society Press.

Costa, M. H. M. (1983). Writing on dirty paper. *IEEE Transactions on Information Theory*, 29(3), 439–441. doi:10.1109/TIT.1983.1056659

Cousot, P., & Cousot, R. (1977). Abstract interpretation: a unified lattice model for static analysis of programs by construction or approximation of fixpoints. In *Proceedings of the 4th ACM Sigact-Sigplan Symposium on Principles of Programming Languages* (pp. 238–252). ACM.

Couto, D., Aguayo, D., Bicket, J., & Morris, R. (2003). *A High-Throughput Path Metric for Multi-Hop Wireless Routing*. Paper presented at the 9th Annual International Conference on Mobile Computing and Networking. San Diego, CA.

Cowan, C., Pu, C., Maier, D., Walpole, J., Bakke, P., & Beattie, S. … Hinton, H. (1998). StackGuard: Automatic adaptive detection and prevention of buffer-overflow attacks. In *Proceedings of the 7th USENIX Security Symposium*. San Antonio, TX: USENIX.

Crossbow Technologies, I. (2003, March). *Mote In-Network Programming User Reference*. Retrieved 11 13, 2009, from http://www.tinyos.net/tinyos-1.x/doc/Xnp.pdf

Dally, W. J., & Towles, B. (2001). Route packets, not wires: on-chip inteconnection networks. In *Proceedings of the 38th conference on Design automation*, (pp. 684–689). Academic Press.

Dally, W. J., & Poulton, J. W. (1998). *Digital Systems Engineering*. Cambridge University Press. doi:10.1017/CBO9781139166980

Damen, M. O., El Gamal, H., & Caire, G. (2003). On maximum-likelihood detection and the search for the closest lattice point. *IEEE Transactions on Information Theory*, 49(10), 2389–2402. doi:10.1109/TIT.2003.817444

Das, K., & Wymeersch, H. (2010). Censored cooperative positioning for dense wireless networks. In *Proceedings of IEEE 21st International Symposium on Personal, Indoor and Mobile Radio Communications Workshops* (pp. 262-266). Instanbul: IEEE.

Das, S., Tokunaga, C., Pant, S., Ma, W.-H., Kalaiselvan, S., & Lai, K. … Blaauw, D.T. (2009). Razor II: In Situ Error Detection and Correction for PVT and SER Tolerance. JSSC, 44(1), 32-48.

Davidson, J. A. (2008). On the Architecture of Secure Software-Defined Radios. In *Proceedings of the IEEE Military Communications Conference* (MILCOM 2008). IEEE.

Davies, J., & Woodcock, J. (1996). *Using Z: Specification, Refinement and Proof*. Prentice Hall International Series in Computer Science.

De Angelis, M., Fantacci, R., Menci, S., & Rinaldi, C. (2005). Analysis of Air Traffic Control Systems Interference Impact On Galileo Aeronautics Receivers. In *Proceedings of the 2005 National Technical Meeting of The Institute of Navigation* (pp. 346-357). San Diego, CA: Academic Press.

De Angelis, G., Baruffa, G., & Cacopardi, S. (2013). GNSS/Cellular Hybrid Positioning System for Mobile Users in Urban Scenarios. *IEEE Transactions on Intelligent Transportation Systems*, 14(1), 313–321. doi:10.1109/TITS.2012.2215855

de Souza Carvalho, E. L., Calazans, N. L. V., & Moraes, F. G. (2010). Dynamic Task Mapping for MPSoCs. *Design & Test of Computers*, 27(5), 26–35. doi:10.1109/MDT.2010.106

Del Re, E. (2011). Integrated satellite-terrestrial NAV/COM/GMES system for emergency scenarios. In *Proceedings of the 2nd International Conference on Wireless Communication, Vehicular Technology, Information Theory and Aerospace & Electronic Systems Technology* (pp. 1-5). Chennai: IEEE.

Deng, J., Han, R., & Mishra, S. (2006). Secure code distribution in dynamically programmable wireless sensor networks. In *Proceedings of Information Processing in Sensor Networks* (pp. 292–300). IPSN. doi:10.1145/1127777.1127822

Denks, H., Steingaß, A., Hornbostel, A., & Chopard, V. (2009). GNSS Receiver Testing by Hardware Simulation with Measured Interference Data from Flight Trials. In *Proceedings of the 22nd International Technical Meeting of The Satellite Division of the Institute of Navigation (ION GNSS 2009)* (pp. 1-10). Savanna, GA: ION GNSS.

Dennis, J. B. (1974). First Version of a Data Flow Procedure Language. In *Proceedings of Programming Symposium*, (pp. 362 – 376). Springer.

Dhillon, D. (2011, July/August). Developer-Driven Threat Modeling: Lessons Learned in the Trenches. *IEEE Security and Privacy*.

Diamos, G., Yalamanchili, S., & Duato, J. (2007). STARS: A system for tuning and actively reconfiguring SoCs link. In *Proceedings of DAC Workshop on Diagnostic services in NoCs*. DAC.

Dick, R. P., Rhodes, D. L., & Wolf, W. (1998). TGFF: Task Graphs For Free.

Dietrich, F. A. (2008). *Robust Signal Processing for Wireless Communications*. Berlin, Germany: Springer.

Dietterle, D. (2009). *Efficient Protocol Design Flow for Embedded Systems*. Brandenburg University of Technology. Retrieved from http://systems.ihp-microelectronics.com/uploads/downloads/diss_dietterle.pdf

Diggelen, F. (2009). *A-GPS: Assisted GPS, GNSS, and SBAS*. Norwood, MA: Artech House.

Divis, D. A. (2013). GPS Spoofing Experiment Knocks Ship off Course – University of Texas at Austin team repeats spoofing demonstration with a superyacht. *InsideGNSS news*. Retrieved December 5, 2013, from http://www.insidegnss.com/node/3659.

Do, T.-N., Nguyen, V.-H., & Poulet, F. (2008). Speed up SVM algorithm for massive classification tasks. In *Proceedings of the 4th International Conference on Advanced Data Mining and Applications* (ADMA 2008) (LNCS), (vol. 5139, pp. 147–157). Springer.

Dong, W., Mo, B., Huang, C., Yunhao, L., & Chen, C. (2013). R3: Optimizing relocatable code for efficient reprogramming in networked embedded systems. In *Proceedings of INFOCOM, 2013 Proceedings IEEE* (pp. 315-319). IEEE.

Dong, W., Liu, Y., Chen, C., Bu, J., Huang, C., & Zhao, Z. (2013). R2: Incremental Reprogramming Using Relocatable Code in Networked Embedded Systems. *IEEE Transactions on Computers*, 62(9), 1837–1849. doi:10.1109/TC.2012.161

Dovis, F., & Musumeci, L. (2011). Use of Wavelet Transform for Interference Mitigation. In *Proceedings of the 2011 International Conference on Localization and GNSS (ICL-GNSS)* (pp. 116-121). Tampere, Finland: GNSS.

Dovis, F., Chen, X., Cavaleri, A., Khurram, A., & Pini, M. (2011). Detection of Spoofing Threats by Means of Signal Parameters Estimation. In *Proceedings of the 24th International Technical Meeting of The Satellite Division of the Institute of Navigation (ION GNSS 2011)* (pp. 416-421). Portland, OR: GNSS.

Dovis, F., Musumeci, L., & Samson, J. (2012). Performance Comparison of Transformed Domain Techniques for Pulsed Interference Mitigation. In *Proceedings of the 25th International Technical Meeting of the Satellite Division of The Institute of Navigation (ION GNSS 2012)* (pp. 3530-3541). Nashville, TN: GNSS.

Dunkels, A., Finne, N., Eriksson, J., & Voigt, T. (2006). Run-Time Dynamic Linking for Reprogramming Wireless sensor Networks. In *Proceedings of the Fourth ACM Conference on Embedded Networked Sensor Systems (SenSys 2006)*. ACM.

Dunkels, A., Gronvall, B., & Voigt, T. (2004). Contiki - a lightweight and flexible operating system for tiny networked sensors. In *Proceedings of Local Computer Networks, 2004: 29th Annual IEEE International Conference on* (pp. 455-462). IEEE.

Duros, E., & Dabbous, W. (1996). *Handling of Unidirectional Links with OSPF*. Retrieved from http://tools.ietf.org/html/draft-ietf-ospf-unidirectional-link-00

Durupt, J., Bertrand, F., Beroulle, V., & Robach, C. (2006). A DFT Architecture for Asynchronous Networks-on-Chip. *Eleventh IEEE European Test Symposium ETS06* (pp. 219-224). IEEE.

Duzellier, S. (2005). Radiation effects on electronic devices in space. *Aerospace Science and Technology, 9*(1), 93–99. doi:10.1016/j.ast.2004.08.006

Eggers, S. J., & Katz, R. H. (1988). *Evaluating the Performance of Four Snooping Cache Coherency Protocols* (Tech. Rep. UCB/CSD-88-478). University of California, Berkeley, EECS Department.

Eker, J., & Janneck, J. (2003). CAL Language Report. *Technical Report ERL Technical Memo UCB/ERL M03/48*. University of California at Berkeley.

Elrabaa, M. E. S. (2006). An all-digital clock frequency capturing circuitry for NRZ data communications. In *Proceedings of IEEE Int. Conf. on Electronics, Circuits and Systems*, (pp. 106-109). IEEE.

ELVEES R&D Center. (2005). *SpaceWire switch MCK-01*. Retrieved from http://multicore.ru/index.php?id=850

Enge, P., & Misra, P. (2006). *Global Positioning System: Signal Measurements and Performance*. Ganga-Jamuna Press.

Erlandson, R. J., Kim, T., Hegarty, C., & Van Dierendonck, A. J. (2004). Pulsed RFI Effects on Aviation Operations Using GPS L5. In *Proceedings of the 2004 National Technical Meeting of The Institute of Navigation* (pp. 1063–1076). San Diego, CA: Academic Press.

Ernst, D. Kim, Das, S., Pant, S., Rao, R., Pham, … Mudge, T. (2003). Razor: A low-power pipeline based on circuit-level timing speculation. In *Proceedings of 36th Annual IEEE/ACM International Symposium on Microarchitecture*, (pp. 7-18). IEEE/ACM.

Ersfolk, J., Roquier, G., Jokhio, F., Lilius, J., & Mattavelli, M. (2011). Scheduling of dynamic dataflow programs with model checking. In Proceedings of Systems (SiPS), (pp. 37 – 42). SiPS.

Ersfolk, J., Roquier, G., Lilius, J., & Mattavelli, M. (2012). Scheduling of dynamic dataflow programs based on state space analysis. In *Proceedings of IEEE International Conference on Acoustics, Speech, and Signal Processing*. IEEE.

Eshraghian, K. (2006). SoC Emerging Technologies. *Proceedings of the IEEE, 94*(6), 1197–1213. doi:10.1109/JPROC.2006.873615

European Radiocommunication Committee within the European Conference of Postal and Telecommunication Administrations (CEPT). (2002). *The European Table of Frequency Allocations and Utilizations covering the frequency range 9–275 GHz*. Lisboa, Portugal: Author.

Event-B and the Rodin Platform. (2008). Retrieved from http://www.event-b.org/

Ezick, J., & Springer, J. (2011). The Benefits of Static Compliance Testing for SCA Next. In *Proceedings of the SDR'11 Technical Conference and Product Exposition, 2011*. SDR.

Ezick, J., Springer, J., Litvinov, V., & Wohlford, D. (2010. A Path Toward Cost-Effective SCA Compliance Testing. In *Proceedings of the SDR'10 Technical Conference and Product Exposition*. SDR.

Faiz-ul-Hassan. Cheng, B., Vanderbauwhede, W., & Rodriguez, F. (2009). Impact of device variability in the communication structures for future synchronous SoC designs. In *Proceedings of International Symposium on System-on-Chip*, (pp. 68-72). Academic Press.

Falletti, E., Fantino, M., Linty, N., Parizzi, F., & Torchi, A. (2012). *Italian Patent No. TO2012A000408*.

Fan, X., Krstić, M., & Grass, E. (2009). Analysis and Optimization of Pausible Clocking based GALS Design. In *Proceedings of International Conference on Computer Design*, (pp. 358-365). ICCD.

Fan, X., Krstic, M., Wolf, C., & Grass, E. (2010). A GALS FFT Processor with Clock Modulation for Low-EMI Applications. In *Proceedings of 21st IEEE International Conference on Application-specific Systems Architectures and Processors* (ASAP) (pp. 273 - 278). ASAP.

Fantino, M., Molino, A., & Nicola, M. (2010). N-Gene: a Complete GPS and Galileo Software Suite for Precise Navigation. In *Proceedings of the 2010 International Technical Meeting of The Institute of Navigation* (pp. 1075-1081). San Diego, CA: Academic Press.

Fantino, M., Molino, A., Mulassano, P., Nicola, M., & Rao, M. (2009). Signal Quality Monitoring: Correlation mask based on Ratio Test metrics for multipath detection. In *Proceedings of International Global Navigation Satellite Systems Society (IGNSS) Symposium 2009*. Surfers Paradise, Australia: IGNSS.

Farabet, C., Poulet, C., & LeCun, Y. (2009). An FPGA-based stream processor for embedded real-time vision with convolutional networks. In *Proceedings of Fifth IEEE Workshop on Embedded Computer Vision* (ECV'09), (pp. 878–885). IEEE.

Farrag, M., Fengler, W., Streitferdt, D., & Fengler, O. (2010). Test Case Generation for Product Lines based on Colored State Charts. In *Proceedings of 3rd Workshop on Model-based Testing in Practice* (MoTiP 2010) (pp. 31–40). MoTiP.

Federal Communication Commission. (2011). *Light-Squared Technical Working Group final report*. Washington, DC: Academic Press.

Ferguson, A. J. (2005). Fostering e-mail security awareness: The West Point carronade. *EDUCAUSE Quarterly, 28*(1).

Fincke, U., & Pohst, M. (1985). Improved methods for calculating vectors of short length in a lattice, including a complexity analysis. *Mathematics of Computation, 44*(170), 463–471. doi:10.1090/S0025-5718-1985-0777278-8

FIPS PUB 140-2. (2001). *Federal Information Processing Standards Publication 140-2 - Security Requirements for Cryptographic Modules*. Information Technology Laboratory, National Institute of Standards and Technology.

FIPS PUB 140-3. (2009). *Federal Information Processing Standards Publication 140-3 (Revised Draft 09/11/09) - Security Requirements for Cryptographic Modules*. Information Technology Laboratory, National Institute of Standards and Technology.

Fischer, R. F. H., Windpassinger, C., Lampe, A., & Huber, J. B. (2002). Space-time transmission using Tomlinson-Harashima precoding. In *Proceedings of the 4th ITG Conference on Source and Channel Coding* (pp. 139 – 147). Frankfurt, Germany: VDE Verlag.

Fitton, J. J. (2002). Security Considerations for Software-Defined Radios. In *Proceedings of the SDR 02 Technical Conference and Product Exposition, 2002*. SDR.

Frankel, S., Eydt, B., Owens, L., & Scarfone, K. (2007). *Establishing Wireless Robust Security Networks: A Guide to IEEE 802.11i - Recommendations of the National Institute of Standards and Technology*. NIST Special Publication 800-97.

Fu, X., Graham, B., Bettati, R., & Zhao, W. (2003). Active Traffic Analysis Attacks and Countermeasures. In *Proceedings of the 2003 International Conference on Computer Networks and Mobile Computing* (ICCNMC'03). ICCNMC.

Fuketa, H., Hashimoto, M., Mitsuyama, Y., & Onoye, T. (2009). Trade-off analysis between timing error rate and power dissipation for adaptive speed control with timing error prediction. In *Proceedings of ASP-DAC*, (pp. 266-271). ASP-DAC.

Gade, T. (2013). *Acknowledgment Strategies for Efficient Asymmetric Routing in Energy Harvesting Wireless Sensor Networks*. (MS Thesis). Texas Tech University.

Gag, M., Wegner, T., & Timmermann, D. (2010). System Level Power Estimation of System-on-Chip Interconnects in Consideration of Transition Activity and Crosstalk. *Proceedings of the 20th international conference on Integrated circuit and system design: power and timing modeling, optimization and simulation* (pp. 21-30). Springer.

Gallery, E. M., & Mitchell, C. J. (2006). Trusted computing technologies and their use in the provision of high assurance SDR platform. In *Proceedings of the Software-Defined Radio Technical Conference*. Orlando, FL: Academic Press.

Gallo, R., Kawakami, H., & Dahab, R. (2009). On Device Establishment and Verification. In *Proceedings of the EuroPKI 2009*. EuroPKI.

Gallo, R., Kawakami, H., & Dahab, R. (2011). FORTUNA – A Probabilistic Framework for Early Design Stages of Hardware-Based Secure Systems. In *Proceedings of 5th International Conference on Network and System Security* (NSS 2011). NSS.

Gallo, R., Kawakami, H., Dahab, R., & Azevedo, R. Lima, S., & Araujo, G. (2010). T-DRE: A Hardware Trusted Computing Base for Direct Recording Electronic Vote Machines. In *Proceedings of the 26th Annual Computer Security Applications Conference* (ACSAC'10). ACSAC.

Gamma, E., Helm, R., Johnson, R., & Vlissides, J. M. (1994). *Design Patterns: Elements of Reusable Object-Oriented Softwaresystemen*. Addison-Wesley Professional.

Ganesan, D., Estrin, D., Woo, A., Culler, D., Krishnamachari, B., & Wicker, S. (2002). *Complex Behavior at Scale: An Experimental Study of Low-Power Wireless Sensor Networks* (Technical Report No. CSD-TR 02-0013). University of California at Los Angeles.

Gao, G. X. (2007). DME/TACAN Interference and its Mitigation in L5/E5 Bands. In *Proceedings of the 20th International Technical Meeting of the Satellite Division of The Institute of Navigation (ION GNSS 2007)* (pp. 1191-1200). Fort Worth, TX: ION GNSS.

Garcia-Hernandez, C. F., Ibarguengoytia-Gonzalez, P. H., & Perez-Diaz, J. A. (2007). Wireless Sensor Networks and Applications: A Survey. *International Journal of Computer Science and Network Security, 7*(3), 264–273.

Garibotti, R., Ost, L., Busseuil, R., & Kourouma, M. Adeniyi-Jones, Sassatelli, G., & Robert, M. (2013). Simultaneous Multithreading Support in Embedded Distributed Memory MPSoCs. In *Proceedings of Design Automation Conference (DAC)* (pp. 1-7). DAC. doi:10.1145/2463209.2488836

GEMSOS Security Kernel. (2012). Retrieved from http://www.aesec.com/

Gesbert, D., Kountouris, M., Heath, R., & Chae, C. (2007). From single user to multi user communications: Shifting the MIMO paradigm. *IEEE Signal Processing Magazine, 24*(5), 36–46. doi:10.1109/MSP.2007.904815

Giffin, J., Christodorescu, M., & Kruger, L. (2005). Strengthening Software Self-Checksumming via Self-Modifying Code. In *Proceedings of the 21st Annual Computer Security Application Conference* (ACSAC'05). ACSAC.

Gilabert, F., Ludovici, D., Medardoni, S., Bertozzi, D., Benini, L., & Gaydadjiev, G. N. (2009). Designing Regular Network-on-Chip Topologies under Technology, Architecture and Software Constraints. In *Proceedings of International Conference on Complex, Intelligent and Software Intensive Systems*, (pp. 681-687). Academic Press.

Gill, A. (1962). *Introduction to the Theory of Finite-state Machines*. New York: McGraw-Hill.

Gillet, M. (2008). Hardware/software co-simulation for conformance testing of embedded networks. In *Proceedings of 6th Seminar of Finnish-Russian University Cooperation in Telecommunications (FRUCT) Program*. Retrieved October 31, 2008, from http://fruct.org/index.php?option=com_content&view=article&id=68&Itemid=73

Gipper, J. (2007). SystemC the SoC system-level modeling language. *Embedded computing Design*. Retrieved from www.embedded-computing.com/pdfs/OSP2.May07.pdf

Gomes, G., & Sarmento, H. (2009). Indoor Location System Using ZigBee Technology. In *Proceedings of Third International Conference on Sensor Technologies and Applications* (pp. 152–157). Athens, Greece: IEEE.

Gondalia, A. K., & Kathiriya, D. R. (2012). Performance Tuning for Geographic Routing in Wireless Networks. *International Journal of Management, IT and Engineering, 2*(5).

González, A., Carlos, R., Dietrich, C. B., & Reed, J. H. (2009). Understanding the software communications architecture. *IEEE Communications Magazine, 47*(9).

Goodchild, J. (2010). *Social Engineering: The Basics*. Retrieved from http://www.csoonline.com/article/514063/social-engineering-the-basics

Gorin, J., Wipliez, M., Prêteux, F., & Raulet, M. (2011). LLVM-based and scalable MPEG-RVC decoder. *Journal of Real-Time Image Processing*. doi:10.1007/s11554-010-0169-2

Goyal, P., Guo, X., & Vin, H. M. (1996). A Hierarchical CPU Scheduler for Multimedia Operating Systems. In *Proceedings of Symposium on Operating Systems Design and Implementation* (pp. 107-121). New York: ACM. doi: 10.1145/238721.238766

Grecu, C., Ivanov, A., Saleh, R., & Pande, P. P. (2006). NoC interconnect yield improvement using crosspoint redundancy. In *Defect and Fault Tolerance in VLSI Systems* (pp. 457–465). Academic Press. doi:10.1109/DFT.2006.46

Guang, L. (2012). *Hierarchical Agent-Based Adaptation for Self-Aware Embedded Computing Systems*. (Doctoral Dissertation). University of Turku, Turku, Finland. Retrieved from http://www.doria.fi/handle/10024/86210

Guang, L., Jafri, S., Yang, B., Plosila, J., & Tenhunen, H. (2012). Embedding Fault-Tolerance with Dual-level Agents in Many-Core Systems. In *Proceedings of MEDIAN Workshop* (pp. 41-44). MEDIAN.

Guang, L., Nigussie, E., Plosila, J., Isoaho, J., & Tenhunen, H. (2012). Coarse and fine-grained monitoring and reconfiguration for energy-efficient NoCs. In *Proceedings of 2012 International Symposium on System on Chip (SoC)*, (pp. 1-7). SoC.

Guang, L., Yang, B., Plosila, J., Latif, K., & Tenhunen, H. (2010). Hierarchical power monitoring on NoC - A case study for hierarchical agent monitoring design approach. In *Proceedings of NORCHIP* (pp. 1 - 6). Tampere, Finland: IEEE.

Guang, L., Nigussie, E., Plosila, J., Isoaho, J., & Tenhunen, H. (2012). Survey of Self-Adaptive NoCs with Energy-Efficiency and Dependability. *International Journal of Embedded and Real-Time Communication Systems*, *3*(2), 1–22. doi:10.4018/jertcs.2012040101

Guang, L., Nigussie, E., Rantala, P., & Isoaho, J. (2010). Hierarchical Agent Monitoring Design Approach towards Self-Aware Parallel Systems-on-Chip. *ACM Transactions on Embedded Computing Systems*, *9*(3), 1–24. doi:10.1145/1698772.1698783

Gupta, M. S., Oatley, J. L., & Joseph, R. Wei, & Brooks, D.M. (2007). Understanding Voltage Variations in Chip Multiprocessors using a Distributed Power-Delivery Network. In Proceedings of Design, Automation & Test in Europe Conference & Exhibition (pp. 1--6). Academic Press.

Gu, R., Janneck, J., Bhattacharyya, S., Raulet, M., Wipliez, M., & Plishker, W. (2009). Exploring the concurrency of an MPEG RVC decoder based on dataflow program analysis. *IEEE Transactions on Circuits and Systems for Video Technology.*

Gurkaynak, F., Oetiker, S., Villiger, T., Felber, N., Kaeslin, H., & Fichtner, W. (2003). On the GALS design methodology of ETH Zurich. In *Proceedings of 1st Intl. Workshop Formal Methods for Globally Asynchronous Locally Synchronous Architecture* (FMGALS). FMGALS.

Habendorf, R., & Fettweis, G. (2007). Vector precoding with bounded complexity. In *Proceedings of the IEEE Workshop on Signal Processing Advances in Wireless Communications* (pp. 186-190). New York, NY: IEEE.

Hagedorn, A., Starobinski, D., & Trachtenberg, A. (2008). Rateless Deluge: Over-the-Air Programming of Wireless Sensor Networks Using Random Linear Codes. *International Conference on Information Processing in Sensor Networks, IPSN '08* (pp. 457-466). St. Louis, MO: IPSN.

Harashima, H., & Miyakawa, H. (1972). Matched-transmission technique for channels with intersymbol interference. *IEEE Transactions on Communications*, *CM-20*(4), 774–780. doi:10.1109/TCOM.1972.1091221

Harel, D., & Thiagarajan, P. S. (2003). *Message Sequence Charts*. Retrieved March 17, 2010 from http://www.comp.nus.edu.sg/~thiagu/public_papers/surveymsc.pdf

Haroud, M., & Biere, A. (2005). SDL Versus C Equivalence Checking. In *Proceedings of 12th International SDL Forum* (LNCS), (vol. 3530, pp. 323-339). Springer.

Haroud, M., & Blazevic, L. (2006). HW accelerated Ultra Wide Band MAC protocol using SDL and SystemC. In *Proceedings of Fourth IEEE International Conference on Pervasive Computing and Communications Workshops (PERCOMW'06)*. IEEE. Retrieved from http://fmv.jku.at/papers/HaroudBlazevicBiere-RAWCON04.pdf

Hashemi, H. (1993). The Indoor Radio Propagation Channel. *Proceedings of the IEEE, 81*(7), 943–968. doi:10.1109/5.231342

Hay, S., & Harle, R. (2009). Bluetooth Tracking without Discoverability. *Lecture Notes in Computer Science, 5561*, 120–137. doi:10.1007/978-3-642-01721-6_8

Heath, S. (2003). *Embedded Systems Design* (2nd ed.). Newnes.

Hegarty, C., Van Dierendonck, A. J., Bobyn, D., Tran, M., & Grabowski, J. (2000). Suppression of Pulsed Interference through Blanking. In *Proceedings of the IAIN World Congress and the 56th Annual Meeting of The Institute of Navigation* (pp. 399-408). San Diego, CA: IAIN.

Heissenbuttel, M., Braun, T., Walchli, M., & Bernoulli, T. (2007). Evaluating the Limitations and Alternatives in Beaconing. *Elsevier Ad Hoc Networks, 5*(5).

Henderson, H. V., & Searle, S. R. (1981). On deriving the inverse of a sum of matrices. *SIAM Review, 23*(1), 53–60. doi:10.1137/1023004

Henkel, J., Bauer, L., Becker, J., Bringmann, O., Brinkschulte, U., & Chakraborty, S. ... Wunderlich, H. (2011). Design and architectures for dependable embedded systems. In *Proceedings of the 9th International Conference on Hardware/Software Codesign and System Synthesis (CODES+ISSS)*, (pp. 69-78). CODES+ISSS.

Henkel, J., Ebi, T., Amrouch, H., & Khdr, H. (2013). Thermal management for dependable on-chip systems. In *Proceedings of ASP-DAC*, (pp. 113-118). ASP-DAC.

Hernández, C., Roca, A., Silla, F., Flich, J., & Duato, J. (2010). Improving the Performance of GALS-Based NoCs in the Presence of Process Variation. In *Proceedings of Fourth ACM/IEEE International Symposium on Networks-on-Chip* (NOCS), (pp. 35-42). ACM/IEEE.

Hess, C., Wenk, M., Burg, A., Luethi, P., & Studer, C. Felber, N., & Fichtner, W. (2007). Reduced-complexity MIMO detector with close-to ML error-rate performance. In *Proceedings of the ACM Great Lakes Symposium on VLSI* (pp. 200-203). New York, NY: ACM.

Hill, J., Szewczyk, R., Woo, A., Hollar, S., Culler, D., & Pister, K. (2000). System architecture directions for networked sensors. *SIGPLAN Not., 35*(11), 93–104. doi:10.1145/356989.356998

Hoare, C. A. R. (1985). *Communicating Sequential Processes*. Prentice Hall.

Hochwald, B. M., Peel, C. B., & Swindlehurst, A. L. (2005). A vector-perturbation technique for near-capacity multiantenna multiuser communication: perturbation. *IEEE Transactions on Communications, 53*(3), 537–544. doi:10.1109/TCOMM.2004.841997

Hoerl, A. E., & Kennard, R. W. (1970). Ridge regression: Biased estimation for nonorthogonal problems. *Technometrics, 12*, 55–67. doi:10.1080/00401706.1970.10488634

Hölzenspies, P. H. (2008). Run-time spatial mapping of streaming applications to a heterogeneous multi-processor system-on-chip (MPSoC). *Proceedings of the conference on Design, automation and test in Europe*, (pp. 212-217).

Honkavirta, V., Perälä, T., Ali-Löytty, S., & Piché, R. (2009). Location Fingerprinting Methods in Wireless Local Area Network. In *Proceedings of the 6th workshop on Positioning, Navigation and Communication (WPNC'09)* (pp. 243-251). Hannover, Germany: IEEE.

Höppner, S., Walter, D., Eisenreich, H., & Schüffny, R. (2010). Efficient compensation of delay variations in high-speed network-on-chip data links. In *Proceedings of International Symposium on System on Chip* (SoC), (pp. 55-58). Academic Press.

Horn, P. (2001). *Autonomic computing: IBM's perspective on the state of information technology (Tech. Report)*. IBM.

Horn, R., & Johnson, C. R. (1985). *Matrix analysis*. Cambridge University Press. doi:10.1017/CBO9780511810817

Howard, M., & LeBlanc, D. (2007). *Writing Secure Code for Windows Vista*. Microsoft Press.

Hsu, F.-H., Guo, F., & Chiueh, T.-C. (2006). Scalable Network-based Buffer Overflow Attack Detection. In *Proceeding of ACM/IEEE Symposium on Architectures for Networking and Communications Systems 2006*. ACM/IEEE.

Hsu, D., Kakade, S., Langford, J., & Zhang, T. (2009). Multi-label prediction via compressed sensing. *Advances in Neural Information Processing Systems, 22*, 772–780.

Hu, J., & Marculescu, R. (2003). Energy-aware mapping for tile-based NoC architectures under performance constraints. In *Proceedings of Design Automation Conference* (pp. 233-239). IEEE.

Hui, J. W. (2005, July 28). *Deluge 2.0 - TinyOS Network Programming*. Retrieved November 13, 2009, from http://www.cs.berkeley.edu/~jwhui/deluge/deluge-manual.pdf

Hui, J. W., & Culler, D. (2004). The dynamic behavior of a data dissemination protocol for network programming at scale. In *Proceedings of the 2nd international conference on Embedded networked sensor systems* (pp. 81-94). New York: Academic Press.

Humphreys, T. E., Ledvina, B. M., Psiaki, M. L., O'Hanlon, B. W., & Kintner, P. M., Jr. (2008). Assessing the Spoofing Threat: Development of a Portable GPS Civilian Spoofer. In *Proceedings of the 2008 ION GNSS Conference*. GNSS.

Hung, W., Addo-Quaye, C., Theocharides, T., Xie, Y., Vijaykrishnan, N., & Irwin, M. J. (2004). Thermal-Aware IP Virtualization and Placement for Networks-on-Chip Architecture. *IEEE International Conference on Computer Design (ICCD'04)* (pp. 430-437). ACM Press.

IBM. (2006). *IBM PowerPC 405 CPU Core Product Overview*. Retrieved January 19, 2012, from https://www-01.ibm.com/chips/techlib/techlib.nsf/techdocs/3D7489A3704570C0872571DD0065934E/$file/PPC405_Product_Overview_20060902.pdf

IBM. (2009). *SDL Suite and TTCN Suite Help*. IBM Rational SDL and TTCN Suite.

Ienne, P., & Leupers, R. (Eds.). (2006). *Customizable Embedded Processors: Design Technologies and Applications (Systems on Silicon)*. San Francisco, CA: Morgan Kaufmann Publishers.

Intel. (2010). *Single-chip cloud computer*. Retrieved from http://techresearch.intel.com/articles/Tera-Scale/1826.htm

International Telecommunication Union. (2011). *Recommendation Z.100. Specification and Description Language (SDL)*. Geneva: Author.

Isoaho, J., Virtanen, S., & Plosila, J. (2010). Current challenges in embedded communication systems. *International Journal of Embedded and Real-Time Communication Systems, 1*(1), 1–21. doi:10.4018/jertcs.2010103001

ITRS. (2007). *International Technology Roadmap for Semiconductors*. Retrieved from http://www.itrs.net/Links/2007ITRS/Home2007.htm

Jagatic, T., Johnson, N., Jakobsson, M., & Menczer, F. (2007). Social Phishing. *Communications of the ACM, 50*(10). doi:10.1145/1290958.1290968

Janin, L., Li, S., & Edwards, D. (2010). Integrated design environment for reconfigurable HPC. In *Proceedings of 6th international conference on Reconfigurable Computing: architectures, Tools and Applications* (ARC'10) (pp. 406-413). ARC.

Janneck, J. W. (2011). A machine model for dataflow actors and its applications. In Proceedings of Signals, Systems and Computers (ASILOMAR) (pp. 756 – 760). ASILOMAR.

Jantsch, A. (2004). *Modeling Embedded Systems and SoCs*. Stockholm: Morgan Kaufmann Publishers.

Jantsch, A., & Tenhunen, H. (Eds.). (2003). *Network on Chips*. Boston, MA: Kluwer Academic Publishers.

JEDEC. (2009, March). Failure Mechanisms and Models for Semiconductor Devices. *JEDEC publication JEP122F*.

Jevring, M., Groote, R., & Hesselman, C. (2008). Dynamic Optimization of Bluetooth Networks for Indoor Localization. In *Proceedings of the 5th international conference on Soft computing as transdisciplinary science and technology* (pp. 663-668). Paris, France: ACM.

Jin, Y., & Makris, Y. (2008). Hardware Trojan detection using path delay fingerprint. In *Proceedings of the 2008 IEEE International Workshop on Hardware-Oriented Security and Trust*. IEEE.

Joham, M., Brehmer, J., & Utschick, W. (2004). MMSE approaches to multiuser spatio-temporal Tomlinson-Harashima precoding. In *Proceedings of the International ITG Conference on Source and Channel Coding* (pp. 387- 394). New York, NY: IEEE.

Joham, M., Utschick, W., & Nossek, J. A. (2005). Linear transmit processing in MIMO communications systems. *IEEE Transactions on Signal Processing, 53*(8), 2700–2712. doi:10.1109/TSP.2005.850331

Johnson, D. B., & Maltz, D. A. (1996). Dynamic Source Routing in Ad Hoc Wireless Networks. *Mobile Computing, 5*, 153–181. doi:10.1007/978-0-585-29603-6_5

Johnston, W. M., Paul Hanna, J. R., & Millar, R. J. (2004). Advances in dataflow programming languages. *ACM Computing Surveys, 36*(1), 1–34. doi:10.1145/1013208.1013209

Jose, A. P., Patounakis, G., & Shepard, K. L. (2005). Near speed-of-light on-chip interconnects using pulsed current-mode signalling. *Digest of Technical Papers Symposium on VLSI Circuits*, 108- 111.

Joshi, R., Kanj, R., Adams, C., & Warnock, J. (2013). Making reliable memories in an unreliable world. In *Proceedings of 2013 IEEE International Reliability Physics Symposium* (pp. 3A.6.1-3A.6.5). IEEE.

Jozawa, T., Huang, L., Sakai, E., Takeuchi, S., & Kasslin, M. (2006). Heterogeneous Co-simulation with SDL and SystemC for Protocol Modeling. In *Proceedings of IEEE Radio and Wireless Symposium 2006*, (pp. 603–606). Retrieved from http://research.nokia.com/node/5789

Kahn, G. (1974). The semantics of a simple language for parallel programming.[IFIP.]. *Proceedings of IFIP, 74*, 471–475.

Kakoee, M. R., Loi, I., & Benini, L. (2010). A New Physical Routing Approach for Robust Bundled Signaling on NoC Links. In *Proceedings of GLSVLSI10*. GLSVLS.

Kamali, M., Petre, L., Sere, K., & Daneshtalab, M. (2011). Formal Modelling of Multicast Communication in 3D NoCs. In *Proceedings of 14th Euromicro Conference on Digital System Design (DSD 2011)* (pp. 634-642). Oulu, Finland: IEEE.

Kamal, R. (2008). *Embedded systems: architecture, programming and design* (2nd ed.). New Delhi: Tata McGraw-Hill Publishing Company Limited.

Kamran, A., & Navabi, Z. (2013). Online periodic test mechanism for homogeneous many-core processors. In *Proceedings of 2013 IFIP/IEEE 21st International Conference on Very Large Scale Integration* (VLSI-SoC), (pp. 256-259). IEEE.

Kamsties, E., Reuys, A., Pohl, K., & Reis, S. (2004). Testing Variabilities in Use Case Models. In *Software Product-Family Engineering* (pp. 6–18). Berlin: Springer. doi:10.1007/978-3-540-24667-1_2

Kang, K. C., Lee, J., & Donohoe, P. (2002). Feature-oriented product line engineering. *Software, IEEE, 19*(4), 58–65. doi:10.1109/MS.2002.1020288

Kantz, F., Ruschival, T., Nenninger, P., & Streitferdt, D. (2009). Testing with Large Parameter Sets for the Development of Embedded Systems in the Automation Domain. In *Proceedings of the 2nd International Workshop on Component-Based Design of Resource-Constrained Systems at the 33rd Annual IEEE International Computers, Software and Applications Conference*. IEEE.

Kaplan, E. D. (Ed.). (1996). *Understanding GPS: Principles and Applications*. Norwood, MA: Artech House.

Kaplan, E., & Hegarty, C. (2005). *Understanding GPS Principles and Applications* (2nd ed.). Artech House.

Karabegov, A., & Ter-Mikaelyan, T. (1993). *Introduction to the SDL language*. Moscow: Radio and Communication.

Karl, E., Sylvester, D., & Blaauw, D. (2005). Timing error correction techniques for voltage-scalable on-chip memories. In *Proceedings of IEEE International Symposium on Circuits and Systems*, (vol. 4, pp. 3563-3566). IEEE.

Karp, B., & Kung, H. T. (2000). *Greedy Perimeter Stateless Routing for Wireless Networks*. Paper presented at the 6th Annual International Conference on Mobile Computing and Networking (MobiCom). Boston, MA.

Karri, R., Rajendran, J., Rosefeld, K., & Tehranipoor, M. (2010). Trustworthy Hardware: Identifying and Classifying Hardware Trojans. *IEEE Computer, 43*(10), 39–46. doi:10.1109/MC.2010.299

Kaushik, S. S. (2011). Preprocessing-based run-time mapping of applications on NoC-based MPSoCs. *IEEE Computer Society Annual Symposium on VLSI*, (pp. 337-338).

Keane, J., & Kim, C. (2011). Transistor aging. *IEEE Spectrum*. Retrieved Dec. 18, 2013, from http://spectrum.ieee.org/semiconductors/processors/transistor-aging

Kelly, D., McLoone, S., & Dishongh, T. (2008). A Bluetooth-Based Minimum Infrastructure Home Localization System. in *Proceedings of 5th IEEE International Symposium on Wireless Communication Systems* (pp. 638-642). Reykjavik, Iceland: IEEE.

Kempf, T., Ascheid, G., & Leupers, R. (2011). *Multiprocessor Systems on Chip: Design Space Exploration*. Springer. doi:10.1007/978-1-4419-8153-0

Kenyon, C., Kornfeld, A., Kuhn, K., Liu, M., Maheshwari, A., Shih, W., … Zawadzki, K. (2008). Managing Process Variation in Intel's 45nm CMOS Technology. *Intel Technology Journal*.

Khajekarimi, E., & Hashemi, M. R. (2012). Communication and Congestion Aware Run-Time Task Mapping on Heterogeneous MPSoCs. In *Proceedings of International Symposium on Computer Architecture and Digital Systems* (pp. 127-132). doi: 10.1109/CADS.2012.6316432

Khatib, I. A., Bertozzi, D., Poletti, F., Benini, L., Jantsch, A., Bechara, M., & Jonsson, S. (2006). MPSoC ECG biochip: A multiprocessor system-on-chip for real-time human heart monitoring and analysis. In *Proceedings of Conference on Computing Frontiers* (pp. 21-28). New York: ACM.

Kihara, M., Ono, S., & Eskelinen, P. (2003). *Digital Clocks for Synchronization and Communications*. Artech House, Inc.

Kim, C., Burger, D., & Keckler, S. W. (2002). An adaptive, non-uniform cache structure for wire-delay dominated on-chip caches. In *Proceedings of ACM SIGPLAN*, (pp. 211–222). ACM.

Kim, D., Toh, C. K., & Chou, Y. (2000). *RODA: A New Dynamic Routing Protocol Using Dual Paths to Support Asymmetric Links in Mobile Ad Hoc Networks*. Paper presented at the 9th International Conference on Computer Communications and Networks (ICCCN). Las Vegas, NV.

Kim, Y. J., Govidan, R., Karp, B., & Shenker, S. (2004). *Practical and Robust Geographic Routing in Wireless Networks*. Paper presented at the 2nd International Conference on Embedded Networked Sensor Systems (SenSys). Baltimore, MD.

Kim, K. K., Wang, W., & Choi, K. (2010). On-Chip Aging Sensor Circuits for Reliable Nanometer MOSFET Digital Circuits. *IEEE Transactions on Circuits and Systems*, *57*(10), 798–802. doi:10.1109/TCSII.2010.2067810

Kistler, M., Perrone, M., & Petrini, F. (2006). Cell Multiprocessor Communication Network: Built for speed. *IEEE Micro*, *26*(3), 10–23. doi:10.1109/MM.2006.49

Kiszka, J., & Wagner, B. (2003). Domain and Type Enforcement for Real-Time Operating Systems. In *Proceedings of the 9th IEEE International Conference On Emerging Technologies and Factory Automation*. IEEE.

Kiszka, J., & Wagner, B. (2007). Modelling Security Risks in Real-Time Operating Systems. In *Proceedings of the 5th IEEE International Conference on Industrial Informatics*. IEEE.

Kjærgaard, M. B. (2007). A taxonomy for radio location fingerprinting. In *Location-and context-awareness* (pp. 139–156). Berlin: Springer. doi:10.1007/978-3-540-75160-1_9

Kloos, J., & Eschbach, R. (2009). Generating System Models for a Highly Configurable Train Control System Using A Domain-Specific Language: A Case Study. In *Proceedings of AMOST'09 - 5th Workshop on Advances in Model Based Testing*. Denver, CO: AMOST.

Knight, J. C. (2002). Dependability of embedded systems. In *Proceedings of the 24th International Conference on Software Engineering*, (pp. 685-686). IEEE.

Ko, Y., Lee, S., & Lee, J. (2004). *Ad-hoc Routing with Early Unidirectionality Detection and Avoidance*. Paper presented at the International Conference on Personal Wireless Communications (PWC). Delft, The Netherlands.

Kobayashi, K., & Onodera, H. (2008). Best ways to use billions of devices on a chip - Error predictive, defect tolerant and error recovery designs. In *Proceedings of ASP-DAC*, (pp. 811-812). ASP-DAC.

Kocher, P. (1996). Timing Attacks on Implementations of Diffie-Hellman, RSA, DSS and Other Systems. *Lecture Notes in Computer Science*, *1109*, 104–113. doi:10.1007/3-540-68697-5_9

Kocher, P., Jaffe, J., & Jun, B. (1999). Differential Power Analysis: Leaking Secrets. In *Proceedings of Crypto '99 (LNCS)* (Vol. 1666, pp. 388–397). Berlin: Springer Verlag.

Kohout, P., Ganesh, B., & Jacob, B. (2003). Hardware Support for Real-Time Operating Systems. In *Proceedings of IEEE/ACM/IFIP International Conference on Hardware/Software Codesign and System Synthesis* (pp. 45-51). New York: ACM. doi: 10.1145/944645.944656

Konovaltsev, A., & Cuntz, M. Haettich, C., & Meurer, M. (2013). Autonomous Spoofing Detection and Mitigation in a GNSS Receiver with an Adaptive Antenna Array. In *Proceedings of the 26th International Technical Meeting of The Satellite Division of the Institute of Navigation (ION GNSS 2013)* (pp. 2937-2948). Nashville, TN: ION GNSS.

Krstic, M., Piz, M., Ehrig, M., & Grass, E. (2008). OFDM Datapath Baseband Processor for 1 Gbps Datarate. In *Proceedings of IFIP/IEEE VLSI-SoC Conference 2008* (pp. 156-159). IEEE.

Krstic, M., Fan, X., Grass, E., Benini, L., Kakoee, M. R., & Heer, C. et al. (2012). Evaluation of GALS Methods in Scaled CMOS Technology: Moonrake Chip Experience. [IJERTCS]. *International Journal of Embedded and Real-Time Communication Systems*, 3(4), 1–18. doi:10.4018/jertcs.2012100101

Krstic, M., Grass, E., Gürkaynak, F., & Vivet, P. (2007). Globally Asynchronous, Locally Synchronous Circuits: Overview and Outlook. *IEEE Design & Test of Computers*, 24(5), 430–441. doi:10.1109/MDT.2007.164

Krum, A. (2000). Thermal Management. In F. Keith (Ed.), *The CRC Handbook of Thermal Engineering* (pp. 1–92). Boca Raton: CRC Press.

Kuhn, D., Wallace, D., & Gallo, A. M. J. (2004). Software fault interactions and implications for software testing. *IEEE Transactions on Software Engineering*, 30(6), 418–421. doi:10.1109/TSE.2004.24

Kulkarni, S., & Wang, L. (2005). MNP: Multihop Network Reprogramming Service for Sensor Networks.[IEEE.]. *Proceedings of Distributed Computing Systems*, 2005, 7–16.

Kuorilehto, M., Kohvakka, M., Suhonen, J., Hämäläinen, P., Hännikäinen, M., & Hämäläinen, T. D. (2007). Ultra-Low Energy Wireless Sensor Networks. In *Practice: Theory, Realization and Deployment*. John Wiley.

Kurdziel, M., Beane, J., & Fitton, J. J. (2005). An SCA security supplement compliant radio architecture. In *Proceedings of the Military Communications Conference*. IEEE.

Kursun, E. C.-Y. (2006). Investigating the effects of task scheduling on thermal behavior. *In 3rd Workshop on Temperature-Aware Computer Systems*, (pp. 1-12).

Kurz, L., Tasdemir, E., Bornkessel, D., Noll, T. G., Kappen, G., Antreich, F., et al. (2012). An Architecture for an Embedded Antenna-array Digital GNSS Receiver using Subspace-based Methods for Spatial Filtering. In *Proceedings of the 6th ESA Workshop on Satellite Navigation Technologies and European Workshop on GNSS Signals and Signal Processing (NAVITEC 2012)* (pp. 1-8). Noordwijk, The Netherlands: GNSS.

Kusume, K., Joham, M., Utschick, W., & Bauch, G. (2005). Efficient Tomlinson-Harashima precoding for spatial multiplexing on flat MIMO channel. In *Proceedings of the IEEE International Conference on Communications* (pp. 2021-2025). New York, NY: IEEE.

Kusume, K., Joham, M., Utschick, W., & Bauch, G. (2007). Cholesky factorization with symmetric permutation applied to detecting and precoding spatially multiplexed data streams. *IEEE Transactions on Signal Processing*, 55(6 II), 3089- 3103.

L3Nav. (n.d.). *GPS Toolbox*. Retrieved from http://www.l3nav.com/gps_toolbox.htm

Laddaga, R. (1997). *Self-adaptive software* (Tech. Rep. 98-12). DARPA BAA.

Laitinen, H., Juurakko, S., Lahti, T., Korhonen, R., & Lahteenmaki, J. (2007). Experimental evaluation of location methods based on signal-strength measurements. *IEEE Transactions on Vehicular Technology*, 56(1), 287–296. doi:10.1109/TVT.2006.883785

Lammers, D. (2010). The Era of Error-Tolerant Computing. *IEEE Spectrum*, 47(11), 15. doi:10.1109/MSPEC.2010.5605876

Langendoen, K., Baggio, A., & Visser, O. (2006). Murphy loves potatoes: experiences from a pilot sensor network deployment in precision agriculture. In *Proceedings of Parallel and Distributed Processing Symposium*. IPDPS.

Langford, J., Smola, A. J., & Zinkevich, M. (2009). Slow learners are fast. In Advances in Neural Information Processing Systems. MIT Press.

Lanoix, A. (2008). Event-B Specification of a Situated Multi-Agent System: Study of a Platoon of Vehicles. In *Proceedings of International Symposium on Theoretical Aspects of Software Engineering* (pp. 297 - 304). Nanjing: IEEE.

Laukkarinen, T., Suhonen, J., & Hännikäinen, M. (2013). An embedded cloud design for internet-of-things. *International Journal of Distributed Sensor Networks*, 13.

Lee, E. (2006). The problem with threads. *Computer*, 39(5), 33–42. doi:10.1109/MC.2006.180

Lee, E. A., & Parks, T. M. (1995). Dataflow Process Networks. *Proceedings of the IEEE*, 83(5), 773–801. doi:10.1109/5.381846

Lee, E., & Messerschmitt, D. (1987). Synchronous data flow. *Proceedings of the IEEE*, 75(9), 1235–1245. doi:10.1109/PROC.1987.13876

Lee, J. Y., & Scholtz, R. A. (2002). Ranging in a dense multipath environment using an UWB radio link. *IEEE Journal on Selected Areas in Communications*, 20(9), 1677–1683. doi:10.1109/JSAC.2002.805060

Leem, L., Cho, H., Bau, J., Jacobson, Q., & Mitra, S. (2010). ERSA: Error Resilient System Architecture for Probabilistic Applications. In *Proceedings of the Conference on Design, Automation and Test in Europe* (pp. 1560-1565). Academic Press.

Lehtonen, T., Wolpert, D., Liljeberg, P., Plosila, J., & Ampadu, P. (2010). Self-Adaptive System for Addressing Permanent Errors in On-Chip Interconnects. *IEEE Transactions on Very Large Scale Integration Systems*, 18(4), 527–540. doi:10.1109/TVLSI.2009.2013711

Lemaire, R., & Thonnart, Y. (2010). *Magali, A Reconfigurable Digital Baseband for 4G Telecom Applications based on an Asynchronous NoC*. Paper Presented at ACM/IEEE International Symposium on Networks-on-Chip. New York, NY.

Lenstra, A. K., Lenstra, H. W., & Lovász, L. (1982). Factoring polynomials with rational coefficients. *Mathematische Annalen*, 216(4), 513–534.

Leonard, J. J., & Durrant-whyte, H. F. (1991). Simultaneous Map Building and Localization for an Autonomous Mobile Robot. In *Proceedings of IEEE/RSJ International Workshop on Intelligent Robots and Systems, Intelligence for Mechanical Systems* (pp. 1442–1447). IEEE.

Le, T., Sinha, P., & Xuan, D. (2010). Turning Heterogeneity into an Advantage in Wireless Ad-Hoc Network Routing. *Elsevier Ad Hoc Networks*, 8(1), 108–118. doi:10.1016/j.adhoc.2009.06.001

Levis, P., Patel, N., Culler, D., & Shenker, S. (2004). Trickle: a self-regulating algorithm for code propagation and maintenance in wireless sensor networks. In *Proceedings of the 1st conference on Symposium on Networked Systems Design and Implementation*. USENIX Association.

Levis, P., & Culler, D. (2002, December). Maté: A Tiny Virtual Machine for Sensor Networks. *SIGOPS Oper. Syst. Rev.*, 36(5), 85–95. doi:10.1145/635508.605407

Limberg, T., Winter, M., Bimberg, M., Klemm, R., Tavares, M., & Ahlendorf, H. … Fettweis, G. (2009). A Heterogeneous MPSoC with Hardware Supported Dynamic Task Scheduling for Software Defined Radio. In *Proceedings of DAC/ISSCC Student Design Contest*. doi: 10.1.1.150.6363

Lin, G., & Noubir, G. (2003). Low Power DoS Attacks in Data Wireless LANs and Countermeasures. In *Proceedings of ACM MobiHoc*. ACM.

Lindqvist, U., & Jonsson, E. (1997). How to Systematically Classify Computer Security Intrusions. In *Proceedings of the 1997 IEEE Symposium on Security and Privacy*. IEEE.

Lines, A. (2004). Asynchronous interconnect for synchronous SoC design. *IEEE Micro*, 24(1), 32–41. doi:10.1109/MM.2004.1268991

Lippett, M. (2004). An IP Core Based Approach to the On-Chip Management of Heterogeneous SoCs. In *Proceedings of IP Based SoC Design Forum & Exhibition*. Academic Press.

Liu, J., & Chen, R. (2011). *Smartphone Positioning Based on a Hidden Markov Model*. Tech Talk Blog. Retrieved March 2011 from http://gpsworld.com/tech-talk-blog/smartphone-positioning-based-a-hidden-markov-model-11385-0

Liu, R. P., Rosberg, Z., Collings, I. B., Wilson, C., Dong, A., & Jha, S. (2008). *Overcoming Radio Link Asymmetry in Wireless Sensor Networks*. Paper presented at the International Symposium on Personal, Indoor and Mobile Radio Communications. Cannes, France.

Liu, H., Darabi, H., Banerjee, P., & Liu, J. (2007). Survey of wireless indoor positioning techniques and systems. *IEEE Transactions on Systems, Man and Cybernetics. Part C, Applications and Reviews*, *37*(6), 1067–1080. doi:10.1109/TSMCC.2007.905750

Liu, J. (2012). Hybrid Positioning with Smartphones. In *Ubiquitous Positioning and Mobile Location-Based Services in Smart Phones* (pp. 159–194). IGI Global. doi:10.4018/978-1-4666-1827-5.ch007

Liu, J., Chen, R., Chen, Y., Pei, L., & Chen, L. (2012). iParking: An Intelligent Indoor Location-Based Smartphone Parking Service. *Sensors (Basel, Switzerland)*, *12*(11), 14612–14629. doi:10.3390/s121114612 PMID:23202179

Liu, J., Chen, R., Pei, L., Guinness, R., & Kuusniemi, H. (2012). A Hybrid Smartphone Indoor Positioning Solution for Mobile LBS. *Sensors (Basel, Switzerland)*, *12*(12), 17208–17233. doi:10.3390/s121217208 PMID:23235455

Liu, W., Calimera, A., Nannarelli, A., Macii, E., & Poncino, M. (2010). On-chip Thermal Modeling Based on SPICE Simulation.[Springer.]. *PATMOS*, *5953*, 66–75.

Loghi, M., Poncino, M., & Benini, L. (2006). Cache Coherence Tradeoffs in Shared-Memory MPSoCs. *ACM Transactions on Embedded Computing Systems*, *5*(2), 383–407. doi:10.1145/1151074.1151081

Loi, I., Angiolini, F., & Benini, L. (2008). Developing Mesochronous Synchronizers to Enable 3D NoCs. In Design, Automation and Test in Europe, (pp. 1414-1419). Academic Press.

Low, Y., Gonzalez, J., Kyrola, A., Bickson, D., Guestrin, C., & Hellerstein, J. M. (2010). Graphlab: A new framework for parallel machine learning. In *Proceedings of the 26th Conference on Uncertainty in Artificial Intelligence* (UAI 2010). UAI.

Ltd, A. R. M. (2013). *AMBA On-Chip Connectivity*. Retrieved December 15, 2013, from http://www.arm.com/products/system-ip/amba/amba-open-specifications.php

Ludovici, D., Strano, A., Bertozzi, D., Benini, L., & Gaydadjiev, G. N. (2009). Comparing tightly and loosely coupled mesochronous synchronizers in a NoC switch architecture. In *Proceedings of 3rd ACM/IEEE International Symposium on Networks-on-Chip*, (pp. 244-249). ACM/IEEE.

Ludovici, D., Strano, A., Gaydadjiev, G. N., Benini, L., & Bertozzi, D. (2010). Design Space Exploration of a Mesochronous Link for Cost-Effective and Flexible GALS NOCs. In Proceedings of Design, Automation & Test in Europe Conference & Exhibition (DATE) (pp. 679-684). DATE.

Määttä, L., Suhonen, J., Laukkarinen, T., Hämäläinen, T., & Hännikäinen, M. (2010). Program image dissemination protocol for low-energy multihop wireless sensor networks. In Proceedings of System on Chip (SoC), 2010 International Symposium on (pp. 133-138). Tampere: IEEE.

Ma, Y. (2009). *Deliverable D2.2 version 1.0 of the WHERE project: Cooperative positioning (intermediate report)*. Retrieved 2009 from http://www.kns.dlr.de/where/documents/Deliverable22.pdf

Macabiau, C., Julien, O., & Chartre, E. (2001). Use of Multicorrelator Techniques for Interference Detection. In *Proceedings of the 2001 National Technical Meeting of The Institute of Navigation (ION GNSS 2013)* (pp. 353-363). Long Beach, CA: GNSS.

Maccone, C. (2010). The KLT (Karhunen-Loève Transform) to extend SETI Searches to Broad-band and Extremely Feeble Signals. *Acta Astronautica*, *67*(11-12), 1427–1439. doi:10.1016/j.actaastro.2010.05.002

MacDonald, J. T., & Roberson, D. A. (2007). *Spectrum Occupancy Estimation in Wireless Channels with Asymmetric Transmitter Powers*. Paper presented at the 2nd International Conference on Cognitive Radio Oriented Wireless Networks and Communications (CrownCom). Orlando, FL.

Magnusson, P., Christensson, M., Eskilson, J., Forsgren, D., Hallberg, G., & Hogberg, J. et al. (2002). Simics: A full system simulation platform. *Computer*, *35*(2), 50–58. doi:10.1109/2.982916

Manadhata, P. K., & Wing, J. M. (2011). An Attack Surface Metric. *IEEE Transactions on Software Engineering*, *37*(3), 371–386. doi:10.1109/TSE.2010.60

Mandal, A., Khatri, S. P., & Mahapatra, R. N. (2013). Exploring topologies for source-synchronous ring-based Network-on-Chip. In *Proceedings of Design* (pp. 1026–1031). Automation & Test in Europe Conference & Exhibition.

Marina, M. K., & Das, S. K. (2002). *Routing Performance in the Presence of Unidirectional Links in Multihop Wireless Networks*. Paper presented at the 3rd International Symposium on Mobile Ad Hoc Networking and Computing. Lausanne, Switzerland.

Martinez, E., & Atienza, D. (2009). Inducing Thermal-Awareness in Multicore Systems Using Networks-on-Chip. *2009 IEEE Computer Society Annual Symposium on VLSI* (pp. 187-192). IEEE.

Mattavelli, M., Amer, I., & Raulet, M. (2010). The Reconfigurable Video Coding Standard. *IEEE Signal Processing Magazine*, *27*(3), 159–167. doi:10.1109/MSP.2010.936032

McGregor, J. D. (2001). *Testing a software product line* (Tech. Rep. CMU/SEI-2001-TR-022). Pittsburgh, PA: Carnegie Mellon University.

Memik, S.O., & Mukherjee, R., Ni, & Long. (2008). Optimizing Thermal Sensor Allocation for Microprocessors. *IEEE Transactions on Computer-Aided Design of Integrated Circuits and Systems*, *27*(3), 516–527. doi:10.1109/TCAD.2008.915538

Merdadoni, S., Lajolo, M., & Bertozzi, D. (2008). Variation tolerant NoC design by means of self-calibrating links. In Proceedings of Design, Automation and Test in Europe, (pp. 1402-1407). Academic Press.

Métayer, C., Abrial, J.-R., & Voisin, L. (2005, May 31). *Deliverables*. Retrieved from http://rodin.cs.ncl.ac.uk/deliverables/D7.pdf

Meurer, M., Konovaltsev, A., Cuntz, M., & Hättich, C. (2012). Robust Joint Multi-Antenna Spoofing Detection and Attitude Estimation using Direction Assisted Multiple Hypotheses RAIM. In *Proceedings of the 25th International Technical Meeting of The Satellite Division of the Institute of Navigation (ION GNSS 2012)* (pp. 3007-3016), Nashville, TN: GNSS.

Microchip Technology, I. (2008, February 10). *PIC18F8722 Product Page*. Retrieved December 7, 2009, from http://www.microchip.com/

Microchip Technology, I. (2009). *MPLAB C Compiler for PIC18 MCUs*. Retrieved December 13, 2009, from http://www.microchip.com/

Miller, C., & Poellabauer, C. (2008). PALER: A Reliable Transport Protocol for Code Distribution in Large Sensor Networks. In *Proceedings of Sensor, Mesh and Ad Hoc Communications and Networks, 2008: SECON '08. 5th Annual IEEE Communications Society Conference on* (pp. 206-214). IEEE.

Miller, K., Morell, L., Noonan, R., Park, S., Nicol, D., Murril, B., & Voas, J. (1992). Estimating the probability of failure when testing reveals no failures. *IEEE Transactions on Software Engineering*, *18*(1), 33–43. doi:10.1109/32.120314

MIL-STD-188-110B. (2000). *Interoperability and Performance Standards for Data Modems*. United States Department of Defense Interface Standard.

Misra, R., & Mandal, C. R. (2005). *Performance Comparison of AODV/DSR On-Demand Routing Protocols for Ad-Hoc Networks in Constrained Situation*. Paper presented at the IEEE International Conference on Personal Wireless Communications (ICPWC). New Delhi, India.

Mitch, R. H., Dougherty, R. C., Psiaki, M., Powell, S., O'Hanlon, B., Bhatti, J., & Humphreys, T. (2011). Signal Characteristics of Civil GPS Jammers. In *Proceedings of the 24th International Technical Meeting of The Satellite Division of the Institute of Navigation (ION GNSS 2011)* (pp. 1907-1919). Portland, OR: GNSS.

Mitchell, T. M. (1997). *Machine Learning*. McGraw-Hill.

Mitra, P., & Poellabauer, P. (2011). Asymmetric Geographic Forwarding: Exploiting Link Asymmetry in Location Aware Routing. *International Journal of Embedded and Real-Time Communication Systems, 2*(4). doi:10.4018/jertcs.2011100104

Mitschele-Thiel, A. (2001). *System Engineering with SDL.* Chichester, UK: John Wiley & Sons, Ltd. doi:10.1002/0470841966

Mohaisen, M., & Chang, K. (2011). Fixed-complexity sphere encoder for multiuser MIMO systems. *Journal of Communications and Networks, 13*(1), 36–39. doi:10.1109/JCN.2011.6157253

Mondal, M., & Ragheb, T. Wu, Aziz, A., & Massoud, Y. (2007). Provisioning On-Chip Networks under Buffered RC Interconnect Delay Variations. In *Proceedings of 8th International Symposium on Quality Electronic Design,* (pp. 873-878). Academic Press.

Mondal, S., Ali, W., & Salama, K. (2008). A novel approach for K-Best MIMO detection and its VLSI implementation. In *Proceedings of the IEEE International Symposium on Circuits and Systems* (pp. 936-939). New York, NY: IEEE.

Monteiro, T., Moore, T., & Hill, C. (2005). What is the Accuracy of DGPS? *Journal of Navigation, 58*(2), 207–225. doi:10.1017/S037346330500322X

Moore, S. W., Taylor, G. S., et al. (2000). Self Calibrating clocks for globally asynchronous locally synchronous systems. In *Proceedings of IEEE Int. Conf. on Computer Design,* (pp. 73-78). IEEE.

Morgenthaler, S., Braun, T., Zhao, Z., Staub, T., & Anwander, M. (2012). *UAVNet: A Mobile Wireless Mesh Network Using Unmanned Aerial Vehicles.* Paper presented at the 3rd International Workshop on Wireless Networking for Unmanned Autonomous Vehicles. Anaheim, CA.

Morozkin, P., Lavrovskaya, I., Olenev, V., & Nedovodeev, K. (2013). Integration of SDL Models into a SystemC Project for Network Simulation. *Lecture Notes in Computer Science, 7916,* 275–290. doi:10.1007/978-3-642-38911-5_16

Motamedi, K., Ionnides, N., Rümmeli, M., & Schagaev, I. (2009). Reconfigurable Network on Chip Architecture for Aerospace Applications. In *Proceedings of 30th IFAC Workshop on Real-Time Programming and 4th International Workshop on Real-Time Software* (pp. 131-136). IFAC.

Motella, B., Pini, M., & Lo Presti, L. (2012). GNSS Interference Detector Based On Chi-square Goodness-of-fit Test. In *Proceedings of the 6th ESA Workshop on Satellite Navigation Technologies and European Workshop on GNSS Signals and Signal Processing (NAVITEC 2012)* (pp. 1-6). Noordwijk, The Netherlands: GNSS.

Motella, B., Savasta, S., Margaria, D., & Dovis, F. (2009). Assessing GPS Robustness in Presence of Communication Signals. In *Proceedings of the International Workshop on Synergies in Communications and Localization, IEEE International Conference on Communications* (pp. 1-5). Dresden, Germany: IEEE.

Motella, B., Savasta, S., Margaria, D., & Dovis, F. (2011). A Method for Assessing the Interference Impact on GNSS Receivers. *IEEE Transactions on Aerospace and Electronic Systems, 47*(2), 1416–1432. doi:10.1109/TAES.2011.5751267

Moura, D. F. C., da Silva, F. A. B., & Galdino, J. F. (2012). Case Studies of Attacks over Adaptive Modulation Based Tactical Software Defined Radios. *Journal of Computer Networks and Communications.*

Moura, D. F., Salles, R. M., & Galdino, J. F. (2010). Multimedia Traffic Robustness and Performance Evaluation on a Cross-Layer Design for Tactical Wireless Networks. In *Proceedings of the 9th International Information and Telecommunication Technologies Symposium (I2TS'10).* I2TS.

Moura, D. F. C., Salles, R. M., & Galdino, J. F. (2009). Generalized input deterministic service queue model: Analysis and performance issues for wireless tactical networks. *IEEE Communications Letters, 13*(12), 965–967. doi:10.1109/LCOMM.2009.12.091646

Mukhtar, H., Kim, B. W., Kim, B. S., & Joo, S.-S. (2009). An efficient remote code update mechanism for Wireless Sensor Networks. In *Proceedings of Military Communications Conference.* IEEE.

Murali, S., Tamhankar, R., Angiolini, F., Pulling, A., Atienza, D., Benini, L., & De Micheli, G. (2006). Comparison of a Timing-Error Tolerant Scheme with a Traditional Re-transmission Mechanism for Networks on Chips. In *Proceedings of International Symposium on System-on-Chip*, (pp. 1-4). Academic Press.

Murotake, D., & Martin, A. (2004). System Threat Analysis for High Assurance Software Radio. In *Proceedings of the SDR'04 Technical Conference*. SDR.

Murotake, D., & Martin, A. (2009). A High Assurance Wireless Computing System Architecture for Software-Defined Radios and Wireless Mobile Platforms. In *Proceedings of the SDR '09 Technical Conference and Product Exposition*. SDR.

Murtaza, Z., Khan, S. A., Rafique, A., Bajwa, K. B., & Zaman, U. (2006). Silicon Real Time Operating System for Embedded DSPs. In *Proceedings of International Conference on Emerging Technologies* (pp. 188-191). doi: 10.1109/ICET.2006.336032

Musa, J. D. (1993). Operational Profiles in Software-Reliability Engineering. *IEEE Software*, *10*(2), 14–32. doi:10.1109/52.199724

Mushtaq, H., Al-Ars, Z., & Bertels, K. (2011). Survey of fault tolerance techniques for shared memory multicore/multiprocessor systems. In *Proceedings 2011 IEEE 6th Design and Test Workshop (IDT)* (pp. 12 -17). IEEE.

Myagmar, S., Lee, A. J., & Yurcik, W. (2005). Threat Modeling as a Basis for Security Requirements. In *Proceedings of the Symposium on Requirements Engineering for Information Security* (SREIS). SREIS.

Nácul, A. C., Regazzoni, F., & Lajolo, M. (2007). Hardware Scheduling Support in SMP Architectures. In *Proceedings of the Conference on Design, Automation and Test in Europe* (pp. 642-647). San Jose, CA: EDA Consortium. doi: 10.1109/DATE.2007.364666

Nakano, T., Utama, A., Itabashi, M., Shiomi, A., & Imai, M. (1995). Hardware Implementation of a Real-Time Operating System. In *Proceedings of TRON Project International Symposium* (pp. 34-42). doi: 10.1109/TRON.1995.494740

Nakura, T., Nose, K., & Mizuno, M. (2007). Fine-Grain Redundant Logic Using Defect-Prediction Flip-Flops. In *Proceedings of IEEE International Solid-State Circuits Conference*, (pp. 402-403, 611). IEEE.

Narayanan, V., & Xie, Y. (2006). Reliability concerns in embedded system designs. *Computer*, *38*(1), 118–120. doi:10.1109/MC.2006.31

Narayanaswamy, S., Kawadia, V., Sreenivas, R. S., & Kumar, P. R. (2002). *Power Control in Ad-Hoc Networks: Theory, Architecture, Algorithm and Implementation of the COMPOW Protocol*. Paper presented at the European Wireless Conference. Florence, Italy.

NATO. (2000). Profile for High Frequency (HF) Radio Data Communication Ed. 2 (STANAG 5066). North Atlantic Treaty Organization.

Naya, F., Noma, H., Ohmura, R., & Kogure, K. (2005). Bluetooth-Based Indoor Proximity Sensing for Nursing Context Awareness. In *Proceedings of the 9th IEEE International Symposium on Wearable Computers* (pp. 212-213). Osaka, Japan: IEEE.

Nemydrov, V., & Martin, G. (2004). *Systems-on-chip: Design and evaluation problems*. Moscow: Technosphera.

Nethercote, N., & Seward, J. (2007). Valgrind: a framework for heavyweight dynamic binary instrumentation. In *Proceedings of the 2007 ACM SIGPLAN conference on Programming language design and implementation*, PLDI '07, (pp. 89–100). New York, NY: ACM.

Nicola, M., Musumeci, L., Pini, M., Fantino, M., & Mulassano, P. (2010). Design of a GNSS Spoofing Device Based on a GPS/Galileo Software Receiver for the Development of Robust Countermeasures. In *Proceedings of the European Navigation Conference (ENC 2010)*. Braunschweig, Germany: ENC.

Nicopoulos, C., Srinivasan, S., & Yanamandra, A., Park, Narayanan, V., Das, C.R., & Irwin, M.J. (2010). On the Effects of Process Variation in Network-on-Chip Architectures. *IEEE Transactions on Dependable and Secure Computing*, *7*(3), 240–254. doi:10.1109/TDSC.2008.59

Ni, L. M., Liu, Y., Lau, Y. C., & Patil, A. P. (2004). Landmarc: Indoor Location Sensing Using Active RFID. *Wireless Networks*, *10*, 701–710. doi:10.1023/B:WINE.0000044029.06344.dd

Noguchi, J., Saito, T., Ohashi, N., Ashihara, H., Maruyama, H., & Kubo, M. … Hinode, K. (2001). Impact of low-k dielectrics and barrier metals on TDDB lifetime of Cu interconnects. In *Proceedings of 39th IEEE International Reliability Physics Symposium*, (pp. 355-359). IEEE.

Nordic Semiconductors. (2007, July). *nRF24L01 Product Specification*. Retrieved December 7, 2009, from http://www.nordicsemi.com/

Nordström, S., & Asplund, L. (2007). Configurable Hardware/Software Support for Single Processor Real-Time Kernels. In *Proceedings of International Symposium on System-on-Chip* (pp. 1-4). doi: 10.1109/ISSOC.2007.4427426

Nose, & Sakurai. (2000). Optimization of VDD and VTH for low-power and high speed applications. In *Proceedings of the 2000 Asia and South Pacific Design Automation Conference* (pp. 469—474). IEEE.

Object Management Group (OMG). (2000). *Object Constraint Language Specification*. OMG Unified Modeling Language Specification, Version 1.3. Retrieved March 17, 2010, from http://www.omg.org/spec/UML/1.3

Oh, H., Dutt, N., & Ha, S. (2005). Single appearance schedule with dynamic loop count for minimum data buffer from synchronous dataflow graphs. In *Proceedings of the 2005 international conference on Compilers, architectures and synthesis for embedded systems* (CASES '05), (pp. 157 – 165). ACM.

Olenev, V. (2009). Different approaches for the stacks of protocols SystemC modeling analysis. In *Proceedings of the Saint-Petersburg University of Aerospace Instrumentation scientific conference* (pp. 112-113). Saint-Petersburg, Russia: Saint-Petersburg University of Aerospace Instrumentation (SUAI).

Olenev, V., Onishenko, L., & Eganyan, A. (2008). Connections in SystemC Models of Large Systems. In *Proceedings of the Saint-Petersburg University of Aerospace Instrumentation scientific student's conference* (pp. 98-99). Saint-Petersburg, Russia: Saint-Petersburg University of Aerospace Instrumentation (SUAI).

Olenev, V., Rabin, A., Stepanov, A., & Lavrovskaya, I. (2009). SystemC and SDL Co-Modeling Methods. In *Proceedings of 6ᵗʰ Seminar of Finnish-Russian University Cooperation in Telecommunications (FRUCT) Program* (pp. 136-140). Saint-Petersburg, Russia: Saint-Petersburg University of Aerospace Instrumentation (SUAI).

Olenev, V., Sheynin, Y., Suvorova, E., Balandin, S., & Gillet, M. (2009). SystemC Modeling of the Embedded Networks. In *Proceedings of 6ᵗʰ Seminar of Finnish-Russian University Cooperation in Telecommunications (FRUCT) Program* (pp. 85-95). Saint-Petersburg, Russia: Saint-Petersburg University of Aerospace Instrumentation (SUAI).

Olimpiew, E. M., & Gomaa, H. (2005). Model-based testing for applications derived from software product lines. In *Proceedings of the 1st international workshop on Advances in model-based testing*. New York, NY: AMOST.

OMG CORBA Security Service v1.8. (2002). Retrieved from http://www.omg.org/spec/SEC/1.8/PDF/

OMG DDS Portal. (2012). Retrieved from http://portals.omg.org/dds/

OMG Web site. (2012). Retrieved from http://www.omg.org/

Open SystemC Initiative (OSCI). (2005). *IEEE 1666™-2005 Standard for SystemC*. Retrieved from http://www.systemc.org

Oresko, J. J., Jin, Z., Cheng, J., Huang, S., Sun, Y., Duschl, H., & Cheng, A. C. (2010). A wearable smartphone-based platform for real-time cardiovascular disease detection via electrocardiogram processing. *IEEE Transactions on Information Technology in Biomedicine, 14*, 734–740. doi:10.1109/TITB.2010.2047865 PMID:20388600

Orshansky, M., & Keutzer, K. (2002). A general probabilistic framework for worst case timing analysis. In *Proceedings 39th Design Automation Conference*, (pp. 556-561). Academic Press.

Ostroumov, S., & Tsiopoulos, L. (2011). VHDL Code Generation from Formal Event-B Models. In *Proceedings of Digital System Design (DSD), 2011 14th Euromicro Conference* (pp. 127-134). Oulu, Finland: IEEE.

Ostroumov, S., & Tsiopoulos, L. (2012). Formal Development of Hierarchical Agent-Based Monitoring Systems for Dynamically Reconfigurable NoC Platforms. *International Journal of Embedded and Real-Time Communication Systems*, *3*(2), 40–72. doi:10.4018/jertcs.2012040103

Ostroumov, S., Tsiopoulos, L., Plosila, J., & Sere, K. (2013). Formal Approach to Agent-Based Dynamic Reconfiguration in Networks-On-Chip. *Journal of Systems Architecture*, *59*(9), 709–728. doi:10.1016/j.sysarc.2013.06.001

P400. (n.d.). *Data Sheet*. Retrieved 2011 from http://www.timedomain.com/p400.php

Paci, G., Bertozzi, D., & Benini, L. (2009). Effectiveness of adaptive supply voltage and body bias as post-silicon variability compensation techniques for full-swing and low-swing on-chip communication channels. In Proceedings of Design, Automation & Test in Europe, (pp. 1404-1409). Academic Press.

Pahlavan, K., Akgul, O. F., Heidari, M., Hatami, A., Elwell, M. J., & Tingley, D. R. (2006). Indoor Geolocation in the Absence of Direct Path. *IEEE Transactions on Wireless Communications*, *13*(6), 50–58. doi:10.1109/MWC.2006.275198

Painkras, E., Plana, L., Garside, J., Temple, S., Davidson, S., & Pepper, J. ... Furber, S. (2012). SpiNNaker: A Multi-Core System-on-Chip for Massively-Parallel Neural Net Simulation. In *Proceedings of IEEE Custom Integrated Circuits Conference* (CICC) (pp. 1-4). CICC.

Panades, I. M., Clermidy, F., Vivet, P., & Greiner, A. (2008). Physical Implementation of the DSPIN Network-on-Chip in the FAUST Architecture. In *Proceedings of International Symposium on Networks-on-Chip* (NoCs) (pp.139-148). Academic Press.

Pant, R. K., & Bagchi, S. (2009). Hermes: Fast and Energy Efficient Incremental Code Updates for Wireless Sensor Networks.[Rio de Janeiro: IEEE.]. *Proceedings - IEEE INFOCOM*, *2009*, 639–647.

Pan, Z., & Wells, B. E. (2008). Hardware Supported Task Scheduling on Dynamically Reconfigurable SoC Architectures. *IEEE Transactions on Very Large Scale Integration (VLSI). Systems*, *16*(11), 1465–1474. doi: doi:10.1109/TVLSI.2008.2000974

Paonni, M., Jang, J. G., Eissfeller, B., Wallnert, S., Rodriguez, J. A., Samson, J., et al. (2010). Innovative Interference Mitigation Approaches, Analytical Analysis, Implementation and Validation. In *Proceedings of the 5th ESA Workshop on Satellite Navigation Technologies and European Workshop on GNSS Signals and Signal Processing (NAVITEC)* (pp. 1-8). Noordwijk, The Netherlands: GNSS.

Parks, T. M. (1995). *Bounded Scheduling of Process Networks*. (Doctoral dissertation). Berkeley, CA.

Park, S., Hong, D.-S., & Chae, S.-I. (2008). A Hardware Operating System Kernel for Multi-Processor Systems. *IEICE Electronics Express*, *5*(9), 296–302. doi:10.1587/elex.5.296

Patel, A., & Ghose, K. (2008). Energy-efficient mesi cache coherence with pro-active snoop filtering for multicore microprocessors. In *Proceeding of the thirteenth international symposium on Low power electronics and design*, (pp. 247–252). Academic Press.

Paul, B.C., Kang, Kufluoglu, H., Alam, M.A., & Roy, K. (2005). Impact of NBTI on the temporal performance degradation of digital circuits. *IEEE Electron Device Letters*, *26*(8), 560–562. doi:10.1109/LED.2005.852523

Paulin, P., Pilkington, C., Langevin, M., Bensoudane, E., Lyonnard, D., & Benny, O. et al. (2006). Parallel Programming Models for a Multiprocessor SoC Platform Applied to Networking and Multimedia. *IEEE Transactions on Very Large Scale Integration (VLSI). Systems*, *14*(7), 667–680. doi: doi:10.1109/TVLSI.2006.878259

Pedram, M., & Nazarian, S. (2006). Thermal Modeling, Analysis, and Management in VLSI Circuits: Principles and Methods. *Proceedings of the IEEE*, *94*(8), 1487–1501. doi:10.1109/JPROC.2006.879797

Peel, C. B., Hochwald, B. M., & Swindlehurst, A. L. (2005). A vector-perturbation technique for near-capacity multianetenna multiuser communication: channel inversion and regularization. *IEEE Transactions on Communications*, *53*(1), 195–202. doi:10.1109/TCOMM.2004.840638

Pei, L., Chen, R., Liu, J., Tenhunen, T., Kuusniemi, H., & Chen, Y. (2010). Inquiry-Based Bluetooth Indoor Positioning via RSSI Probability Distributions. In *Proceedings of the Second International Conference on Advances in Satellite and Space Communications (SPACOMM 2010)* (pp. 151–156). Athens, Greece: IEEE.

Pei, L., Chen, R., Liu, J., Tenhunen, T., Kuusniemi, H., & Chen, Y. (2010). Using Inquiry-based Bluetooth RSSI Probability Distributions for Indoor Positioning. *Journal of Global Positioning Systems, 9*(2), 122–130.

Pei, L., Liu, J., Guinness, R., Chen, Y., Kuusniemi, H., & Chen, R. (2012). Using LS-SVM Based Motion Recognition for Smartphone Indoor Wireless Positioning. *Sensors (Basel, Switzerland), 12*(5), 6155–6175. doi:10.3390/s120506155 PMID:22778635

Pellegrini, A., Bertacco, V., & Austin, T. (2010). Fault-Based Attack of RSA Authentication. In Proceedings of Design, Automation and Test in Europe (DATE). DATE.

Penna, F., Caceres, M. A., & Wymeersch, H. (2010). Cramér-Rao bound for hybrid GNSS-terrestrial cooperative positioning. *IEEE Communications Letters, 14*(11), 1005–1007. doi:10.1109/LCOMM.2010.091310.101060

Peterson, J. L. (1981). *Petri Net Theory and the Modeling of Systems*. Prentice Hall.

Pini, M., Motella, B., & Troglia Gamba, M. (2013). Detection of Correlation Distortions Through Application of Statistical Methods. In *Proceedings of the 26th International Technical Meeting of The Satellite Division of the Institute of Navigation (ION GNSS 2013)* (pp. 3279-3289). Nashville, TN: GNSS.

Pino, J. L., & Lee, E. (1995). Hierarchical static scheduling of dataflow graphs onto multiple processors. In *Proceedings of International Conference on Acoustics, Speech, and Signal Processing* (ICASSP-95), (pp. 2643–2646). ICASSP.

Pinto, E. L., & Galdino, J. F. (2009). Simple and robust analytically derived variable step-size least mean squares algorithm for channel estimation. *Communications, IET, 3*(12), 1832–1842. doi:10.1049/iet-com.2009.0038

Plackett, R. L. (1950). Some theorems in least squares. *Biometrika, 37*(1/2), 149–157. doi:10.2307/2332158 PMID:15420260

Plana, L., Furber, S., Temple, S., Khan, M., Yebin, S., Jian, W., & Shufan, Y. (2007). A GALS Infrastructure for a Massively Parallel Multiprocessor. *IEEE Design & Test of Computers, 24*(5), 454–463. doi:10.1109/MDT.2007.149

Poggio, T., & Smale, S. (2003). The mathematics of learning: Dealing with data.[AMS]. *Notices of the American Mathematical Society, 50*(5), 537–544.

Pohl, K., Böckle, G., & van der Linden, F. (2005). *Software Product Line Engineering: Foundations, Principles, and Techniques*. Springer-Verlag.

Proakis, J. G. (2001). Digital Communications (4th Ed.). McGraw-Hill International Ed.s.

Prowell, S. (2005). Using markov chain usage models to test complex systems. In *Proceedings of the 38th Annual Hawaii International Conference on System Sciences*. Academic Press.

Prowell, S., Trammell, C., Linger, R., & Poore, J. (1999). *Cleanroom Software Engineering: Technology and Process*. Addison-Wesley-Longman.

Prowell, S. J. (2000). TML a description language for Markov chain usage models. *Information and Software Technology, 42*(12), 835–844. doi:10.1016/S0950-5849(00)00123-3

Prowell, S. J., & Poore, J. H. (2003). Foundations of Sequence-Based Software Specification. *IEEE Transactions on Software Engineering, 29*(5), 417–429. doi:10.1109/TSE.2003.1199071

Psiaki, M. L., Powell, S. P., & O'Hanlon, B. W. (2013). GNSS Spoofing Detection using High-Frequency Antenna Motion and Carrier-Phase Data. In *Proceedings of the 26th International Technical Meeting of The Satellite Division of the Institute of Navigation (ION GNSS 2013)* (pp. 2949-2991). Nashville, TN: GNSS.

PTOLEMUS Consulting Group. (2012). *Global Insurance Telematics Study 2012*. Author.

Quach, N. (2000). High Availability and Reliability in the Itanium Processor. *IEEE Micro, 20*(5), 61–69. doi:10.1109/40.877951

Qualcomm, Inc. (2013). *Snapdragon*. Retrieved December 16, 2013, from http://www.qualcomm.com/snapdragon

Radetzki, M., Feng, C. C., Zhao, X. Q., & Jantsch, A. (2013). Methods for fault tolerance in networks-on-chip. *ACM Computing Surveys, 46*(1). doi:10.1145/2522968.2522976

Rantala, P., Isoaho, J., & Tenhunen, H. (2007). Novel Agent-Based Management for Fault-Tolerance in Network-on-Chip. In *Proceedings of Euromicro Conference on Digital System Design Architectures, Methods and Tools* (pp. 551 - 555). Lubeck: IEEE.

Reijers, N., & Langendoen, K. (2003). Efficient code distribution in wireless sensor networks. In *Proceedings of the 2nd ACM international conference on Wireless sensor networks and applications (WSNA '03)* (pp. 60-67). New York: ACM.

Reijers, N., Halkes, G., & Langendoen, K. (2004). *Link Layer Measurements in Sensor Networks*. Paper presented at the 1ˢᵗ IEEE International Conference on Mobile Ad-hoc and Sensor Systems (MASS). Fort Lauderdale, FL.

Reinhold, M. (2003). *Praxistauglichkeit von Vorgehensmodellen: Specification of large IT-Systems – Integration of Requirements Engineering and UML based on V-Model'97*. Nordrhein-Westfalen: Shaker Verlag.

Renauld, M., Standaert, F. X., & Veyrat-Charvillon, N. (2009). Algebraic Side-Channel Attacks on the AES: Why Time also Matters in DPA. In *Proceedings of the Workshop on Cryptographic Hardware and Embedded Systems 2009* (CHES 2009). CHES.

Retscher, G., & Fu, Q. (2007). Integration of RFID, GNSS and DR for Ubiquitous Positioning in Pedestrian Navigation. *Journal of Global Positioning Systems, 6*(1), 56–64. doi:10.5081/jgps.6.1.56

Reuys, A., Kamsties, E., Pohl, K., & Reis, S. (2005). *Model-Based System Testing of Software Product Families. Advanced Information Systems Engineering*. Berlin: Springer.

Rifkin, R., Yeo, G., & Poggio, T. (2003). Regularized least-squares classification. In Advances in Learning Theory: Methods, Model and Applications, (vol. 190, pp. 131–154). IOS Press.

Riley, R., Jiang, X., & Xu, D. (2010). An Architectural Approach to Preventing Code Injection Attacks. *IEEE Transactions on Dependable and Secure Computing, 7*(4), 351–365. doi:10.1109/TDSC.2010.1

Robinson, K. (2010, October 10). *System Modelling & Desing using Event-B*. Retrieved from http://wiki.event-b.org/images/archive/20101010115803!SM%26D-KAR.pdf

RODIN. (2011). Retrieved from http://sourceforge.net/projects/rodin-b-sharp/

Romdhani, B., Barthel, D., & Valois, B. (2012). *Exploiting Asymmetric Links in a Convergecast Routing Protocol*. Paper presented at the 11th International Conference on Ad Hoc Networks and Wireless (ADHOC-NOW). Belgrade, Serbia.

Roosta, T., Menzo, M., & Sastry, S. (2005). *Probabilistic Geographic Routing in Ad Hoc and Sensor Networks*. Paper presented at Wireless Networks and Emerging Technologies (WNET). Banff, Canada.

Rossi, M., Bui, N., Zanca, G., Stabellini, L., Crepaldi, R., & Zorzi, M. (2012). SYNAPSE++: Code Dissemination in Wireless Sensor Networks Using Fountain Codes. *IEEE Transactions on Mobile Computing, 9*(12), 1749–1765. doi:10.1109/TMC.2010.109

RTCA DO-292. (2004). *Assessment of Radio Frequency Interference Relevant to the GNSS L5/E5A Frequency Band*.

Ruschival, T., Nenninger, P., Kantz, F., & Streitferdt, D. (2009). Test Case Mutation in Hybrid State Space for Reduction of No-Fault-Found Test Results in the Industrial Automation Domain. In *Proceedings of the 2nd International Workshop on Industrial Experience in Embedded Systems Design (IEESD 2009) at the 33rd Annual IEEE International Computers, Software and Applications Conference*. IEEE.

Salehie, M., & Tahvildari, L. (2009). Self-adaptive software: Landscape and research challenges. *ACM Trans. Auton. Adapt. Syst., 4*(2).

Sang, L., Arora, A., & Zhang, H. (2007). *On Exploiting Asymmetric Wireless Links via One-way Estimation*. Paper presented at the 8th International Symposium on Mobile Ad Hoc Networking and Computing. Montreal, Canada.

Sarangi, S. R., Greskamp, B., Teodorescu, R., Nakano, J., Tiwari, A., & Torrellas, J. (2008). VARIUS: A Model of Process Variation and Resulting Timing Errors for Microarchitects. *IEEE Transactions on Semiconductor Manufacturing, 21*(1), 3–13. doi:10.1109/TSM.2007.913186

Savage, J. E. (1998). *Models of computation: Exploring the power of computing.* Addison-Wesley Pub.

Savasta, S. (2010). *GNSS Localization Techniques in Interfered Environment.* (Unpublished Ph.D. Dissertation). Politecnico di Torino, Italy.

SCA - Software Communications Architecture Specification, v.2.2. (2001). Retrieved from http://jtnc.mil/sca/Pages/sca1.aspx

SCA - Software Communications Architecture Specification, v.2.2.2. (2006). Retrieved from http://jtnc.mil/sca/Pages/sca1.aspx

SCA - Software Communications Architecture Specification, v.4.0. (2012). Retrieved from http://jtnc.mil/sca/Pages/sca1.aspx

SCA Next Specification v. 1.0. (2011). Retrieved from http://jtnc.mil/sca/Pages/sca1.aspx

SCA Security – Security Supplement to the Software Communications Architecture Specification. (2001). Retrieved from http://jtnc.mil/sca/Pages/sca1.aspx

Schmid, K., & Santana de Almeida, E. (2013). Product Line Engineering. *Software, IEEE, 30*(4), 24–30. doi:10.1109/MS.2013.83

Schmidt, D., Joham, M., & Utschick, W. (2008). Minimum mean square error vector precoding. *European Transactions on Telecommunications, 19*(3), 219–231. doi:10.1002/ett.1192

Schnorr, C. P., & Euchner, M. (1994). Lattice basis reduction: Improved practical algorithms and solving subset sum problems. *Mathematical Programming, 66*(2), 181–199. doi:10.1007/BF01581144

Schürmans, S., Zhang, D., Auras, D., Leupers, R., Ascheid, G., Chen, X., & Wang, L. (2013). Creation of ESL Power Models for Communication Architectures using Automatic Calibration. In *Proceedings of Design Automation Conference (DAC)* (pp. 1-6). doi: 10.1145/2463209.2488804

Scott, L. (2003). Anti-spoofing and Authenticated Signal Architectures for Civil Navigation Systems. In *Proceedings of the 16th International Technical Meeting of the Satellite Division of The Institute of Navigation* (pp. 1543-1552). Portland, OR.

Scott, L. (2013). Spoofing – Upping the Anti. *Inside-GNSS, 4.*

Scott, L. (2004). *GPS and GNSS RFI and Jamming Concerns IV. Navtech Tutorial 410D.* Long Beach, CA: ION GNSS.

Seidel, H. (2006). *A Task-level Programmable Processor.* Duisburg, Germany: WiKu-Verlag.

Shabbir, A., Kumar, A., Mesman, B., & Corporaal, H. (2011). Distributed Resource Management for Concurrent Execution of Multimedia Applications on MPSoC Platforms. In *Proceedings of International Conference on Embedded Computer Systems* (pp. 132-139). doi: 10.1109/SAMOS.2011.6045454

Shang, L., Peh, L., Kumar, A., & Jha, N. K. (2004). Thermal Modeling, Characterization and Management of On-Chip Networks. In *Proceedings of the 37th International Symposium on Microarchitecture* (pp. 67-78). IEEE.

Sheng, Z., & Pollard, J. K. (2006). Position Measurement Using Bluetooth. *IEEE Transactions on Consumer Electronics, 52*(2), 555–558. doi:10.1109/TCE.2006.1649679

Shotton, J., Fitzgibbon, A., Cook, M., Sharp, T., Finocchio, M., Moore, R., et al. (2011). Real-time human pose recognition in parts from single depth images. In *Proceedings of IEEE Computer Vision and Pattern Recognition (CVPR) 2011.* IEEE.

Silva, F. A. B., Moura, D. F., & Galdino, J. F. (2012). Classes of attacks for tactical software defined radios. [IJERTCS]. *International Journal of Embedded and Real-Time Communication Systems, 3*(4), 57–82. doi:10.4018/jertcs.2012100104

Simone, M., Lajolo, M., & Bertozzi, D. (2008). Variation tolerant NoC design by means of self-calibrating links. In Proceedings of Design, Automation and Test in Europe DATE '08, (pp. 1402--1407). DATE.

Singh, A. K. (2010). *Run-time mapping of multiple communicating tasks on MPSoC platforms* (pp. 1019–1026). Procedia Computer Science.

Skadron, K., Stan, M., W., H., Velusamy, S., Sankaranarayanan, K., & Tarjan, D. (2003). Temperature-Aware Microarchitecture: Exended Discussion and Results. *Computer Engineering, (CS-2003-08)* .

Sloss, A., Symes, D., & Wright, C. (2004). *ARM System Developer's Guide: Designing and Optimizing System Software*. San Francisco, CA: Morgan Kaufmann Publishers.

Son, D., Helmy, A., & Krishnamachari, B. (2004). The Effect of Mobility-induced Location Errors on Geographic Routing in Ad Hoc Networks: Analysis and Improvement using Mobility Prediction. *IEEE Transactions on Mobile Computing*, 233–245. doi:10.1109/TMC.2004.28

Sottile, F., Caceres, M. A., & Spirito, M. A. (2011). *A Simulation tool for hybrid-cooperative positioning*. Paper presented at IEEE International Conference on Localization and GNSS. Tampere, Finland.

Sottile, F., Vesco, A., Scopigno, R., & Spirito, M. A. (2012). *MAC layer impact on the performance of real-time cooperative positioning*. Paper presented at IEEE Wireless Communications and Networking Conference. Paris, France.

Sottile, F., Wymeersch, H., Caceres, M. A., & Spirito, M. A. (2011). *Hybrid GNSS-ToA cooperative positioning based on particle filter*. Paper presented at IEEE Global Communication Conference. Houston, TX.

Spirent. (n.d.). *Wi-Fi Positioning Access Point Simulator*. Retrieved from http://www.spirent.com/~/media/Datasheets/Positioning/GSS5700.ashx

Srinivasan, J., & Adve, S. (2003). *RAMP: A Model for Reliability Aware MicroProcessor Design*. IBM Research Report, RC23048.

Srinivasan, K., & Levis, P. (2006). *RSSI is Under Appreciated*. Paper presented at the 3rd Workshop on Embedded Networked Sensors. Cambridge, MA.

Srinivasan, K., Dutta, P., Tavakoli, A., & Levis, P. (2006). *Understanding the Causes of Packet Delivery Success and Failure in Dense Wireless Sensor Networks*. Paper presented at the 4th International Conference on Embedded Networked Sensor Systems. Boulder, CO.

Standaert, F.-X., Malkin, T. G., & Yung, M. (2009). A Unified Framework for the Analysis of Side-Channel Key Recovery Attacks. *Lecture Notes in Computer Science*, *5479*, 443–461. doi:10.1007/978-3-642-01001-9_26

Standards, I. E. E. E. (2008, August 31). *Part 15.4: Wireless Medium Access Control (MAC) and Physical Layer (PHY) Specifications for Low-Rate Wireless Personal Area Networks (WPANs)*. Retrieved December 16, 2010, from http://standards.ieee.org/getieee802/download/802.15.4a-2007.pdf

Stango, A., & Prasad, N. R. (2009). Policy-Based Approach for Secure Radio Software Download. In *Proceedings of the SDR '09 Technical Conference and Product Exposition*. SDR.

Stathopoulos, T., Heidemann, J., Estrin, D., & Sensing, C. U. (2003). *A remote code update mechanism for wireless sensor networks*. Citeseer.

Stepanov, A., Lavrovskaya, I., & Olenev, V. (2010). SDL and SystemC Co-Modelling: The Protocol SDL Models Tester. In *Proceedings of the 8th Conference of Open Innovation Framework Program FRUCT*. Saint-Petersburg, Russia: Saint-Petersburg University of Aerospace Instrumentation (SUAI).

Stepanov, A., Lavrovskaya, I., Olenev, V., & Rabin, A. (2010). SystemC and SDL Co-Modelling Implementation. In *Proceedings of the 7th Conference of Finnish-Russian University Cooperation in Telecommunications (FRUCT) Program*. Saint-Petersburg, Russia: Saint-Petersburg University of Aerospace Instrumentation (SUAI).

Stergiou, S., Angiolini, F., Carta, S., Raffo, L., Bertozzi, D., & De Micheli, G. (2005). Xpipes Lite: a synthesis oriented design library for networks on chips. In Proceedings, Design, Automation and Test in Europe, (pp. 1188-1193). Academic Press.

STMicroelectronics & CEA. (2010). *Platform 2012: A Many-core Programmable Accelerator for Ultra-Efficient Embedded Computing in Nanometer Technology* (White Paper). Retrieved December 10, 2013, from http://www.2parma.eu/index.php/documents/publications.html

Strano, A., Ludovici, D., & Bertozzi, D. (2010). A Library of Dual-Clock FIFOs for Cost-Effective and Flexible MPSoCs Design. In *proceedings of International Conference on Embedded Computer Systems* (SAMOS) (pp. 20-27). SAMOS.

Strasser, M., Popper, C., Capkun, S., & Cagalj, M. (2008). Jamming-resistant Key Establishment using Uncoordinated Frequency Hopping. In *Proceeding of the 2008 IEEE Symposium on Security and Privacy*. IEEE.

Suhonen, J., Kuorilehto, M., Hännikäinen, M., & Hämäläinen, T. (2006). Cost-Aware Dynamic Routing Protocol for Wireless Sensor Networks - Design and Prototype Experiments. In *Proceedings of Personal, Indoor and Mobile Radio Communications, 2006 IEEE 17th International Symposium on* (pp. 1-5). IEEE.

Sullivan, H., & Bashkow, T. R. (1977). A large scale, homogeneous, fully distributed parallel machine. In *Proceedings of the 4th annual symposium on Computer architecture*, (pp. 105–117). Academic Press.

Sun, C.-C., & Jan, S.-S. (2011). GNSS Interference Detection and Excision Using Time-Frequency Representation. In *Proceedings of the 2011 International Technical Meeting of The Institute of Navigation* (pp. 365-373). San Diego, CA: Academic Press.

Sun, Y., & Cheng, A. C. (2012). Machine learning on-a-chip: A high-performance low-power reusable neuron architecture for artificial neural networks in ECG classifications. *Computers in Biology and Medicine, 42*(7), 751–757. doi:10.1016/j.compbiomed.2012.04.007 PMID:22595230

Sutherland, W. R. (1966). *The on-line graphical specification of computer procedures*. (PhD Thesis). MIT, Cambridge, MA.

Suvorova, E. (2007). A Methodology and the Tool for Testing SpaceWire Routing Switches. In *Proceedings of the first International SpaceWire Conference*. Retrieved September 19, 2007, from http://spacewire.computing.dundee.ac.uk/proceedings/Papers/Test and Verification 2/suvorova2.pdf

Suvorova, E., & Sheynin, Y. (2003). *Digital systems design on VHDL language*. Saint-Petersburg, Russia: BHV-Petersburg.

Suykens, J., Van Gestel, T., De Brabanter, J., De Moor, B., & Vandewalle, J. (2002). *Least Squares Support Vector Machines*. World Scientific Pub. Co.

Swan, S. (2003). *A Tutorial Introduction to the SystemC TLM Standard*. Retrieved July 7, 2008, from http://www-ti.informatik.uni-tuebingen.de/~systemc/Documents/Presentation-13-OSCI_2_swan.pdf

Swanson, M., Bartol, N., & Moorthy, R. (2010). *Piloting Supply Chain Risk Management Practices for Federal Information Systems*. Draft NIST IR 7622.

Swere, E. A. (2008). *Machine Learning in Embedded Systems*. (PhD thesis). Loughborough University.

Swiderski, F., & Snyder, W. (2004). *Thread Modeling*. Microsoft Press.

Sylvester, D., Blaauw, D., & Karl, E. (2006). ElastIC: An Adaptive Self-Healing Architecture for Unpredictable Silicon. *IEEE Design & Test of Computers, 23*(6), 484–490. doi:10.1109/MDT.2006.145

Synopsys, Inc. (2013). *Synopsys Platform Architect*. Retrieved December 06, 2013, from http://www.synopsys.com/Systems/ArchitectureDesign/Pages/PlatformArchitect.aspx

Synopsys, Inc. (2013). *Synopsys Processor Designer*. Retrieved December 06, 2013, from http://www.synopsys.com/Systems/BlockDesign/ProcessorDev/Pages/default.aspx

Synopsys. (2007). *PrimeTime VX Application Note - Implementation Methodology with Variation-Aware Timing Analysis, Version 1.0*. Author.

Syrjärinne, J. (2001). *Studies on Modern Techniques for Personal Positioning*. (Ph.D. Thesis). Tampere University of Technology, Tampere, Finland.

Szumski, A. (2011). Finding the Interference, Karhunen-Loève Transform as an Instrument to Detect Weak RF Signals. *InsideGNSS, 3*, 56-64.

Tanenbaum, A. S., & David, J. W. (2011). *Computer Networks* (5th ed.). Prentice Hall.

TEMPEST. (1982). *Tempest fundamentals, NSA-82-89, NACSIM 5000, National Security Agency*. Retrieved from http://cryptome.org/jya/nacsim-5000/nacsim-5000.htm

TEMPEST. (1995). *NSTISSAM TEMPEST/2-95, Red/Black Installation Guidance*. Retrieved from http://cryptome.org/jya/tempest-2-95.htm

Texas Instruments, Inc. (2013). *OMAP*. Retrieved January 25, 2013, from http://focus.ti.com/docs/prod/folders/print/omap3530.html

Thonnart, Y., Vivet, P., & Clermidy, F. (2010). A Fully Asynchronous Low-Power Framework for GALS NoC Integration. In Proceedings of Design, Automation & Test in Europe Conference & Exhibition (DATE) (pp. 33-38). DATE.

Tockhorn, A., Cornelius, C., Saemrow, H., & Timmermann, D. (2010). Modeling temperature distribution in Networks-on-Chip using RC-circuits. *13th IEEE Symposium on Design and Diagnostics of Electronic Circuits and Systems* (pp. 229-232). IEEE.

Tomasevic, M., & Milutinovic, V. (1994). Hardware Approaches to Cache Coherence in Shared-Memory Multiprocessors, Part 1. *IEEE Micro, 14*(5), 52–59. doi:10.1109/MM.1994.363067

Tomasevic, M., & Milutinovic, V. (1994). Hardware Approaches to Cache Coherence in Shared-Memory Multiprocessors, Part 2. *IEEE Micro, 14*(6), 61–66. doi:10.1109/40.331392

Tomlinson, M. (1971). New automatic equaliser employing modulo arithmetic. *Electronics Letters, 7*(5-6), 138–139. doi:10.1049/el:19710089

Tran, A. T., Truong, D. N., & Baas, M. B. (2009). A low-cost high-speed source-synchronous interconnection technique for GALS Chip multiprocessors. In *Proceedings of ISCAS*, (pp. 996-999). ISCAS.

Troglia Gamba, M., Falletti, E., Rovelli, D., & Tuozzi, A. (2012). FPGA Implementation Issues of a Two-pole Adaptive Notch Filter for GPS/Galileo Receivers. In *Proceedings of the 25th International Technical Meeting of The Satellite Division of the Institute of Navigation (ION GNSS 2012)* (pp. 3549-3557). Nashville, TN: GNSS.

Troglia Gamba, M., Motella, B., & Pini, M. (2013). Statistical Test Applied to Detect Distortions of GNSS Signals. In *Proceedings of the 2013 International Conference on Localization and GNSS (ICL-GNSS)* (pp. 1-6). Torino, Italy: GNSS.

Unsal, O. S., Tschanz, J. W., Bowman, K., De, V., Vera, X., Gonzalez, A., & Ergin, O. (2006). Impact of Parameter Variations on Circuits and Microarchitecture. *IEEE Micro, 26*(6), 30–39. doi:10.1109/MM.2006.122

Utting, M., Pretschner, A., & Legeard, B. (2012). A taxonomy of model-based testing approaches. *Software Testing. Verification & Reliability, 22*(5), 297–312. doi:10.1002/stvr.456

van de Ven, A. (2005, July). Limiting buffer overflows with ExecShield. *Red Hat Magazine*.

Van der Linden, F. J., Schmid, K., & Rommes, E. (2007). *Software Product Lines in Action: The Best Industrial Practice in Product Line Engineering*. Berlin: Springer.

van Dijk, M., Rhodes, J., Sarmenta, L. F. G., & Devadas, S. (2007). Offline Untrusted Storage with Immediate Detection of Forking and Replay Attacks. In *Proceedings of the The Second ACM Workshop on Scalable Trusted Computing* (STC'07). STC.

Vangal, S., Howard, J., Ruhl, G., Dighe, S., Wilson, H., Tschanz, J., et al. (2007). An 80-tile 1.28tflops network-on-chip in 65nm cmos. In *Proceedings of IEEE International Solid-State Circuits Conference ISSCC 2007*, (pp. 98–589). IEEE.

Verbitsky, D., Dobkin, R., & Ginosar, R. (2011). A Four-Stage Mesochronous Synchronizer with Back-Pressure and Buffering for Short and Long Range Communications. In Technical Report.

Verdú, S. (1998). *Multiuser Detection*. Cambridge, UK: Cambridge University Press.

Viterbo, E., & Boutros, J. (1999). A universal lattice code decoder for fading channels. *IEEE Transactions on Information Theory, 45*(5), 1639–1642. doi:10.1109/18.771234

Vitkovskiy, A., Soteriou, V., & Nicopoulos, C. (2012). A Dynamically Adjusting Gracefully Degrading Link-Level Fault-Tolerant Mechanism for NoCs. *IEEE Transactions on Computer-Aided Design of Integrated Circuits and Systems, 31*(8), 1235–1248. doi:10.1109/TCAD.2012.2188801

Vivet, P., Lattard, D., Clermidy, F., Beigne, E., & Bernard, C. … Varreau, D. (2006). FAUST, an Asynchronous Network-on-Chip based Architecture for Telecom Applications. In Proceedings of Design, Automation & Test in Europe Conference & Exhibition (DATE). DATE.

Von Platen, C., Eker, J., Nilsson, A., & Arzen, K.-E. (2012). *Static Analysis and Transformation of Dataflow Multimedia Applications (Technical report)*. Lund University.

Waldén, M. (1998). Layering Distributed Algorithms within the B-Method. In Proceedings of Second International {B} Conference (pp. 243-260). Springer-Verlag.

Wang, G., Ji, Y., & Turgut, D. (2004). *A Routing Protocol for Power Constrained Networks with Asymmetric Links*. Paper presented at the 1st International Workshop on Performance Evaluation of Wireless Ad Hoc, Sensor, and Ubiquitous Networks (PE-WASUN). Venice, Italy.

Wang, Q., Zhu, Y., & Cheng, L. (2006, June). Reprogramming wireless sensor networks: Challenges and approaches. *Network, IEEE, 20*(3), 48–55. doi:10.1109/MNET.2006.1637932

Wang, S., Hu, J., & Ziavras, G. S. (2008). Self-Adaptive Data Caches for Soft-Error Reliability. *IEEE Trans. on CAD of Integrated Circuits and Systems, 27*(8), 1503–1507. doi:10.1109/TCAD.2008.925789

Wegner, T., Cornelius, C., Gag, M., Tockhorn, A., & Uhrmacher, A. (2010). Simulation of Thermal Behavior for Networks-on-Chip. *28th NORCHIP Conference* (pp. 1-4). IEEE Xplore.

Wen, H., Huang, P. Y., Dyer, J., Archinal, A., & Fagan, J. (2005). Countermeasures for GPS Signal Spoofing. In *Proceedings of the 18th International Technical Meeting of the Satellite Division of The Institute of Navigation (ION GNSS 2005)* (pp. 1285-1290). Long Beach, CA: GNSS.

Wenk, M. (2010). MIMO-OFDM Testbed: Challenges, Implementations, and Measurement Results. Zürich, Switzerland: Eidgennössische Technische Hochschule (ETH).

Wenk, M., Zellweger, M., Burg, A., Felber, N., & Fichtner, W. (2006). K-Best MIMO detection VLSI architectures achieving up to 424 Mbps. In *Proceedings of the IEEE International Symposium on Circuits and Systems* (pp. 1150- 1154). New York, NY: IEEE.

Wentzlaff, D., Griffin, P., Hoffmann, H., Bao, L., Edwards, B., & Ramey, C. et al. (2007). On-Chip Interconnection Architecture of the Tile Processor. *IEEE Micro, 27*(5), 15–31. doi:10.1109/MM.2007.4378780

Wesson, K. D., Shepard, D. P., Bhatti, J. A., & Humphreys, T. E. (2011). An Evaluation of the Vestigial Signal Defense for Civil GPS Anti-Spoofing. In *Proceedings of the 24th International Technical Meeting of The Satellite Division of the Institute of Navigation (ION GNSS 2013)* (pp. 2646-2656). Portland, OR: GNSS.

Wesson, K., Rothlisberger, M., & Humphreys, T. (2011). A Proposed Navigation Message Authentication Implementation for Civil GPS Anti-spoofing. In *Proceedings of the 24th International Technical Meeting of The Satellite Division of the Institute of Navigation (ION GNSS 2011)* (pp. 3129-3140). Portland, OR: GNSS.

Wesson, K., Rothlisberger, M., & Humphreys, T. (2012). Practical Cryptographic Civil GPS Signal Authentication. *Navigation. Journal of The Institute of Navigation, 59*(3), 177–193. doi:10.1002/navi.14

Wildemeersch, M., Rabbachin, A., Cano, E., & Fortuny, J. (2010). Interference assessment of DVB-T within the GPS L1 and Galileo E1 band. In *Proceedings of the 5th ESA Workshop on Satellite Navigation Technologies and European Workshop on GNSS Signals and Signal Processing (NAVITEC 2010)* (pp. 1-8). Noordwijk, The Nederlands: GNSS.

Windpassinger, C., Fischer, R. F. H., & Huber, J. B. (2004). Lattice-reduction-aided broadcast precoding. *IEEE Transactions on Communications, 52*(12), 2057–2060. doi:10.1109/TCOMM.2004.838732

Wipliez, M. (2010). *Compilation Infrastructure for Dataflow Programs*. (Ph.D. Dissertation). National Institute of Applied Sciences (INSA).

Wipliez, M., & Raulet, M. (2010). Classification and Transformation of Dynamic Dataflow Programs. In *Design and Architectures for Signal and Image Processing (DASIP)*. Academic Press. doi:10.1109/DASIP.2010.5706280

Wireless Innovation Forum. (2002). *Requirements for Radio Software Download for RF Reconfiguration, Approved Document SDRF-02-S-007-V1.0.0, November 2002*. Author.

Wireless Innovation Forum. (2010). *Securing Software Reconfigurable Communications Devices, Approved Document WINNF-08-P-0013, Version 1.0.0, July 2010.* Author.

Wireless Innovation Forum. (2011). *International Radio Security Services API Task Group. IRSS API Specification WINNF-09-S-0011-V1.0.0.* Author.

Woo, A., Tong, T., & Culler, D. (2003). *Taming the Underlying Challenges of Reliable Multi-hop Routing in Wireless Networks.* Paper presented at the 1st International Conference on Embedded Networked Sensor Systems. Los Angeles, CA.

Worm, F., Ienne, P., Thiran, P., & De Micheli, G. (2005). A robust self-calibrating transmission scheme for on-chip networks. *IEEE Transactions on Very Large Scale Integration (VLSI). Systems, 13*(1), 126–139.

Wübben, D., Böhnke, R., Rinas, J., Kühn, V., & Kammeyer, K. (2001). Efficient algorithm for decoding layered space-time codes. *IEEE Electronic Letters, 37*(22), 1348–1350. doi:10.1049/el:20010899

Wymeersch, H., Lien, J., & Win, M. Z. (2009). Cooperative localization in wireless networks. *Proceedings of the IEEE, 97*(2), 427–450. doi:10.1109/JPROC.2008.2008853

X.731 - ITU X.731 ISO/IEC10164-2 State Management. (2012). Retrieved from http://www.itu.int/rec/T-REC-X.731-199201-I/en

Xilinx. (2006). Retrieved from http://www.xilinx.com/support/documentation/application_notes/xapp441.pdf

Yan, T., He, T., & Stankovic, J. (2003). Differentiated surveillance for sensor networks. In *Proceedings 1st International Conference on Embedded Networked Sensor Systems* (pp. 51–62). Los Angeles, CA: ACM.

Yang, C., & Nguyen, T. (2009). Self-calibrating position location using signals of opportunity. In *Proceedings of the 22nd International Technical Meeting of The Satellite Division of the Institute of Navigation* (pp. 1055-1063). Savannah, GA: The Institute of Navigation.

Ye, T. (2004). Packetization and routing analysis of on-chip multiprocessor networks.[Elsevier North-Holland, Inc.]. *Journal of Systems Architecture, 50*(2-3), 81–104. doi:10.1016/j.sysarc.2003.07.005

Yin, A., Guang, L., Liljeberg, P., Nigussie, E., Isoaho, J., & Tenhunen, H. (2009). Hierarchical Agent Based NoC with Dynamic Online Services. In Proceedings of Industrial Electronics and Applications (pp. 434 - 439). Taichung: IEEE.

Yu, Z., & Baas, B. M. (2006). Implementing tile-based chip multiprocessors with GALS clocking style. In *Proceedings of ICCD*, (pp. 174-179). ICCD.

Yviquel, H., Casseau, E., Wipliez, M., & Raulet, M. (2011). Efficient multicore scheduling of dataflow process networks. In *Proceedings of IEEE Workshop on Signal Processing Systems* (SiPS), (pp. 198 – 203). IEEE.

Zamalloa, M. Z., & Krishnamachari, B. (2007). An Analysis of Unreliability and Asymmetry in Low-Power Wireless Links. *ACM Transactions on Sensor Networks (TOSN), 3*(2).

Zander, J., & Schieferdecker, I. (2010). Model-based Testing of Embedded Systems Exemplified for the Automotive Domain. In Behavioral Modeling for Embedded Systems and Technologies: Applications for Design and Implementation (pp. 377-412). Idea Group Inc (IGI).

Zanini, F., Atienza, D., & G., D. M. (2009). A Control Theory Approach for Thermal Balancing of MPSoC. *Proc. of 14th Asia and South Pacific Design Automation Conference (ASP-DAC 2009)* (pp. 37-42). IEEE.

Zebelein, C., Falk, J., Haubelt, C., & Teich, J. (2008). Classification of General Data Flow Actors into Known Models of Computation. In *Proceedings of MEMOCODE*. Anaheim, CA: MEMOCODE.

Zhang, J., & Kim, K. J. (2005). Near-capacity MIMO multiuser precoding with QRD-M algorithm. In *Proceedings of the Asilomar Conference on Signals, Systems, and Computers* (pp. 1498–1502). New York, NY: IEEE.

Zhang, C., Yu, Q., Huang, X., & Yang, C. (2008). An RC4-Based Lightweight Security Protocol for Resource-constrained Communications. In *Proceedings of Computational Science and Engineering Workshops* (pp. 133–140). IEEE. doi:10.1109/CSEW.2008.28

Zhang, D., Lu, L., Castrillon, J., Kempf, T., Ascheid, G., Leupers, R., & Vanthournout, B. (2013). Efficient Implementation of Application-Aware Spinlock Control in MPSoCs. *International Journal of Embedded and Real-Time Communication Systems, 4*(1), 64–84. doi:10.4018/jertcs.2013010104

Zhang, D., Zhang, H., Castrillon, J., Kempf, T., Vanthournout, B., Ascheid, G., & Leupers, R. (2011). Optimized Communication Architecture of MPSoCs with a Hardware Scheduler: A System-Level Analysis. *International Journal of Embedded and Real-Time Communication Systems, 2*(3), 1–20. doi:10.4018/jertcs.2011070101

Zhao, Y. J., & Govidan, R. (2003). *Understanding Packet Delivery Performance in Dense Wireless Sensor Network.* Paper presented at the 1st International Conference on Embedded Networked Sensor Systems. Los Angeles, CA.

Zhao, Z., Rosario, D., Braun, T., Cerqueira, E., Xu, H., & Huang, L. (2013). *Topology and Link Quality-aware Geographical Opportunistic Routing in Wireless Ad-hoc Networks.* Paper presented at the 9th International Wireless Communications and Mobile Computing Conference. Sardinia, Italy.

Zhdanov, F., & Kalnishkan, Y. (2010). An identity for kernel ridge regression. In *Proceedings of the 21st international conference on Algorithmic learning theory* (LNCS) (vol. 6331, pp. 405–419). Berlin: Springer-Verlag.

Zhou, G., He, T., Krishnamurthy, S., & Stankovic, J. A. (2004). *Impact of Radio Irregularity on Wireless Sensor Networks.* Paper presented at the 2nd International Conference on Mobile Systems, Applications, and Services. Boston, MA.

Zhou, G., He, T., Krishnamurthy, S., & Stankovic, J. A. (2006). Models and Solutions for Radio Irregularity in Wireless Sensor Networks. *ACM Transaction on Sensor Networks, 2,* 221–262. doi:10.1145/1149283.1149287

Zimmermann, F., Eschbach, R., Kloos, J., & Bauer, T. (2009). Risiko-basiertes statistisches Testen (in German). In TAV group Meeting of the GI (Gesellschaft für Informatik). Dortmund, Germany: GI.

Zimmermann, F., Kloos, J., Eschbach, R., & Bauer, T. (2009). Risk-based Statistical Testing: A refinement-based approach to the reliability analysis of safety-critical systems. In *Proceedings of EWDC'09, European Workshop on Dependable Systems.* Toulouse, France: EWDC.

Zinkevich, M., Smola, A., & Langford, J. (2009). Slow learners are fast. *Advances in Neural Information Processing Systems, 22,* 2331–2339.

About the Contributors

Seppo Virtanen received his BSc in applied physics, MSc in electronics and information technology (1998), and DSc (Tech.) in communication systems (2004) from the University of Turku (Finland). Since 2009, he has been Adjunct Professor of Embedded Communication Systems at University of Turku. He was the first Editor-in-Chief of the *International Journal of Embedded and Real-Time Communication Systems* (IGI Global), serving from 2009 to 2014. He is a senior member of the IEEE. His published academic research has been in the areas of hardware acceleration for protocol processing, protocol processor architectures, and hardware/software codesign methodologies for embedded communication systems. He has supervised more than 50 Master's theses to completion. In the past few years, his research interests have been focused on platforms capable of handling the processing of communication protocols, DSP routines, and software-defined radio algorithms and applications in parallel on a parameterizable hardware platform, as well as information security and network security related topics especially in the embedded systems domain.

* * *

Antti Airola received his M.Sc. degree in software engineering and the D.Sc. (Tech) degree in information and communication technology from the University of Turku, Finland in 2006 and 2011, respectively. He is currently a university teacher at the University of Turku. His research interests include both basic and applied research in machine learning, with focus on regularized kernel methods.

Gerd Ascheid received his Diploma and Ph.D. degrees in electrical engineering (communications eng.) from RWTH Aachen University, Aachen, Germany, in 1977 and 1984. In 1988, he started as a co-founder of CADIS GmbH. The company has successfully brought the system simulation tool COS-SAP to the market. In 1994 CADIS GmbH was acquired by Synopsys, a California-based EDA market leader where his last position was Senior Director (executive management), wireless and broadband communications service line, Synopsys Professional Services. Since April 2003, he heads the Chair for Integrated Signal Processing Systems (ISS) of the Institute for Communication Technologies and Embedded Systems (ICE), RWTH Aachen University. He has published several scientific papers in reputed journals and conferences. He is also the chairman of the cluster of excellence in "Ultra-high speed Mobile Information and Communication (UMIC)" at RWTH Aachen University.

Sergey Balandin is a Founder and President of FRUCT Oy company. He also has positions of General Chair in Open Innovations Association FRUCT and Adjunct Professor of the Department of Communications Engineering at the Tampere University of Technology. Sergey has received the first M.Sc. degree in Computer Science from St.-Petersburg State Electrotechnical University "LETI" (Russia), then M.Sc. in Telecommunications from Lappeenranta University of Technology and M.Sc. in Business Administration (IBMA program) from Haaga-Helia University of Applied Sciences in 2012. In 2003, Sergey Balandin graduated from Ph.D. School of Nokia Research Center and St. Petersburg State Electrotechnical University "LETI" with PhD degree in Telecommunications and Control Theory. From 1999-2011, he worked for Nokia Research Center. His last position was Principal Scientist of Ubiquities Architectures team and leader of universities cooperation program in Russian and CIS. Sergey has authored or co-authored over 70 papers and 29 patents. He co-edited several proceedings books. For 3 years, he worked in embedded networks field and took part in MIPI UniPro standardization. Currently main research interest of Sergey is in development of smart systems and future service solutions for m-Health and e-Tourism, by using advanced technologies, such as IoT, smart spaces, embedded networks, etc.

Maitane Barrenechea was born in Aretxabaleta, Spain in 1983. She received the BS and MS degrees in Telecommunications Systems' Engineering from the University of Mondragon, Mondragon, Spain, in 2004 and 2007, respectively. In 2007, she joined the Department of Electronics and Computer Sciences at the University of Mondragon, where she obtained her PhD degree in February 2012, focusing on vector precoding systems for the multiuser broadcast channel from both implementation and theoretical perspectives. She is a lecturer and researcher at the Signal Theory and Communications Area of the University of Mondragon. She has been a visiting researcher at The University of Edinburgh (2008), Queen's University Belfast (2009), Technische Universität München (TUM)(2009) and École Polytechnique Fédérale de Lausanne (EPFL) (2011 and 2012). Her areas of interest include low-complexity lattice-search techniques and implementation issues of complex algorithms.

Thomas Bauer has been working as a researcher at the department of Embedded Systems Quality Assurance at the Fraunhofer Institute for Experimental Software Engineering (IESE) after finishing his studies of software engineering at the University of Potsdam (Germany, 1999-2005). His research focus comprises the cost-efficient application of quality assurance, test automation, and model-based statistical testing to complex software-intensive systems.

Luca Benini is Full Professor at the ETH Zuerich and at University of Bologna. He received a PhD degree in electrical engineering from Stanford University in 1997. He is a Fellow of the IEEE and a member of the Academia Europaea. Dr. Benini's research interests are in energy-efficient system design and Multi-Core SoC design. He is also active in the area of energy-efficient smart sensors and sensor networks for biomedical and ambient intelligence applications. On these topics he has published more than 600 papers in peer-reviewed international journals and conferences, four books, and several book chapters.

Davide Bertozzi got his PhD in Electrical Engineering from the University of Bologna (Italy) in 2003. Since 2005 he has been an Assistant Professor at the University of Ferrara (Italy), where he leads the Research group on Multi-Processor Systems-on-Chip and on Networks-on-Chip. He has been a visiting researcher at international academic institutions (Stanford University) and large semiconductor companies (NEC America Labs, USA; NXP Semiconductors, Holland; STMicroelectronics, Italy; Samsung Electronics, South Korea). Bertozzi was the Program Chair of the Int. Symposium on Networks-on-Chip (2008) and of the NoC track at DATE conference (2010-2012). He is a member of the Editorial Board of the IET-CDT, Springer DAEM and ACM TODAES journals. Bertozzi is a member of the Hipeac NoE and was/is involved in EU funded initiatives (Galaxy, NaNoC, vIrtical).

Andreas Burg was born in Munich, Germany, in 1975. He received his Dipl.-Ing. degree in 2000 from the Swiss Federal Institute of Technology (ETH) Zurich, Switzerland. He then joined the Integrated Systems Laboratory of ETH Zurich, from where he graduated with the Dr. sc. techn. degree in 2006. From 2006 to 2007, he held positions as postdoctoral researcher at the Integrated Systems Laboratory and at the Communication Theory Group of the ETH Zurich. In 2007, he co-founded Celestrius, an ETH-spinoff in the field of MIMO wireless communication, holding the position of Director for VLSI. In January 2009, he joined ETH Zurich as SNF Assistant Professor and as head of the Signal Processing Circuits and Systems group at the Integrated Systems Laboratory. Since 2011, he is a Tenure Track Assistant Professor at the École Polytechnique Fédérale de Lausanne (EPFL) where he leads the Tele-communications Circuits Laboratory in the School of Engineering.

Emmanuel Casseau received the M.S. degree in Electrical Engineering in 1990 and the Ph.D degree in Electrical and Computer Engineering from the Universite de Bretagne Ouest, France, in 1994. From 1994 to 1996, he was a research engineer in the French National Telecom School, ENST Bretagne, France. From 1996 to 2006, he was an Associate Professor in the Electronic Department at the Universite de Bretagne Sud, France. He is currently a Professor in IRISA/INRIA (French National Institute for Research in Computer Science and Control), Universite de Rennes1, France. His research interests include system design, high-level synthesis, SoCs design methodologies, and reconfigurable architectures for multimedia applications.

Jeronimo Castrillon received the Electronics Engineering degree with honors from the Pontificia Bolivariana University in Colombia in 2004, the master degree from the ALaRI Institute in Switzerland in 2006 and the Ph.D. degree (Dr.-Ing.) on Electric Engineering and Information Technology with honors from the RWTH Aachen University in Germany in 2013. From 2009 to 2013, he was the chief engineer of the research institute for Software for Systems on Silicon in the RWTH Aachen University, where he was enrolled as research staff since late 2006. In June 2014 he joined the Dresden University of Technology as an Assistant Professor in the Computer Science Faculty. His research interests lie on MPSoC programming: automatic code partitioning, code generation from abstract parallel programming models, compile time mapping and scheduling as well as HW/SW support for run-time systems.

Liang Chen is a Senior Research Scientist at the Department of Navigation and Positioning at the Finnish Geodetic Institute. He also holds a postdoctoral researcher position of the Academy of Finland from 2011 to 2014. Before joining the Finnish Geodetic Institute in the beginning 2011 he worked as a postdoctoral research associate at Tampere University of Technology, Department of Mathematics, Finland. He received his PhD in signal and information processing from Southeast University, China, in 2009. He is the author of more than 30 scientific publications and 4 pending patents. His research interests include wireless positioning and navigation, statistical signal processing, software defined radio, and sensor fusion.

Ruizhi Chen is an endowed chair and professor at the Conrad Blucher Institute of Surveying & Science, Texas A&M University Corpus Christ. Dr. Chen is the editor of the book entitled *Ubiquitous Positioning and Mobile Location-Based Services in Smart Phones*. He is an author/co-author of 131 scientific papers and 5 book chapters. His research results have been selected twice as cover stories in *GPS Worlds* – the World's most popular magazine in satellite navigation. He has supervised 4 Ph.D. dissertations, and examined 4 PhD theses from Finland, Spain and Hong Kong. His students have received 7 international awards in best student paper competitions, including 3 student-winning papers in the Institute of Navigation for 2010, 2012 and 2013. Dr. Chen is the general chair of the IEEE conferences "Ubiquitous Positioning, Indoor Navigation and Location-based Services" 2010, 2012 and 2014. He is the associate editor of *Journal of Navigation*, member of editorial boards *Journal of Global Positioning Systems* and *International Journal of Information Engineering*. Dr. Chen is the chair of the council of principle investigators and research administrators of Texas A&M University Corpus Christi (2013-2014), and member of the Provost's Leadership Team of the same university (2013-2014). Dr. Chen was President of the International Association of Chinese Professionals in Global Positioning Systems (2008) and board member of the Nordic Institute of Navigation (2009-2012). Dr. Chen's research interests are satellite navigation, mobile geospatial computing and ubiquitous sensing.

Yuwei Chen is a research manager of Department of Remote Sensing and Photogrammetry, FGI, where he led remote sensing electronics group and mainly developed hardware for various remote sensing and ubiquitous positioning research. He joined FGI in 2005 as a post-doc research. He is the author of more than 70 scientific publications and holds 8 patents (5 of them are granted). His research interests cover: mobile laser scanning, Spaceborne/Airborne LiDAR, hyperspectral Lidar, Seamless Indoor/Outdoor navigation with multi-sensor multi-network device.

Fabrício Alves Barbosa da Silva is a Researcher in Applied Mathematics at Oswaldo Cruz Foundation (FIOCRUZ) and a Professor at the Military Institute of Engineering (IME), both located at Rio de Janeiro, Brazil. Formerly he was a computer scientist at the Army Technological Center, Rio de Janeiro, Brazil, and a visiting professor at the Informatics Department, University of Lisbon. He obtained his doctoral degree in Computer Science from Université Pierre et Marie Curie (Paris VI) in 2000. His current research interests are Advanced Communication Systems, Big Data Analysis, Modeling and Simulation of Biological Systems.

Fabio Dovis' research addresses many field of signal processing and communications, and it is currently mainly focused on the satellite navigation systems (GPS and Galileo). He works at the Electronics and Telecommunications Department of Politecnico di Torino. He strongly contributed to the creation of the Navigation Signal Analysis and Simulation (NavSAS), specializing the telecommunication skills in the satellite navigation field. His research interests in the field of satellite navigation are focused on: design of fully software and software radio navigation receivers, multipath and interference detection and mitigation, study of optimized signals for positioning and navigation and augmentation systems based on high altitude platforms. His past research interests include communication systems based on HAPs, wavelet signals for communications and array signal processing.

Robert Eschbach works at ITK Engineering AG as Senior Software Quality Engineer and Technical Senior Consultant for Software Engineering, with a focus on (Agile) Project Management, Software Quality, and Software Economics. He works in different industrial sector like automotive, commercial vehicles, agricultural sector, medical device sector, and rail. He has a broad expertise for regulatory requirements in these sectors, as well as practical experiences in process improvement (SPICE, CMMI, TPI). Formerly, he was the Embedded Systems Quality Assurance department head at Fraunhofer IESE, where he was mainly engaged in projects with safety-critical and software intensive systems. His research in software and systems engineering centered on modeling theory, model-based testing, dependability engineering, and formal methods. He is the author of more than 80 refereed publications. Dr. Eschbach has been a principal investigator in numerous research and industrial development projects, including the standardization of an important formal description technique for the telecommunication sector. He has given more than 50 industrial talks at different important industrial workshops and conferences. Dr. Eschbach earned his Ph.D. in Informatics at the University of Kaiserslautern, Germany.

Xin Fan received the BSc degree in Communication Engineering from Sichuan Univ. China, in 2000, and the MEng. degree in Signal and Information Processing from Peking Univ., China, in 2004. From July 2004 to May 2008, he was working in Shanghai Jade Technologies Corp., as a senior ASIC engineer. In June 2008, he joined IHP Microelectronics, Germany, as a research associate. Currently, his work focuses on the asynchronous circuits design and applications.

Martin Gag received his Bachelor and Master degree in Information Technology from the University of Rostock, Germany in 2007 and 2009, respectively. Since 2009, he has been with the Institute of Applied Microelectronics and Computer Engineering at the University of Rostock, where Mr. Gag is currently working towards his Ph.D. His research interests include performance, power and reliability evaluation of on-chip communication networks as well as signal integrity issues in modern nanoscale technologies.

Juraci Ferreira Galdino received the B.Sc. degree in Electronic Engineering from the Federal University of Paraiba (UFPB), Brazil, in 1991. He received the M.Sc. degree in Electrical Engineering from the Military Institute of Engineering (IME), Brazil, in 1998 and was awarded with the D.Sc. degree from the Federal University of Campina Grande (UFCG), Brazil, in 2002. His research interests include adaptive transmission, soft defined radio, adaptive filtering, cross-layer design, and wavelets. Since 2003, he has been a professor and researcher at IME. He has authored more than 80 articles published in journals and conference proceedings of national and international scientific societies. His biography is included in the following books: Who's Who in the World, Who's Who in Science and Engineering, 2000 Outstanding Intellectuals of the 21st Century, and Leading Engineers of the World. Dr. Galdino is a Brazilian army lieutenant-colonel and member of the Brazilian Telecommunications Society (SBrT).

Alberto Ghiribaldi is a PhD student at University of Ferrara (Italy). His research interests concern asynchronous circuits and design toolflows, and CAD tool support in general for the physical synthesis of multi-core systems on sub-40nm technologies. He has been visiting researcher at Columbia University (USA) and at Intel Mobile Communications (Germany), where he developed an asynchronous interconnect fabric and a NoC debugging infrastructure respectively. He was lecturer at the NaNoC summer school in 2012 (held in Munich), with a contribution on fault-tolerant systems.

Michel Gillet is an architect at Nokia Devices in Helsinki, Finland. Gillet has a MSc in computer science from University of Liege, Belgium, 1999, with the title "Ethernet ISDN gateway for VoIP." From 1999 until 2008, he was in the Nokia Research Center with the following research activities: high-speed serial links, signal integrity, electromagnetic simulations, sensor integration and embedded networks. In 2008, he joined Nokia Devices to drive the productization of embedded network technologies. As the main Nokia representative, he participated in the MIPI UniPro standardization. His research interests include on-chip, off-chip and die-to-die embedded networks, system architecture, system modeling, and distributed systems.

Jérôme Gorin is a postdoctoral scholar at Telecom ParisTech, Paris, France. He works with the Signal and Image Processing department [TSI] group. Before joining the TSI in 2013, he received his Ph.D. from the University of Pierre and Marie Curie. His main topic of research includes the transport and the parallel processing of digital multimedia contents.

Eckhard Grass received the PhD degree in Microelectronics from the Humboldt-University Berlin, Germany, in 1992. He was a Visiting Research Fellow at Loughborough University, UK, from 1993 to 1995, and a Senior Lecturer in Microelectronics at the University of Westminster, London, U.K., from 1995 to 1999. Since 1999, he has been with IHP GmbH, Frankfurt (Oder), Germany, leading a number of projects on the implementation of wireless broadband communication systems in the 5 GHz and 60 GHz bands. Since 2008, E. Grass is group leader of the Wireless Broadband Communications Group at IHP. Furthermore, he is Professor at the Department of Computer Science at Humboldt-University Berlin since 2011. E. Grass has published a number of papers – mainly in the areas of circuits for wireless communication systems and asynchronous circuit design.

Liang Guang is a post-doctoral senior researcher in the laboratory of Embedded Computer and Electronic Systems at the Department of Information Technology, University of Turku, Finland. He holds a D.Sc. (PhD Tech.) degree from the same department, a MBA certificate in innovation and growth from Business and Innovation Development (BID) unit of University of Turku, and a M.Sc. degree from Royal Institute of Technology (KTH), Sweden. He has a wide interest in embedded system design, adaptive systems, on-chip communication architecture, low power, and dependable computing. His current research is the integration of heterogeneous computing and communication systems, in particular concerning the collective adaptivity and timing predictability. Dr. Guang has served as PC members, session chairs and organizers of many international conferences, for instance HPCS2013, CASEMANS 2012, PECCS2013, etc. He is on the editorial board of IJERTCS. He has published over 40 papers and articles in international conferences and journals.

Christoph Heer is Division Vice President at Intel Mobile Communications, and responsible for "Design System & IP" within the company. Dr. Heer received a Diploma degree in solid-state electronics from Aachen University of Technology in 1990 and a PhD degree in electrical engineering from Ulm University in 1995. Dr. Heer has published more than 50 papers in peer-reviewed international journals and conferences. He has been member of the technical program committees of various international conferences including DAC, DATE, and ASYNC.

Carles Hernández received the M.S. degree in telecommunications and PhD in computer sciences from Universitat Politècnica de València, in 2006 and 2012, respectively. He is currently researcher at the Barcelona Supercomputing Center. His area of expertise includes network-on chip and reliability-aware processor design. He has participated in NaNoC, parMERASA, PROXIMA FP7, and VeTeSS ARTEMIS projects. In 2012, he was intern at Intel Mobile Communications Munich.

Tanvir Hussain is working as a pilot engineer at The Mathworks GmbH. He is working in the area of verification, validation. In 2008, he completed his PhD in Electrical and Computer Engineering from TU Kaiserslautern, Germany.

Marko Hännikäinen received his MSc in information technology in 1998, and PhD in information technology in 2002 from the Tampere University of Technology, Finland. He is currently a research fellow at Tampere University of Technology and he was nominated an adjunct professor in 2005. He has authored over 140 publications and holds several patents. He has supervised eight PhD thesis and his teaching experience includes project working, scientific publishing, and wireless sensor networks. He has also been founding several spin-off companies resulting from research projects. His research interests include wireless sensor networks and mobile Web services.

M. R. Kakoee received his MSc degree in computer architecture from University of Tehran, Iran, in 2003, and his PhD degree in electronics from University of Bologna, Italy, in 2012. He was a post-doctoral fellow at the same university working in the laboratory of Professor Luca Benini. From June 2010 until December 2010, he was a visiting scholar in University of Michigan, Ann Arbor, USA. His research interests include MPSoC design and reliability, variability, and low power design. He has published several papers on these topics in journals and visible conferences. Currently, he holds a staff engineer position at Qualcomm (USA).

Florian Kantz is a Scientist at the ABB Corporate Research Center in Ladenburg, Germany. He joined ABB in 2008 with a background in automated testing of networked embedded systems. Currently he focuses on technologies for the integration of embedded devices in automation systems. His background is in electronic engineering and information technology. He received his Dipl.-Ing. from the Technical University of Darmstadt in 2002.

Holger Kaul is a Scientist at the ABB AG Corporate Research Center in Ladenburg, Germany since 2006. He is working in the area of embedded systems, focusing in hardware and software design. His research topics are software and hardware architectures, ultra low power design and autonomous measurement systems. Holger studied electrical engineering at the University of Applied Sciences in Mannheim, spent a half year of his studies at the Ngee Ann Poly in Singapore (work topic: FPGA design in the field of Software Defined Radio) and a half year at company Renishaw (thesis topic: Development of a Motor Control Analyzer).

Miloš Krstić received the Dr-Ing. degree in electronics from Brandenburg University of Technology, Cottbus, Germany in 2006. Since 2001, he has been with IHP Microelectronics, Frankfurt (Oder), Germany, where he currently holds the position of a Team leader for Design & Test Methodology in the Systems Department. Dr. Krstic was leading the IHP activities in many national and international projects including FP7 EU-funded project GALAXY, FP7 project VHiSSi, Eurostars project IC-NAO, DAAD Project ESD, various industry projects and grants. His work was followed with around 80 publications and 6 registered patent applications mainly in the area of low power and fault-tolerant digital design methods for wireless and embedded applications and globally-asynchronous locally-synchronous (GALS) methodologies for digital systems integration.

Heidi Kuusniemi is an acting Head of Department and Professor at the Department of Navigation and Positioning at the Finnish Geodetic Institute (FGI). She is also a Lecturer in GNSS Technologies at the Department of Surveying Sciences at Aalto University, Finland, where she is also an Adjunct Professor. Her research interests cover various aspects of GNSS and sensor fusion for seamless outdoor/indoor positioning including quality control, software defined receivers, as well as interference and error mitigation. She is the President of the Nordic Institute of Navigation.

Teemu Laukkarinen received the M.Sc. degree in computer science from Tampere University of Technology (TUT) in 2010. He is currently pursuing towards Ph.D. in the Department of Pervasive Computing at TUT. He has researched wireless sensor networks from 2006 at TUT. He has published several international peer reviewed journal and conference articles about wireless sensor networks. His teaching experience covers computer architecture, microcontroller systems and wireless sensor network applications. He has taken a part in launching a spin-off company from the wireless sensor network research at TUT. His research interests include operating systems, applications and high-level abstractions in wireless sensor networks.

Irina Lavrovskaya is a Junior Researcher in a SUAI-Nokia Embedded Computing for Mobile Communications Lab of the Saint-Petersburg State University of Aerospace Instrumentation (SUAI). She holds M. Sc. in Computer Science from Saint-Petersburg State University of Aerospace Instrumentation, the main research subject: "Modeling of formally specified embedded systems communication protocols." Irina Lavrovskaya has over 6 years of industrial experience: she participated in the development of such standards as UniPro and SpaceWire. The last project she took part in was SpaceWire-RT project funded under the EU's Seventh Framework Programme. Her main research interests are embedded systems, modeling, SDL and SystemC modeling languages and development of communication protocols for on-board and spacecraft systems.

Rainer Leupers received the M.Sc. and Ph.D. degrees in Computer Science with honors from the Technical University of Dortmund, Germany, in 1992 and 1997. From 1997 to 2001, he was the chief engineer at the Embedded Systems chair at TU Dortmund. He joined RWTH Aachen University as a professor in 2002, where he currently heads the Chair for Software for Systems on Silicon (SSS) of the Institute for Communication Technologies and Embedded Systems (ICE). He published numerous books and technical papers, and served as a program committee member and topic chair of leading international conferences, including DAC, DATE, and ICCAD. He has been a co-founder of LISATek, an EDA tool provider for embedded processor design, acquired by Synopsys Inc. His research and teaching activities comprise software development tools, processor architectures and electronic design automation for embedded systems, with emphasis on multiprocessor system-on-chip design tools.

Pasi Liljeberg received his M.Sc. and Ph.D. degrees in electronics and information technology from the University of Turku, Turku, Finland, in 1999 and 2005, respectively. He is an Adjunct Professor in embedded computing architectures at the University of Turku, Embedded Computer Systems laboratory. He is the author of over 160 peer reviewed publications. His current research interests include Internet-of-Things, embedded computing platforms, fault tolerant and energy aware system design, 3D multiprocessor system architectures, cyber physical systems, intelligent network-on-chip communication architectures and reconfigurable system design. He has established and is leading a research group focusing on reliable and fault tolerant self-timed communication platforms for multiprocessor systems.

Nicola Linty received the "diplôme d'ingénieur" at INP, in Grenoble (France) and the Master of Science in Communication Engineering at Politecnico di Torino, summa cum laude, in 2010. In 2010, he spent six months in the PLAN group, at the University of Calgary, where he developed his thesis. Afterwards, he worked as a researcher at ISMB (Torino), in the navigation technologies area. At present, he is pursuing his Ph.D. at the Department of Electronics and Telecommunications at Politecnico di Torino, in the frame of which he spends six months as a student trainee at the European Space Agency (The Netherlands). His research interests cover the field of signal processing and simulation, applied to telecommunication and satellite positioning and navigation areas. In particular, his work is mainly focused on the design and development of innovative architectures, techniques and algorithms for GPS and Galileo receivers, both professional high performance and commercial mass-market.

Jingbin Liu is a senior research scientist in the Department of Remote Sensing and Photogrammetry at the Finnish Geodetic Institute (FGI), and is working on ubiquitous positioning technologies. Dr. Liu has been working on navigation and positioning, and obtained his M.Sc. degree (2004) and Doctoral degree (2008) in Geodesy from Wuhan University, China. Before joining the FGI, he worked in industry for more than four years at SiRF Technology Corporation (Shanghai). As a staff engineer and the group manager, Dr. Liu led the team of research and development of GPS receiver software, and has one patent issued. Currently, his research interests cover various aspects of outdoor/indoor seamless navigation, including GNSS precise positioning, integrated GNSS/inertial sensor positioning, indoor positioning, data fusion of multiple sensors and multiple signals, software defined GNSS receiver technology, and GNSS-based meteorology. He has been developing a hybrid positioning solution, named HIPE (Hybrid Indoor/outdoor Positioning Engine), for mobile devices, e.g. smartphones and tablets, to enable LBS in seamless indoor and outdoor environments. He is now the author of 50 scientific publications, including two chapters in the Book *Ubiquitous Positioning and Mobile Location-Based Services in Smart Phones*, IGI Global, 2012, an issued patent in GPS receiver technology (US 2007/0152876 A1), and a Finnish patent under application. He is a member of working group "Ubiquitous Positioning Systems" in the International Association of Geodesy (IAG).

Lasse Määttä received the M.Sc. degree in computer science from Tampere University of Technology (TUT) in 2010. He has researched wireless sensor network protocols, application development, debugging, and application testing at Department of Computer Systems in TUT from 2006 to 2010. He has taken a part in launching a spin-off company from the wireless sensor network research at TUT. Since then, he has worked three years in research and development of wireless sensor network protocols, products and applications. His research interests include software design for low-power wireless sensor networks.

Mikel Mendicute was born in Eibar, Spain, in 1979. He received the BS, MS (Kutxa Prize) and PhD degrees in electrical engineering from the University of Mondragon in 2000, 2003 and 2008, respectively. He was a member of the Communications Area of Ikerlan-IK4 research center, from 1998 to 2003, where he worked on the development of communications systems. From 2003 on, he has been with the Signal Theory and Communications Area, Department of Electronics and Computer Science, of the University of Mondragon, where he is now a senior lecturer and researcher. He was a visiting researcher at the Institute for Digital Communications of the University of Edinburgh (2004) and at the Telecommunications Circuits Laboratory of the Swiss Federal Institute of Technology in Lausanne (2011). His current research interests include signal-processing algorithms, multiuser MIMO communications, cognitive radio interfaces and the implementation of embedded systems.

Gabriele Miorandi got his master degree in 2013 at University of Ferrara (Italy). He currently holds a PhD position at the same university on low-power asynchronous component design for cyber-physical systems. He won the best paper award at the Int. Symposium on Systems-on-Chip 2013 with a work on in-bus memory management architectures. He has been visiting researcher at Escola Politecnica da Universidade De Sao Paulo (Brasil), where he developed a routing framework for runtime reconfigurable multicore systems.

Pramita Mitra is a Senior Solutions Architect at the Vehicle Design and Infotronics Department at Ford Research and Innovation Center in Dearborn, Michigan. Her research interests include Mobile Computing and Wireless Communications, Human-Machine Interaction, In-Vehicle Infotainment, Cloud-Connected Services, and Parallel Computing. Pramita received her BE in Computer Science and Engineering from Jadavpur University, India; and her MS and PhD, also in Computer Science and Engineering from University of Notre Dame, Indiana. In between her Masters and Doctoral programs, she held a yearlong research intern position at the Oak Ridge National Laboratory where she worked on the implementation of novel and massively parallel FFT algorithms. Pramita is a member a member of IEEE, ACM, and USENIX.

Pavel Morozkin is a 5th year student of the Saint-Petersburg State University of Aerospace Instrumentation (SUAI). He joined a SUAI-Nokia Embedded Computing for Mobile Communications Lab as engineer in 2011 and continues working there up to the present time. His research interests are: embedded systems design, SDL and SystemC languages, software engineering, and model-driven development.

David Fernandes Cruz Moura received the B.Sc. degree in communications engineering from the Military Institute of Engineering (IME), Brazil, in 1997. He received the M.Sc. degree in electrical engineering from the Federal University of Rio de Janeiro (UFRJ), in 2003, and was awarded with the D.Sc. degree in defense engineering from IME in 2011, after holding a position of Visiting Scholar at the Virginia Polytechnic Institute and State University (Virginia Tech, USA), in 2010. Dr. Moura is a Brazilian Army Major and his research interests include cross-layer design, wireless tactical networks, game theory, and operations research applications in telecommunications systems. Since 2012, he has been a Professor at IME and a Researcher at the Brazilian Army Technological Center. He was awarded in 2012 with the Marechal-do-Ar Casimiro Montenegro Filho Prize, offered by the Brazilian Strategic Affairs State, and with the Oscar Niemeyer Prize, offered by the Rio de Janeiro Engineering Council.

Luciano Musumeci is a Post-doc researcher at the Department of Electronics and Telecommunications of Politecnico di Torino. He has received his Ph.D. diploma in 2014 from Politecnico di Torino, and since January 2011 is part of the Navigation Signal Analysis and Simulation (NavSAS) group. In the context of the GNSS signal-processing framework, his research activity focuses on the development and study of innovative receiver based interference countermeasures for GNSS applications. The last part of his Ph.D. studies has been focused on the investigation and performance analysis of advanced signal processing techniques which over-perform the traditional interference countermeasures and which increase the GNSS receivers' sensitivity and robustness in extremely harsh and denied environments expected for GNSS aviation receivers. Recently, he has been also fully involved in an European Space Agency (ESA) funded project, the LUNAR GNSS, investigating the design of a high sensitivity acquisition and tracking scheme for weak GNSS signals reception on the moon.

Philipp Nenninger is the global technology manager for flow measurement products in ABB. His background is in electronic engineering and industrial information technology. He received his Dipl.-Ing. and Dr.-Ing. in 2003 and 2007, respectively, both from the University of Karlsruhe (now KIT).

Valentin Olenev is Senior Researcher of the Saint-Petersburg State University of Aerospace Instrumentation (SUAI). He holds M. Sc. in Computer Science from Saint-Petersburg State University of Aerospace Instrumentation, the main research subject: "Research and development of a system for Wi-Fi networks security increase"; and Ph.D in Mathematical support for computers, complexes and computer networks from Saint-Petersburg State University of Aerospace Instrumentation, the main research subject: "Design of communication protocols models for embedded systems." Valentin Olenev has about 10 years of industrial experience and his main research interests are: networking, embedded systems, modeling, SDL and SystemC modeling languages, models architecture, Petri Nets, SpaceWire, on-board systems. He has over 30 scientific publications. Currently Valentin Olenev is a project leader at SUAI-Nokia Embedded Computing for Mobile Communications Lab and assistant at Aerospace Computer and Software Systems Department at SUAI.

Sergey Ostroumov is a PhD student at Åbo Akademi University and TUCS – Turku Centre for Computer Science. He received his MSc with honours in Computer Systems and Networks from National Aerospace University "KhAI," Kharkiv, Ukraine in 2008. His research interests include methods for designing dependable and efficient embedded systems. He is specifically interested in agent-based management systems for many-core platforms such as network-on-chip. In addition, he explores methods for deriving fault-free implementations into hardware description languages by applying formal methods to hardware designs.

Tapio Pahikkala currently acts as a Professor of Intelligent Systems in University of Turku, Finland. He received his Doctoral degree from University of Turku in 2008 and held an Academy of Finland postdoctoral position during 2010-2012. His research focuses on machine learning, pattern recognition, algorithmics, and computational intelligence. He has authored more than 80 peer reviewed scientific publications and served in program committees of numerous scientific conferences.

Claudio Pastrone received the M.Sc. degree in telecommunications engineering from the Politecnico di Torino, Turin, Italy, in 2002. In 2004, he joined the Electronics Department, Politecnico di Torino, where he pursued his previous activity. In February 2005, he joined the Istituto Superiore Mario Boella, Turin, Italy. His research interests include short-range wireless communications technologies, with a focus on cognitive Wireless Sensor Networks (WSNs), network management, and e-security protocols. Mr. Pastrone won an annual research grant from CNIT to study mobility and security issues in wireless networks in 2003.

Ling Pei received his Ph.D degree in test measurement technology and instruments from the Southeast University, China, in 2007, joining the Finnish Geodetic Institute as a senior research scientist at the same year. He worked as a specialist research scientist in the Navigation and Positioning Department at the Finnish Geodetic Institute until 2013. He is now an associate professor in School of Electronics, Information and Electrical Engineering at Shanghai Jiao Tong University. He has authored or co-authored over 50 scientific papers and book chapters. He is also an inventor of 6 patents and pending patents. His research interests include indoor/outdoor seamless positioning, ubiquitous computing, wireless positioning, mobile computing, context-aware applications and location-based services. Nowadays, he has been session chairs, research chair, advisory chair, and technical program committee members of some international conferences. Moreover, he is an editorial board member of the *International Journal of Advanced Robotic Systems*, a referee for *IEEE Pervasive Computing, Sensors, International Journal of Geographical Information Science*, and various IEEE international conferences.

Marco Pini heads the Navigation Technology research area at Istituto Superiore Mario Boella (ISMB). He received the Ph.D. in Electronics and Communication Engineering on December 2006 from Politecnico di Torino, working on the software radio design of GNSS receivers. His major interests cover the field of baseband signal processing, algorithms for new GNSS signals, interference and spoofing mitigation. As a result of the experience gained on GNSS receiver technology, Marco Pini has been responsible for the activities of several European and National projects, is co-author of papers on journals and international conferences and has an appointment with the European GNSS Agency as technical reviewer of R&D projects.

Juha Plosila is an Associate Professor in Embedded Computing and an Adjunct Professor in Digital Systems Design at the University of Turku (UTU), Department of Information Technology, Finland. He received a PhD degree in Electronics and Communication Technology from UTU in 1999. Dr. Plosila is the leader of the Embedded Computer and Electronic Systems (ECES) research unit and a co-leader of the Resilient IT Infrastructures (RITES) research programme at Turku Centre for Computer Science (TUCS). He is the coordinator of the Embedded Systems master's program at the EIT ICT Labs Master School and is a management committee member of the EU COST Actions IC1103 (MEDIAN: Manufacturable and Dependable Multicore Architectures at Nanoscale) and IC1202 (TACLe: Timing Analysis on Code Level). Dr. Plosila is an Associate Editor of *International Journal of Embedded and Real-Time Communication Systems* (IGI Global). His current research deals with adaptive multiprocessor systems at different abstraction levels. This includes e.g. specification, development, and verification of self-aware, multi-agent monitoring and control architectures for massively parallel systems, as well as applications of autonomous energy-efficient architectures to new computational challenges in the cyber-physical systems domain.

Christian Poellabauer received his Dipl.Ing. degree from the Vienna University of Technology, Austria and his Ph.D. degree from the Georgia Institute of Technology, Atlanta, GA, both in Computer Science. He is currently an Associate Professor in the department of Computer Science and Engineering at the University of Notre Dame. His research interests are in the areas of wireless sensor networks, mobile computing, ad-hoc networks, pervasive computing, and mobile healthcare systems. He has published more than 90 papers in these areas and he has co-authored a textbook on Wireless Sensor Networks. His research has received funding through the National Science Foundation (including a CAREER award in 2006), Army Research Office, Office of Naval Research, IBM, Intel, Toyota, and Motorola Labs. He is a senior member of ACM and IEEE.

Alexey Rabin is Associate Professor of St.-Petersburg State University of Aerospace Instrumentation. Alexey Rabin is graduated engineer in Computer Science from St.-Petersburg State University of Aerospace Instrumentation, the main research subject: "Implementation of the switches in systems with arithmetical code division of channels," and holds Ph.D. in System Analysis, Control and Information Processing from St.-Petersburg State University of Aerospace Instrumentation, the main research subject: "Use of orthogonal coding for increase of noise immunity of information transmission systems." Alexey Rabin has over 11 years of industrial experience. His main research interests are: information theory, noise immunity coding, data transmission systems and networks, data transmission protocols, embedded systems modelling, embedded computing and digital signal processing.

Mickaël Raulet received a Ph.D. degree in 2006 from INSA in electronic and signal processing in collaboration with Mitsubishi Electric ITE (Rennes – France). He is currently in the research Institute of Electronics and Telecommunications of Rennes (IETR) where he is a research engineer in rapid prototyping of standard video and he is a project leader of several French and European projects. He is also a member of a new research institute IRT B-COM (http://b-com.org). His particular interests include dataflow programming, signal processing systems and reconfigurable video coding. Since 2007, he is particularly involved in the ISO/IEC MPEG standardization activities as a Reconfigurable Video Coding Expert. He is the author of 3 book chapters and more than 80 international conferences and journal papers. Dr. Mickaël Raulet serves as a member of the technical committee of the Design and Implementation of Signal Processing Systems (DISPS) of the IEEE Signal Processing Society.

Laura Ruotsalainen is the deputy Head of the Department of Navigation and Positioning and a specialist research scientist at the Finnish Geodetic Institute. She received her M.Sc. degree in 2003 from the Department of Computer Science, University of Helsinki, Finland and doctoral degree in 2013 from the Department of Computer Systems, Tampere University of Technology, Finland. Her doctoral studies were partly conducted at the Department of Geomatics Engineering at the University of Calgary, Canada. The doctoral research was focused on vision-aided seamless indoor/outdoor pedestrian navigation. Her current research interests cover, in addition to vision-aided navigation, GNSS interference detection and mitigation and various aspects of GNSS and sensor fusion. She is also lecturing a course on GNSS technologies at the Department of Surveying Sciences at Aalto University, Finland.

Thomas Ruschival worked as Scientist at ABB Corporate Research Center Germany. He is currently working as Embedded Software Engineer for carrier grade Ethernet switches at Datacom Telematica in Porto Alegre, Brazil. Thomas Ruschival graduated as Dipl.-Ing. in Electrical Engineering at University of Stuttgart in 2008.

Tapio Salakoski received the M.Sc. degree in computer science in 1987 and the Ph.D. degree in computer science and bioinformatics in 1997 from the University of Turku, Turku, Finland. He is currently a Professor of computer science and the Head of the Department of Information Technology, University of Turku, Turku, Finland. His research interests include machine learning and language technology in the BioHealth domain, especially bioinformatics and bioNLP, as well as educational informatics, especially technology-enhanced learning of mathematics and programming. He has more than 170 scientific publications in international journals and conference proceedings. He has led many research projects and held numerous positions of trust involving both academia and industry.

Birgit Sanders is Program Manager Design System & IP at Intel Mobile Communications and is responsible for advanced deep sub-micron CMOS library & flow development. Birgit received her Diploma degree in electrical engineering from the Technical University München, Germany, in 1997. Birgit has published on Static Timing Analysis and holds two patents in modeling of digital libraries and crosstalk calculation.

Stefan Schürmans received his diploma degree in computer science from RWTH Aachen University in 2005. After working as a software developer in industry for three years, he started his PhD studies at Institute for Communication Technologies and Embedded Systems, RWTH Aachen University in 2008. He is working on methodologies and software tools for MPSoC architecture and software exploration on Electronic System Level. Within this area, his main research focus is on power estimation on high abstraction levels.

Federico Silla received the MS and PhD degrees in Computer Engineering from the Technical University of Valencia, Spain, in 1995 and 1999, respectively. He is currently an associate professor at the Department of Computer Engineering (DISCA) at that university, and external contributor of the Advanced Computer Architecture research group at the Department of Computer Engineering at the University of Heidelberg. He is also member of the Advanced Technology Group of the HyperTransport Consortium, whose main result to date has been the development and standardization of an extension to HyperTransport (High Node Count HyperTransport Specification 1.0). Furthermore, he worked for two years at Intel Corporation, developing on-chip networks. His research addresses high performance on-chip and off-chip interconnection networks as well as distributed systems. He has published numerous papers in peer-reviewed conferences and journals, as well as several book chapters. He has been member of the Program Committee in several of the most prestigious conferences in his area, including IPDPS, HiPC, SC, etc.

Francesco Sottile received the MS degree in telecommunications engineering from Politecnico di Torino, Turin, Italy, in 2002. In November 2002, he joined Istituto Superiore Mario Boella, Torino, Italy, where he currently works as senior researcher and project manager in the Pervasive Technologies

Research Area. From 2002 to 2005, his main research activities were focused on channel coding, where he developed several algorithms for the design of LDPC codes. His current research interests include distributed and cooperative localization based on Bayesian approaches, Wireless Sensor Network (WSN) applications and the Internet of Things (IoT). Moreover, his research activities also encompass the design of hybrid GNSS-terrestrial positioning algorithms.

Alessandro Strano is senior engineer at Intel Mobile Communications with a focus on virtual platform prototype design. Alessandro received a master degree in Electronic and Telecommunication Engineering in 2009 and a PhD degree in Electronic Engineering in 2013 from the University of Ferrara, Italy. He has been visiting researcher in Thales Research and Technology, Paris, France and EPFL, Lausanne, Switzerland. Alessandro has published more than 20 papers in peer-reviewed international journals and conferences on synchronization issues and fault-tolerant communication architectures.

Detlef Streitferdt is the head of the Software Architectures and Product Lines group at the Ilmenau University of Technology since August 2010. The research fields are the efficient development of software architectures and product lines, their analysis and their assessment as well as software development processes and model-driven development. Before that, he was Principal Scientist at the ABB AG Corporate Research Center in Ladenburg, Germany, where he started 2005. He was working in the field of software development for embedded systems. Detlef studied Computer Science at the University of Stuttgart and spent a year of his studies at the University of Waterloo in Canada and graduated 1999. He received his doctoral degree from the Technical University of Ilmenau in the field of requirements engineering for product line software development in 2004.

Jukka Suhonen received his M.Sc. in 2004 from Tampere University of Technology (TUT), and D.Sc. degree from TUT in 2012. He currently acts as a researcher at the Department of Pervasive Computing at TUT. He has authored two books about wireless sensor networks and has published over 20 international peer reviewed journal and conference publications. He has participated in several research projects on wireless networks and he has taken part on launching spin-off companies resulting from the research. His research focus is on wireless sensor networks, communication protocols, quality of service, and software architectures.

Jian Tang is now a visiting scholar at Department of Remote Sensing and Photogrammetry, Finnish Geodeitic Institute, Finland.

Hannu Tenhunen is chair professor of Electronic Systems at Royal Institute of Technology (KTH), Stockholm, Sweden. Prof. Tenhunen has hold professor position as full professor, invited professor or visiting honorary professor in Finland (TUT, UTU), Sweden (KTH), USA (Cornel U), France (INPG), China (Fudan and Beijing Jiatong Universities), and Hong Kong (Chinese University of Hong Kong). He has contributed to over 150 journal papers, over 650 reviewed international conference papers, over 170 non-reviewed papers, and 9 international patents granted in multiple countries. Most lately, he has served at EU-level as being the Education Director in European Institute of Technology and Innovations, and its Knowledge and Innovation Community EIT ICT Labs.

John Thompson received the B.Eng. and PhD degrees from the University of Edinburgh, U.K., in 1992 and 1996, respectively. From July 1995 to August 1999, he was a postdoctoral researcher funded by the U.K. Engineering and Physical Sciences Research Council (EPSRC) and Nortel Networks. Since September 1999, he has been with the School of Engineering and Electronics, University of Edinburgh, where he is now Professor. His research interests include signal-processing algorithms for wireless systems, antenna array techniques, multihop wireless, and green communications. He has published approximately 200 papers to date including a number of invited papers, book chapters, and tutorial talks, as well as coauthoring an undergraduate textbook on digital signal processing. Dr. Thompson is the founding Editor-In-Chief of the *IET Signal Processing Journal*. He has been a Technical Programme Co-Chair for IEEE Globecom 2010 and the IEEE International Conference on Communications (ICC) 2007.

Dirk Timmermann studied Electrical Engineering at the University of Dortmund (Germany) and graduated in 1984 as Dipl.-Ing. Afterwards, he worked until 1989 as a Ph.D. student at the Fraunhofer Institute of Microelectronic Circuits and Systems. In July 1990, he received his Ph.D. from the Department of Electrical Engineering at the University of Duisburg and worked afterwards as PostDoc and project-leader at the Fraunhofer Institute of Microelectronic Circuits and Systems. In 1993, he became Professor for Computer Engineering at the University of Paderborn. Since 1994, he is a full University Professor for Computer Engineering (School for Computer Science and Electrical Engineering) at the University of Rostock. Concurrently, he is head of the Institute for Applied Microelectronics and Computer Engineering. In 2004 and 2007, he has been elected vice dean of the School for Computer Science and Electrical Engineering. Currently, he serves as dean. His research focuses on energy aware and robust digital CMOS-circuits and systems, embedded sensor networks, and self-organizing systems.

Leonidas Tsiopoulos received his B. Eng in Electronic Engineering and Technology Management from the University of Huddersfield, England, in 2001, his MSc in Software Engineering from Åbo Akademi University, Finland, in 2003 and his PhD in Computer Science from Åbo Akademi University, Finland, in 2010. Currently, he is a postdoctoral researcher at Åbo Akademi University. The focus of his research has been on applying formal methods to the design of many-core systems and exploiting new approaches based on formal methods for efficient and dependable safety-critical, mixed-critical, and massively parallel application development and mapping to many-core platforms. Moreover, the efficient mapping of state-of-the-art parallel programming languages and frameworks to higher-level modelling and specification paradigms is of high interest.

Bart Vanthournout received the electrical engineering degree from the Katholieke Universiteit Leuven, Heverlee, Belgium, in 1990. Subsequently, he joined the VLSI Systems Design Group of IMEC, Leuven, Belgium, and was involved in developing scheduling and assignment techniques as part of a behavioral synthesis environment. From 1994 to 1996, he was a member of the Hardware and Software Systems group, developing a synthesis and simulation environment for Hardware and Software codesign. Since 1997, he continued this work at CoWare, of which he is one of the cofounders. At CoWare he held several positions managing applications and engineering groups. He joined Synopsys in 2010 after the acquisition of CoWare. Most recently, he is responsible for product definition and research activities as product architect. Currently, his areas of interest are Transactional Modeling (TLM) and MPSOC design methodologies.

Marina Waldén is a Docent and Senior University Lecturer in distributed systems at the Department of Information Technologies at Åbo Akademi University, Finland. She received her PhD degree in Computer Science from Åbo Akademi Universty in 1998 and her docent title at Åbo Akademi University in 2005. Dr Waldén is a co-leader of the Distributed System research unit (DS-Lab) and a leader of the Integrated Design of Quality Systems group. Her research interests include formal methods and their application on industrial strength systems. She is especially interested in making the formal design process more efficient, flexible and maintainable for developing complex, dependable software systems that are correct by construction. Currently she is a responsible leader of the projects ADVICeS (Adaptive Integrated Formal Design of Safety-Critical Systems) financed by the Academy of Finland (2013-2017), as well as of the projects DiHy and Digihybrid financed by TEKES programme EFFIMA (Energy and life cycle efficient machines) (2011-2014) coordinated by the FIMECC SHOK.

Tim Wegner received his Bachelor and Master degree in Information Technology from the University of Rostock, Germany in 2007 and 2009, respectively. Since 2009, he has been with the Institute of Applied Microelectronics and Computer Engineering at the University of Rostock, where he is currently working towards his Ph.D. His research focuses on reliability and robustness of on-chip communication networks as well as energy awareness of nanoelectronic systems.

Matthieu Wipliez is CTO at Synflow SAS, where he works on designing a new programming language called C~ for hardware design along with the development environment and compiler for that language. This work leverages the experience he gained while doing his thesis. Indeed, Matthieu previously worked in academia for five years at IETR, France, where he did a PhD on the compilation and analysis of dataflow programs. His contributions included the design of an Intermediate Representation, classification, and sequential software code generation of dataflow programs. During his thesis, he also created the Open RVC-CAL Compiler (Orcc) project that is still being used for research about dataflow.

Zhoubing Xiong is a PhD student in the department of electronics and telecommunications at Politecnico di Torino (Italy). He received MSc degree in telecommunications engineering from Politecnico di Torino in 2010. Since September 2012, he is working as a researcher under the grant at the Istituto Superiore Mario Boella (ISMB) in the Pervasive Technologies area. His main research interest is focused on the design and development of hybrid positioning architectures and algorithms. Moreover, his research activities also include channel coding and Internet of Things (IoT).

Thomas Canhao Xu received his Doctor of Science degree from University of Turku, Finland in 2012, and Master of Engineering degree from Zhejiang University, China in 2007, respectively. He worked as a system engineer, project manager and system analyst for several projects since 2005. During 2006-2008, he has been teaching the National Certification of Information Engineer (NCIE) courses. Since September 2008, he has been working in the Computer Systems Laboratory, University of Turku and Turku Center for Computer Science (TUCS) as a researcher. He has been involved in several projects of Academy of Finland. Currently he is working as an university lecturer (assistant professor) at University of Turku. His research interests include hardware/software co-design, parallel systems, three-dimensional multiprocessor architecture design, energy efficient cache/memory architectures and software engineering. He has authored and co-authored over 40 international referred papers. He is a member of IEEE, ACM, and ISOC since 2007.

Hervé Yviquel is currently a postdoctoral scholar in the research Institute of Electronics and Tele-communications of Rennes (IETR), where he works on the implementation of video codecs on multi-core platforms using dataflow modeling. He is also one of the main developer of the Orcc project, a development environment dedicated to dynamic dataflow programming. Before in 2013, he received a Ph.D in computer science from the University of Rennes 1. His research interests include the modeling and implementation of data-intensive applications.

Diandian Zhang received his Diploma in electrical engineering from the Chair for Integrated Signal Processing Systems (ISS) of the Institute for Communication Technologies and Embedded Systems (ICE), RWTH Aachen University, Germany, in 2006. From 2006 to 2013, he was a Ph.D. candidate in electrical engineering at RWTH Aachen University. Currently he is a design engineer at Spansion International Inc. in Germany. His research interests are development of Application-Specific Instruction-Set Processors (ASIPs), communication architecture exploration and high-level power estimation in Multi-Processor Systems-on-Chip (MPSoCs) in the embedded domain.

Index

A

Abstract Interpretation 290-291

Agent-Based Management System 302-303, 308, 330-331, 333

Aging 1-6, 9-11, 13, 17, 21, 419

Application-Specific Instruction-Set Processor (ASIP) 335, 338, 367

Asynchronous 399, 401-403, 419, 421, 423-427, 432-433, 442-443, 447

Automation Domain 234-235, 240, 242, 261

Autonomous Robot 164

B

Batch Learning 263, 267, 269, 281

Bayesian Framework 177

Bluetooth 162-164, 166, 169-174, 177

Bootloader 86, 91, 95-96, 106

Broadcast Channel 136-139, 161

C

Cache Coherency 364, 367

Classification 24-25, 28, 55-56, 58, 264, 266-267, 272-273, 283, 289, 291, 295

Communication Architecture 335-336, 341, 343-345, 350-352, 356-358, 360-364, 367, 397

Co-Modeling 206-207, 213-219, 223-225, 230-231, 233

Computer Science 98-99, 204, 236, 263

Configuration Parameter 261

Cycle Computer 234-240, 242, 246, 251, 254, 256, 261

D

Dataflow Models of Computation 282, 287, 300

Dataflow Program 283-284, 288

Delay Variations 399-400, 412, 419

Delta-Cycle 216-217, 233

Dependability 1-6, 10, 13-18, 21, 302-303, 330

Direct Memory Access (DMA) 336, 367

Dissemination Protocol 85-86, 88, 102-103, 106

Down-Link Node 118-120, 122-124, 133

Dynamic Dataflow Model 285, 301

Dynamic Reconfiguration 302-303, 329, 333

Dynamic Scheduling 301, 337, 364, 367

E

Embedded Network 209, 233

Embedded Networked System 106

EMI 420-421, 424-425, 439-440, 444, 447

Error Resilient Computing 1, 17, 21

Event-B Decomposition 333

F

Fault Tolerance 3, 21, 371

Feature Model 234-235, 238, 241, 246-247, 254-255, 261

Fingerprinting 162-165, 173-174, 177

Firmware 85-91, 96-97, 102, 106, 170, 340

Fixed Sphere Encoder 135, 138, 144, 161

G

Globally Asynchronous Locally Synchronous (GALS) 399-403, 405, 407, 416, 419-429, 432-435, 439-441, 444, 447

GNSS 54-66, 68-72, 74-77, 83, 163, 170, 178-180, 187, 204

GPS 31, 35, 54, 56, 64, 66-67, 73, 76, 83, 108, 162-163, 170, 179-180, 187, 191, 204

H

Hello 113-115, 118-126, 128, 133

High Efficiency Video Coding 282, 301

CPSIA information can be obtained at www.ICGtesting.com
Printed in the USA
BVOW06*1508310514

354891BV00007B/75/P